AA002910

31st Ethylene Producers Conference 2019

Topical Conference at the 2019 AIChE Spring Meeting
and 15th Global Congress on Process Safety

New Orleans, Louisiana, USA
31 March - 4 April 2019

ISBN: 978-1-5108-8685-8

Printed from e-media with permission by:

Curran Associates, Inc.
57 Morehouse Lane
Red Hook, NY 12571

CURRAN ASSOCIATES INC. proceedings.com

Some format issues inherent in the e-media version may also appear in this print version.

Printed by Curran Associates, Inc. (2019)

For permission requests, please contact AIChE
at the address below.

AIChE
120 Wall Street, FL 23
New York, NY 10005-4020

Phone: (800) 242-4363
Fax: (203) 775-5177

www.aiche.org

Additional copies of this publication are available from:

Curran Associates, Inc.
57 Morehouse Lane
Red Hook, NY 12571 USA
Phone: 845-758-0400
Fax: 845-758-2633
Email: curran@proceedings.com
Web: www.proceedings.com

TABLE OF CONTENTS

Title: Open Process Automation: Unlocking Value Capture by Leveraging Advances in Digital Technology

Authors: Bradley G. Houk, David L. DeBari, Tyler A. Soderstrom; ExxonMobil Research and Engineering

Meeting: American Institute of Chemicals Engineering 2019 Spring Meeting, Ethylene Producers Conference

Location and Date: New Orleans, LA. March 31 – April 4, 2019

Abstract

Distributed Control Systems (DCS) are critical for the safe and reliable operation of today's ethylene plants. The DCS also provides the foundation for our advanced process control and real-time optimization applications. Over the years these applications have become a key component for maximizing the value from our steam crackers. Advances in digital technology are now making it possible to consider a move from the traditional closed proprietary industrial control systems to an open architecture based upon industry standards. Such an open system has the potential to unlock value capture by creating an architecture which supports interoperability and portability for both hardware and software. Examples could include a new state of the art sensor or a recently developed data analytics application. An open system will lower the barrier for innovation and accelerate the development and deployment of new advanced applications to generate greater value from the production of ethylene. This paper will highlight the research done by ExxonMobil to create a proof of concept system which is leading to a prototype system to demonstrate the technical feasibility of an open process control architecture. The work of the industry consortium operating as the Open Process Automation Forum™ to create the required standards will also be discussed. Expected benefits supported by a move to an open process automation architecture will be highlighted.

AIChE Paper Number:

For Robust, Maintainable Advanced Process Control, KEEP IT SIMPLE ("Simple" Does Not Mean Easy!)

Lou Ann Perry
Sr. Process Control Engineer
Chevron Phillips Chemical Company LP

Prepared for Presentation at the 2019 Spring National Meeting
New Orleans, LA, April 1 – 4, 2019

For Robust, Maintainable Advanced Process Control, KEEP IT SIMPLE ("Simple" Does Not Mean Easy!)

Lou Ann Perry
Sr. Process Control Engineer
Chevron Phillips Chemical Company LP

Abstract:

The trend in Advanced Process Control (APC) implementation over the past several decades has been toward more automated project implementation and complex controllers. In addition, demographic changes have resulted in fewer experienced process control engineers available to implement and maintain APC in many operating facilities. These trends have contributed to the proliferation of advanced control solutions that are challenging to operate, troubleshoot, and maintain.

"Keep it Simple" can be used as a guiding principle for robust, maintainable APC. Steps can be taken that will provide a stable platform for APC, a workable scope, and lifecycle planning for training and support.

Introduction

The trend in Advanced Process Control (APC) implementation over the past several decades has been toward more automated project implementation and complex controllers. In addition, demographic changes have resulted in fewer experienced process control engineers available to implement and maintain APC in many operating facilities. These trends have contributed to the proliferation of advanced control solutions that are challenging to operate, troubleshoot, and maintain. Additionally, the maturation of the APC industry has made it increasingly likely that process control engineers will be expected to maintain complex legacy APC applications without the benefit of knowledge accrued during the implementation phase.

The requirements for robustness and maintainability should ideally be defined prior to project scope development, whether the project is to be internally executed or contracted. "Keep it Simple" can be used as a guiding principle at all phases of project execution, revamp, and maintenance.

Simplify the Problem!

If you give your APC software the simplest possible control problem to solve, the result will be predictable and stable control. Achieving this state often requires much more time and effort than the APC implementation itself.

The first step is to provide a stable platform of distributed control system (DCS) regulatory controls, instrumentation, and equipment. Tune the regulatory controls affecting the entire target area, not just those expected to receive APC manipulated variable setpoints. This effort may encompass entire heat integrated systems, as well as other users on common fuel gas, heating medium or cooling medium systems. Valves may need repair or replacement, transmitters may need to be re-ranged, and instruments may need calibration.

Evaluate whether the DCS control configuration is appropriate for the APC solution. Often the regulatory control can be simplified because APC will reduce the need for non-critical cascades, overrides, feed-forwards, etc. Take the opportunity to evaluate and potentially remove control loops that have fallen into disuse.

Consider whether to break manipulated variable cascade loops and write setpoints to the lowest level regulatory control or even directly to controller outputs or valves. In many cases, this will result in lower order dynamic responses, simpler models, and therefore a simpler problem for the APC software.

Simplify the Solution!

Limit the scope of the APC to that which is necessary to achieve the targeted benefits. Often a robust, stable, 80% solution will provide more benefits than an attempted 95% solution, due to increased uptime and operator acceptance.

Limit the number of instruments used in inferred property control variables (CVs). Every additional variable used in an application is a possible point of failure that can affect APC uptime. Additional inputs can also result in complex dynamic responses which are difficult to control.

Use linear transformed CV's sparingly! Transforms are often mis-applied and overused, and they can be challenging to troubleshoot. Many non-linear variables (column dPs, analyzers) can be modeled linearly within the normal operating range and near the constraint limit. Always try the simplest solution first.

Simplify the Hardware/Software Solution!

Implement supporting calculations on robust platforms at the lowest practical level, preferably on the DCS, so that they are available to operators even when the APC is offline. Calculations that are only necessary for the APC may be implemented within the APC software itself. Limit the number of data paths and interfaces, as each one is a potential source of failure.

For multiple similar APC applications (such as ethylene cracking furnaces), develop a procedure and tools to implement modifications consistently and correctly, such as text substitution to create APC configuration files. Use dynamic model gain multipliers and logic instead of multiple configuration files to accommodate operational mode changes or feed type changes.

Bigger is Not Always Better!

The larger the APC controller, the more difficult it will be to operate, troubleshoot, and maintain. It's far simpler to modify or troubleshoot several small controllers than a large one. Often a control matrix needs to be only large enough to encompass the feed system up to the primary constraint to feed maximization (where utilized).

Limit the number of control variables to necessary targets and current process constraints. Avoid the mindset of "we may need this in the future." As processes are modified or debottlenecked, revisit the process constraints and remove unnecessary CVs.

Leave a Trail of Breadcrumbs!

Any APC effort, whether a new project, revamp, or minor modification, should not be considered complete without leaving a record behind. Documentation doesn't need to be extensive or complex, but it does need to be an ongoing effort. It can consist of a simple overview of the APC goals and constraints, and a map pointing to the components of the application, each containing more detailed notes. Detailed comments are appropriate with the calculations themselves, and should include a revision history (date, name of reviser, description of modification).

The Customer is Always Right!

Poor APC online metrics, or clamping MV limits to prevent poor control action, is an indication that your customers (operators) are not finding your product to be useful. This is usually indicative of a design problem, dynamic model issue, or model matrix issue. "Tuning" the controller cannot alleviate these problems. Automated step testing and modelling may not alleviate the problems. If all the easy fixes have been tried and have failed, it's time to call in an expert for an evaluation.

Develop customized operator displays, if necessary, but keep customizations to a minimum. Add custom features to reduce operator error or highlight the need for operator action, such as:

- toggle switches for redundant CVs
- initialize operator limits according to feed type or operating mode
- send messages/alarms to the DCS when APC sheds or cannot achieve its goals (controller give-up)

Training is an ongoing effort due to staff turnover. Create a training program for incoming operators and engineers, and keep it updated as APC applications evolve over time. Invite operators to improve the training program with their own learnings and tips.

Don't Buy a Ferrari if You Don't Know a Good Mechanic!

Complex APC solutions should be maintained by experienced staff who plan to be in the position long-term. In the planning stages, cater the complexity of an APC solution to the expected experience level of the support staff, both current and future. Advocate for a technical mentorship program to provide expertise where more complex controllers exist.

The APC implementation phase is a learning opportunity that shouldn't be squandered due to expediency. Make the training of support staff and operations a critical measure of project success. APC vendors should be expected to allow hands-on configuration, modeling and tuning by the future support staff (which may include operators).

Conclusions

"Keep it Simple" can be used to guide decisions at every stage of APC implementation and support. Simplicity starts with the base layer of control – the DCS. The DCS-level control must be steady and should be simplified where possible to increase stability in your APC and simplify the control problem your APC must solve. This base level is where you should store supporting calculations. The key to robustness and stability of the controller is in simplifying not only the problem but also the solution. Limit CVs and be careful with transforms. Use a matrix that is adequate but abbreviated. Reduce confusion for those who inherit your work. Leave a trail of information regarding your changes and embrace all training opportunities. Use logic to supplement where multiple configuration files may otherwise be necessary. Review the quality of your solution by observing controller uptime and maintainability. Keeping it simple is something that takes time to perfect. When possible, implement mentorship programs to foster learning and set the team up for future success.

AIChE Paper Number: 4c

TROUBLESHOOTING PROCESS CONTROL PROBLEMS TO IMPROVE PERFORMANCE OF ETHYLENE AND OTHER PETROCHEMICAL PLANTS

Brian K. Sullivan
Director of Aftermarket Services
McDermott - Lummus Technology

Prepared for Presentation at the 2019 Spring National Meeting
New Orleans, Louisiana April 1, 2019

TROUBLESHOOTING PROCESS CONTROL PROBLEMS TO IMPROVE PERFORMANCE OF ETHYLENE AND OTHER PETROCHEMICAL PLANTS

Brian K. Sullivan
Director of Aftermarket Services
McDermott - Lummus Technology

ABSTRACT

This paper will address common, yet often hidden, base level process control problems that can adversely affect the overall performance of a petrochemical plant, such as an ethylene plant. These problems can usually be corrected without complicated software or highly specialized expertise. A checklist of typical control system problems that have been observed at numerous petrochemical plants will be presented with examples and suggestions for avoiding and/or correcting them. Many of these problems exist from initial plant startup, never get corrected, and become part of the "that never worked" lore of the plant. These base level process control problems can also limit the ability of an advanced control system to achieve its full potential benefit. In fact, some of the benefit that is often attributed to advanced control technology is actually the result of correcting base level problems during the implementation of an advanced control project.

INTRODUCTION

The performance that is theoretically attainable by petrochemical plants, like ethylene plants, is closely aligned with the technology applied and the equipment installed during the design and construction of the plant. Achieving the full potential performance of a petrochemical plant is dependent upon proper operation of that equipment, which depends in part on the effectiveness of the plant's control system. For decades, the industry has tried to close the gap between actual and potential plant performance with the application of advanced control, but there are also numerous opportunities available to improve plant operation through the rectification of the basic regulatory control system. In fact, some of the benefit that is often attributed to advanced control technology is actually the result of correcting these regulatory control problems during the implementation of an advanced control project.

As a minimum, improving the performance of the regulatory control system in this manner will smooth the plant operation and make the advanced control applications that are layered on top of the regulatory control system even more effective. Even if you are not looking to apply advanced controls, a periodic review and rectification of regulatory control system performance can result in cost effective improvements for the operation of your plant.

This paper intended to be a tutorial to assist chemical process/plant engineers, who are not necessarily process control experts, with identifying and troubleshooting problematic regulatory control loops. Common, yet often hidden, base level process control problems are explained and addressed in this paper. Checklists for identifying problematic control loops and for troubleshooting typical control system problems are also presented, with examples and suggestions for avoiding and/or correcting them. If you have a specific problematic loop that you want to rectify, then skip right to the sections on troubleshooting. If you want to perform a general review of your regulatory control system, start with the following section (i.e. "Where to Start? Look for These Symptoms").

TERMINOLOGY

Since process control terminology can vary between systems and between organizations, the terminology used in this paper is defined below and illustrated in Figure 1.

- Advanced Process Control (APC) = Any high level software that is capable of handling process control issues such as long dead time and time lag dynamics, variable interactions, and multiple constraint limitations to improve plant performance. This can include multivariable control, model-based control, etc.
- Automatic mode = A control setting that causes the control algorithm to execute and prevents the operator from changing the value of the controller output, but allows the operator to enter the value of the set point.

- Cascade mode = A control setting that causes the control algorithm to execute and allows the controller to receive its set point from another controller while preventing the operator from changing the value of the set point or controller output.
- Conditional or tracking status = A control setting that prevents the control algorithm from executing and forces the controller output value and possibly the set point value to track the value of other variables for the purpose of bumpless initialization.
- Controller Output (CO) or Manipulated Variable (MV) = the output signal from the controller that is sent to downstream controller or final control element.
- Control loop = a combination of instrumentation, control functions, and final control elements that serve to maintain a specific process objective.
- Critical variable = The most important variable in a function block that is used to calculate the process variable for a controller. An example would be the flow measurement in a flow compensation calculation.
- DCS = Distributed Control System
- Final control element = A field device (i.e. actuator, positioner, control valve, etc.) that implements the mechanical action requested by the controller output signal.
- Instrumentation = field devices that measure the value of the process variable (see flow element example in Figure 1).
- Integral wndup or reset windup = an accumulation of the output value of a controller due to the integral term continually adding to the controller output due to the presence of a steady error and improper control processing functionality.
- Manual mode = A control setting that prevents the control algorithm from executing and allows the operator to directly set the value of the controller output.
- Non-critical variable = A variable of lesser importance in a function block that is used to calculate the process variable for a controller. An example would be the temperature or pressure measurement in a flow compensation calculation.
- PLC = Programmable Logic Controller
- Primary controller = The top level controller in a cascade control strategy that determines the set point of the lower level controller in the strategy.
- Process Variable (PV) or measurement = the value of the measured variable, such as flow, temperature, pressure, etc.
- Regulatory control system = Basic level of control equipment and functionality required to safely operate a process plant.
- Secondary controller = The lower level controller in a cascade control strategy that receives its set point from the primary controller in the strategy.
- Service Factor = Percentage of time that a controller is operating in its optimal mode (i.e. automatic or cascade, depending upon the strategy).
- Set Point (SP) = target value for the controller.
- Split-range controller = A controller with an output that sequentially adjusts more than one downstream controller or valve in series based upon the value of its controller output signal.

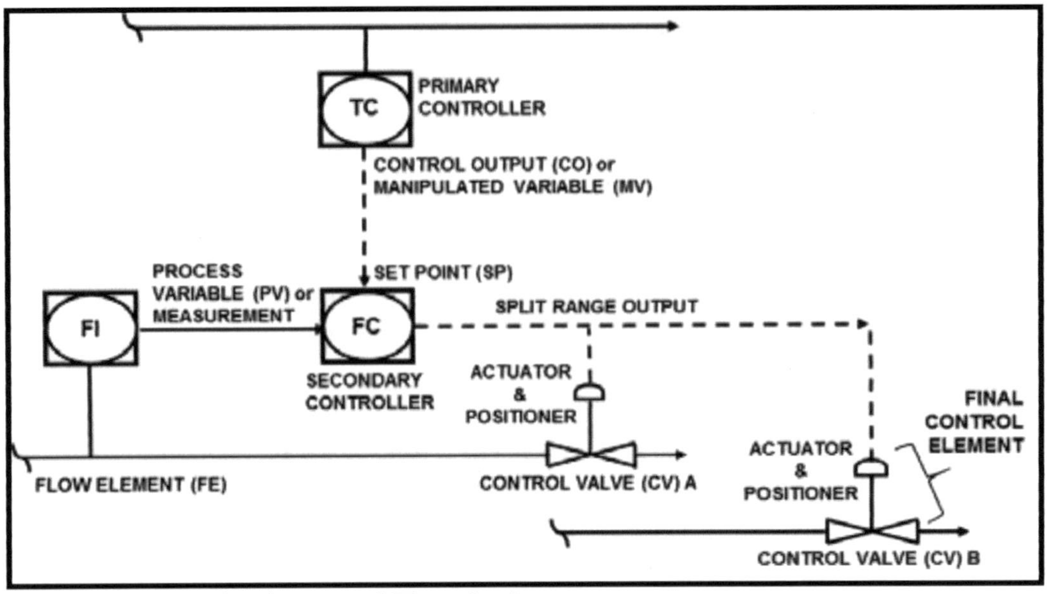

Figure 1. Standard Control Terminology

WHERE TO START? LOOK FOR THESE SYMPTOMS

While some control loop problems may be well known to the plant personnel, others may exist and adversely impact the overall performance of the plant in relative obscurity. Therefore, the first step in the process of rectifying a regulatory process control system is identifying the loops that are problematic. Asking the following questions can lead to identification of potentially poor regulatory control implementation and problematic control loops:

- Which control loops have a low service factor (see below for suggested ranges for poor, satisfactory, and good service factors)?
  - Are there control loops that are continuously operated in manual mode? Operators will quickly lose patience with controllers that do not work well, so this is a key indicator of a problem. However, keep in mind that there are some controllers that are only used sparingly, such as start-up controllers or regeneration controllers, so a low service factor does not always mean there is a problem.
  - Are there control loops that normally work well in automatic mode, but are periodically/repetitively placed in manual mode by operators to handle large disturbances, implement set point changes, or squelch unstable control responses? In ethylene plants, this has been known to occur during cracking heater swaps, dryer switches, and feed property swings.
  - It is also important to note that in certain cases, like cascade loops, overrides, and others, there are primary controllers that remain in automatic mode, even though they are in "tracking" or "conditional" status because the downstream secondary controller or function block is not in the proper mode or position. Therefore, simply looking at the mode is not sufficient. It is also necessary to look at the status of the controller to confirm whether or not it is in use.

- Which controllers perform poorly, even though they have a satisfactory service factor?
    - Are there controllers that typically exhibit a large degree of variability or oscillatory/cyclic behavior?
    - Are there controllers that typically require a very long time to reach their set point following a disturbance, process excursion, or set point change?
    - Are there controllers that frequently have their set points changed? This especially applies to level controllers (which should hardly ever have their set points changed) and temperature controllers (which typically don't have their set points changed often). When operators frequently change the set point of a controller, it can be an indication that they are trying to "help" the controller to respond to a disturbance, but of course the controller should be able to respond to a disturbance without help.

While the above list looks formidable, many of these questions can be answered easily by importing the relevant control parameters (i.e. mode, status (i.e. track, conditional, open input, etc.), process variable/measured value, set point, controller output, alarm settings, and clamp values) from the control system into a spreadsheet and performing common statistical analyses upon them. For example, one minute data for a week or one hour data for a year could be imported for the key parameters for all the control loops of interest, depending upon the dynamics of the control loops you are investigating. However, if you are pulling the data from your historian and not directly from your control system, make sure that the data compression is not set too broadly (to save storage capacity), which can distort the statistical analysis by interpolating between threshold values.

When determining service factor, the mode and status states can be converted to a numerical value (e.g. manual = 0 and other modes = 1, etc.) and the average value should be calculated for both parameters. The average values for the converted mode and status states will indicate the service factor for the controller. If either average value is between 0 to 50%, then the service factor is poor and the controller is normally not used. If either parameter is between 50 and 90%, then the service factor is non-optimal and the controller may be adversely affected by particular disturbances. If both parameters are > 90%, then the service factor of the controller is good. Of course, you can modify the bracketing values to meet your preferences.

When analyzing controller performance, multiple calculations are required, but they are easily performed in a spreadsheet. Start with calculating the difference between the set point and process variable of the controller for each available data point (note that the result can be positive or negative). Then calculate the standard deviation for the full array of those difference values. This standard deviation value, divided by the engineering range of the controller (i.e. if the range is -50 F to + 50 F, then the range is 100 F), provides an excellent normalized indication of the performance of the controller. Simply sort the final list of controllers and their performance standard deviations and focus on the controllers with the highest values (start with the top 10, but it can be more if you wish) to identify the controllers with the most significant performance issues.

If you wish, similar statistical analyses can be performed by calculating the difference between the measured value and the alarm value and by calculating the difference between the controller output value and the output clamp value. In these cases however, you will be looking for the smallest value as the indication of a controller that is operating near its constraining value.

For certain controllers, such as levels and temperatures, you may wish to look at the variance of their set points as an indication of whether or not the operators are frequently changing the values.

In the other cases, where the question is not conducive to this statistical approach, then having discussions with a number of the plant operators will often identify additional problematic control loops. The following questions are good starting points for such discussions:

- Are there any controllers that you frequently have to operate in manual mode?
- Are there any controllers in which their output value frequently saturates to its end of scale or clamp position?
- Are there any transmitters/control inputs or analyzers that are frequently out of service?
- Do any of your temperature measurements drift?
- Are there any valves that frequently experience erosion?
- Are there any base level control strategies that do not perform well?
- Is there any part of your advanced control system that does not perform as well as it should?
- Do you have any GC analyzers that are directly used in a control loop?
- In general, are there any other instrumentation or control problems that come to mind?
- Are there any control loops that are associated with causing occasional trips or flaring incidents?
- Are there control loops that used to perform well, but now perform poorly?

When doing so, it is best to discuss these questions with operators from several different shifts because different shifts can have different approaches to the same operating and control issues.

Once you have a short-list of control loops that are not performing as well as they should, then you can move on to troubleshooting the cause and developing the solution. There are three general types of troubleshooting, which are:

- Systematic review of the general control functionality of the controller. This includes a check of the characteristics that are common to most control loops.
- Investigation of particular aspects of specific control loop structures.

- High level assessment of the control strategy/philosophy.

These three methods of troubleshooting are addressed in the next three sections of the paper.

TROUBLESHOOTING GENERAL CONTROL FUNCTIONALITY OF PROBLEMATIC CONTROL LOOPS

This section will address the systematic review of the general control function characteristics of the problematic control loops. Methodically reviewing problematic control loops using the suggestions below has helped control engineers uncover many issues and systematically eliminate potential causes for control system problems. Some of the suggestions are simple ideas, whereas others are not as obvious, but each step is worth investigating. Don't stop when you find one problem, as more than one often occurs in a problematic loop.

TUNING

Controller tuning is mentioned first here because it is often blamed for the majority of regulatory process control problems, although often it is not the culprit. It is most likely that someone has already tried to tune the problematic controller, but did not solve the problem. However, it is best to confirm that tuning is not the source of the problem before moving on to a more in-depth investigation. There are many articles in the literature on the topic of controller tuning, so it will not be covered in depth in this paper, although some tips are provided.

Tuning is often more of an art than a science. While there are always exceptions, the following guidelines have been used successfully in many petrochemical plants to determine if controller tuning constants were reasonable. Note that the suggested gains are on a dimensionless basis and the integral and derivative suggestions are based upon typical process dynamics, but individual applications can vary.

- Flow controllers should generally be tuned for a rapid response without excessively throttling the control valve. Typically, a dimensionless gain between 0.1 and 0.2 with an integral time between 3 to 6 seconds provides good flow control.
- Level controllers should generally be tuned for disturbance rejection without excessive oscillation. Typically a dimensionless gain between 0.8 and 1.2 with an integral time between 3 to 10 minutes (less for small volumes, more for large volumes) is used.
- Temperature controllers should generally be tuned for smooth response to set point change. Typically a dimensionless gain of 0.4 to 0.6, with an integral time between 3 to 10 minutes (depending upon the time constant of the process response) is used.
- The dynamics of pressure controllers can vary quite a bit depending upon the application. Direct pressure relief is generally tuned quickly, whereas pressure control by condensing is generally tuned much more slowly.

- Note that any controllers that exhibit highly oscillatory behavior may be tuned with overly aggressive gain.

It is also important to understand that different vendor control systems use different control equations, so transferring PID controller tuning constants that work well in one system may require adjustment to work well in another system that uses a different control equation. The above values may need to be adjusted for units (i.e. proportional band instead of gain, resets per minute instead of integral time per reset, time units of minutes vs seconds, etc.) used by the control system installed in your plant prior to using them for comparison. The optimal tuning constants will depend upon several factors, including the engineering range of the instrument and controller, control valve size and characteristics, PID equation, etc.). The suggestions above are only provided as a benchmark of reasonableness. If the tuning of the problem control loop deviates significantly from those values (i.e. a factor of three or more after adjusting for your system's equations), then investigating the tuning may be warranted.

If your tuning constants look reasonable, then consider if you need to apply adaptive tuning (also called gain scheduling). If a controller appears to work well in some scenarios and not well in others, then you should consider applying adaptive tuning to the controller. In an ethylene plant, a good example of this type of application is the coil outlet temperature (COT) controller on the cracking heater. The COT controller is typically tuned during normal cracking operation, when an endothermic cracking reaction is absorbing much of the fired duty. However, during decoking there is an exothermic combustion reaction occurring in the radiant coils. As a consequence, the response of the COT to a step change in the fuel firing is quite different between those two operating modes. Adaptive tuning can be used to automatically change the gain of the COT controller during decoking operations to account for this change in the process response of the temperature.

There are other cases in which adaptive tuning might be applicable, including:

- Split-range controllers
- Cascade loops with multiple secondary controllers
- Controllers in non-linear processes in which the process gain changes significantly as the plant load changes (i.e. 100% capacity operation vs turndown at 70%, etc.).

Alternatively, you might want to consider if you need to apply derivative action tuning. An ethylene plant is full of significant, non-continuous disturbances (i.e. furnace switches, feed quality switch, plant load change, dryer switch, etc.) that cause controllers that are normally well behaved to exhibit problems. This can be due to the magnitude and speed of the disturbance or the non-linear nature of the process in the operating region. If a controller's performance is poor during this type of disturbance or if it is frequently placed in manual mode by operators to handle these types of disturbances, then you may wish to consider the use of derivative action. Too much derivative action can lead to instability, but up to 20 seconds of derivative time (in extreme cases) has been useful. Make certain that your control system uses derivative action on the "process" signal only and not on the "error." Otherwise, the derivative action will respond aggressively to set point changes and possibly introduce instability.

INSTRUMENT RELIABILITY, RANGE AND CALIBRATION

For controllers that are mostly or always in manual mode (or tracking/conditional mode), the first thing to check is whether or not the measured process variable is reliable. Trend the measured process variable while the controller is in manual mode and the associated control valve is at a constant opening. The following checklist of measurement characteristics is an indication of a problem with the measurement instrumentation:

- The measured value is very close to or at its low or high end of scale value.
  - If it is constantly frozen at its low or high end of scale value, then the instrument may not be scaled or ranged properly or may be installed incorrectly.
  - If it is close to its low or high end of scale value, but is fluctuating in a reasonable manner, then there is a risk that the controller is operating in a non-linear region of the process or valve response.
- The value exhibits high frequency noise with large amplitude.
- The value appears to be relatively constant and then exhibits large jumps in value.

These instrument problems can include installation mistakes, such as inversion of the flow orifice, insufficient space between flow element and valve, etc.

For controllers that perform poorly, but can remain in the proper mode, then consider that there may be a mismatch or calibration issue. The transmitter range of the measured process value should be checked and compared against the engineering unit range programmed in the DCS or PLC. This can occur with any instrument, as sometimes the instrument is purchased after the control system configuration specification is prepared and any resulting deviation does not make it back into the control system configuration management of change documentation.

A common example of instrument/controller mismatch occurs when there is a difference between the installed thermocouple type and the transmitter calibration type. Another common example of this issue is with density-based level measurements. While a level measurement is almost always scaled from 0 to 100%, if the level measurement is calibrated for a density that doesn't match the liquid in the vessel, then the measured level can be off significantly. This can result in blow through when the level is lower than the reading or carryover when the level is higher than the reading.

FINAL CONTROL ELEMENTS (i.e. control valve, positioner, actuator, etc.)

For controllers that exhibit oscillatory or highly variable performance when in automatic mode, a check of the control valve and its associated hardware is warranted. In particular, if the controller output appears as a saw tooth pattern and the process variable exhibits a square wave, like shown in Figure 2, then valve stiction, which leads to such instability, could be the problem.

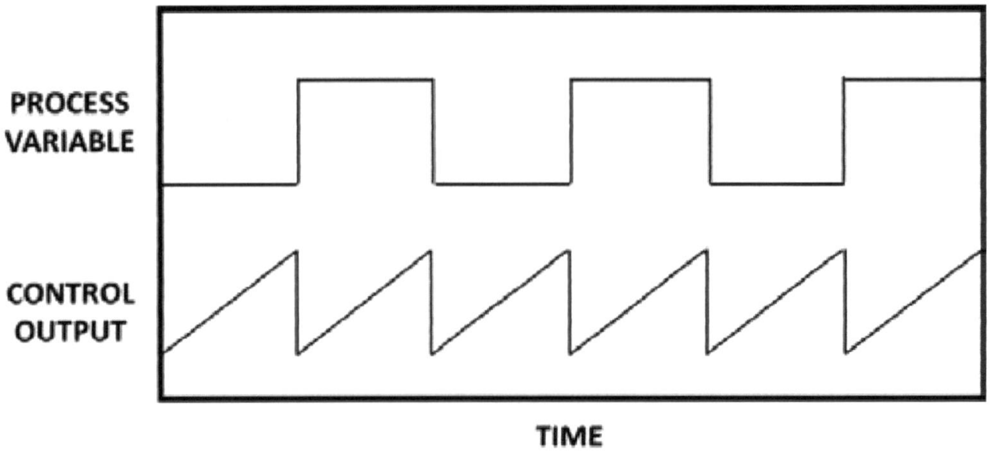

Figure 2. Symptoms of Valve Stiction

There is a quick test for valve stiction. If the measured variable stabilizes when the controller is placed in manual mode with a constant opening of the associated control valve, then control valve stiction could be the source of the problem. There are several possible issues that need to be considered:

- The valve could be sticking. In this case, the valve would have to have a bypass to allow maintenance to be performed on the valve.
    - The packing could be too tight, which restricts the movement of the valve stem.
    - Friction in other sliding services, combined with an underpowered actuator, could result in sticking also.
- Alternatively, the final control element (i.e. actuator, positioner, etc.) could have a deadband that does not initiate a valve movement until a threshold value is reached. In some systems, this deadband is referred to as a hysteresis factor. This can usually be easily adjusted.

If that test does not resolve the issue, then you can try a stroke test. However, it is best to make small changes when performing the stroke test because large changes will usually apply enough force to move the stickiest valve, which will mask the problem. At each increment of the controller output confirm that the valve has indeed moved a matching amount in the field.

A checklist of other possible issues with the final control element includes:

- Mis-calibration or mismatch between the controller output signal and the actual valve position can result in a loss of control range for the valve. If the problem continues to reoccur after recalibration, then installation of a smart positioner might be warranted.
- Each valve has a particular type of trim that provides for more reliable (i.e. linear) response in a particular region of the valve opening (see examples in Figure 3). If the valve has the wrong trim for the application or if the process can operate in multiple regions of the valve opening, then the valve may not be operating in its optimal region and the controller performance can degrade if the valve is operating in a region for which it was not tuned.

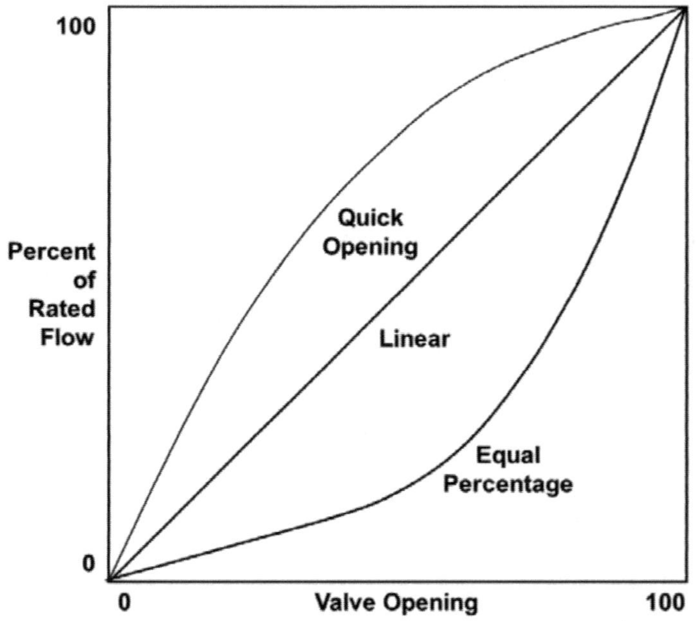

Figure 3. Sample Control Valve Characteristic Curves

CONTROL EQUATION

As noted above, different vendor control systems use different control equations. Some examples of the different basic forms of the PID control algorithm used in different systems are shown below.

Interactive Gain

$$Output = K_c\left[e + \frac{1}{T_i}\int edt + T_d\frac{de}{dt}\right]$$

Independent Gain

$$Output = K_p e + K_i\int edt + K_d\frac{de}{dt}$$

Proportional Band

$$Output = \frac{100}{PB}\left[e + \frac{1}{T_i}\int edt - T_d\left(\frac{de}{dt}\right)\right]$$

For an individual control system the same "structure" is generally used for all controllers, so the differences between systems are not generally a problem – as long as you keep the particular structure of your control system in mind when tuning your loops. However, with the advent of Foundation Fieldbus technology, by which smart devices from multiple vendors perform control logic in the field rather than in a common location, a variety of possible control equations structures may be employed at a single plant site due to the different smart device vendors using different control equations. If this is the case in your plant, take the time to confirm which control equations are used in which devices.

To further complicate the situation, within a given vendor's control system, it may be possible to select from one of several control equations. Although these equations have the same structure, they contain subtle differences that impact the control loop response. In the sample algorithms listed above, the proportional (Kp), integral (Ki), and derivative (Kd) terms are all acting upon a function of the controller error (i.e. set point – measured value). Alternatively, there are PID equations by which the proportional and derivative terms act upon the process variable instead of the error. In most situations, the preferred equation uses the proportional and integral term acting upon the error and the derivative term acting on the process variable for the following reasons:

- If the proportional term acts upon the process variable instead of the error, then there will not be a "proportional kick" whenever the operator changes the set point. This can lead to very sluggish response to set point changes.
- If the derivative term acts upon the error, then the derivative action (which responds to rate of change) can overreact to set point changes (which cause a rapid rate of change in the error).

CONTROL ACTION and VALVE FAILURE MODE

If a controller is always in manual mode and is unstable when you attempt to use it, then one easy thing to do is to check that the control action is configured properly in the control system (i.e. ensure that it matches the design and good engineering sense - in case there is a typo/error in the design information). Controllers are either specified as DIRECT or REVERSE, which defines whether the controller output increases (DIRECT) or decreases (REVERSE) when the measured process variable increases. If the controller is designated with the wrong control action, it will become unstable almost immediately (within minutes in most cases).

It is common for a few controllers out of the hundreds in the plant to be initially configured with the wrong control action prior to the original plant startup. If this check is overlooked, then the controller just remains in disuse, despite such a simple fix being available.

A second thing to check, if a controller is always in manual mode and is unstable when you attempt to use it, is the valve failure mode in combination with the control output processing. All output signals from the control system to valves in the field are normally displayed such that 100% represents a fully open valve and 0% represents a fully closed valve. In the case of valves that use a signal to close (i.e. fail open valves), the inversion of the controller output is performed in a manner that is invisible to the operator. The inversion can commonly be accomplished by configuration of the positioner or by configuration of an output processing block in the control system (see Figure 4 for an example). If this inversion is not properly implemented, it will have the same essential effect on control performance as incorrectly configuring the control action (i.e. it will cause the control to be unstable).

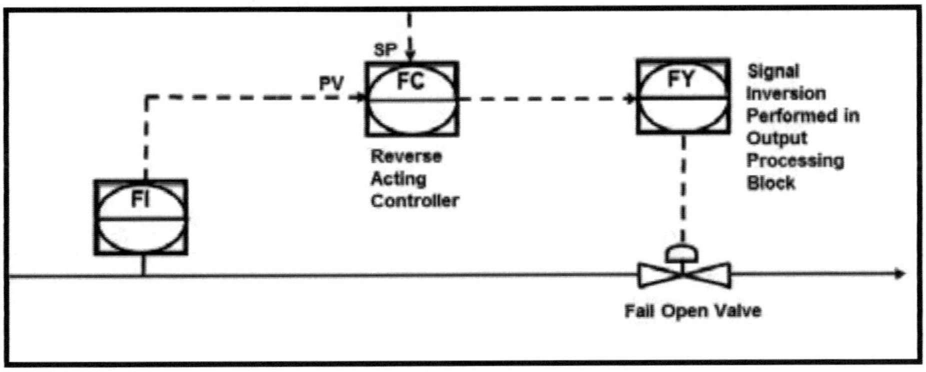

Figure 4. Output Signal Inversion

TRACKING AND INITIALIZATION

If a controller exhibits a significant disturbance whenever it is placed in service, but eventually stabilizes, then it is possible that the controller's tracking and initialization was not configured properly.

Initialization is primarily an issue with cascade and other multi-level control strategies. In these cases, output of the higher level controller is set equal to the equivalent value of the set point of the lower level controller when the lower level controller is not in the mode that allows it to receive a set point (see Figure 5). Most control systems automatically provide this type of initialization when control functions are connected to each other, although there are some special cases in which this does not happen (see the section on "Cascade Strategies with One Primary Controller and Multiple Secondary Controllers" for an example).

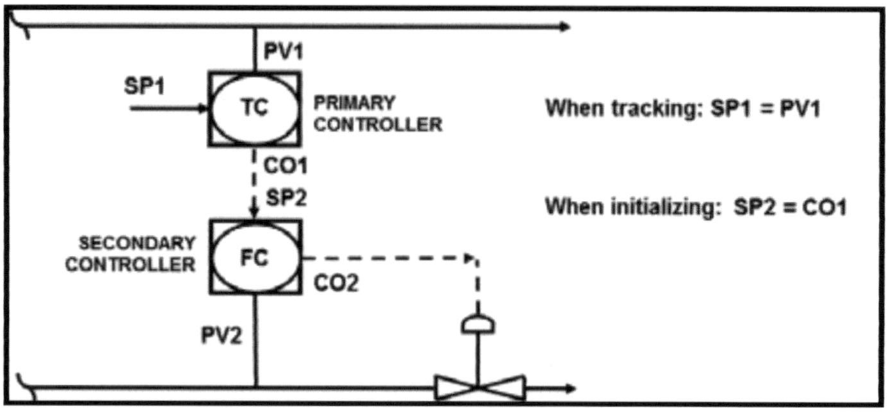

Figure 5. Initialization and Tracking

The more common problem is tracking, which refers to whether or not the value of the set point is automatically adjusted to equal the value of the process variable when a controller is not in service. If tracking is performed, then when the controller is placed in service, the set point and the process variable will be equal and therefore the error will equal a value of zero and the controller output will not instantly jump to a new value. In most cases, the controllers should be set for the set point to track its process variable. However there are some special cases in which the set point tracking should not be used. These include the following cases:

- Override controllers. An example is a high level override that prevents carryover of liquid into a vapor stream.
- Relief controllers. An example is a high pressure relief controller on a tower that sends the overhead stream to flare if the pressure is too high.
- Equipment protection controllers. An example is a minimum flow protection controller on a pump.

In each of these cases, the correct set point of the controller is based upon the safety of the equipment, which in turn depends upon its design. Therefore, once determined, the set point should never change. If set point tracking is used on these types of controllers, then the set point will reset to the current process variable value the first time that the controller is placed in manual mode. As an example, in the case of an override controller, if the override controller is placed back in service with its set point equal to the process variable value, then the controller will immediately override its primary controller.

Even if these special controllers are not configured for tracking, it is still important to make certain that the value of their set points (which should not be changed) have been properly determined according to the requirements of the equipment that they are protecting.

INPUT FILTERING and VARIABLE CLAMPING

If a controller is highly oscillatory or extremely sluggish, it could be due to improper input filtering. A measured process variable exhibiting high frequency signal noise, such as shown in Figure 6, can cause its associated controller to oscillate by accelerating the proportional action on your controller. Filtering can improve this situation, but care must be used. Filtering should only be used to alleviate high frequency noise. If not enough filtering is used, then the proportional action on the controller may exacerbate the situation by responding to the noise and moving the valve needlessly. If too much filtering is used, the true process response will be masked and the control will have to be tuned too sluggishly.

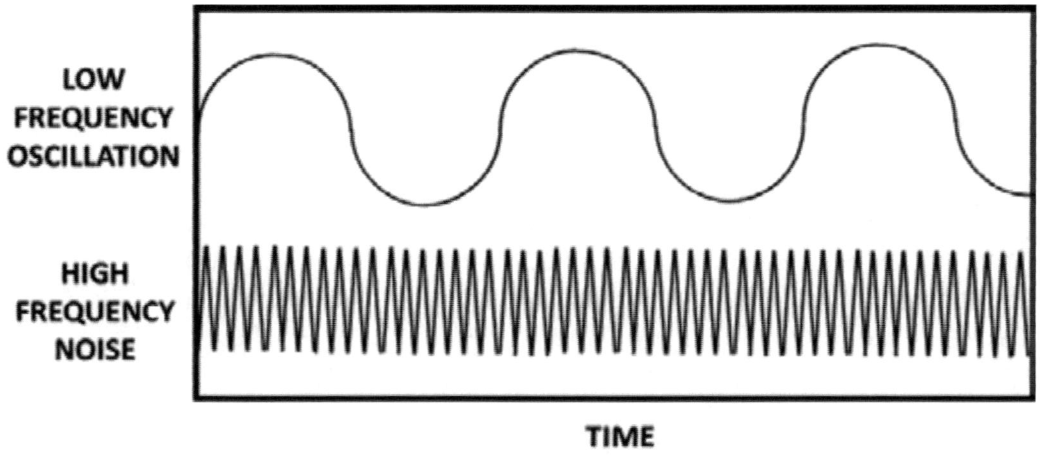

Figure 6. High Frequency Noise vs. Low Frequency Oscillation

On the other hand, if the control response is sluggish, watch out for double dipping on the filtering. Some transmitters have filtering applied in the field. If the signal is filtered again in the DCS or PLC, then the signal may be made too sluggish. A first order filtering constant of 1 to 5 seconds is normally sufficient to squelch high frequency noise.

Do not try to use filtering to attenuate a low frequency oscillation, such as is shown in Figure 6. This will only expand/elongate the oscillation and make it more difficult to control.

Finally, it should also be noted that use of clamps can be problematic in some instances. For example, a set point rate of change clamp is a form of filtering that prevents the operator from changing the value of the set point too rapidly. However, if the clamp rate is set too small, it has the same effect as tuning the controller too sluggishly for set point changes.

In addition, hard clamps (i.e. high and low limits) that are placed on set points and controller outputs can limit the range of the controller if they are not set properly. Confirm that these clamps are not limiting your problematic controller from operating in the full range that is required.

TROUBLESHOOTING SPECIFIC CONTROL LOOP STRUCTURES/ALGORITHMS

While problems can occur in any loop, control loops with the following features could be more prone to non-optimal implementation that can lead to poor performance and therefore should warrant special attention when trying to identify problematic loops, regardless of whether or not they exhibit the generic problem characteristics noted in the previous section.

- Primary controllers with multiple cascaded secondary controllers
- Control loops with overrides
- Split-range controllers
- Level controllers with gap action
- Controllers that use calculated input values
- Controllers that use inputs from GC analyzers
- Fired heat duty controllers
- Heat exchanger heat duty controllers (including reboilers and condensers)

If your problematic loop falls into one of the special categories listed above, then there are specific aspects of the controller that you should check because there are common errors that can occur with the configuration of these types of control loops. It may be more efficient to check these items before proceeding with troubleshooting the more general control functionality issues addressed in the previous section, if your problem controller fits into one of these categories.

CASCADE STRATEGIES WITH ONE PRIMARY AND MULTIPLE SECONDARY CONTROLLERS

A standard cascade control loop has one primary controller and one secondary controller, where the output of the primary controller sets the set point of the secondary controller, as shown in Figure 7. When configuring a control system, if two control algorithms are connected in this type of standard cascade configuration, the system generally automatically performs the proper initialization such that when the secondary controller is placed in cascade mode, the output of the primary controller is set to a value that is equivalent to the current set point of the secondary controller so that no initialization bump occurs when the connection is made.

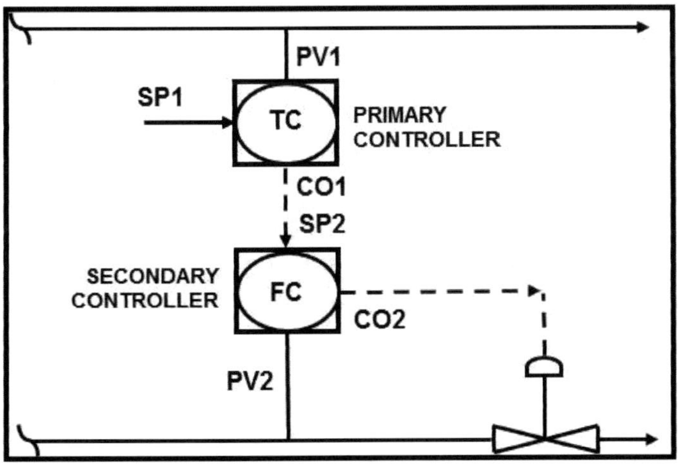

Figure 7. Standard Cascade Loop

However, when there is a cascade structure with one primary controller and multiple secondary controllers, as shown in Figure 8, the initialization mechanism is non-standard and requires customization, so it is more likely to be implemented incorrectly. In this case, the output of the primary controller cannot track all of secondary controller set points simultaneously (i.e. the primary controller only has one output value), but it will send the same output to each of them when it initializes, thereby bumping some of them, assuming that they did not all start with the same value for their set point.

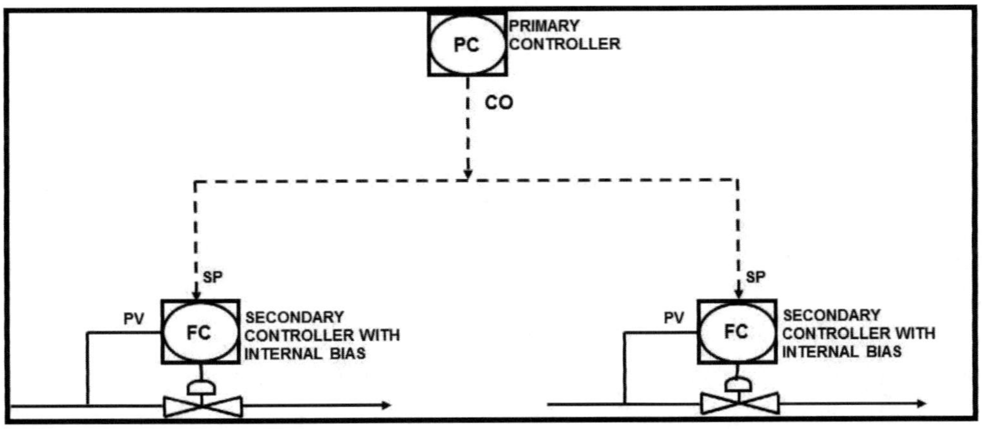

Figure 8. Primary Controller with Multiple Secondary Controllers

If you have such a cascade structure and it exhibits an initialization problem, this problem can be overcome with the use of internal biases in the secondary controllers. The internal biases would be applied to the set points of each of the secondary controllers to compensate for the difference between the primary controller output value and the individual secondary controller set point value.

It is also important to note that the gain of the primary controller will change, depending upon how many secondary controllers are in service. In the example shown in

Figure 8, when the primary pressure controller output signal changes by 1%, then flow to the condensers will change by 1% when both of the condenser flow controllers are in service (i.e. both secondary flow controllers are in cascade mode), but it will only change by 0.5% if one of the two condensers is not in cascade with the primary pressure controller. Therefore the gain of the control response is significantly different depending upon how many of the secondary flow controllers are in cascade mode. This can be addressed by applying adaptive tuning logic (also called gain scheduling) to the primary pressure controller based upon the number of secondary flow controllers that are in cascade mode.

CONTROL LOOPS THAT INCORPORATE OVERRIDES

A simple override control loop has one primary controller and one override controller, plus a signal selector that selects the higher or lower of the two controller output signals (depending upon the objective of the override), as shown in Figure 9.

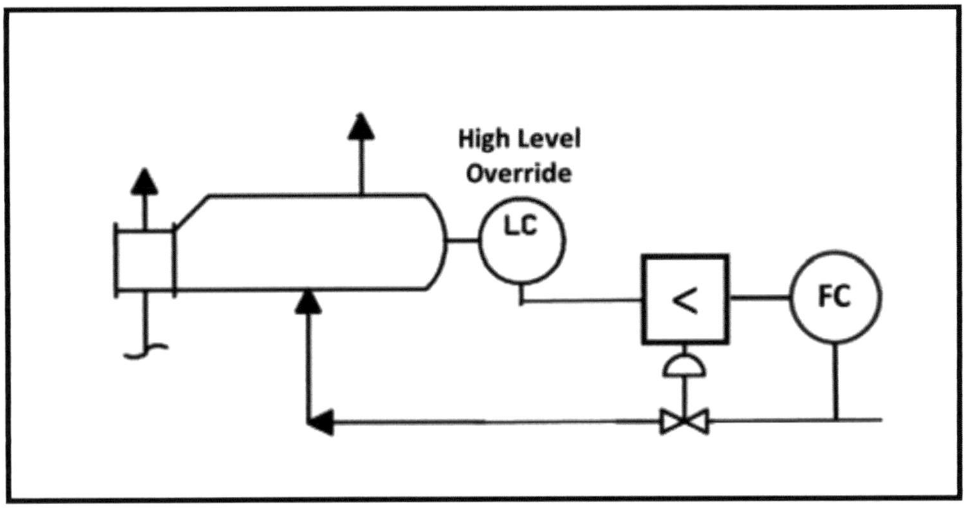

Figure 9. Simple High Level Override Control Strategy

An override controller normally operates with an offset between its set point and measured variable. Therefore, the non-selected controller would have a tendency to windup and saturate at its minimum or maximum controller output value unless the system is configured with external feedback from the selector output to each of the controllers in the override loop, as shown in Figure 10. Some DCS or PLC systems automatically incorporate this logic when a control function is configured, while others require the configuration engineer to specifically specify it. In the latter case, when it is not done properly, the ensuing performance issue often results in operators placing the override controller in manual mode and setting its output to either 0 or 100% to avoid any interference with the primary loop. This essentially negates the existence of the override, which was most likely placed there to protect the equipment or process in some manner.

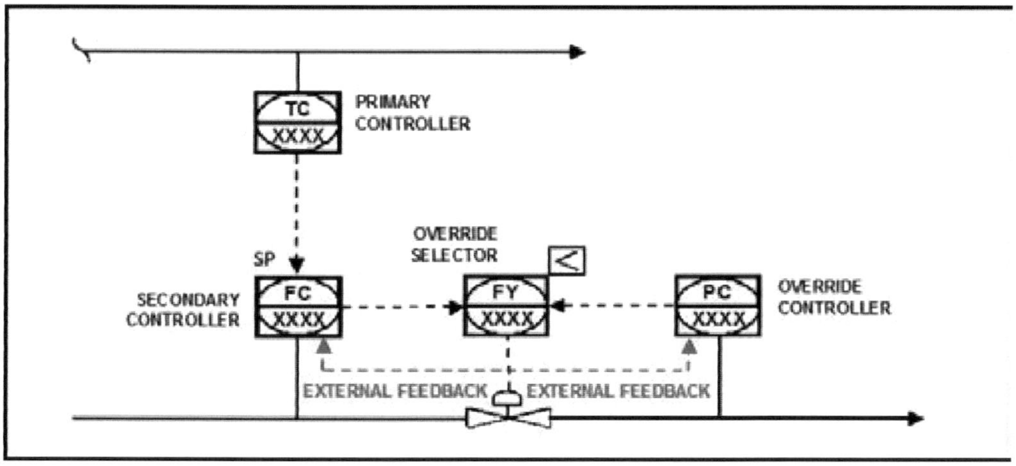

Figure 10. Override Configuration with External Feedback

If the external feedback is properly configured, but the controller output values from your override and normal controllers are still so close that they interfere with each other in the selector block, then you may need to apply more integral action in your override controller.

SPLIT-RANGE CONTROLLERS

A split-range controller is one with an output that sequentially adjusts more than one downstream controller or valve in series based upon the value of its control output signal. As shown in the example in Figure 11, the output of the split-range temperature controller is split and sent to two control valves. As the temperature increases, the A valve is closed from 100% open to 0% open as the controller output value changes from 0 to 50%. Then the B valve opens from 0% open to 100% open as the controller output value changes from 50 to 100%. These controllers are notorious for working well some of the time, but not others. There are several potential reasons (and solutions) for this. Some of the issues that might arise include the following:

- If the process response of the two valves is different (i.e. changing one of the valves by 1% changes the value or dynamics of the temperature differently than a 1% change in the other valve), then a single set of tuning constants may not be optimal for performance across the entire range of operation and adaptive tuning/gain scheduling may be required.
- If the installed valves have very little sensitivity at the ends of their valve opening range (i.e. if there is very little change in flow between a valve opening of 0 to 5% or 95 to 100%), then there will be a deadband in the response of the split-range controller in the transition region between the two valves. In this case, it may be necessary to overlap the action of the two valves (i.e. in the example, start opening the B valve at a controller output value of 45% and do not fully close the A valve until reaching a controller output value of 55%).

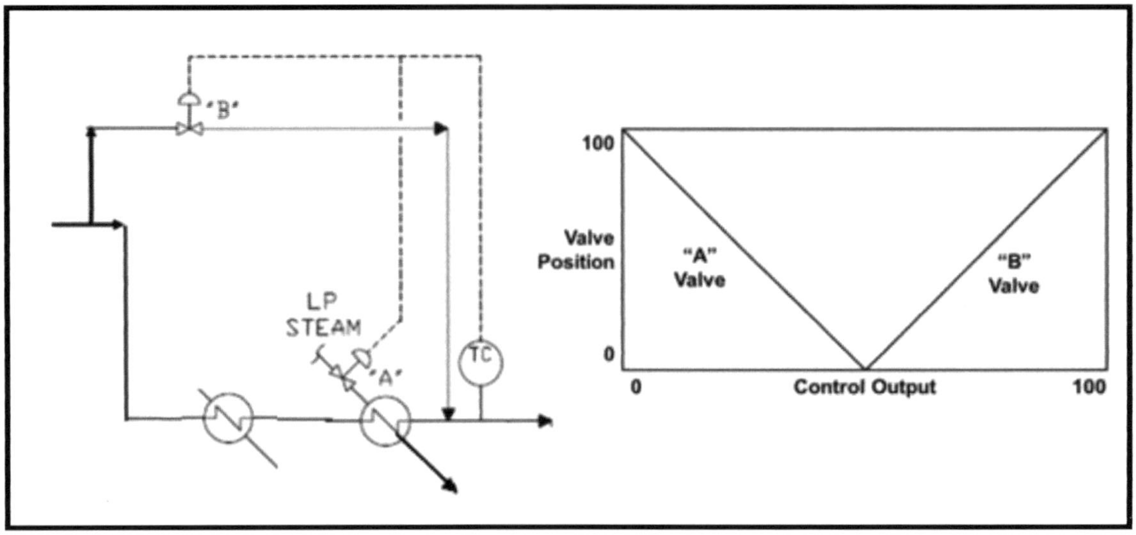

FIGURE 11. Example of a Split-Range Controller

LEVEL CONTROLLERS WITH GAP ACTION

In control jargon, level is not a "self-regulating" variable, but rather an integrating variable. In other words, if the measured level is at a steady-state value and you make a step change in the feed flow rate to the vessel and no change to the output from from the vessel, then the measured level would not reach a new steady-state value, but rather it would eventually overflow or empty the vessel (see Figure 12).

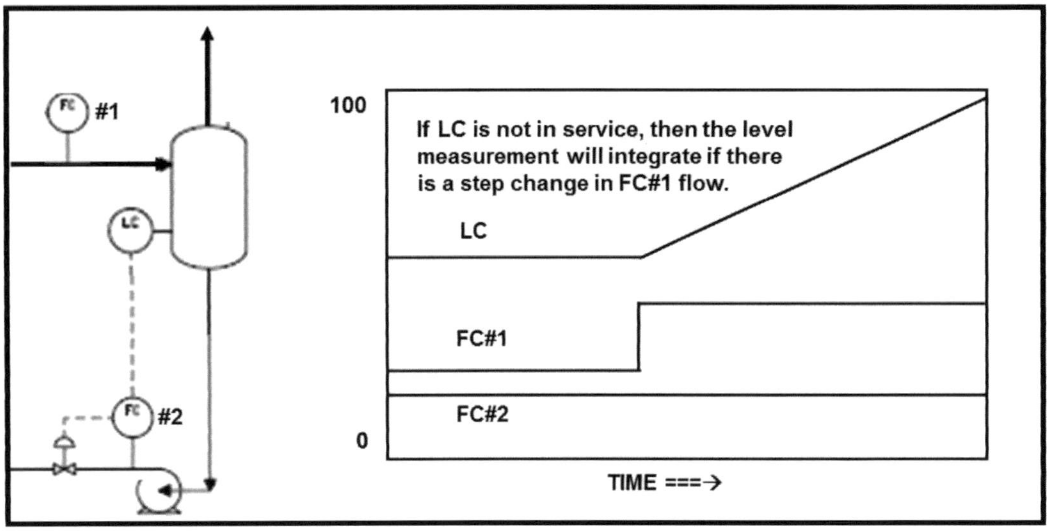

Figure 12. A Simple Level Control Loop

A gap action level control strategy allows the level to float within high and low limits without taking any control action. This type of strategy is actually trickier than it would appear and is often applied in a manner that adds to instability rather than detracts from it, as it was intended. If you use gap-action level control with a zero gain inside the gap, with the intention of letting the level measurement float around in an acceptable band without changing the flow to the downstream equipment, then you may be subject to this integrating effect. As a result you might end up bouncing your measured level between the gap limits and creating large changes in the flow to the downstream equipment when crossing the gap limits. It may be more prudent to pass a one time bump in flow down through the plant than to continually propagate an oscillation in this manner. Alternatively, you could consider using a smaller, non-zero gain inside the gap to minimize the effect on downstream equipment and eliminate the continual oscillation caused by frequently bouncing between the two gap limits. A larger gain would be used outside of the gap limits to handle severe disturbances that would occur much less frequently.

CONTROLLERS THAT USE CALCULATED PROCESS VARIABLES (i.e. flow compensation, heat duty control, etc.)

The power of the DCS has encouraged the implementation of many loops that use a calculated value as their process variable. The basic calculation and control is normally quite straightforward, but there are hidden traps. Consider the simple example of a controller for a flow that is compensated for temperature and pressure (see Figure 13). The trap occurs when the non-critical variable (i.e. temperature or pressure) signal quality drifts out of a reasonable range or obtains a bad status for some reason. Depending upon how the configuration was performed, one of several things could happen, including:

- An inaccurate calculated flow reading
- A bump in the calculated flow value that causes the controller to respond erroneously (i.e. either when the status of the non-critical variable (e.g. temperature or pressure) goes bad or returns to good status).
- An automatic change to manual mode operation for the controller

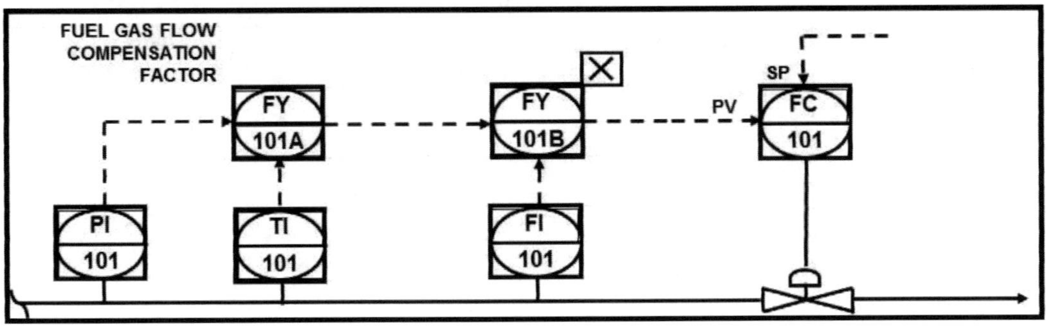

Figure 13. Typical Flow Compensation

These problems can be mitigated by applying the following logic:

- Clamping the non-critical inputs within reasonable bounds to protect against signal drift.

- Installing logic to continue to use the last good value of a non-critical input if its status becomes bad – so that the controller can continue to operate based upon the value of the critical input.

Similar "bumping" can occur when the input to the controller is switchable. When two or more potential input values pass through a switch prior to being used in a controller, the inputs that are not in use will likely have a different value than the one that is in use. Therefore, switching the inputs without properly initializing the controller will "bump" the control response – even though the process hasn't actually changed.

Another trap common with this type of loop is the mismatch in execution or communications timing between the various processing blocks in the control system configuration. Depending upon how the control system is organized, the various inputs to the calculation may come from different "boxes" or may be processed at different frequencies. Therefore, at the time that the calculation is performed, the inputs may not be synchronized with each other. Check with your control system configuration engineer to confirm that all of the inputs and the controller function blocks are synchronized properly.

CONTROLLERS WITH MEASURED PROCESS VARIABLES FROM GC ANALYZERS (i.e. non-continuous inputs)

Use of inputs from gas chromatographs (i.e. GCs) in controllers requires special processing, which may have been neglected during implementation. A GC only sends a new analysis value on an intermittent basis (e.g. once every 3 to 5 minutes, or longer if the analyzer is multi-streamed). If the controller that uses that signal is not synchronized with the update frequency of the analyzer input, several problems can occur, as listed below:

- If the controller that is using the analyzer input is running continuously (i.e. executing once a second), then it will be subject to windup. In this case, it will either result in oscillatory behavior or a sluggish response (if it has been "detuned" to compensate for the mismatch in execution timing).
- If the controller is run intermittently, on the same frequency as the analyzer, other problems can occur.
    - If the signals are not synchronized, then additional dead time will be introduced into the loop, which will require detuning and sluggish response.
    - If the analyzer freezes (i.e. stops sending an updated value), then the controller output will wind up (i.e. its output will continue to change and drive the controller output and the process variable to an out of range value).

To avoid these problems, it is best (if possible) to trigger the execution of the controller whenever a new analyzer value is sent. This can be done directly if the analyzer also sends a digital bit indicating that a new value has been sent or indirectly by continually monitoring the analyzer for a change (to several significant digits) and generating your own digital bit trigger.

If your DCS or PLC functionality will not allow such triggering of controllers, then watchdog timer logic should be used to determine if the analyzer value has frozen and if so, change the controller to manual mode.

RATIO CONTROLLERS

Ratio controllers are very simple, so they are often overlooked during troubleshooting because "What could go wrong with a ratio controller?" While unlikely, mistakes have been made. The key thing to check in this case is which type of function block was used for the ratio controller because several different techniques are available.

The best technique is to use a ratio control block. This uses a simple ratio equation (i.e. set point x process variable (as called wild input) = controller output), with no dynamic adjustment, but includes all of the normal control processing functionality (i.e. tracking, initialization, etc.). A problem can occur when one of the following techniques is used instead of the ratio control block:

- If a ratio calculation block is used instead of a ratio control block, then none of the control processing functionality will be in place, which will lead to initialization problems.
- If the ratio to be controlled is calculated as a process variable (see section on "Controllers That Use Calculated Process Variables") and then used as an input to a PID type controller, an unnecessary dynamic lag will be introduced into the control loop which will make it more sluggish than it needs to be. However, please note that depending upon the ratio that needs to be controlled, it may be necessary to use a calculated ratio.

FLOODBACK EXCHANGERS/REBOILERS

An example of a floodback reboiler is shown in Figure 14. The level of condensate is maintained in the reboiler between the top and bottom of the tube bundle at a point of equilibrium between the steam condensing pressure and the available surface area in the reboiler to the extent that the level in the exchanger will provide just the right amount of surface to condense the steam at a pressure sufficient to enter the LP condensate header. (Generally, the condensate pot will always operate completely filled.)

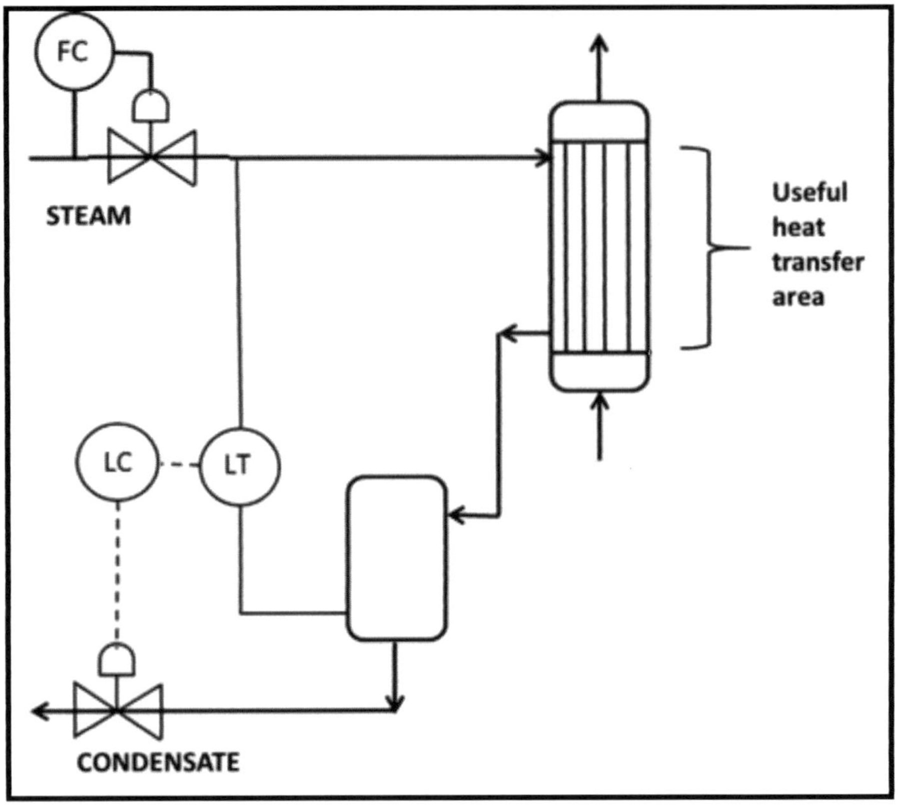

Figure 14. Floodback Exchanger Hookup

Note that the top tap is located above the top of the tube bundle and the bottom tap is located at the bottom of the condensate pot. Therefore, the usable level (i.e. the elevation between the top and bottom of the tube bundle), is a more narrow range than the full 0 to 100% scale of the level indicator.

To prevent a non-linear response by operating outside of the useful heat transfer surface area, consideration should be given to alarming the level measurement and clamping the level set point at the level values that are equivalent to the top and bottom of the useful heat transfer surface area. Note that elevation information will be required to calculate these values.

FIRED FUEL HEAT DUTY CONTROLLERS

Controllers used to maintain the fired duty from combusting fuel generally work well. In an ethylene plant, they are used to control the fired duty in the cracking heaters. Please note that these controllers are a subset of the category of "Controllers That Use Calculated Process Variables" that was mentioned earlier in this paper (refer to that section for general troubleshooting concerns for this type of controller).

However, there are some additional concerns to consider in this case. Many of the fired fuel heat duty controllers use a Wobbe Index analyzer. You must take care when configuring your system to properly handle the engineering units and the flow compensation when using a Wobbe Index value in the calculation. The Wobbe Index value and the fuel flow are used to calculate the fired heat duty using the equation given below, when the flow is measured with a differential pressure device such as an orifice plate:

$$Q = FI \times \sqrt{\frac{P}{P_B}\frac{T_B}{T}} \times \sqrt{SG_B} \times \text{Wobbe Index}$$

Where:

FI = measured flow in standard volumetric units per hour
P = pressure of fuel gas, in absolute units
P_B = base pressure, in absolute units, at which the flow meter was calibrated
SG = specific gravity of the gas (ratio of the gas molecular weight to that of air)
SG_B = base gas specific gravity at which the meter was calibrated
T = temperature of fuel gas, in absolute units
T_B = base temperature in absolute units, at which the flow meter was calibrated
Q = fired heat duty per hour

$$\text{Wobbe Index} = \frac{\text{Heating Value}}{\sqrt{\text{SG}}}$$

The flow measurement is corrected for temperature and pressure, but not fuel gas molecular weight because the molecular weight correction is part of the Wobbe Index equation. Please note also that the flow measurement must be in volumetric units, not mass units, for that compensation to work properly.

Alternatively, this type of controller might use the heating value that is correlated from the molecular weight of the fuel (based upon a continuous molecular weight analyzer signal). This works well on a volumetric basis, but the correlation is non-linear on a mass basis and can lead to problems around the edges of your operating range (see Figure 15).

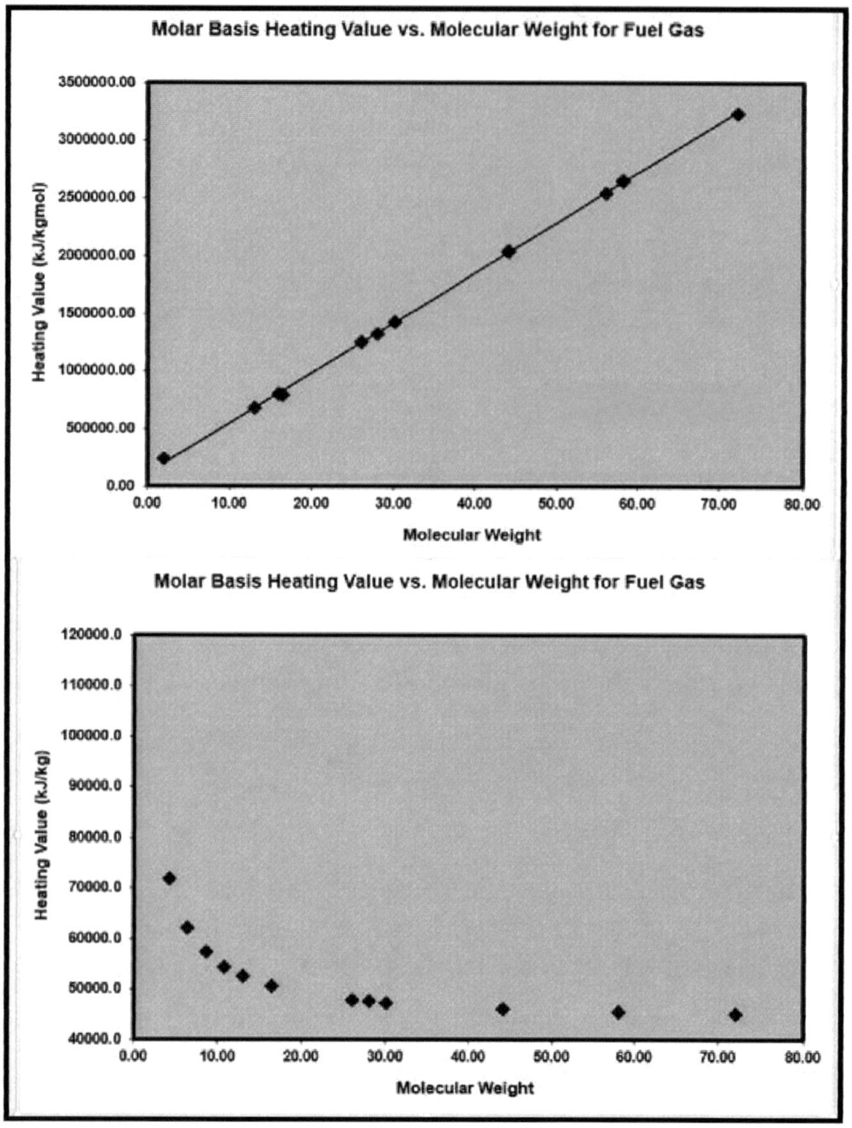

Figure 15. Fuel Heating Value vs Molecular Weight

Of course, the heating value of the fuel should not be calculated from a GC analyzer. The reason is that the GC does not update continuously (i.e. it might update every 3 to 5 minutes or longer if it is multi-streamed). Therefore, each time it updates, it will bump the calculated value rather than continuously update the value as when a Wobbe meter or continuous molecular weight analyzer is used.

HEAT EXCHANGER HEAT DUTY CONTROLLERS (including reboilers and condensers)

As opposed to fuel fired heat duty controllers, these controllers maintain the heat transfer that occurs via heat exchange. There are two basic types. The first type is simple heat exchange between two process streams, while the second type is condensing/vaporizing heat exchanger with utilities like steam and refrigeration. Please note that these controllers are a subset of the category of "Controllers That Use Calculated Process Variables" that was mentioned earlier in this paper (refer to that section for general troubleshooting concerns).

However, there are some additional concerns to consider in this case. While these controllers would appear to be quite simple and intuitive, they can actually be quite troublesome. In the first case, where they are used in exchangers with no change in phase and the heat duty is calculated based upon the temperature difference across the exchanger (and the flow and heat capacity of the fluids), there is often a dynamic lag of several minutes between the time at which the inlet temperature changes and the time at which the outlet temperature reaches its corresponding new steady-state value. The larger the exchanger, the bigger the dynamic offset. Unless this dynamic is properly compensated, the temperature differential calculation will infer a disturbance that is really only a dynamic blip. The controller will respond when it shouldn't and add to the instability of the loop. This can lead to detuning the controller and resultant sluggish performance. It is often more effective to control the flow of the fluid and, if the fluid properties change often, add a feedforward adjustment for the inlet conditions of the fluid.

In the second case, the dynamic lag in the outlet temperature is not as much of a concern, since the condensation or vaporization generally takes place at a constant temperature. However, the utility can have varying degrees of superheating or sub-cooling, which can impact the total heat transfer. Moreover, if the pressure of the supplied utility varies, so will its latent heat. These factors should be considered when developing the heat duty calculation.

TROUBLESHOOTING CONTROL STRATEGY/PHILOSOPHY AT A HIGHER LEVEL

Are you still having a problem with your control loop? After checking the control functionality and structure as noted above, then it is time to take a more global look at the control strategy and philosophy. Here is a checklist of things to consider:

- Make sure you are working on the true problem. While the cause for most problematic loops is self-contained, sometimes the obvious symptom is masking a more deeply hidden problem. For a simple example, if a tower temperature controller is performing poorly, sometimes it is related to poor level control in the upstream tower that is causing abrupt or oscillatory feed disturbances in the tower that the temperature controller cannot handle. Another typical example occurs when an oscillation in cracking heater COT is traced to an oscillation in the fuel gas drum pressure control. See if the poor performance in your problematic loop correlates to poor performance further upstream.
- Keep in mind that controllers interact with each other, so what may appear to be poor performance of one controller can actually be caused by a disturbance due to inappropriate tuning of a related controller. When tuning a group of associated controllers, start with the fastest loops (usually flow and pressure differential) and move

toward the slowest loops (usually temperature and level). Review the control strategy to confirm that the proper variables have been paired in the loop.

  - In some cases, the interacting controllers may have the same process dynamics. In that case, the more important controller should be tuned for the appropriate response and the less important controller should be detuned.

- Interference between normal control strategies and logic functions such as dryer switch sequencing can create interactions that disturb normal control functions. This interaction can interfere with the operation of the normal control loops. Look for these types of interference.
- The performance of the standard PID control algorithm is less effective when the process response dead time and time lag is long. If you can find no other fault and the process control response has a very long dead time or time constants, then you may want to consider the application of advanced control technology, which is better at handling long dynamics and interactions between variables.
- If a controller performs worse at dawn and at dusk, consider if it could be impacted by the cooling water temperature. As the sun rises and sets, ambient temperature changes more rapidly, which can impact the cooling water temperature on a daily cyclic basis.
- Misunderstanding of control philosophy/objectives by operations personnel can lead to improper actions that appear as bad control performance. Reviewing the control strategy philosophy for the problem loop with operations personnel can alleviate many problems.
- If a controller performs worse when a rain storm occurs, consider if it could be impacted by damaged or absent insulation, which can result in rapid cooling of the piping and equipment.
- When a pump runs out toward the end of its characteristic curve, the head produced will decrease. This may affect the response of any control valves that are downstream, as the upstream pressure on the control valve (i.e. driving force) will be reduced, possibly causing the valve to open wider and approach a non-linear region of operation.
- Consider if there are manual operations, such as opening or closing manual valves, that will affect the performance of your process, but that will not show up in a statistic review or trend of operating data.

GENERAL CONTROL LOOP TROUBLESHOOTING ADVICE

In addition to the specific suggestions made in this paper, the following general advice may prove helpful during your troubleshooting efforts.

- Keep an open mind and consider all possible issues including the process control philosophy, DCS/PLC configuration, instrumentation, etc. Don't try to force the solution to a preconceived notion.
- Discuss the control loop behavior with operations. Be careful to sort the fact from the opinion and the "lore" of the operations history.

- To the extent possible, when troubleshooting, stabilize the operation around the area of interest to prevent masking the true problem with external influences.
- Prepare sufficient real-time and historical data trending with suitable resolution when performing your analysis.
- Have contact information for the appropriate vendor personnel available so that you can call them if you get stumped or are uncertain about the functionality of their system (i.e. DCS/PLC configuration engineers, analyzer specialists, etc.)
- Have all pertinent documentation readily available at site for reference when troubleshooting. For example, /PLC configuration manual and loop diagrams (i.e. configuration drawings), instrument specification sheets, override and relief set point settings, etc.
- Be aware of operational plans, such as furnace or feed switches, dryer switches, change in plant load, etc. that may impact your troubleshooting efforts.
- Use extreme care when changing the DCS/PLC configuration online to implement a solution. One system crash can use up a lot of accumulated good will. In addition, if not performed correctly, valves can be slammed shut leading to potentially dangerous consequences.

SUMMARY

Achieving the full potential performance of a petrochemical plant is dependent upon proper operation of the process equipment, which is closely aligned with the effectiveness of the plant's control system. Prior to investing in advanced process control (APC) technologies to close the gap between actual and potential plant performance, a rectification of the basic regulatory control system may produce a number opportunities for improved performance at very low cost. Moreover, periodic reviews of the basic regulatory control system will identify more opportunities throughout the life cycle of the plant.

There are straightforward statistical analyses that can be performed on data, which can be easily extracted from the control system, to identify potentially problematic control loops that need attention or investigation. Once the target control loops are identified, there are several common faults that can be investigated quickly to determine if a simple solution is available. Those potential faults (which are described in this paper) fall into one of three categories:

- General control function characteristics
- Specific control loop structure.
- Control strategy philosophy

Hopefully, some of these tips will help you achieve improved performance of your regulatory control system and your plant. Good luck!

AIChE Paper Number: 4D

ON-LINE MONITORING OF WATER AT SUB-PPM LEVELS IN CRACKED GAS AT THE OUTLET OF MOLECULAR SIEVE DRYER VESSELS USING TUNABLE DIODE LASER ABSORPTION SPECTROSCOPY ANALYZERS

W. Gary Engelhart
Product Line Marketing Manager
SpectraSensors, Inc.

Dr. Xiang (Sherry) Liu
Director of R&D
SpectraSensors, Inc.

Luc Broekaert
Manager – Center of Competence (CoC)
SpectraSensors, Inc.

Prepared for Presentation at the 2019 Spring National Meeting
New Orleans, LA, April 1 – 4, 2019

ON-LINE MONITORING OF WATER AT SUB-PPM LEVELS IN CRACKED GAS AT THE OUTLET OF MOLECULAR SIEVE DRYER VESSELS USING TUNABLE DIODE LASER ABSORPTION SPECTROSCOPY ANALYZERS

W. Gary Engelhart
Product Line Marketing Manager
SpectraSensors, Inc.
Dr. Xiang (Sherry) Liu
Director of R&D
SpectraSensors, Inc.
Luc Broekaert
Manager – Center of Competence (CoC)
SpectraSensors, Inc.

Abstract:

Molecular sieve dehydration of cracked gas is a critical step for efficient separation and recovery of ethylene. Breakthrough of elevated concentrations of water from molecular sieve adsorbent beds can lead to freeze-up of cold box heat exchangers and curtailment or stoppage of ethylene production.

Traditional analytical techniques for measuring water in dry cracked gas include aluminum oxide (Al_2O_3) capacitance probes and quartz crystal microbalances (QCMs). Both techniques employ an indirect measurement principle with a non-specific molecular response to water. Consequently, these analyzers respond to water and some other compounds present in cracked gas. The resulting measurement bias complicates detection of H_2O breakthrough and efficient operation of molecular sieve dehydration systems.

Tunable diode laser absorption spectroscopy (TDLAS) analyzers employ a high-resolution laser that emits a near infrared (NIR) wavelength that is selective and specific for water molecules in complex hydrocarbon gas streams. This design characteristic supports accurate and reliable trace-level measurement of water in cracked gas following molecular sieve dehydration. Data demonstrating TDLAS analyzer detection sensitivity at sub-ppm concentrations of H_2O in cracked gas, long-term measurement stability and repeatability, and results obtained from field installations will be presented.

Introduction

Production of high-purity ethylene (C_2H_4) involves steam cracking of naphtha or ethane followed by a series of unit operations to remove or convert contaminants in the resulting cracked gas stream as depicted in Figure 1. Cracked gas exiting the quench tower is compressed by a multi-stage compressor and fed to a caustic wash tower to remove CO_2 and H_2S by contacting the gas with a countercurrent stream of aqueous sodium hydroxide (NaOH). Cracked gas exiting the caustic wash tower is saturated with water vapor. Water must be removed before the cracked gas undergoes cryogenic fractionation to avoid formation of hydrates and ice (1). Gas treated in the caustic wash tower is compressed and then cooled to remove as much entrained water as possible before it enters the molecular sieve dryers. Compressing and cooling the gas reduces the water load on the molecular sieve adsorbent beds. This reduces the frequency of bed regeneration which extends the overall service life of molecular sieve beds. Decreasing regeneration frequency also lowers energy consumption and operating costs associated with regeneration.

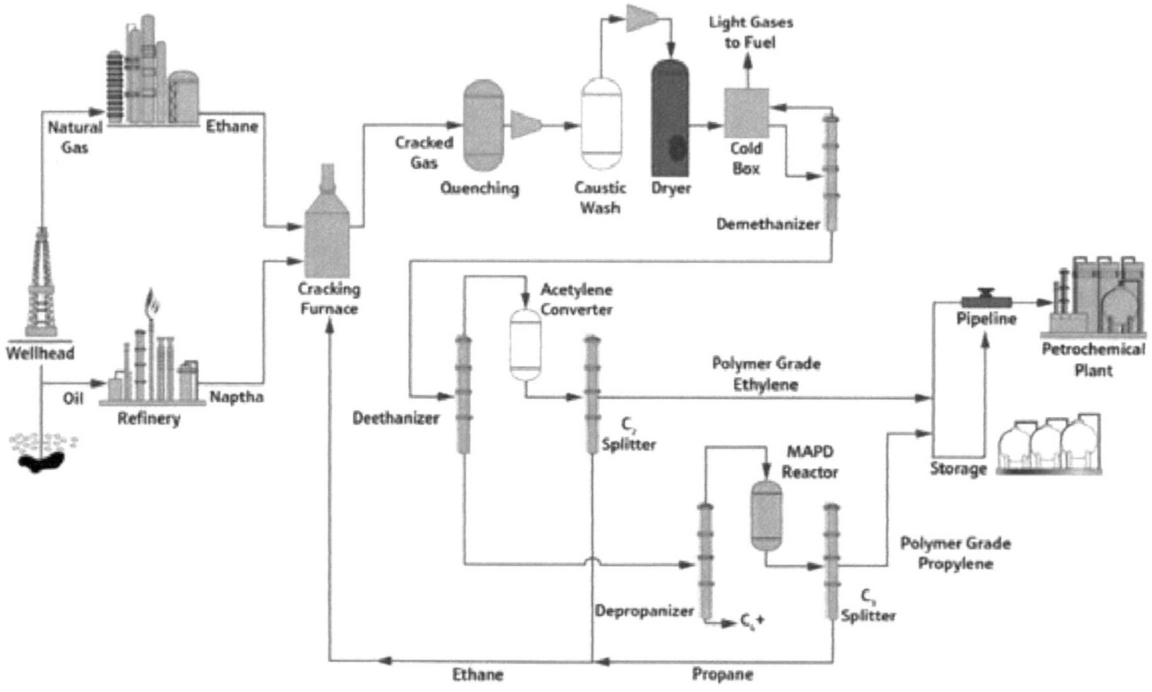

Figure 1. Simplified Diagram of Unit Operations in the Ethylene Production Process

Molecular sieve dehydration is used to dry cracked gas down to < 1 ppm_v, before it is introduced into the cold box heat exchanger to remove hydrogen and light gases and sent on to the fractionation columns.

Molecular Sieve Dehydration

The use of molecular sieve adsorbents to dry cracked gas was patented in 1964 (2). The process supplanted the use of amorphous silica gel and aluminosilicate (zeolite) adsorbents. Amorphous adsorbents contain a broad range of pore sizes (from 20Å to 10,000Å) (3). Consequently, they exhibit no selectivity to separate molecules based upon size. Molecular sieves contain a network of precisely controlled pore size openings that facilitate separation of molecules by size exclusion. Molecular sieves with a 3Å nominal pore size are typically employed for dehydration of cracked gas (4). Figure 2 contains a table listing the molecular size (in Angstroms) of some molecules found in cracked gas, and underscores why 3Å molecular sieves are used for dehydration.

Molecule	Size Å (Angstroms)
Water	**2.6**
Carbon Dioxide	3.3
Acetylene	3.3
Hydrogen Sulfide	3.6
Ethylene	3.9
Propylene	4.0
1,3 Butadiene	4.0
Ethane	4.2
Propane	4.3
i-Butane	5.1

Figure 2. Relative Size of Molecules Found in Cracked Gas (in Angstroms)

Cracked gas is introduced at the top of a molecular sieve dryer vessel. In operation, water molecules entrained in the cracked gas diffuse into the pores formed by the tetrahedron crystalline structure of molecular sieves. As water in the gas is adsorbed the mass transfer zone where active adsorption is occurring progressively moves down the adsorbent bed as depicted in Figure 3. Adsorbent that is saturated with water forms an equilibrium zone within the bed in which no further adsorption occurs. As the mass transfer zone advances down the adsorbent bed and the equilibrium zone expands, the unused and active adsorbent capacity decreases. At some point traces of water begin to "breakthrough" from the adsorbent bed and are present in gas exiting the dryer vessel.

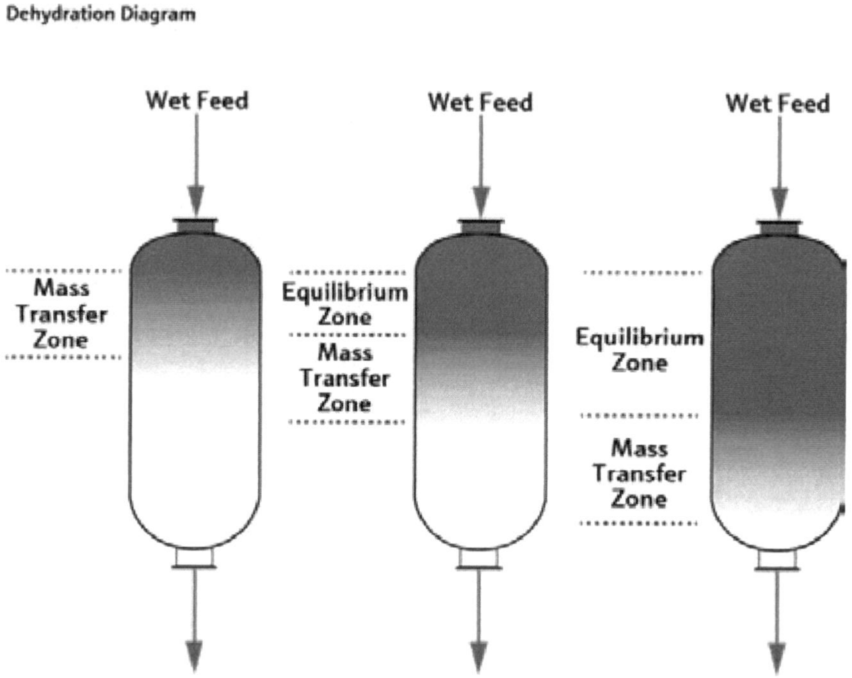

Figure 3. Depiction of Mass Transfer of H_2O During Molecular Sieve Dehydration

Monitoring and detecting moisture breakthrough from molecular sieve cracked gas dryer vessels is of critical importance. A publication by one molecular sieve supplier underscores this point stating; "Operation of a cracked-gas dryer may be improved by continuously monitoring the effluent gas to detect moisture breakthrough. Moisture breakthrough is defined as that point in the drying cycle when the first trace of moisture reaches the point of measurement. This procedure detects any upset in the drying operation, permitting the dryer to be switched before icing can occur in the downstream equipment and a shutdown becomes necessary" (3).

A process flow diagram of a 3-bed molecular sieve dehydration system with locations of possible measurement points is shown in Figure 4.

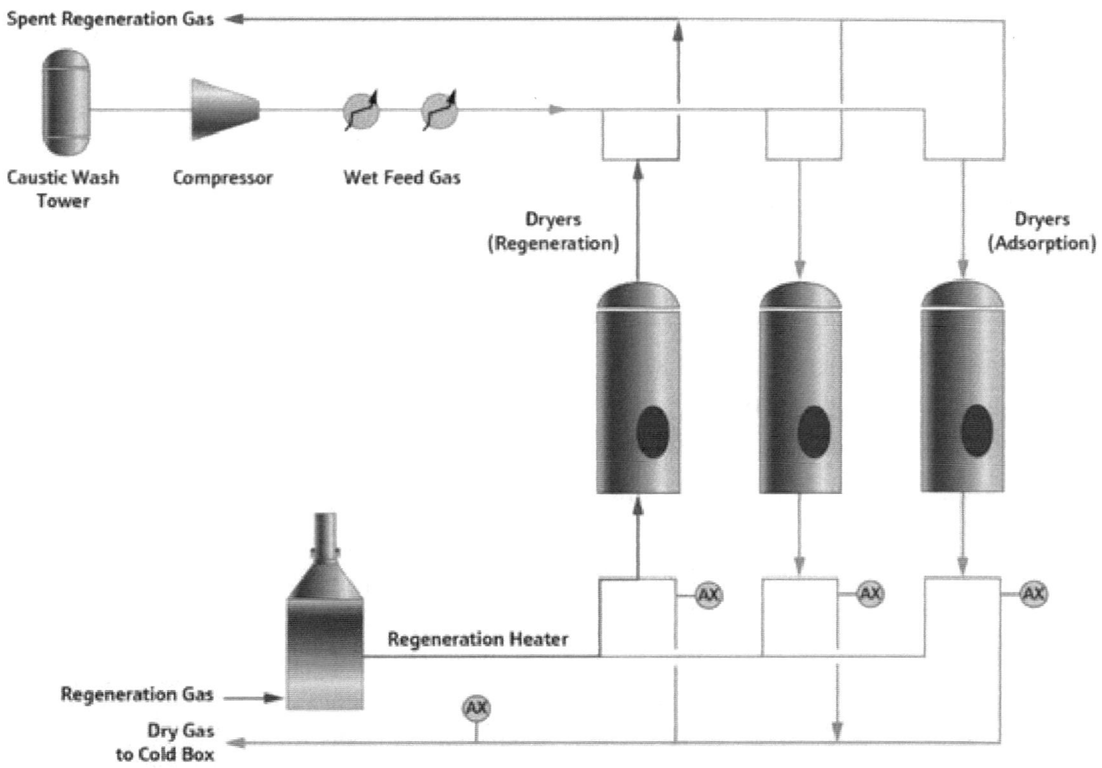

Figure 4. Process Flow Diagram of a 3-Bed Molecular Sieve Cracked Gas Dehydration System Showing Locations of Possible H_2O Measurement Points

Traditional Techniques for Monitoring Moisture in Cracked Gas

The development and commercial application of molecular sieves for dehydration of process gas streams started 50+ years ago. Aluminum oxide (Al_2O_3) capacitance sensors and quartz crystal microbalances (QCMs) available during this time-period became the traditional technologies used for trace level moisture measurement in gas streams. Both techniques employ an indirect measurement principle with a non-specific molecular response to water. Consequently, these analyzers respond to water and other compounds present in cracked gas. This can result in a positive measurement bias that complicates detection of H_2O breakthrough from molecular sieve dehydration systems. Because the operating principle of both techniques requires that the sensor element be in direct contact with the process gas stream, any measurement bias arising from other components in the gas stream cannot be avoided.

Union Carbide adopted QCM technology for water measurement in cracked gas, commenting; "To effectively monitor the dryer effluent quality, the water analyzer should have a minimum sensitivity of at least one part per million by volume (1 ppmv), and should be unaffected by the presence of olefins. Experience in commercial installations has shown the

DuPont Model 510 Sorption Hygrometer to be a suitable analyzer for this application. This commercial analyzer has been modified by Union Carbide to improve its ability to detect water in cracked gas...Note that precautions have been taken to prevent liquid water or hydrocarbons from reaching the analyzer and rendering it inoperable" (3).

Cracked gas is a particularly challenging matrix for QCM analyzers. The hygroscopic coating on the sensor element in contact with cracked gas will change and degrade over time affecting measurement performance. A review of trace water vapor analysis techniques noted; "Users should also be aware that the absorption of water vapor by the QCM sensor is dependent on the material properties of the polymer film. Although the sensors are quite rugged, these properties can change over time or can be affected by exposure to different sample gases. Therefore, periodic calibration of the sensor, as well as detection limit confirmation is required" (5).

TDLAS Analyzer Design & Performance Characteristics

TDLAS analyzers possess design and performance characteristics that make them especially well-suited for monitoring trace-level (low ppm to sub-ppm) concentrations of H_2O in hydrocarbon gas streams.

The high-resolution laser used in TDLAS H_2O analyzers emits a wavelength of near-infrared (NIR) light that is selective and specific for H_2O molecules in a gas stream facilitating concentration measurements based upon molar absorptivity.

The solid-state laser and detector components in a TDLAS analyzer are isolated by a window from the process gas flowing continuously through the measurement cell as depicted in Figure 5. This design allows measurements to be performed with absolutely no contact between the process gas (and entrained contaminants) and these critical analyzer components.

Cracked gas contains highly reactive compounds such as acetylene, olefins, and diolefins. These compounds can react on the catalytically active surface of the macro-porous binder material of molecular sieve adsorbents (4). Adsorbed hydrocarbons polymerize forming "green oil", a viscous mixture of C_4 – C_{20} oligomers (1,4). Droplets of green oil are entrained in cracked gas exiting molecular sieve dryer vessels and sample streams directed to H_2O analyzers. Deposition of the green oil on the sensing element of Al_2O_3 capacitance probes and QCM analyzers in direct contact with the cracked gas stream is unavoidable and decreases their on-stream factor (availability). Installing coalescing and demisting filters mitigates fouling of these direct sensor technologies but doesn't prevent it. In-situ cleaning of the sensing elements is not possible, so fouling leads to measurement biases and errors. The non-contact laser spectroscopy principle of TDLAS analyzers avoids measurement biases and errors and provides better analyzer availability.

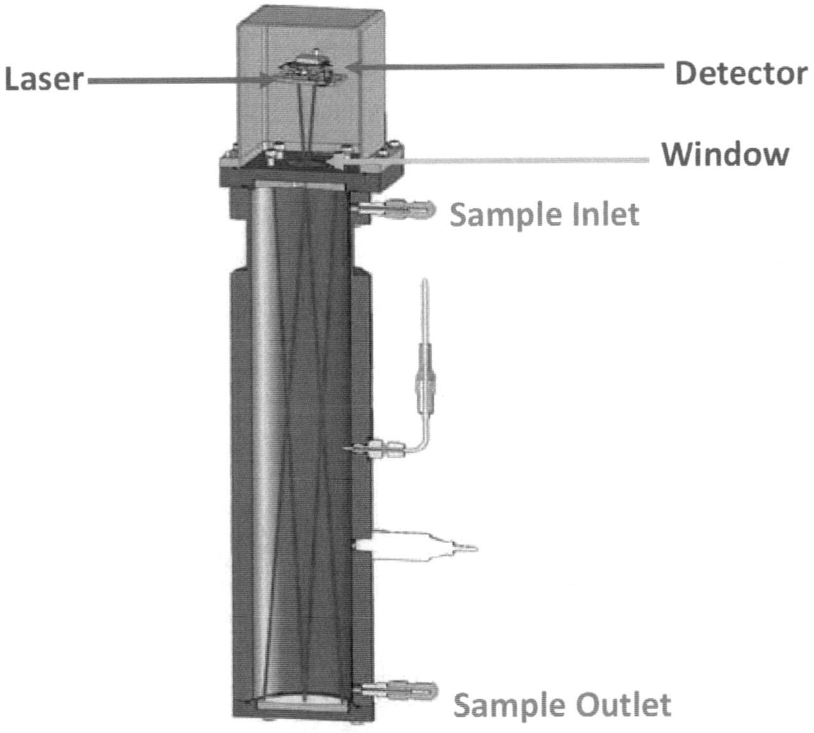

Figure 5. TDLAS Analyzer Flow Cell for Trace H_2O Measurements Showing Positions of Laser, Detector, and Window

Molecular sieve dehydration is a highly efficient means of removing H_2O from cracked gas, ethylene, and propylene. Under normal operating conditions gas exiting a molecular sieve dryer vessel contains no detectable H_2O, or only trace-level (sub-ppm) concentrations of H_2O.

Cracked gas is composed of a complex mixture of hydrocarbons that varies based on the feedstock being used, plant design, and operating parameters. Some hydrocarbons present in cracked gas absorb NIR energy at or near the wavelength used to measure H_2O. Light energy measured by the TDLAS analyzer can be affected by this background interference effect.

SpectraSensors developed and patented a spectral subtraction technique (differential spectroscopy) that enables trace-level measurements of H_2O (and H_2S or NH_3) to be made when a process gas stream contains very low levels of an analyte and background interferences (6,7,8,9).

The differential spectroscopy technique is depicted in Figure 6. In operation a TDLAS analyzer performs a sequence of steps to obtain a "zero" spectrum and "process" spectrum that are used to calculate analyte concentration by spectral subtraction. The zero spectrum is obtained by passing the process gas through a high-efficiency dryer which selectively removes the trace H_2O, without altering the process gas composition and background absorbance. The analyzer records the resulting dry spectrum of the process gas and automatically switches the sample gas flow to bypass the dryer and collect the process gas "wet" spectrum. Subtraction of the recorded zero spectrum from the process spectrum generates a differential spectrum of the trace analyte which is free of background interferences. The concentration of H_2O is calculated from the differential spectrum.

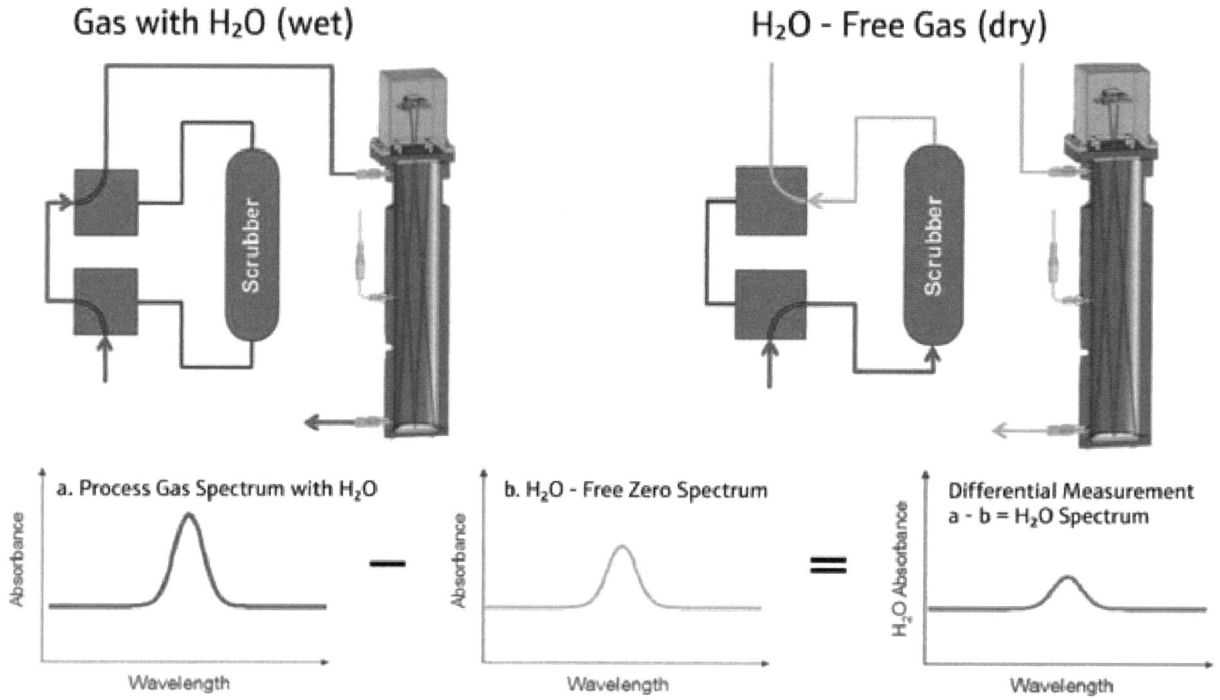

Figure 6. Depiction of Differential Spectroscopy Using Spectral Subtraction

An important advantage of the differential spectroscopy technique for trace-level H_2O in cracked gas streams is that it dynamically corrects for changes in gas stream composition, temperature, or pressure, when spectral distortion from such changes exceed preset values in the analyzer firmware.

Differential spectroscopy is used to perform trace-level measurements of H_2O in cracked gas with a calibration range of 0 – 10 ppm. Figure 7A shows the dry and wet spectra of H_2O in cracked gas, and Figure 7B shows the differential spectrum derived by spectral subtraction.

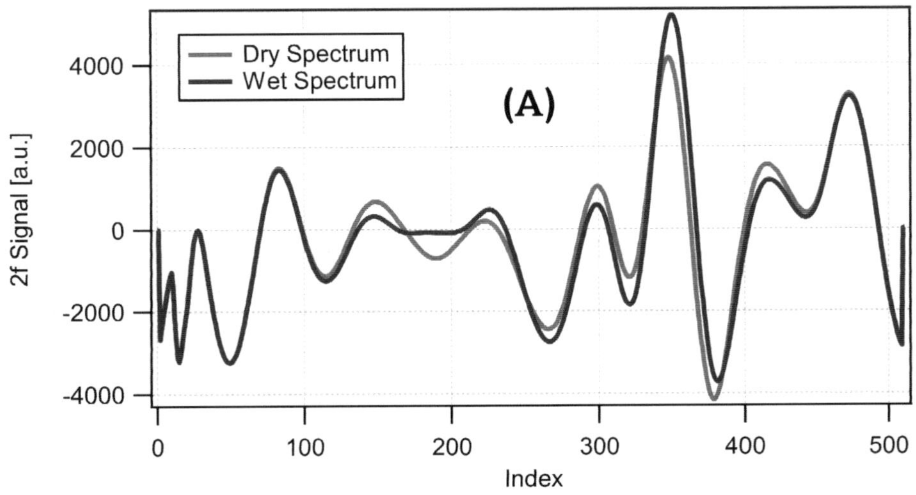

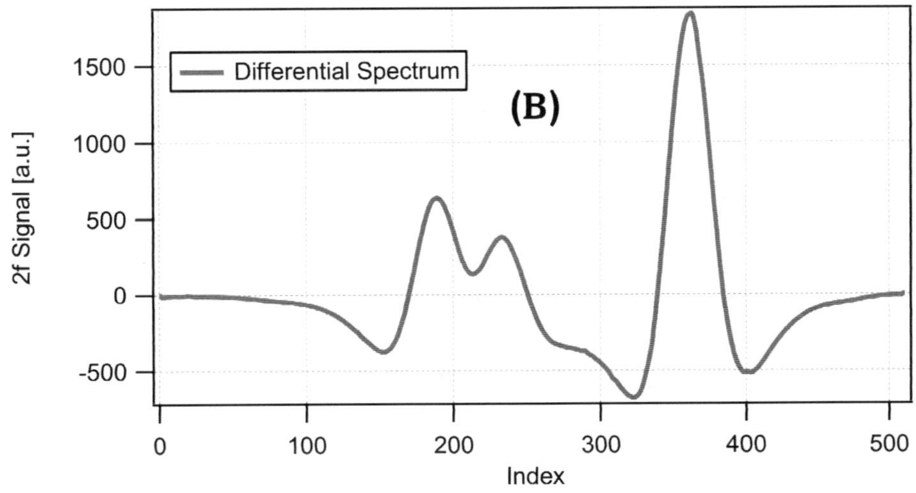

Figure 7. Wet and Dry Spectrum (A) and Differential Spectrum (B) of H_2O in Cracked Gas

Factory Calibration, Performance Testing, and Field Validation

Every TDLAS trace H_2O analyzer is factory tested and calibrated using a test mixture blended to simulate the cracked gas stream composition furnished by the end user. The dilution ratio of the standard and mixing ratio of the background stream gases are controlled by digital mass flow controllers with NIST certifications.

The solid-state laser and detector components used in TDLAS analyzers are intrinsically stable, so no field calibration is required over the lifetime of the analyzer. Users perform periodic validation checks to verify the analyzer is operating properly within its factory-certified calibration range and ensure measurements are accurate for process control purposes.

Validation of analyzers for trace-level measurements of H_2O in gas exiting molecular sieve dryer vessels is particularly important because H_2O will not be present for extended periods of time under normal operating conditions. The analyzer will read zero ppm.

Cylinders of NIST-traceable certified reference gases are not commercially available for trace-level (low to sub-ppm) concentrations of H_2O because they would not be stable. To address this situation TDLAS analyzers for trace-level H_2O measurements are equipped with a permeation tube system to perform automated validation checks at user-designated time intervals. Because reliable H_2O measurements at the outlet of molecular sieve dryer vessels are critical to uninterrupted operation of ethylene, LNG, and natural gas processing plants, TDLAS analyzer validation checks are commonly performed daily in these plants.

The results of a long term (> 12 hour) factory repeatability test of a TDLAS analyzer conducted with 1 ppm H_2O in a simulated cracked gas stream are shown in Figure 8. The 2σ repeatability observed during this test period was ± 12 ppb.

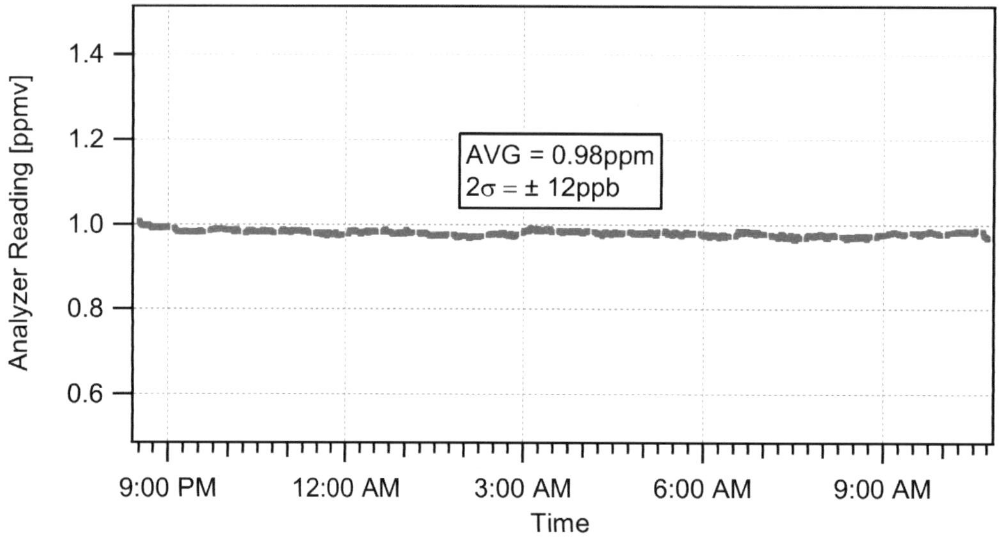

Figure 8. Long Term Repeatability Test Using 1 ppm H_2O in a Simulated Cracked Gas Stream

Figure 9 shows the results of a concentration step change test from 10 ppm down to 0 ppm H_2O in a simulated cracked gas stream. The data demonstrates the exceptionally fast response of the TDLAS analyzer to H_2O concentration changes. This performance characteristic of TDLAS analyzers is especially important for rapidly detecting H_2O breakthrough and tracking the decreasing H_2O concentration following switchover of a molecular sieve adsorbent bed.

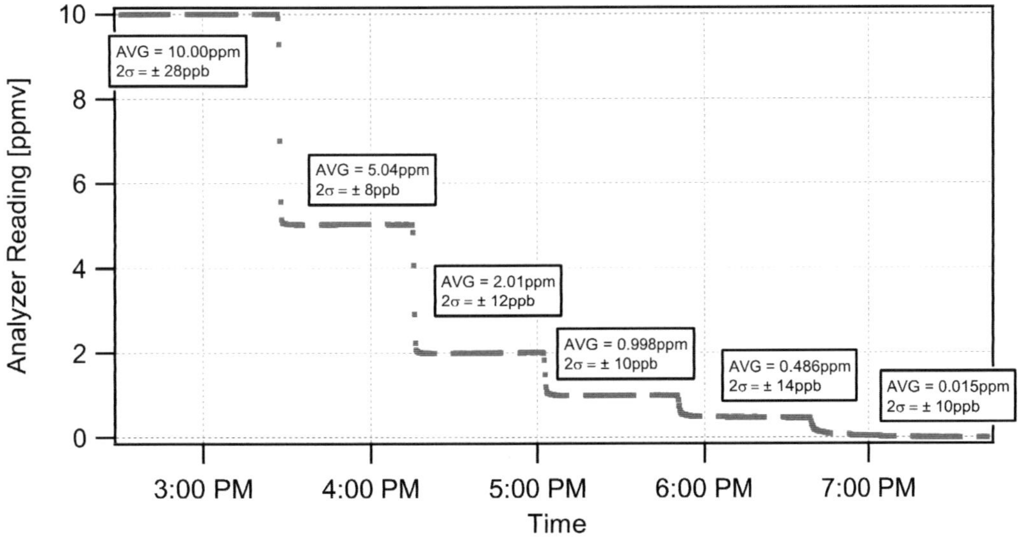

Figure 9. Concentration Step Change Test 10 – 0 ppm H_2O in a Simulated Cracked Gas Stream

Figure 10 shows the excellent linearity (R^2 value of 0.99999) of a TDLAS analyzer over a range of 0 – 10 ppm_v H_2O in a simulated cracked gas stream.

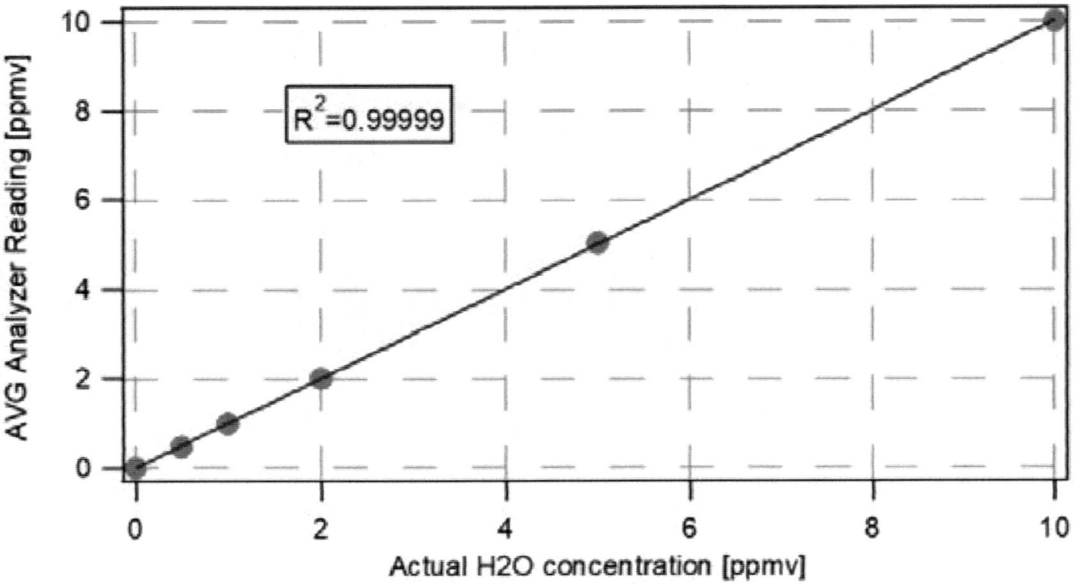

Figure 10. TDLAS Analyzer Linearity 0 – 10 ppm H_2O in a Simulated Cracked Gas Stream

Field Trial Results – TDLAS Versus QCM Analyzers

A comparative field trial of TDLAS and QCM analyzers was conducted in an European ethylene plant operating a 3-bed molecular sieve dehydration system. In this plant, two dryer vessels are on stream while the adsorbent bed in the third vessel is being regenerated. Regeneration occurs approximately once every 60 hours per adsorbent bed. The TDLAS analyzer used in the field trial (shown in Figure 11) was installed in the same sample line as the QCM analyzer.

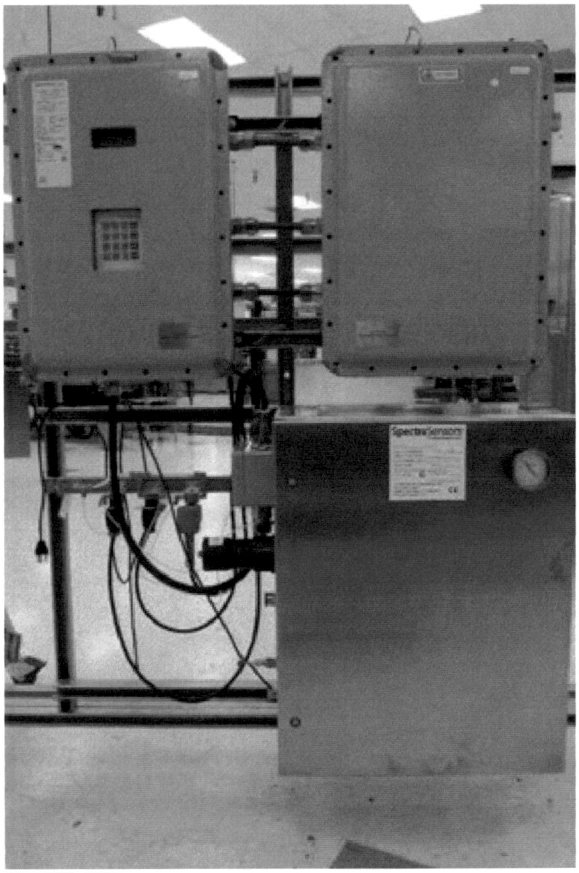

Figure 11. ATEX Explosion-proof TDLAS Analyzer Used in H_2O in Cracked Gas Field Trial

Response of TDLAS and QCM Analyzers During Normal Operation

During normal operation of a molecular sieve dehydration system, no water should be detectable in cracked gas at the outlet of the dryer vessel. Figure 12 shows the H_2O concentration measured during side-by-side testing of QCM and TDLAS analyzers during normal molecular sieve dryer operation. The data plotted in light blue is from the QCM analyzer, the data plotted in dark blue is from the TDLAS analyzer. The QCM analyzer erroneously reports a concentration of 0.629 ppm_v H_2O. The TDLAS analyzer correctly reports a concentration of 0.000 ppm_v H_2O.

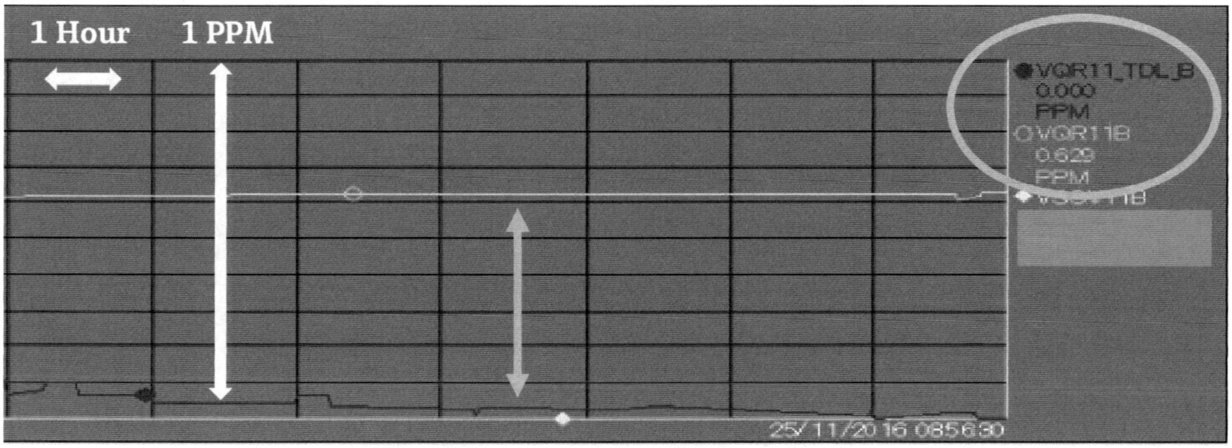

Figure 12. QCM Analyzer (Light Blue Line) Erroneously Reports an H_2O Concentration of 0.629 ppm During Normal Operation When No H_2O is Present While the TDLAS Analyzer (Dark Blue Line) Correctly Reports an H_2O Concentration of 0.000 ppm

Response of TDLAS and QCM Analyzers Following H_2O Breakthrough

The principal purpose of monitoring the H_2O concentration in cracked gas exiting a molecular sieve dryer vessel is to detect moisture breakthrough to trigger switchover to a fresh adsorbent bed when the concentration reaches a control set point. A low limit of detection (LOD) and fast response to H_2O concentration changes are important analyzer attributes in this application.

Figure 13 shows the response of TDLAS and QCM analyzers to a change in H_2O concentration caused by moisture breakthrough from the molecular sieve adsorbent bed. The H_2O concentration plotted in dark blue is from the TDLAS analyzer and the data plotted in light blue is from the QCM analyzer. The TDLAS analyzer detected the breakthrough and increase in H_2O concentration up to the control set point 30 minutes faster than the QCM analyzer.

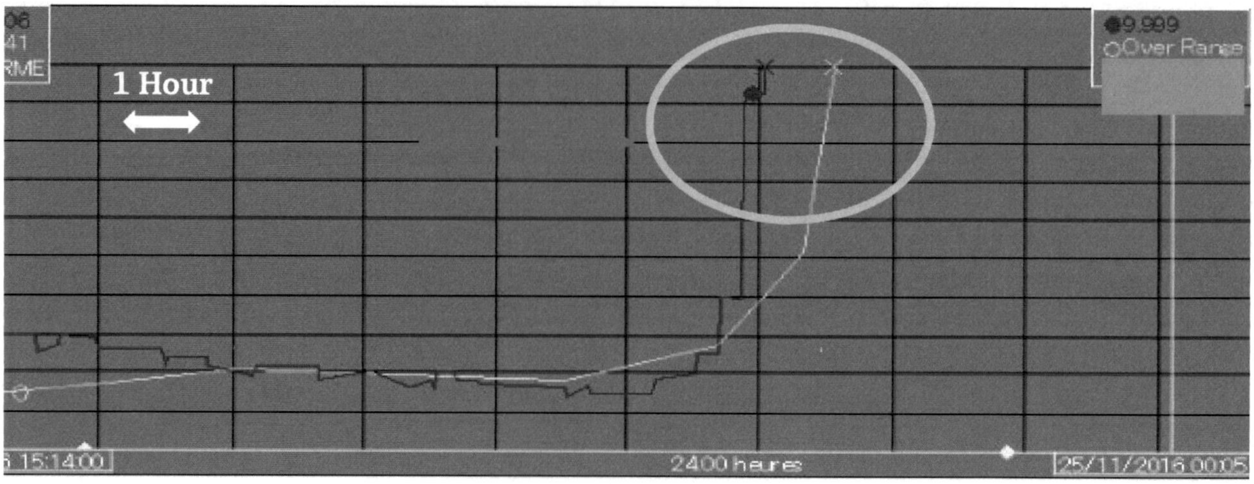

Figure 13. TDLAS Analyzer (Dark Blue Line) Responds to H_2O Breakthrough 30 Minutes Before QCM Analyzer (Light Blue Line)

TDLAS analyzers exhibit an exceptionally fast response to H_2O concentration changes because sample gas flows continuously through the measurement cell, as laser light is introduced to the cell and absorbance measured continuously. Consequently, once the volume of sample gas in the measurement cell is replaced with fresh gas the resulting change in absorbance and H_2O is detected immediately. In contrast, QCM analyzers operate using intermittent cycles in which the sensor element is alternatively exposed to wet sample gas and a dry reference gas stream. In cases where entrained contaminants are present in the process gas stream the analyzer may be operated in a "sensor saver" mode bypassing the analyzer with reference gas flowing to the analyzer to minimize the time the sensor is in contact with the process gas. Operating in the sensor saver mode results in less QCM analyzer on-stream time delaying the analyzer's response to H_2O concentration changes.

Response of TDLAS and QCM Analyzers to Adsorbent Bed Switchover

After breakthrough from an on-stream molecular sieve adsorbent bed is detected the H_2O concentration will increase until it reaches a control set point and triggers switchover to a fresh adsorbent bed. During the time interval required to complete switchover, a measurable concentration of H_2O will be present in the system piping before tapering off to below a detectable concentration. The duration of this time interval is governed by the volume and design of the molecular sieve dehydration system (number of dryer vessels, piping arrangement using a manifold / common outlet, etc.).

Figure 14 shows the response of QCM and TDLAS analyzers following switchover to a fresh molecular sieve adsorbent bed. The H_2O concentration data plotted in dark blue is from the TDLAS analyzer, and the data plotted in light blue from the QCM analyzer. The QCM analyzer response to the change in H_2O concentration following adsorbent bed switchover lags behind the TDLAS analyzer by more than 2 hours.

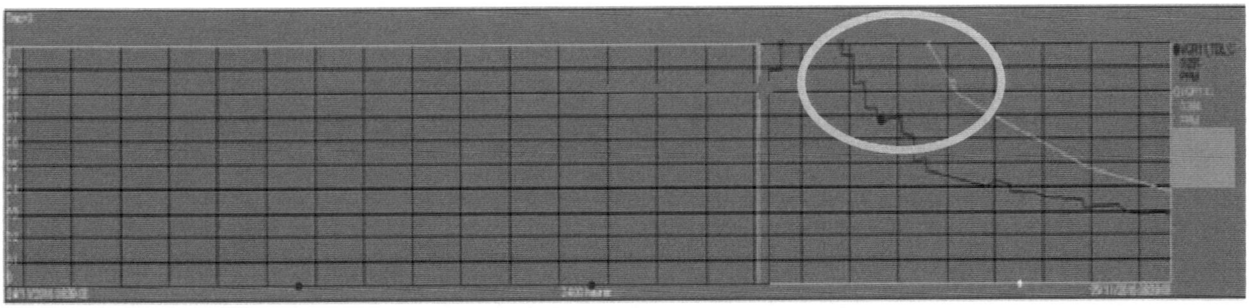

Figure 14. QCM Analyzer Response to Adsorbent Bed Switchover Lags Behind the TDLAS Analyzer by More than 2 Hours

Conclusions

On-line monitoring of water in cracked gas at the outlet of molecular sieve dryer vessels helps to prevent freeze-up of cold box heat exchangers and cryogenic fractionation equipment to avoid production shutdowns in ethylene plants. TDLAS analyzers possess design and performance characteristics that are well-suited for monitoring trace-levels of H_2O in cracked gas at the outlet of molecular sieve dryer vessels and detecting adsorbent bed breakthrough.

TDLAS analyzers employing differential spectroscopy are capable of rapidly detecting H_2O breakthrough from molecular sieve adsorbent beds at low ppb levels and tracking the H_2O concentration up to the control set point for switchover to a fresh adsorbent bed. The solid-state laser and detector components in TDLAS analyzers are isolated and protected from the complex mixture of hydrocarbons and entrained contaminants in cracked gas. This design avoids the fouling, corrosion, measurement errors and biases associated with moisture analyzers using direct contact sensors.

Field testing of a TDLAS analyzer installed and operating concurrently alongside a quartz crystal microbalance (QCM) was conducted in an ethylene plant in Europe. Data from field testing demonstrated the TDLAS analyzer; correctly reported the absence of H_2O during normal operation, responded to adsorbent bed H_2O breakthrough 30 minutes faster than the QCM analyzer, and tracked the decrease in H_2O concentration following adsorbent bed switchover 2 hours faster than the QCM analyzer.

References:

1. Zafirakis, V. and Hoefer, H., "Zeolite Based Dryers in Ethylene Plants", PTQ, Q2 2014, digitalrefining.com/article 1000920
2. Milton, R.M., "Drying Cracked Gas", US Patent 3,130,021, April 21, 1964.
3. Southworth II, J. and Peralta, S.M., "Molecular Sieve Processing in Ethylene Plants", UOP publication F-4070A, 1989.
4. Sicard, J.P. and Richman, P., "Product and Process Improvements in Cracked Gas Drying, UOP publication F-4343C, 1990.
5. Raynor, M.W., Bertness, K.A., Cossel, K.C., Adler, F. and Ye., J, "Trace Water Vapor Analysis in Specialty Gases: Sensor and Spectroscopic Approaches", Chapter 7 in Trace Analysis of Specialty and Electronic Gases, John Wiley & Sons Inc., 2013.
6. Zhou, X., Liu, X., Feitisch, A., and Sanger, G.M., "Reactive Gas Detection in Complex Backgrounds", US Patent 7,704,301 B2, Apr. 27, 2010.
7. Zhou, X., Liu, X., Feitisch, A., and Sanger, G.M., "Reactive Gas Detection in Complex Backgrounds", US Patent 7,819,946 B2, Oct. 26, 2008.
8. Zhou, X., Liu, X., Feitisch, A., and Sanger, G.M., "Reactive Gas Detection in Complex Backgrounds", US Patent 8,500,849 B2, Aug. 6, 2013.
9. Liu, X., Ji, W., and Feitisch, A., "Development of H_2S, H_2O, NH_3, and C_2H_2 TDL Analyzer for Petrochemical Applications in Optically Interfering Hydrocarbon Streams", ISA-AD, Apr. 2010

AIChE PAPER NUMBER: 550575

MULTIDISCIPLINARY TEAMWORK PROPERLY IDENTIFIES A MACHINE PERFORMANCE ISSUE IN AN ETHYLENE PROCESS GAS COMPRESSOR

J. M. (Marty) Novak, Jr. P.E.

Principal Engineer

Huntsman

Jeff Castetter

Global Product Application Engineer

SUEZ Water Technologies & Solutions

Contributors :

Matthew Kress	***Kayla Pompa***	***Brent Gates***
Reliability Manager	***Engineering Intern***	***Account Manager***
Huntsman	***Huntsman***	***SUEZ Water Technologies & Solutions***

Prepared for Presentation at the 2019 Spring National Meeting

New Orleans, Louisiana, April 1-4, 2019

MULTIDISCIPLINARY TEAMWORK PROPERLY IDENTIFIES A MACHINE PERFORMANCE ISSUE IN AN ETHYLENE PROCESS GAS COMPRESSOR

J. M. (Marty) Novak, Jr. P.E.

Principal Engineer

Huntsman

Jeff Castetter

Global Product Application Engineer

SUEZ Water Technologies & Solutions

Contributors:

Matthew Kress	***Kayla Pompa***	***Brent Gates***
Reliability Manager	***Engineering Intern***	***Account Manager***
Huntsman	***Huntsman***	***SUEZ Water Technologies & Solutions***

Abstract: The Process Gas Compressor (PGC) is one of the most critical pieces of operating equipment in the ethylene manufacturing process.

When there is a problem associated with the PGC, the required actions can have significant financial impacts (reducing rates, applying expensive solvents/additives, unplanned outage, etc.).

It is very important to quickly identify machine asset performance issues and apply the proper solutions to the identified issues.

In this case, a multidisciplinary team from Huntsman and SUEZ jointly identified a mechanical issue within the 2nd/3rd stage case by reviewing and evaluating the thermodynamic, aerodynamic, rotor dynamic, auxiliary equipment and mechanical performance versus design.

General Comments/Information/Process

The Huntsman A3 Olefins Unit was built in 1965 as a propane cracker. Modified in 1988 to process FCC tail gas, the unit currently operates with ethane and ethane/propane mix feeding the pyrolysis furnaces. The unit utilizes a 3-stage PGC and includes gas scrubbing and acetylene conversion between the 2nd and 3rd stages.

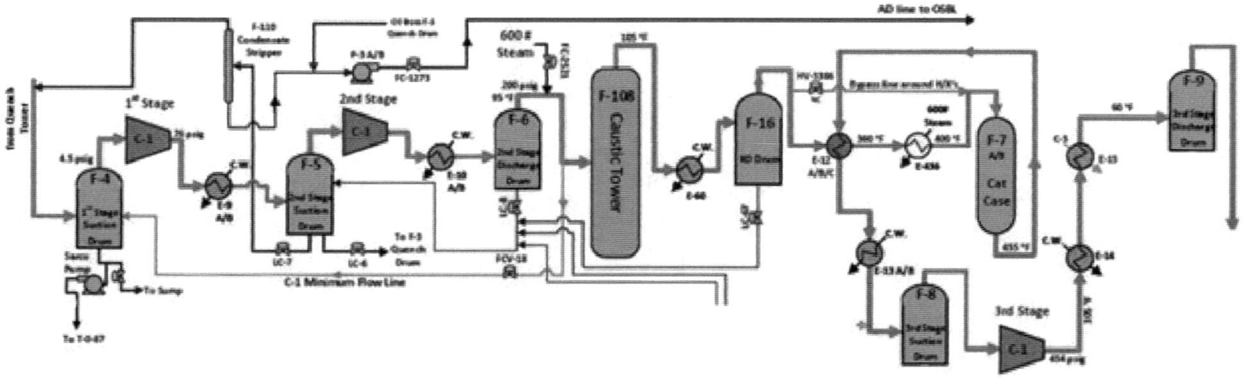

Following a contract change in mid-2015, SUEZ assumed treatment and monitoring of the Huntsman PGC. The SUEZ CompRISE® monitoring and reporting tool identified a concern regarding performance of the 2nd stage around November 2016. Performance remained acceptable and impacts were attributed to identified process conditions and mitigated as opportunities allowed (see key events log).

Following the impacts due to Hurricane Harvey in the fall of 2017 and a boiler failure at the end of the year, Huntsman operations identified a concern that the 2nd stage suction pressure was approaching a pressure relief valve (PRV) setpoint and limiting rates in January 2018. At the same time, the business area requested the Unit turnaround, scheduled for January 2018, be delayed until May.

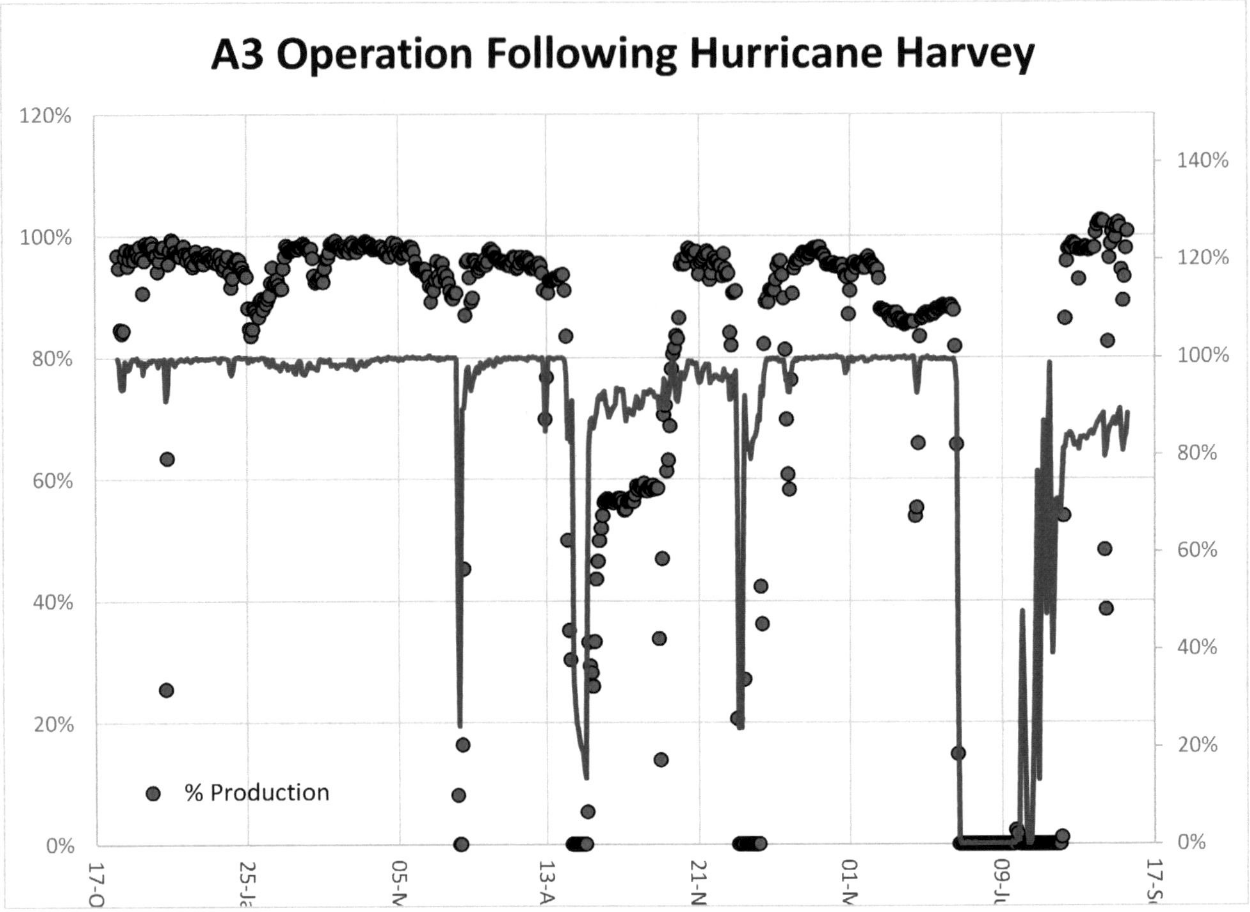

In February 2018, meetings to discuss PGC performance issues began. A multidisciplinary team was formed that included Huntsman engineering, reliability, rotating equipment and operations, along with the SUEZ local account manager and product application subject matter expert.

Theories/Conclusions

As a result, process instruments were checked, pressure and thermodynamic studies were conducted, vibration trends and orbits were reviewed, and performance versus design comparisons were discussed. As the unit was near the end of its normal 5-year turnaround cycle, process exchanger fouling was first assumed to be the cause. When this was disproven, the team identified the following potential causes:

1) 2nd and/or 3rd stage fouling based on vibration orbit changes. Wash oil to 2nd stage was increased until turnaround.

2) Excessive dry seal gas flow. Dry gas seal instruments were checked for calibration and flows were optimized.

3) Worn bushings between 2nd and 3rd stage discharges allowing internal gas recycling. No opportunities to address this possibility were identified until turnaround.

The unit was operated with no observed performance improvement or decline until shutdown for turnaround in mid-May. The PGC was opened for inspection and rebuilt. The inspection of the 1st and 3rd stages and interstage coolers all looked acceptable. Light to moderate fouling was noted on PGC 2nd stage but not at levels to account for the observed performance issues. The 2nd/3rd stage bushings were found to be intact.

No cause of poor performance was identified until the Reliability Manager and the Rotating Equipment Specialist removed the 2nd stage intake piece from the bottom half and inserted it into the lid at the wash slab. As shown below, the mechanical damage formed a gap, allowing 3rd stage suction gas to recycle to 2nd stage inlet.

This defect was not observable from the opened compressor due to the fit against the removed lid. As expected, this check will be made on the compressor in future rebuilds.

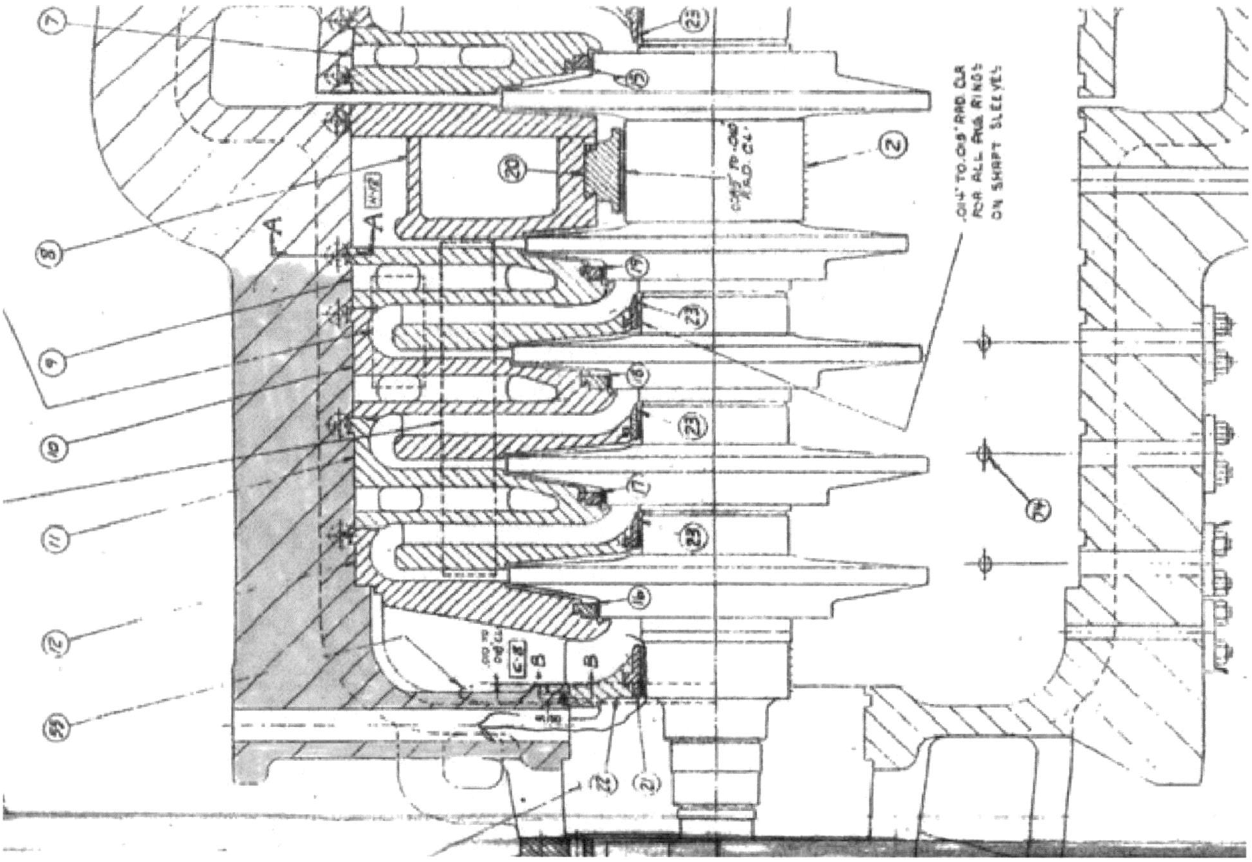

Lessons Learned

The event and its discovery highlight several points:

- Step changes in operation, either production rates, feedstocks, or operating conditions, can provide good opportunities to diagnose the health of an operating unit and to prepare for maintenance activities.
- Diligence must be maintained in determining the cause of changes, either in confirming what is expected (e.g., fouling) or determining another cause.
- In this case, no single discipline would have properly identified the issue. It took a diverse team with multiple tools and experience to hone in on the defect.

Disclaimer

Appendix

Key Events Log

November 2015 - Unit shutdown due to power failure resulted in rebuild of PGC 1st stage. Interstage coolers were also opened and cleaned.

March to May 2016 - Weather events contribute to two separate unit trips. Quench Tower flows manipulated to lower PGC suction temperature. Temperatures were about 5 degrees hotter than historically normal for the month of May. Antifoulant dosage increased on 2nd/3rd stages for the summer to help cope with hotter temps.

June 2016 – Feedstock changes to PGC. Compressor appears to see an increase in volumetric flow but a reduction in pressure ratio. Two of three process interchangers pulled and cleaned in effort to reduce pressure drop between 2nd and 3rd PCG stages. As-found condition was good and third interchanger was not opened.

October 2016 - Process temperatures return to normal and 2nd/3rd stages return to maintenance antifoulant dosages.

November 2016 - 1st/3rd stages have recovered well with seasonal changes, but 2nd has not recovered well. Suspect issues around pressure transmitters or fouling incurred during really hot summer months. 2nd stage does not get as much wash oil as other stages. Recommended blowing down 2nd stage pressure transmitters. Bearing/vibration data still acceptable.

February 2017 - Rebuilt furnace back online. Feedstock changes correspond to increases in compressor efficiency.

March 2017 - Suction temperatures reduced after quench water coolers cleaned. Reduced first stage suction temp from 125 to 98 deg F.

April to June 2017 – Reduced PGC flow identified as impact to production. Pressure drop survey identifies fouling in Acetylene Converter Pre-Heater. Bypass installed to assist with flow. In June, unrelated outage allowed exchanger to be cleaned.

July to August 2017 – PGC performance continues to limit rates. With unit at 6 months from planned turnaround cycle, assumption is fouled interstage coolers combined with summer cooling water temperatures.

August 2017 - Hurricane Harvey result in site shutdown, but the PGC was kept rolling. Storm impacts to downstream units result in Unit restarted at minimum rates.

October 2017 – Rates increased with more downstream units coming on-line, but still below full capacity.

December 2017 - Unit shutdown due to waste heat boiler failure. Business area requests planned turnaround be delayed from January to May 2018 for economic reasons.

January 2017 – Unit re-starts from boiler repairs. In attempting to achieve full rates for first time since summer, the PGC is limited on 2nd stage suction. Pressure drop surveys indicate cause is not exchanger fouling. Huntsman/Suez multidisciplinary team formed to identify cause and ensure machine lasts until new turnaround date.

February 2018 – Team identifies potential causes as:

> 1) 2nd and/or 3rd stage fouling based on vibration orbit changes. Wash oil to 2nd stage was increased until turnaround.
>
> 2) Excessive dry seal gas flow. Dry gas seal instruments were checked for calibration and flows were optimized.
>
> 3) Worn bushings between 2nd and 3rd stage discharges allowing internal gas recycling. No opportunities to address this possibility are identified until turnaround.

March 2018 - Unit feedstock and operating conditions changed for economic reasons. Heavier process stream also reduced temperatures and increased compression ratio.

May to June 2018 - Unit shuts down for 5-year turnaround and compressor is opened for inspection and rebuild. 1st and 3rd stages, interstage coolers all look acceptable. Light-moderate fouling on 2nd stage but still considered acceptable. 2nd/3rd stage bushings intact. No cause of poor performance identified until Reliability Manager and Rotating Equipment Specialist remove the 2nd stage intake piece from bottom half and insert into lid at wash slab, finding gap allowing 3rd stage suction gas to recycle to 2nd stage inlet.

July 2018 – Restart of unit following turnaround at reduced rates awaiting downstream unit start-up.

August 2018 – Return to full rates indicate broken intake piece likely in place for at least last ten years. Analysis of data indicates additional wear during period after Hurricane Harvey resulted in performance change large enough to notice effects.

***** Formatting Issue - Best Available Paper/Graphic *****

DEVELOPMENT OF NEW ON-LINE WASH OIL INJECTION SYSTEM FOR CENTRIFUGAL COMPRESSOR

Masaki Shakuda
Design engineer in Compressor design section
Mitsubishi Heavy Industries Compressor Corporation
6-22, 4-Chome, Kan-On-Shin-Machi, Nishi-Ku, Hiroshima,
733-8553, Japan

Shinichiro Tokuyama
Senior Design engineer in Compressor design section
Mitsubishi Heavy Industries Compressor Corporation
6-22, 4-Chome, Kan-On-Shin-Machi, Nishi-Ku, Hiroshima,
733-8553, Japan

Mr. Ajay Mathew
Machinery Engineering Services
ExxonMobil Manufacturing Engineering Singapore
100 Jurong Island Hwy, Singapore, 627867

Mr. Subramani Elumalai
Machinery Engineering Section
ExxonMobil Research and Engineering
22777 Springwood Village Parkway, Spring, TX 77389

Masaki Shakuda is a design engineer in Compressor Designing Section in the Engineering and Design Division, Mitsubishi Heavy Industries Compressor Corporation, in Hiroshima, Japan. He has 6 years experience in designing compressor and its package. He has B.S and M.S degrees in Mechanical from Tokyo University.

Shinichiro Tokuyama is a senior design engineer in Compressor Designing Section in the Engineering and Design Division, Mitsubishi Heavy Industries Compressor Corporation, in Hiroshima, Japan. He has 12 year experience in designing and developing compressors.

Ajay Mathew is Lead Machinery Engineer of Engineering Services Department at Exxonmobil Chemicals, located in Singapore. He has thirteen years of rotating equipment asset management experience in refining and petrochemical industry. Mr. Ajay received a Bachelor of Engineering degree(Mechanical,2005) from the National Institute of Technology, Suratkal, India.

Elumalai Subramani is Machinery engineer, ExxonMobil Research Engineering, Houston, Texas, USA. He has 20 years of experience in Refinery & Petrochemical plant Machinery Maintenance, Diagnosis, Commissioning and Reliability Improvement projects. He graduated from Coimbatore Institute of Technology, Tamil Nadu, India.

*** Formatting Issue - Best Available Paper/Graphic ***

ABSTRACT

The process gas compressor is the most critical unit in an ethylene plant and several types of contaminates can often foul the compressor flow path, causing deterioration in performance and significant losses in plant production. In order to prevent the efficiency losses due to this fouling during long term operation, as common practice, washing oil is injected through the nozzles installed on the suction piping and return bend of each stage.

However, fouling material has been reported during turnaround with few years of operation even though wash oil injection has been carried out at required intervals. Hence, user and OEM carried out CFD (Computational Fluid Dynamics analysis) to investigate the conventional design process and its practical effect to washing capability. And OEM found the effective approach to develop the new concept improving on-line wash oil injection system.

Optimized oil injection system which includes new injection location, injection hole size, injection angle and oil quantity were evaluated with consideration to prevent any risk including erosion by droplet, any leakage through newly designed part and etc. by means of CFD and FEM(Finite Element Method) analysis. Evaluation through numerical approach was verified through subsequent verification test using actual component and fully assembled machine according to the newly developed design. And every analysis was confirmed to work as intended.

After completion of these investigations, OEM manufactured and delivered newly developed oil injection system in user site where the fouling materials on the compressors were reported with conventional oil injection system. And after a few years commercial operation, User and OEM confirmed the effectiveness of newly developed system at site by checking the compressor performance.

INTRODUCTION

Fouling of the process gas compressor is considered to be caused, as shown in the Fig.1, by the reaction of compressed gas in condition that its temperature exceeds a certain level. This fouling leads performance degradation of compressor and increases user's power consumption to keep intended plant production plan. OEM implements water injection to each of the compressor stages to reduce the gas temperature in order to prevent this fouling, but it is practically impossible to completely eliminate the fouling possibility. Hence, on-line wash oil injection system is also included in the process gas compressor package. This conventional oil injection system installed on the compressor casing is shown in Fig.2.

OEM found that there was the room to improve the washing capability through the verification with the user's turnaround report and OEM developed new concept of on-line wash oil injection system. This paper introduces the detailed process toward establishment of this development.

*** Formatting Issue - Best Available Paper/Graphic ***

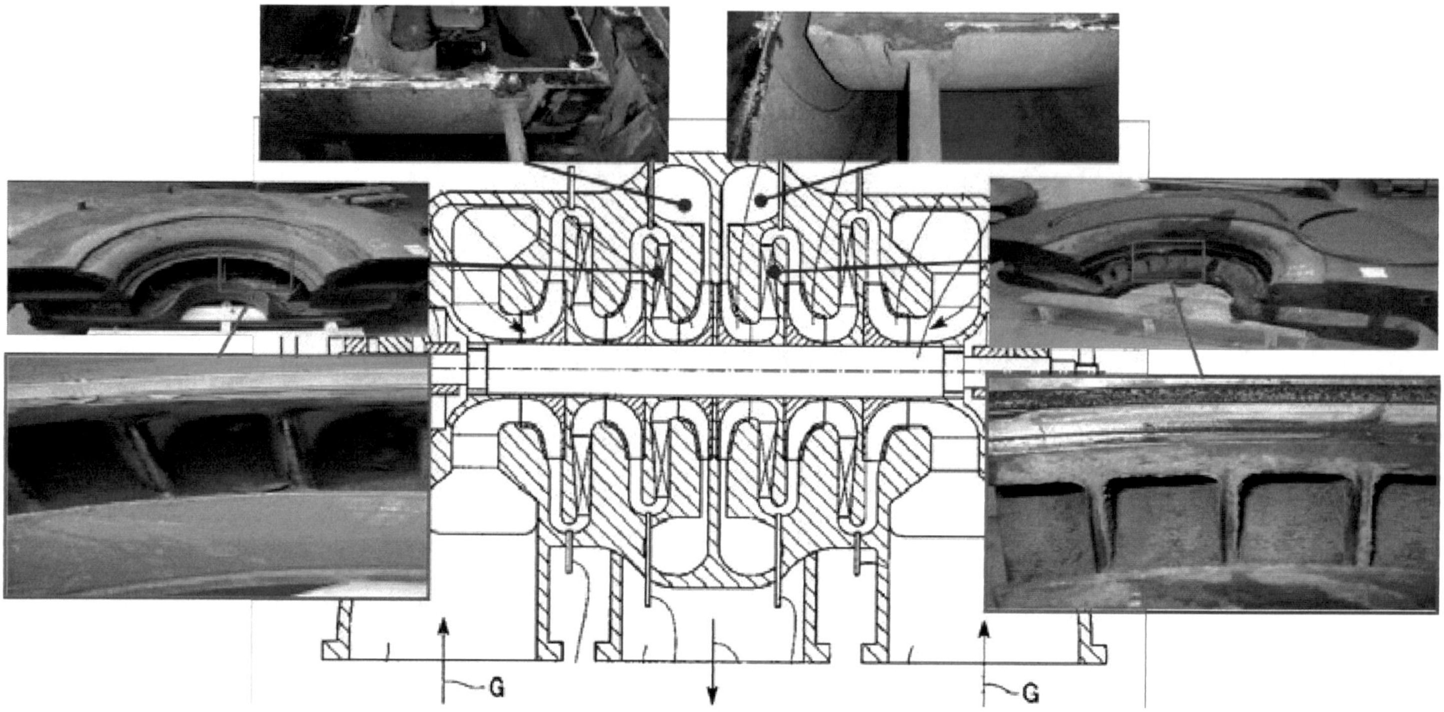

Fig. 1 Fouling condition found at user site

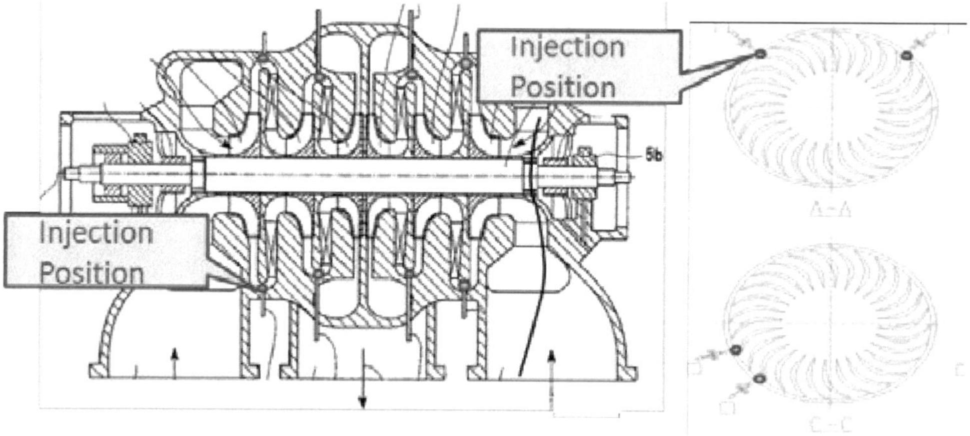

Fig. 2 Conventional wash oil injection system configuration

DEVELOPMENT APPROACH

The step toward establishment of new on-line wash oil system is shown in the Fig.3. Investigation of conventional wash-oil injection system was carried out to understand the current cleaning capability. Based on this study, improved new concept was considered through CFD analysis. Then, new concept of wash oil system was verified by the test using the component test at first and subsequently the whole compressor machine. And new system was assessed from the view point of possible risk at site and its mitigation plan. Finally, the actual washing oil system was implemented in the actual compressor and it was delivered at user plant. The operating data proved that newly developed on-line wash system worked as intended and OEM development was found successfully completed. The more detail of each steps are described in the following paragraphs.

*** Formatting Issue - Best Available Paper/Graphic ***

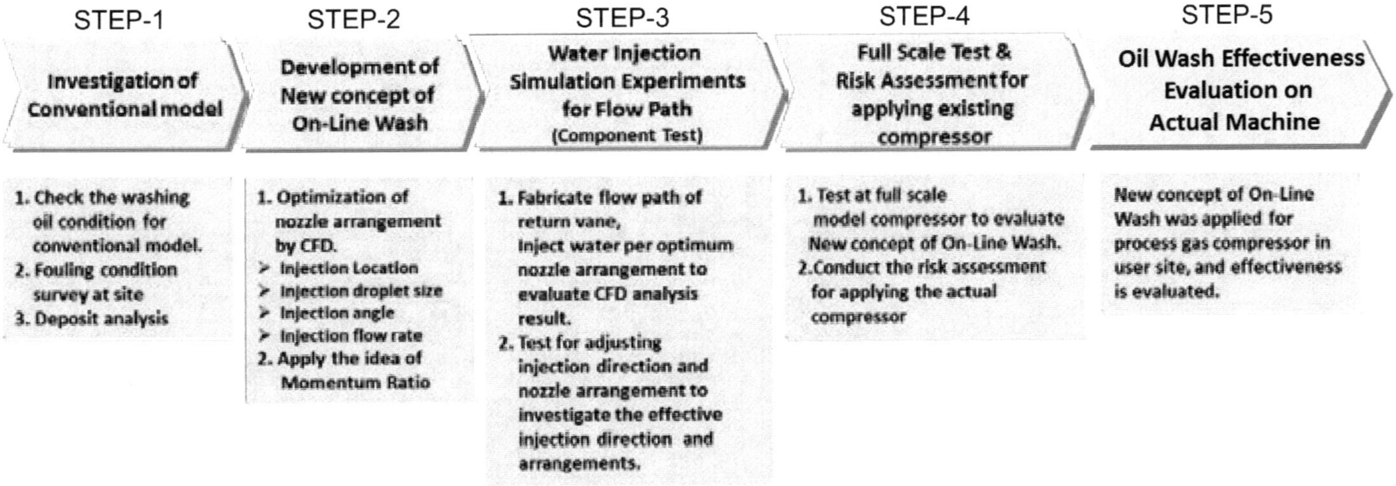

Fig. 3 Step of development of new concept on-line wash oil system

STEP-1. INVESTIGATION OF CONVENTIONAL MODEL

OEM investigated the conventional system through CFD analysis to explain the condition reported by user. As a result, it was found that injected washing oil might not have been equally distributed along the whole gas passage area as shown in Fig.4. Therefore, OEM investigated new design concept to enhance the distribution equality so that washing oil can reach all the gas passage area.

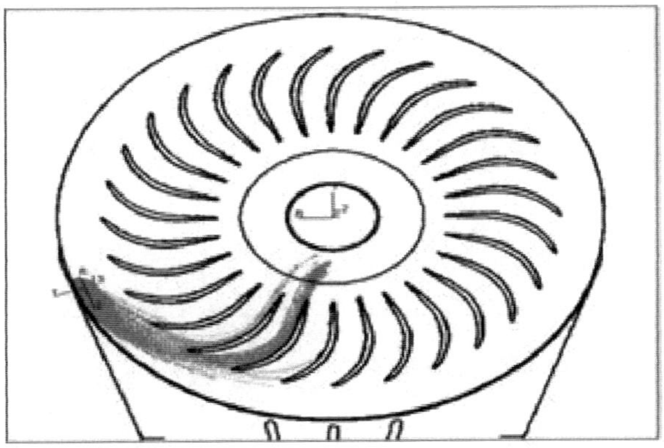

Fig. 4 Expected injected oil flow line from single injection nozzle

STEP-2. DEVELOPMENT OF NEW CONCEPT OF ON-LINE WASH

The case study was carried out to find the optimized design to maximize the cleaning capability of on-line wash oil system. Various injection locations, droplet sizes, nozzle angles and flow rates were taken into consideration as shown in Fig.5. This case study revealed that injection at near impeller tip from impeller hub side, Pos.1, was the best position from the view point that injected oil mist can be distributed equally along with flow path. CFD analysis result of Fig.6 shows improved injected mist distribution compared with conventional system and it can be understood that injected mist reaches approximately 1/12 of flow passage area from single injection point. Hence, OEM decided to implement totally 12 injection holes in the diaphragm as shown in Fig.7 to enable wash oil mist distribution to whole the gas passage area.

*** Formatting Issue - Best Available Paper/Graphic ***

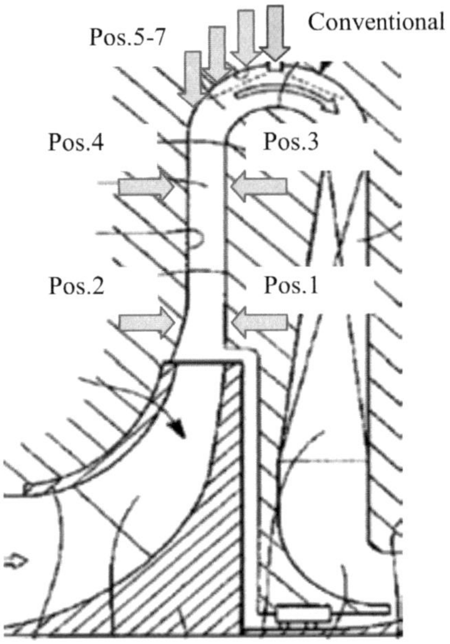

Case	Nozzle Hole Diameter	Velocity	Injection Direction	Location
1 RSCT	1mm	24m/s	Inclined	Diffuser Center
2 RDST	1mm	24m/s	Inclined	Diffuser Shroud
3 RDHT	1mm	24m/s	Inclined	Diffuser Hub
4a DST	1mm	24m/s	Inclined	Diffuser Exit Wall Shroud Surface
4b DHT	1mm	24m/s	Inclined	Diffuser Exit Wall Hub Surface
5 IST	1mm	24m/s	Inclined	Impeller Exit Shroud Surface
6 DHV	1mm	24m/s	Perpendicular	Diffuser Exit Wall Hub Surface
7 IHV	1mm	24m/s	Perpendicular	Diffuser Inlet Wall Hub Surface
8 ISV	1mm	24m/s	Perpendicular	Diffuser Inlet Wall Hub Surface
9 IHT	1mm	24m/s	Perpendicular	Diffuser Inlet Wall Hub Surface
7'IHV	1mm	32m/s	Perpendicular	Diffuser Inlet Wall Hub Surface
7'IHV	1mm	45m/s	Perpendicular	Diffuser Inlet Wall Hub Surface
7'IHV	0.5mm	32m/s	Perpendicular	Diffuser Inlet Wall Hub Surface

Fig. 5 Case study of oil injection for optimization

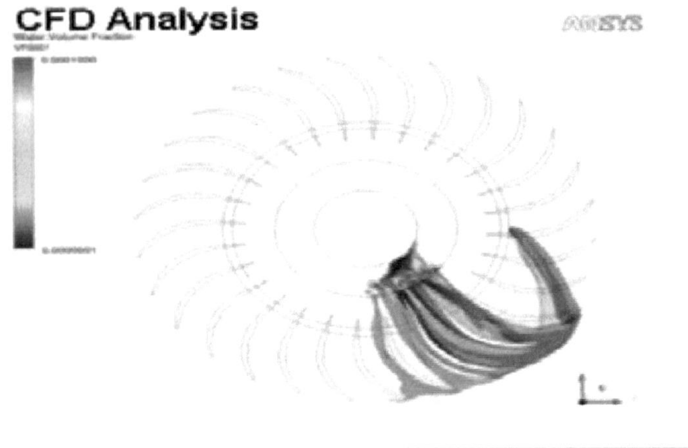

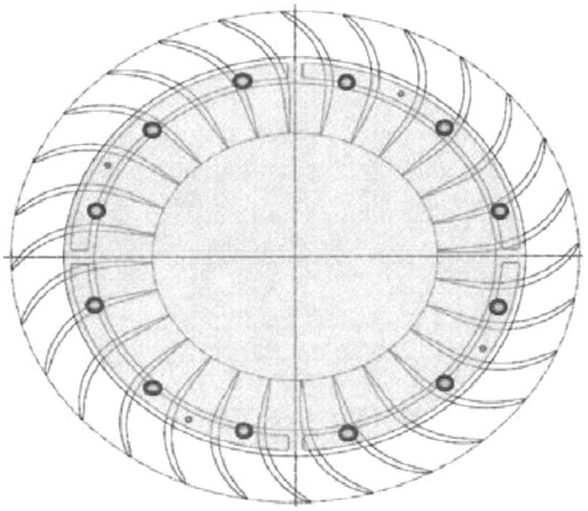

Fig. 6 CFD analysis result of oil injection at Pos.1 (Left)
Fig. 7 Wash oil Injection nozzle arrangement based on newly developed concept (Right)

As for the injection flow rate, it was optimized by adjusting momentum ratio of gas phase and liquid phase. Simplified model is shown in Fig.8. Optimum flow rate was considered from the stand point of capability to equally distribute the oil mist. CFD analysis result showed that higher liquid flow rate to the gas flow rate is better. Meanwhile, OEM simultaneously considered the risk of erosion and performance decrease due to increased total gas flow and finally established the optimal amount as standardized value covering the all of these perspectives.

Bottom of Fig.8 shows the case study result of momentum ratio. Accordance to the increased liquid flow, distribution area was increased but it was found to be saturated at the certain level. OEM decided that low liquid momentum ratio case, in this table, is the optimum condition from the all of the perspective.

< Concept of momentum ratio >

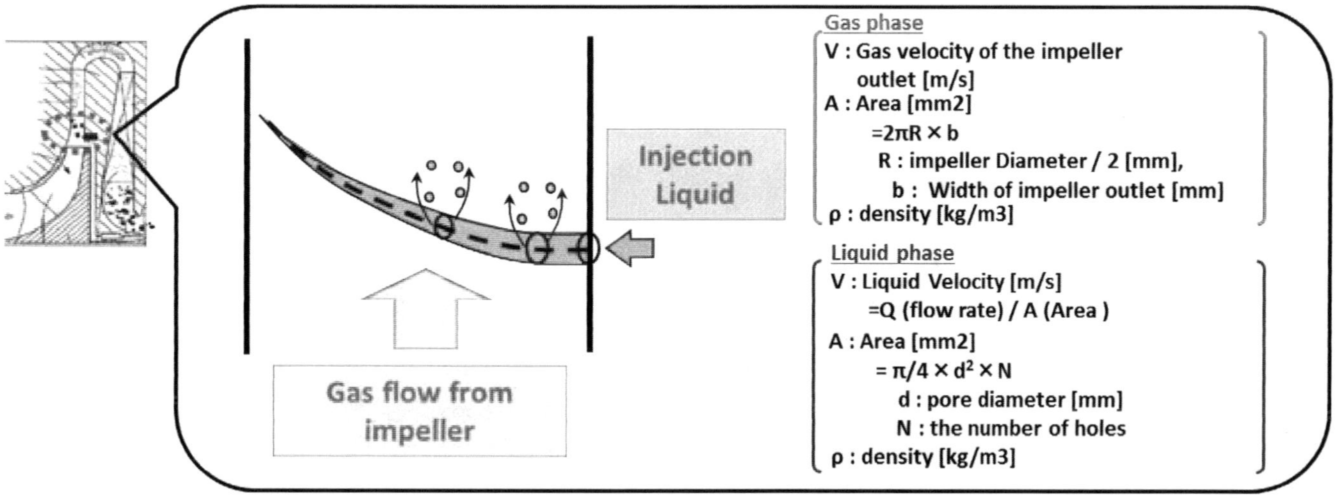

< Case study of momentum ratio >

Process Gas Flow

Liquid Injection Flow

Moment-um ratio	Measurement points			
	Very High	High	Low	Very Low
Very Low	A, 65 degree	A, 65 degree	A, 65 degree	A, 90 degree
Low	A, 90 degree	A, 90 degree	A, 90 degree	A, 130 degree
High	A, 115 degree	A, 115 degree	A, 115 degree	Data not available
Very High	A, 115 degree	Data not available	A, 115 degree	Data not available

Fig. 8 Concept of momentum ratio

STEP-3&4 VERIFICATION TEST OF DEVELOPED NEW CONCEPT

In order to verify CFD analysis result, test by means of components and whole compressor machine was conducted. Fig.9 shows the arrangement of this verification test. Water injection was used to represent injected wash oil and temperature measurement enabled visualization of the mist distribution by sensing of compressed gas temperature decrease

Fig.10 shows the test result showing injection nozzle location and measured mist distribution. At the both of component and full assembly test, approximately 40 degree distribution per injection nozzle was found as highlighted in pink in this figure. CFD analysis result shown in Fig.6 well agreed with this verification test result. Then, OEM confirmed that CFD analysis was calculated under the reasonable condition and result was reliable.

<Injection nozzle installation>

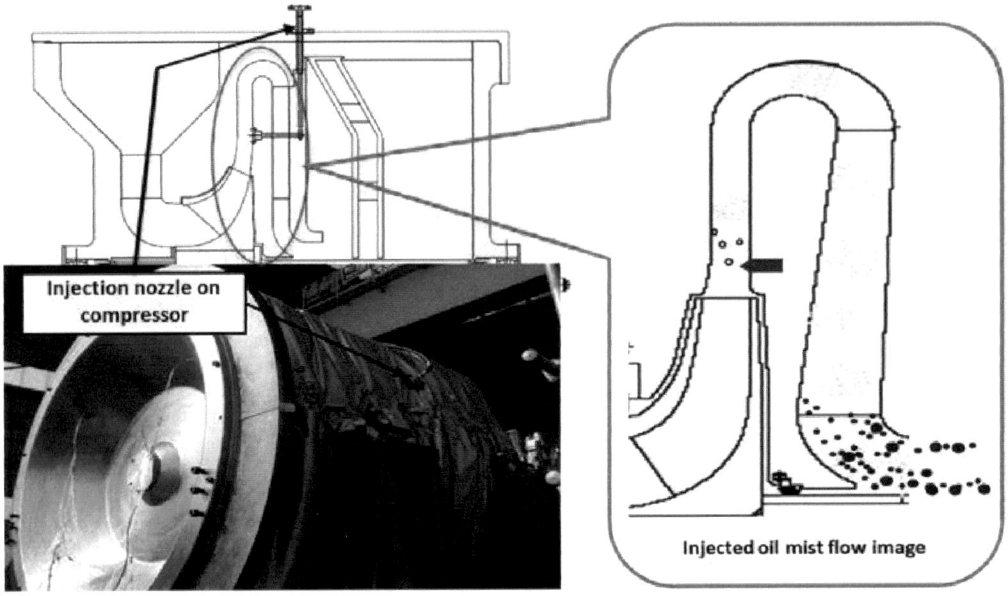

<Temperature sensor location>

Fig. 9 Verification test arrangement

***** Formatting Issue - Best Available Paper/Graphic *****

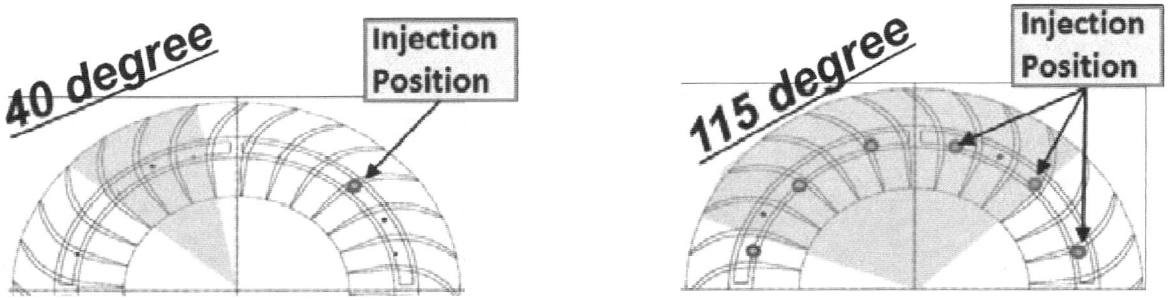

Fig. 10 Verification test nozzle location(pink circle) and distributed area(pink area)

RISK ASSESMENT OF DEVELOPED WASH OIL INJECTION SYSTEM

Any risks related to the components newly developed were evaluated from the perspective but only one point is introduced as example for understanding the procedure which was taken. As shown in Fig.2 and Fig.5, injection location was changed from return bend to the diaphragm of each stage. Due to this change, unintended oil leakage through diaphragm was considered as the possible failure mode. Fig.11 shows the structure of diaphragm at horizontal split surface. This failure mode could be caused by the insufficient sealing at diaphragm and would lead to the increased injection oil flow rate and less plant production due to insufficient washing and increased mass flow rate. This failure mode can be detected by the flow meter implemented on the piping upstream of wash oil injection nozzle. Performance data would also be the means of the detector of this failure mode. In order to prevent such failure, OEM carried out FEM analysis to check stress distribution in diaphragm caused by pressured wash oil. And also OEM conducted in-house testing to verify the sealing is functioning as designed.

Following table shows the summary of the process as described in the above. Every possible risks and risk mitigating plan for each of items were considered and summarized before this new design concept was deployed to the actual product to prevent any trouble at site.

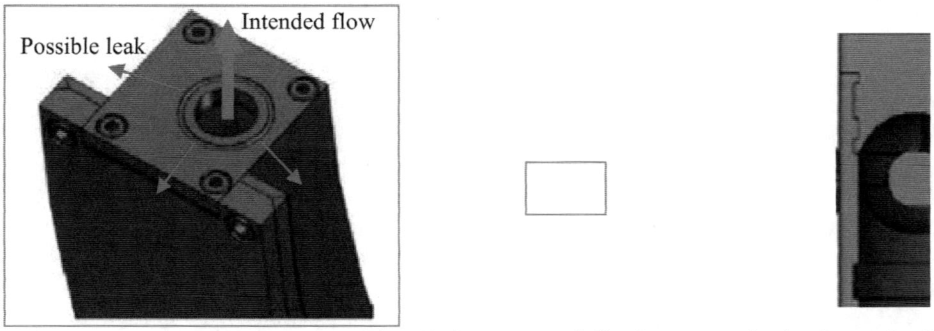

Fig. 11 Structure of diaphragm at the horizontal split surface

Table. Risk assessment summary for leakage at diaphragm horizontal split surface

Component	Fail Mode	Cause	Impact by Fail	Fail Detection	Fail Prevention
Diaphragm	Oil Leakage from horizontal split surface	Insufficient sealing	Increased flow rate	Flow meter	1. Calculation of hydraulic pressure of oil (internal press) and weight of upper side diaphragm(sealing press) 2. Leakage check during in-house hydrostatic test.
			Degraded plant production	Performance data	

*** Formatting Issue - Best Available Paper/Graphic ***

EFFECTIVENESS EVALUATION ON ACTUAL MACHINE

Improved compressor with new wash oil injection system was delivered to user site, where the performance degradation was reported. Fig.12 shows long term operating data of compressor efficiency including before and after replacement with new oil injection system. This trend shows performance getting worse along with the time of operation before modification of wash oil injection system. However, it can be seen that performance improvement after implement of new wash oil injection system and no further performance degradation even after 30 months operation. This trend data ensure that new wash oil injection system is working as intended.

Actual fouling condition is going to be confirmed during the next turnaround of user plant.

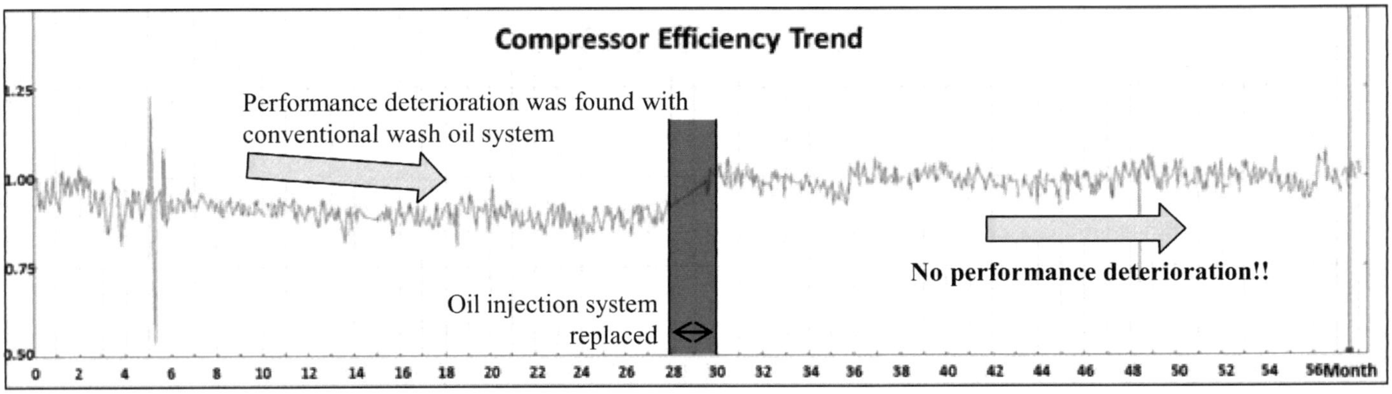

Fig. 12 Efficiency trend data through long term operation (before and after oil injection system replacement)

CONCLUSIONS

CFD analysis was carried out to decide optimized configuration of wash oil injection to facilitate the distribution of oil mist all along the gas passage area inside centrifugal compressor. As a result, OEM found optimized design as below.

1) Injection at the nearest point of impeller tip
2) Injection from the hub side to shroud side
3) Optimum momentum ratio (Gas phase/Liquid phase)

The verification test by means of compressor component & whole machine was carried out to verify the CFD calculation result and OEM confirmed that the verification test result had good compatibility with the founding of CFD calculation. And also, developed system was installed on actual compressor and operating trend data more than one year proved the new wash oil system is working as intended.

As a result of these investigation, OEM successfully established the new concept of enhanced On-Line wash oil injection design.

REFERENCES

- Satoshi Hata, Junichi Horiba, Shari R. Snider, 2010, “Process Gas Compressor Fouling & Washing Mitigation A better understanding of fouling phenomenon & compressor design techniques”, AIChE Paper Number: T4 175313
- Shinichiro Tokuyama, et al, 2018, ATPS-18 Case study_” New Concept of On-Line Wash for Centrifugal Compressor”

ACKNOWLEDGEMENTS

The authors gratefully wish to acknowledge the following individuals for their thoughtful suggestions, contribution and technical assistance in analyzing and sharing valuable experiences.

- A. Mathew of Lead Machinery Engineer of Engineering Services Department at ExxonMobil Chemicals
- E. Subramani of Machinery engineer at ExxonMobil Research Engineering
- A. Nakaniwa of Mitsubishi Heavy Industries Ltd

Ethylene Producers' Committee

Improved Coatings to Efficiently Prevent Fouling in Centrifugal Compressors

Alice Pranzetti

Coating Technology Manager

BHGE

Domenico Di Pietro

Materials Science Engineer

BHGE

Sergio Ghezzi

Centrifugal Compressors Product Manager

BHGE

Francesca Monti

Petrochemicals Platform Manager

BHGE

Marco Romanelli

Engineering Manager - Materials Science and Engineering

BHGE

Prepared for Presentation at the 2019 Spring National Meeting

New Orleans, March 31st-4th April 2019

Abstract:

Charge gas fouling is not a new challenge for ethylene plant operations. A major contributor to the fouling phenomenon is increased process gas temperature which is typically observed in steam crackers with gas feedstock. In addition, the market focus on capital expenditure is also impacting charge gas temperature in both liquid and gas cracking facilities. To drive down rotating equipment costs, the number of compression sections in the train is being reduced. However, fewer sections of compression increases sectional pressure ratio and results in higher gas temperature which exponentially increases the risk of polymerization.

Polymer deposits in the compressor internals are inhibited by water injection and anti-fouling coatings or removed with solvent injections. The evaporative cooling effect of water is essential for temperature control and to prevent polymerization. However, the main consequence of introducing water or oil solvents into the flow path is the erosion phenomenon, caused by liquid droplet impact on compressor internals, damaging components and removing traditional coatings, e.g. traditional organic base and spray fluoropolymers. To minimize the erosion effect of the frequent washing cycles, electroless nickel plating with high phosphorous content (ENP-11% P) has been implemented. This metallic coating, erosion and corrosion resistant, has demonstrated to be an acceptable compromise between durability and fouling resistance.

Fouling resistance is significantly improved with a modification of the ENP-11%P matrix by addition of fluoropolymer nano-particles. The polymer is incorporated into the process and distributed homogeneously in the coating surface, dramatically improving the anti-fouling behavior and enhancing the erosion and corrosion resistance. This coating is applied by an auto-catalytic chemical reaction ensuring a uniform coating thickness over the entire surface of the subject component.

The material properties and fouling-release results for fluoropolymer doped ENP-11%P are discussed herein.

1. Introduction

The cracked-gas compressor (CGC) is one of the most critical components of an ethylene plant. CGCs are high-power compressors that draw gas from the cracking furnaces through the various quench coolers and separation sections of the plant and pressurize the gas for further separation [1]. These machines are characterized by large volumes, high mass flow rates, low pressures, and the need for multiple compression sections (typically three to five across two or three compressor casings) to meet the required discharge pressure. Compressor operation must be fine-tuned for optimized performance and reduced downtime to meet contractual agreements and market demands. Plant operating efficiency and availability are key elements of plant performance to which cracked-gas compressor is critical. The nature of the process presents significant challenges for operations as CGCs are susceptible to fouling due to polymerization [2]. In fact, the cracked gas contains compounds that polymerize and adhere to both stationary and rotating components of the compressor flow path and other equipment in the process loop. Fouling degrades performance, reduces efficiency, and increases power consumption. To recover the degradation, producers must stop production and thoroughly inspect and clean compressor components as often as every six months [3].

The threat of fouling is always present and frequent plant shutdowns are unacceptable. Therefore, both fouling prevention and cure methods are present in most installations. Cures remove fouling after it has occurred, while preventative measures are implemented to create an environment unfavorable for polymer formation or adherence of the foulant to the flow path e.g. temperature control or surface treatments [4]-[5].

The heat of compression elevates cracked-gas temperatures which promotes polymer formation. The process and compressor designs seek to balance risk of polymerization, as well as capital and operational expense of the compressor train. Water injection into the compressor is frequently implemented to further improve economics and as anti-fouling insurance. Atomized water is injected into the gas stream through nozzles located in the return channel at the discharge of each impeller where the temperature is to be controlled.

When the water contacts the hot gas, it evaporates or flashes, removing heat. Therefore, each compressor section can generate more head versus operation without water injection with similar discharge temperature [4]. Operating with water injection may make certain compressor arrangements feasible, but the primary objective of cooling the gas is to prevent the formation of polymers which foul the equipment.

To further prevent fouling, coatings may be applied to the stationary and rotating parts of the compressor. These coatings prevent polymers and other foulants from adhering the coated surface. However, once the polymers have formed and fouling has collected within the flow path, wash oil is used to physically remove the foulant. Wash oil is injected into the inlet piping and intermediate return channels of the compressor. The wash oil dissolves and erodes foulant from the compressor's flow path but the injected fluid causes erosion of both coatings and compressor components [5].

The application of coatings is considered one of the most economical ways to reduce fouling issues. However, current coating solutions, typically fluoropolymer or silicon-based spray type, do not perform well long term. Coating performance degrades quickly under charge gas service conditions and even faster in the presence of liquid injection. Initially the smoothness of the coating surface prevents polymer foulant deposits. However, over time the polymer accumulates, and must be removed with high pressure off-line washing. Online and off-line washing weakens the coating-substrate bond and causes coating detachment. Continuous water injection for cooling purposes removes coating even faster, typically within the first few months of service due to continuous liquid impact [6].

The formulation of coatings capable of resisting both erosion and fouling attack is highly desirable. Electroless nickel plating (ENP) has recently been used instead of fluoropolymer-based organic coatings [6]. Specifically, high phosphorous (10-13%) ENP provides both corrosion and erosion resistance, and decreases surface roughness; thus, it is a suitable solution to avoid frequent coating replacements. The capability to improve ENP characteristics with co-deposition of fine particulate matter, within the electroless metal matrix, has led to a new generation of composite coatings with enhanced properties. These coatings have controllable and reproducible microstructure, uniform thickness and homogeneous dispersion of functional particles. The following describes the characterization of an ENP-based coating doped with polytetrafluoroethylene (ENP + PTFE) particles and compares this solution with the coatings commonly applied in centrifugal compressor applications for the downstream market.

2. Coating solutions description

2.1 Existing Coating Solutions

Current coating solutions used on both stationary and rotating parts of CGC compressors are organic and/or inorganic polymer based, applied in one or more layers. The examples considered herein are two of the most used anti-fouling coatings types in the Oil&Gas industry: PTFE-based polymeric coatings and silicon-based polymeric coatings.

PTFE-based polymeric coating considered in this study combines three types of coatings used to prevent corrosion of metal components: barrier coatings, inhibitive coatings and sacrificial coatings, (Figure 1). The outermost layer of the coating is a high-temperature resistant resin which is a barrier against corrodents present in the environment. Barrier coatings prevent corrosion by sealing the substrate from the environmental effect. However, when the seal is broken corrosion proceeds unchecked at the point of breach in the coating. The intermediate layer of the PTFE-based polymeric system is a durable inhibitive coating. Inhibitive coatings contain pigments (e.g. chromates or complex metal-organic compounds) that prevent corrosion by modifying the chemistry of environmental corrodents contacting the coated surfaces. These reactions change the pH, reactivity, and even the molecular structure of the corrodents. Inhibitive coatings are very effective as long as the film remains intact and inhibitive pigments remain reactive. When the coating is damaged, corrosion of the exposed substrate is slowed until active pigments are depleted. The foundation of the coating system is a tightly adherent layer of a sacrificial aluminum-filled ceramic. Sacrificial or galvanic coatings actively prevent corrosion of the metallic substrate. They are made of a more reactive metal, which when placed in contact with a less reactive (more noble) metal will be consumed entirely by the environment before the more noble material begins to corrode. The final thickness of this coating is approximatively 100-150 microns.

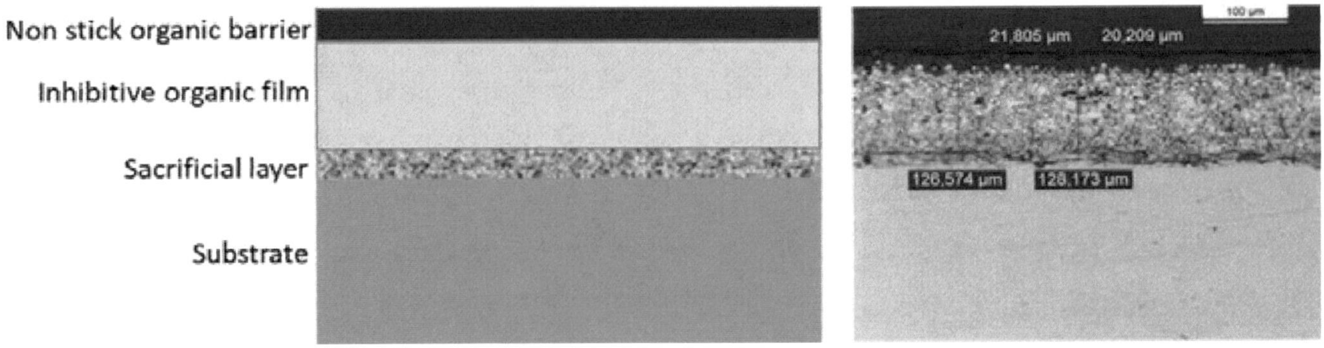

Figure 1 Schematic representation of PTFE-based polymeric coating(left) and cross section light optical microscopy (right).

An alternative solution to PTFE-based polymeric coatings are the silicon-based polymeric coatings. This studied coating is formed by a unique thin layer with an approximate thickness of 50 microns (Figure 2) of a silicon-based polymer enriched with titanium and aluminum oxides. These oxides are added to increase the hardness of the coating providing enhanced erosion resistance to the polymer.

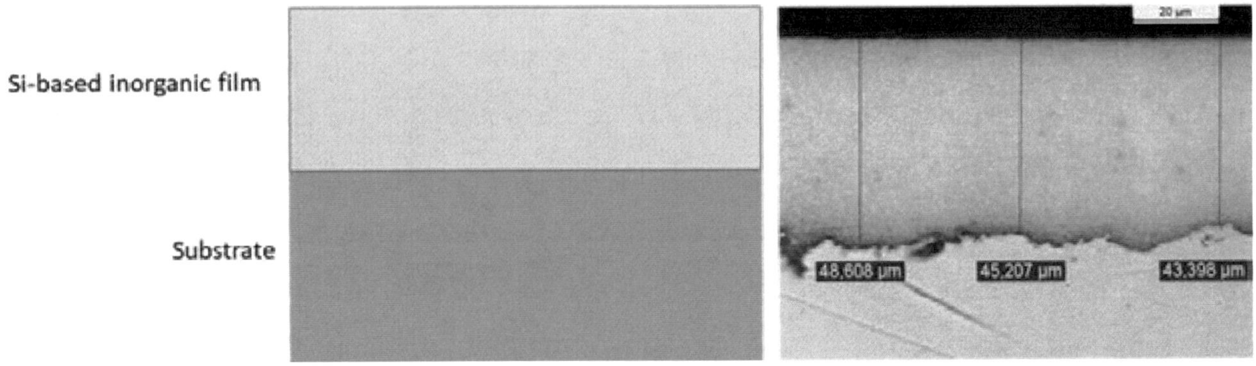

Figure 2 Schematic representation of silicon-based polymeric coating (left) and cross section light optical microscopy (right).

Both, PTFE-based polymeric coatings and silicon-based polymeric coatings are "line of sight", and present significant challenges when coating complex geometries with areas difficult to access. In addition, the application procedures are laborious and involve specific application sequences which may include curing, heating, peening, and blasting.

2.2 Electroless Nickel plating (ENP) solution

Electroless process is an autocatalytic method in which Ni^{2+} ions in solution are deposited through the oxidation of a chemical compound or reducing agent, typically, sodium hypophosphite. This process can effectively and economically improve the properties of steel products by creating a protective layer with high wear and corrosion resistance, low surface roughness, and uniform thickness even when applied to complex geometries. Three classes of ENP coatings can be identified according to their phosphorus content, e.g. Low phosphorus 2 - 5% P. Medium 6 - 9% P. High 10 - 13% P.

Among the three ENP coating classes, the high phosphorus class (> 10 wt.%), shown in Figure 3, exhibits the best corrosion resistance (especially, in acidic environments [7]) and the lowest porosity. These properties are related to the amorphous structure of the coating that make the high-P ENP the preference for corrosion protection in sour and acidic environments.

Figure 3 Scanning Electron Microscope (SEM) image of ENP (11%P) coating, surface (left) and cross section view (right).

2.3 ENP composites solution

Composite coatings containing nano and micro-sized second phase hard particles such as diamond, SiC, Al_2O_3, Si_3N_4, CeO_2, TiO_2, and soft particles, namely PTFE, MoS_2, HBN have been successfully co-deposited in the ENP matrix and exhibit superior properties compared to the base ENP coatings. The ENP-composites are found in many applications ranging from molds for rubber and plastic components, tools for pumps, valves and butterfly valves for oil and gas industry, fasteners, precision instrument parts, aluminum air cylinders, carburetors, choke shafts, etc. Among, the various cited solutions, ENP + PTFE-doped coatings offer a series of desirable characteristics such as: non-stick, non-galling, high lubricity, low friction, precise and uniform torque and tension, good wear and corrosion resistance. The production of a double-layer coating having a high-P-ENP primer and an ENP + PTFE top coat would bring together the corrosion and erosion resistance of the ENP with the non-stick properties of the PTFE (Figure 4).

Figure 4 Schematic representation of ENP+PTFE coating (left) and Scanning Electron Microscope (SEM) image of cross section.

3. Methodology

3.1 Materials

All testing activities were performed on coatings applied on low alloy steel substrate (ASTM A182 F22 grade). The substrates have been machined to reach a surface roughness of 0.8 μm Ra. Tested coatings are shown in Table 1. Samples of uncoated F22 steel were also tested in the same conditions for control purposes.

Coating type	Thickness
Silicon-based polymeric coatings	50
PTFE-based polymeric coatings	100
ENP 11%P	75
ENP-PTFE	30

Table 1 Tested coatings list and nominal thickness.

3.2. Test procedures

3.2.1 Solid Particle Erosion test

The resistance of the tested surfaces to solid particle erosion (SPE) was assessed by following the standard test protocols ISO 20567:2005, Method B [8]. Twelve samples, (three samples for each coating type) were tested against SPE resistance. The samples were produced with the following dimensions: 25 mm long, 25 mm large and 3 mm thick. The analyzed samples were tested in the conditions described in Table 2:

Total grit mass:	500 ± 20 g	**Particle Size:**	4 ÷ 5 mm
Gravelometer air pressure:	200 ± 10 kPa	**Shot duration:**	10 seconds 2 times
Impact distance:	290 ± 1 mm	**Impact angle:**	54 ± 1°
Test temperature:	23±2°C	**Test RH:**	50±5%

Table 2 Experimental conditions for solid particles erosion test.

Samples were weighted before and after the tests to measure the weight loss due to the erosion of the coating surface. The SPE test results refer the thickness loss to the initial coating thickness. The thickness loss values were calculated by comparing the thickness of the unexposed sample to the that same sample at different stages of the test.

3.2.2 Liquid Droplet Erosion test

The resistance of the coated surfaces to liquid droplet erosion (LDE) was assessed by following a procedure based on the standard test protocols of ASTM G73 [9]. The LDE test was set up in a cylindric chamber with a rotating equipment that can accommodate two samples. Liquid droplets are injected in the chamber through a nozzle of adjustable exit diameter (to control droplet size). Ten samples (two for each coating type and two for the F22 bulk) were tested under the following conditions:

Temperature	70°C
Pressure:	30 mbar
Peripheral Speed	250 m/s
Nozzle diameter	200 μm

Table 3 Experimental conditions for liquid droplet erosion test.

The test liquid was distilled water and each sample was subject to five million of impacts. Samples were weighed before and after the tests. The LDE test results compare the weight loss with respect to the initial sample weight.

3.2.3 Wettability

Wettability is the interaction between fluid and solid phases while surface free energy (SFE) is the measure of how a surface interacts with liquids. Each sample was evaluated and compared to understand fouling resistance properties. SFE was evaluated in accordance with the sessile drop technique using Attension Theta equipment by Biolin Scientific. A 1 μ-liter droplet was set on the coated surface at room temperature (25 °C) and the droplet image was recorded with a micro CCD camera which was then analyzed using Attension software to obtain the static contact angle on each of the coatings. The static contact angle is the angle at which the liquid drop surface meets the coated surface, see Figure 4.

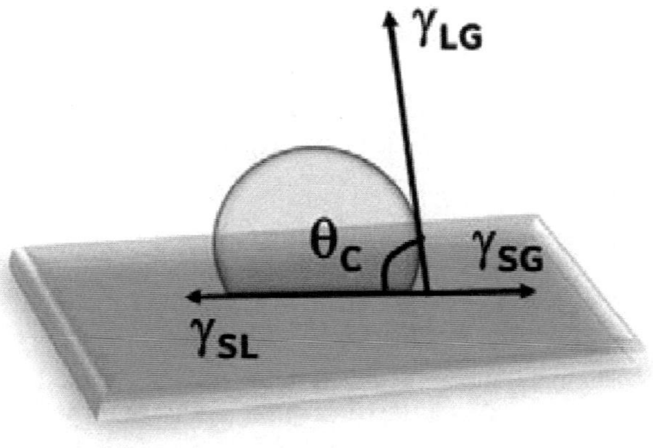

Figure 4 Scheme representing the contact angle (Θ) and the free energy components between gas, liquid and solid phases for a liquid droplet on a solid surface.

The Owens-Wendt model has been considered for SFE calculation. This model is based on the idea that surface free energy has two contributions, polar and dispersive, due to intermolecular interactions.[10] Four different liquids with different polarity were tested on each surface: water, diiodomethane, ethanol and glycerol. The contact angle for each liquid-surface combination was measured over three separate trials, for a total of thirty recorded contact angles.

3.2.4 Fouling Release-Test

Anti-fouling properties were characterized using an in-house developed test apparatus. The coated samples are mounted in the apparatus and rotated at high-speeds (200 rpm). Then, foulant media is injected into the chamber and impacts the surface of the samples at an angle of 45° at high speed. The scheme of the test equipment is shown in Figure 5.

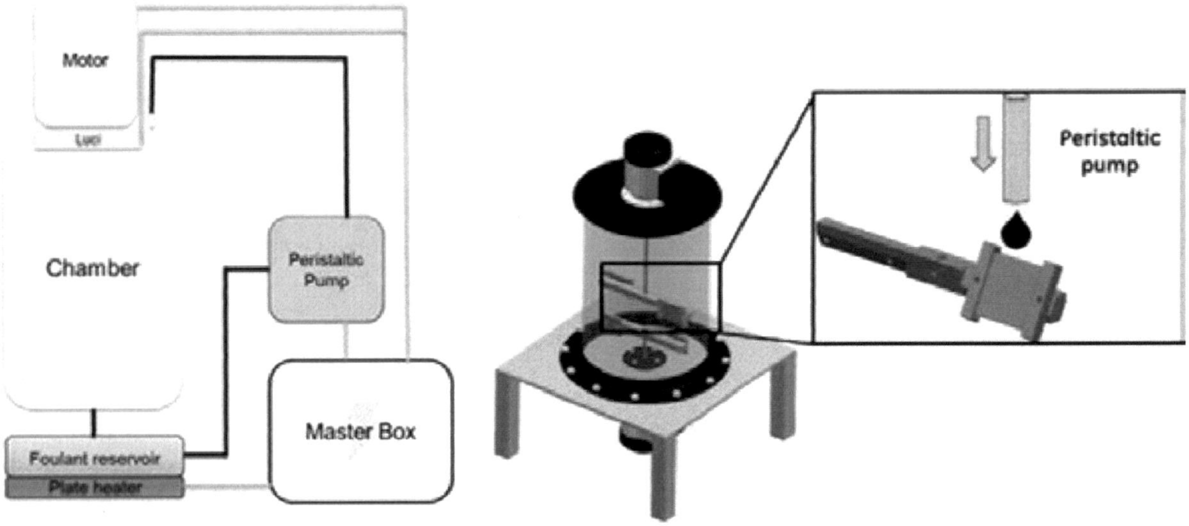

Figure 5 3D drawing of the anti-fouling test rig (center), with detail on the sample holder (right) and illustration of the foulant supply circuit, sample holder drive and system control (left).

The foulant composition is a mixture of asphalt (35% v/v) and high viscosity mineral oil (65% v/v). The foulant media are heated at 100°C through a heating plate and injected in the test chamber by a peristaltic pump (15 ml/min). Samples are weighed before and after the tests. The fouling test results are recorded as the percentage mass gain of the samples with respect to the reference sample (silicon-based polymeric coating) tested under the same conditions.

4. Results and discussion

4.1 Solid Particle Erosion test

Figure 6 summarizes the solid particle erosion resistance (SPE) results obtained by testing the coated samples according to ISO 20567:2005, Method B.

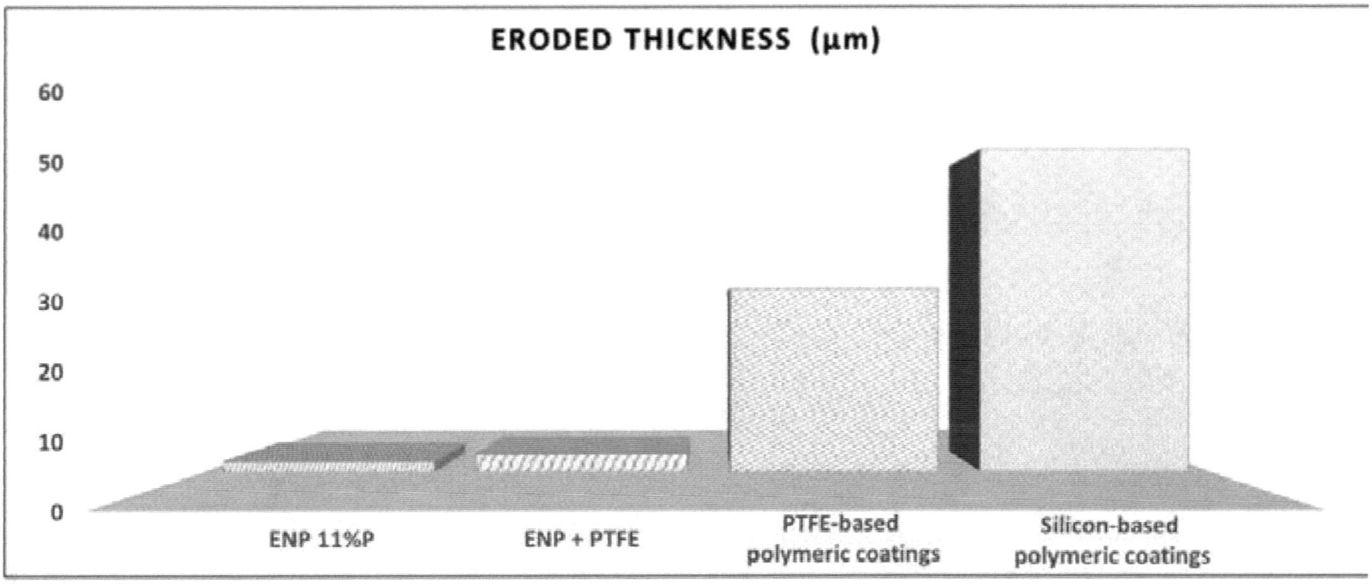

Figure 6 Thickness loss eroded (µm) after samples exposure to two 10 seconds-long blast shots.

The results are expressed as the coating average thickness loss in microns. The highest thickness loss was experienced by the silicon-based polymeric coating which was completely removed from the tested sample. Similarly, PTFE-based polymeric coating experienced a substantial thickness loss exposing large areas of the F22 substrate. The highest erosion resistance was observed on the ENP sample where only a light abrasion of the ENP surface occurred. ENP+PTFE showed higher levels of erosion compared to the ENP sample, but the overall thickness loss was comparable to that of ENP and significantly less than the polymeric coatings. Although the ENP+PTFE hardness (~400 HV) is less than that of the ENP surface (~650 HV), the PTFE doped top layer shows a significant level of resistance to solid particle erosion. This was substantiated by the test results which show the average thickness loss was less than the top coat thickness. Furthermore, the results confirmed that ENP+PTFE is appreciably more erosion resistant relative to PTFE-based and silicon-based polymeric coating.

4.2 Liquid Droplet Erosion test

Figure 7 summarizes the liquid droplet erosion resistance results obtained by testing in accordance with ASTM G73, where the samples were exposed to five million water droplet impacts. Sample weights were taken at regular intervals over the duration of the test. The results are expressed as eroded mass in mg. The trends from the collected data are presented in Figure 7 along with some physical observations.

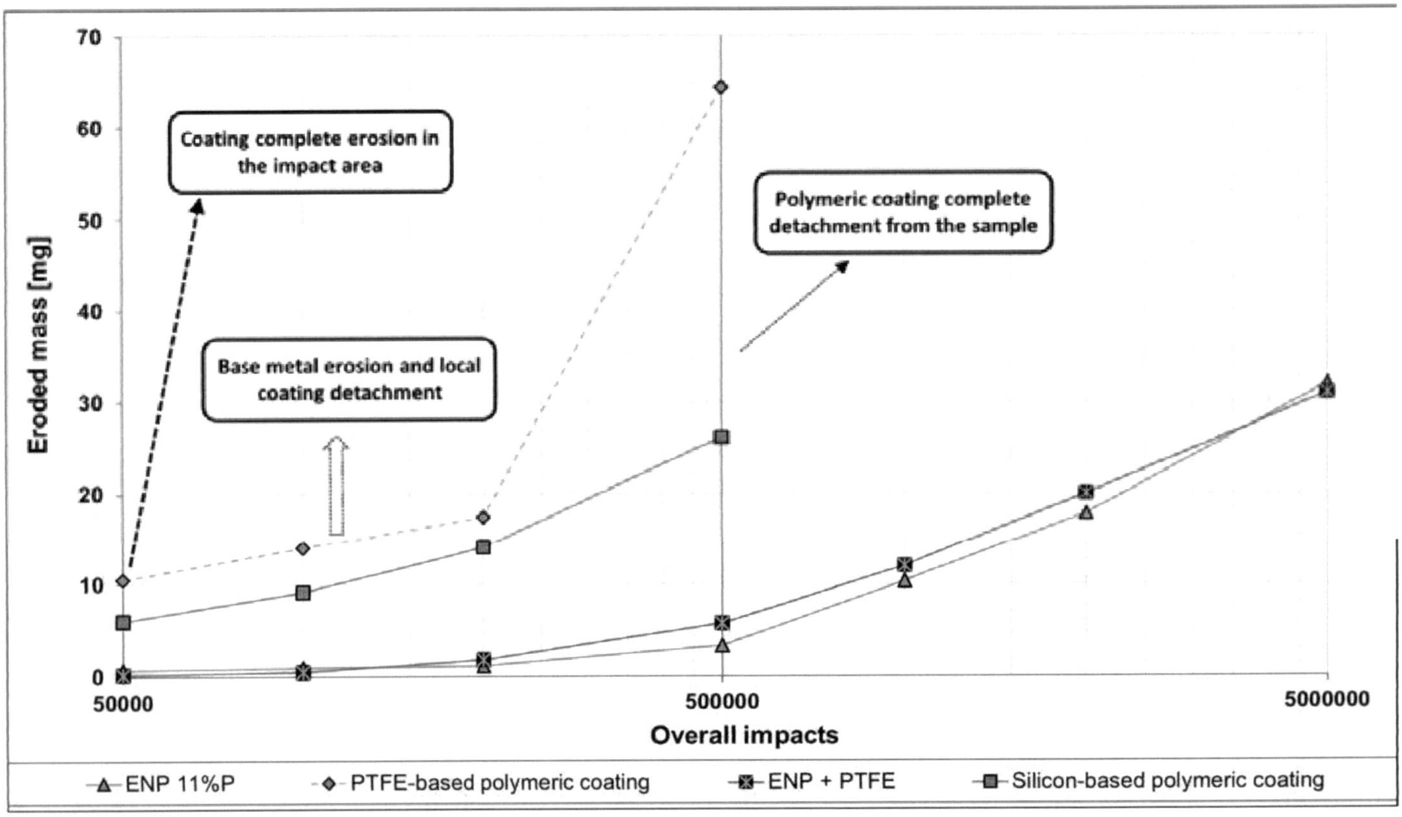

Figure 7 Eroded mass (mg) versus number of water droplet impacts (in log scale).

PTFE-based and silicon-based polymeric coating samples experienced complete erosion in the impact area after the first inspection, at 50,000 impacts, with some incipient attack on the substrate material (see Figure 8 a-b). The higher loss of mass observed for PTFE-based polymeric coating was related to the detachment of the top coat in the area surrounding the impact. The non-linear trends in the data for these coatings, from 50,000 to 200,000 impacts, was due to base material erosion in the impact area and coating detachment in the surrounding area. Beyond 200,000 impacts, the coating completely detached from the sample, (Figure 8) with a consequent sharp mass loss increase. Erosion was even more severe for the PTFE-based polymeric coating which was thicker and denser compared to the silicon-based polymeric coating. The two ENP based coatings showed similar data trends with curves overlapping in the overall test domain. Both coatings showed negligible erosion after 100,000 impacts, then demonstrated excellent durability in the impact area, exposing the base material only after one million impacts where some base material erosion was also observed (Figure 8 c-d). The test continued up to five million impacts, and abrasion proceeded on the base material without perturbing the coating surrounding the impact area. These results confirm the superior liquid droplet impact resistance of the samples coated with ENP and ENP+PTFE compared to the polymeric coatings.

Figure 8 LDE samples after several numbers of impacts: a) PTFE-based polymeric coatings; b) silicon-based polymeric coatings; c) ENP 11%P; d) ENP+PTFE.

4.3 Wetting properties

Table 4 shows the average contact angle measured for drops of water, diiodomethane, ethanol, and glycerol formed on the bare F22 substrate and on the tested coatings surface. The Owens-Wendt equation was used to determine the values of polar and dispersive energy, which are in turn summed to calculate the surface free energy. Energy values of PTFE bulk material [11] are listed for comparison. Note that error for experimental data is less than 5%.

	Contact angle Θ [deg]				Dispersive Energy	Polar Energy	Surface Energy
	H_2O	Glycerol	Et-Gly	Diiodomethane			
F22 steel	84	96	70	69	21.0	5.8	**26.8**
Silicon-based polymeric coating	92	78	65	49	33.3	1.1	**34.4**
PTFE-based polymeric coating	77	88	72	71	18.4	9.7	**28.1**
ENP + PTFE	120	89	81	70	21.5	1.0	**22.5**
ENP 11%P	84	70	71	53	30.8	3.5	**34.4**
PTFE					18.4	1.6	**20**

Table 4 Average contact angles measured with the four liquid, and calculated surface energy components.

Low contact angles (Θ<90°) correspond to a high affinity between the liquid and the surface resulting in the spreading of the liquid over the surface, termed a wettable surface. High contact angles (Θ>90°) indicate a resistance to contact or non-wettable, where the contact area between liquid and surface becomes smaller as the angle increases. Surface free energy is also an indication of surface wettability, the lower the SFE value, the higher the capability of the surface to prevent liquid spreading.

The results listed in Table 4 are better expressed in a graphic form, in what is called the wetting envelope. The calculation requires the numerical resolution of the Owens-Wendt equation, replacing the energy components of the considered surface and setting a wetting condition in terms of contact angle value. In this way, a series of closed contours with a quadratic behaviour was obtained in terms of the dispersive and polar energy components of the liquid.

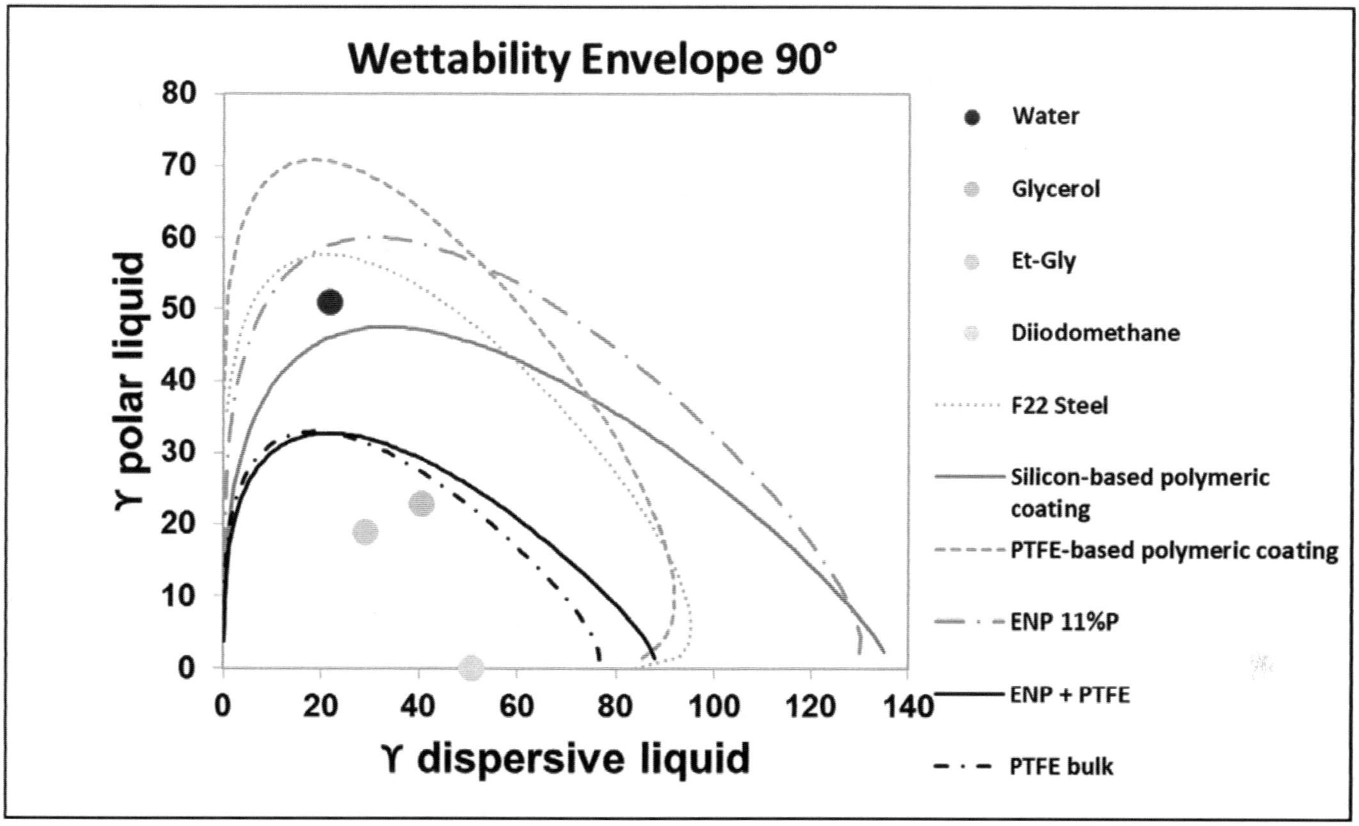

Figure 9 Wetting envelope for the investigated surfaces, at 90° contact angle.

Figure 9 represents the wetting envelopes for the investigated surfaces which are reported together with the polar-dispersive coordinate of water. For water, the contact angle of 90° represents the boundary between hydrophilic and hydrophobic which is why this reference point was selected. If the liquid point is outside the domain representing one of the test surfaces, the contact angle between the liquid drop and the surface will be higher than 90° which indicates low wettability. Furthermore, the domains with smaller areas have low surface free energy, also indicating low wettability. As visible from the graph, ENP+PTFE demonstrates superior properties in terms of wettability: such properties are superior to the other tested materials, and nearly equivalent to bulk PTFE.

4.4 Fouling Release-properties

Figure 10 summarizes the fouling release property results obtained by testing the coated samples with the "purpose-built" apparatus described in section 3.2.2.

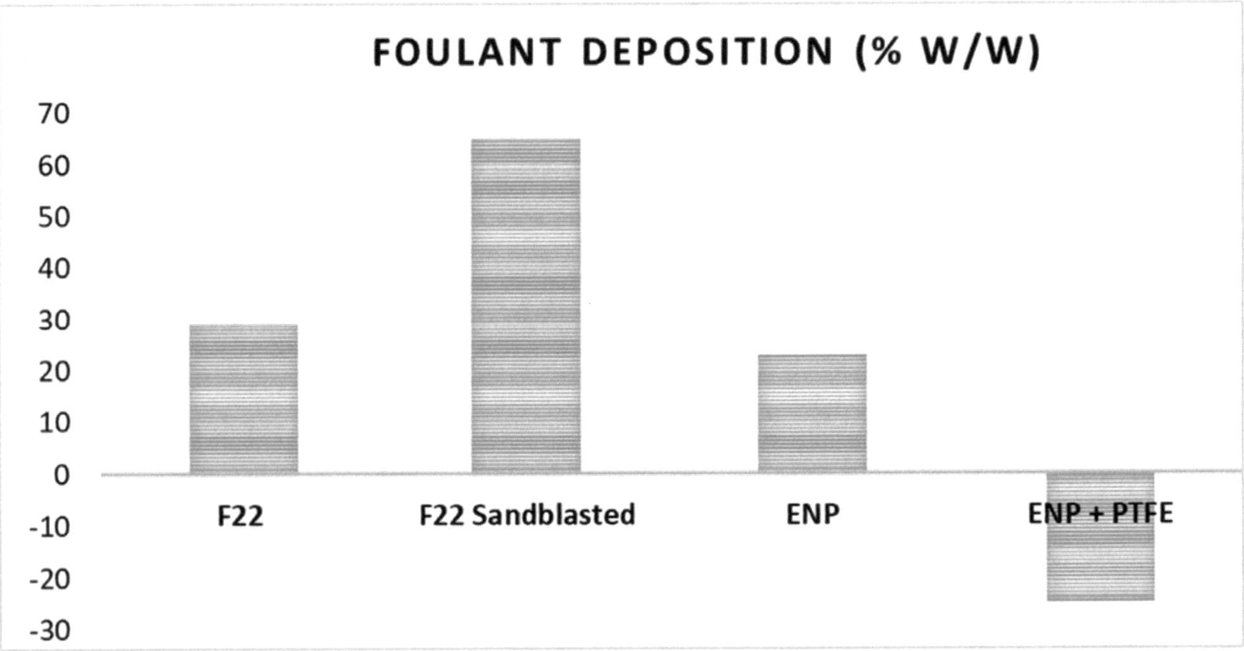

Figure 10 Foulant accumulation on F22 sandblasted, F22, ENP coated and ENP+PTFE coated samples. The baseline (0 value) represents the foulant accumulated (in weight %) on the silicon-based polymeric coating substrate.

Silicon-based polymeric coating sample was selected as the "zero reference" or baseline for the comparison of results. A bare F22 surface with a roughness of 0.8 μm had a 29% mass gain. In order to investigate the effect of surface roughness, a second sample of F22 was sandblasted creating a 3.2 μm surface roughness. When exposed to the foulant, the recorded weight increase was significantly higher than the 0.8 μm finish (> 60% mass gain), confirming the surface roughness effects on fouling resistance. The ENP-coated surface experienced a 23% weight gain, similar to the F22 uncoated sample. The sample coated with an ENP+PTFE showed a significant reduction in fouling accumulation (-25% weight gain) with respect to the reference sample. Since both the ENP and ENP+PTFE maintain the same surface roughness as the substrate, the low degree of mass gain of the ENP+PTFE sample is directly related to the low wettability and the low friction features of the "nano-texture" created by the fluoropolymer particles dispersion in the coating. (see electron microscope image in Figure 11 showing PTFE particles in the coating).

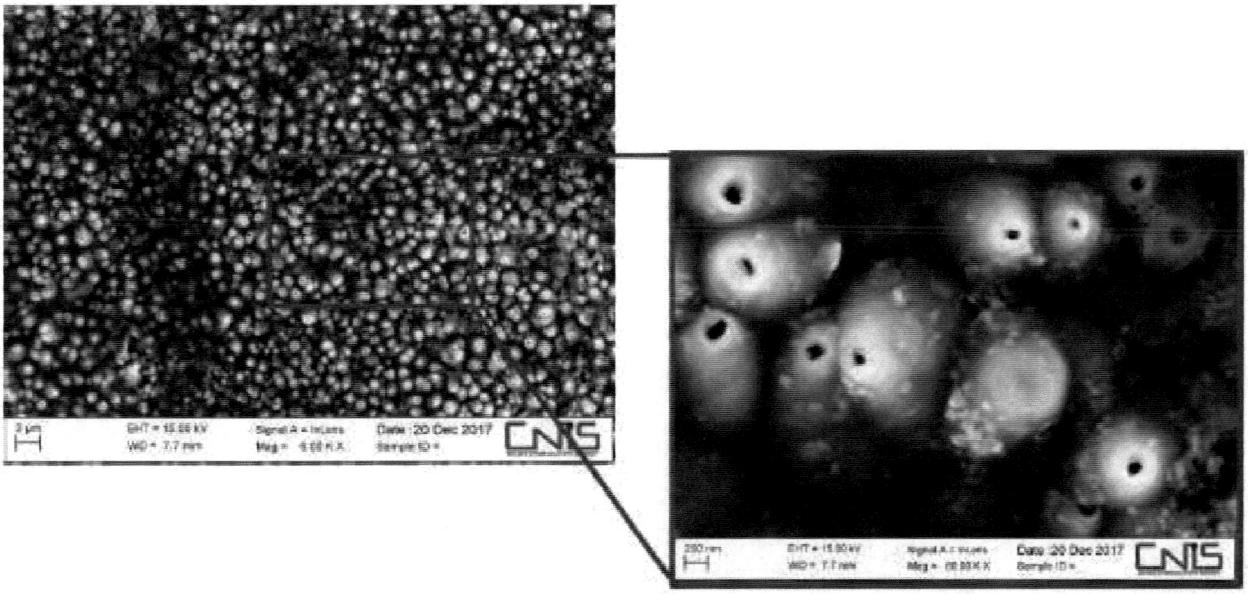

Figure 11 Scanning Electron Microscope (SEM) image of ENP+PTFE coating (top view).

5. Conclusions

Fluorinated and non-fluorinated polymers are an effective solution against fouling and corrosion attack, however, they are not inherently resistant to erosion. In fact, they are rapidly eroded when inorganic particulate is present in the cracked-gas, or when water is injected for cooling purposes. Eventually, fouling occurs on the eroded and damaged areas and on uncoated substrate. The attempt to wash away the foulant, on and off-line, further compromises the coating integrity. Furthermore, the application of polymeric coatings requires line-of-sight to the surface being coated, a characteristic of all spray application processes. These coatings present significant challenges when coating inner surfaces of small diameter bores and other complex geometries causing defects in the coating or variation in the thickness. In this context ENP coating doped with PTFE particles represents a step forward in mitigating fouling issues in the downstream market for cracked gas and other services. The benefit of the electroless plating application process is the removal of geometry limitations. And, the nature of the application which is finely controlled to deliver a high quality, uniform coating, without altering the original surface finish, including super-finished, low roughness surfaces. ENP+PTFE coating quality control coupled with excellent protection from fouling and resistance to corrosion and erosion, results in extended time between maintenance, improved equipment performances decreasing the overall cost of operation.

References

[1] S. Snider, Ethylene plant cracker gas compressor fouling, conference paper, April 2006, EPC Conference 2006, AIChe Spring National Meeting.

[2] R. Shah, Olefins Compressors' Reliability Performance Survey Results, 2008 Ethylene Producers' Conference.

[3] The Impact of Fouling and Corrosion on Ethylene Plant Efficiency, Conference Paper, April 2009, EPC Conference: 2009 AIChE Spring National Meeting.

[4] J. M. Shende, S. Vijayaraghavan, S. Mankin, K. D. Subramaniyam, M. Dorf Ketal, Improve compressor reliability with advanced chemical treatments, Hydrocarbon Processing, August 2014.

[5] B. McCune, Coatings to mitigate fouling on rotating equipment used in ethylene production, Article 36/37F, 1997, Elliot.

[6] W. Wang, P. Dowson, A. Baha, Development of Antifouling and Corrosion Resistant Coatings for Petrochemical Compressors, Proceedings of the Thirty-Second Turbomachinery Symposium - 2003, Turbomachinery Laboratory, Texas A & M University, College Station, Texas, pp. 91-97.

[7] Electroless Nickel Coatings, NACE International Publication 6A287 (1997 Revision).

[8] ISO 20567-1:2005 - Paints and varnishes — Determination of stone-chip resistance of coatings — Part 1: Multi-impact testing.

[9] ASTM G73 - 10(2017) - Standard Test Method for Liquid Impingement Erosion Using Rotating Apparatus.

[10] D.K. Owens and R.C. Wendt, L. Appl. Polymer Sci. 13, 1741 (1969).

[11] H.A. Guleç, K. Sarioglu, M. Mutlu, J. Food Engineering 75, 187 (2006).

AIChE Paper Number: 550265

CONSIDERATION & RECOMMENDATIONS FOR ETHYLENE PRODUCERS WHEN SPECIFYING API STANDARD 692

F. Alex Schaefer

Senior Engineer

Elliott Group

Belle Vernon, PA

Michael Shaffer

Technical Services Engineer

Elliott Group

Jeannette, PA

Prepared for Presentation at the 2019 Spring National Meeting

New Orleans, LA, April 1, 2019

CONSIDERATION & RECOMMENDATIONS FOR ETHYLENE PRODUCERS WHEN SPECIFYING API STANDARD 692

F. Alex Schaefer

Senior Engineer

Elliott Group

Belle Vernon, PA

Michael Shaffer

Technical Services Engineer

Elliott Group

Jeannette, PA

Prepared for Presentation at the 2019 Spring National Meeting

New Orleans, LA, April 1, 2019

INTRODUCTION

The summer of 2018 saw the highly anticipated release of API Standard 692, Dry gas Sealing Systems for Axial, Centrifugal, Rotary Screw Compressors and Expanders. Now what? Inside the front cover of the standard and under "Special Notes" the fourth paragraph frames this question most eloquently;

"API publications are published to facilitate the broad availability of proven, sound engineering and operating practices. These publications are not intended to obviate the need for applying sound engineering judgement regarding when and where these publications should be utilized. The formulation and publication of API publications is not intended in any way to inhibit anyone from using any other practices."

Several parties are involved in the design, manufacture, and commissioning of a dry gas sealing system. If these parties do not communicate (or worse yet - miscommunicate) critical information then process reliability will be diminished.

DATASHEETS, OPERATING CONDITIONS & THEIR IMPACT

Datasheets are a generally accepted method to communicate vital information between parties in a uniform, consistent manner. Let's emphasize that this transfer of information is intended to be bi-directional. Datasheets should be attached to the purchase order with as much information filled out as possible. The supplier in turn completes the datasheets and returns them to the purchaser[1] for approval/comment and record. Figure 1 is only a typical example of this flow of information between the parties involved.

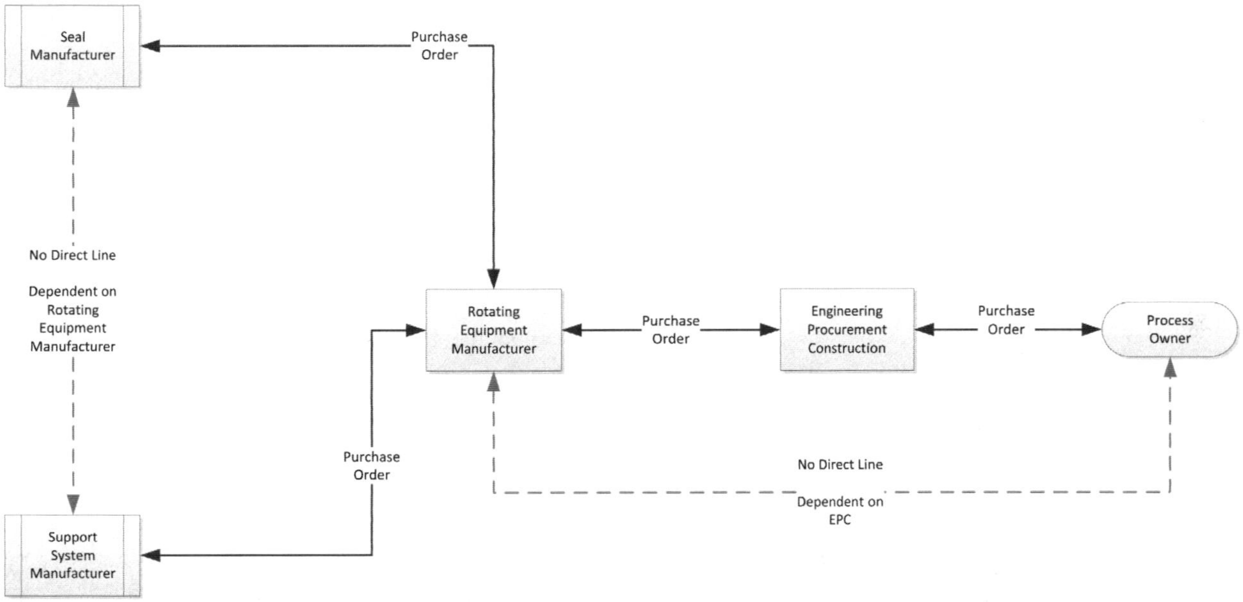

Figure 1: Flow of Project Information

It is clear that the opportunity for lack of communication or even misinformation is present. Therefore datasheets can be valuable tool in maintaining and controlling this flow of data.

So, what information should be included on these datasheets? The whole objective of the contract is to deliver a stable process operating at a given condition. The guaranteed[2] and normal operating conditions are typically filled out in great detail on API 617 datasheets. Often there are several alternate operating points defined as well. The more thought and detail included at the front end of the project will help

---

[1] API defines purchaser as "Agency that issues the order and specifications to the vendor." The purchaser can be; the owner of the facility in which the equipment is installed, the owner's appointed agent, or the equipment manufacturer purchasing components.

[2] API 692, 1st edition defines the certified point as the "Point to which the performance tolerances will be applied."

control costs and prevent changes later in the schedule. Figure 2 is a familiar abstraction of this principle showing the control of costs in relation to project schedule.

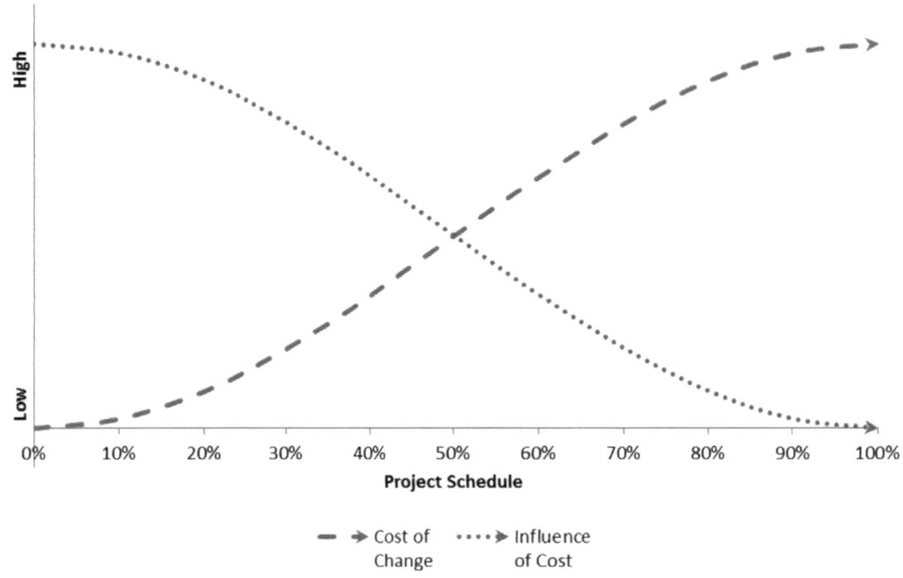

Figure 2: Control Over Project Costs

But what about alternate conditions? Processes are designed to operate in a steady state[3] condition; however boundary conditions such as standby[4], transient[5], slow-roll[6] , turning gear[7], and settle-out[8] cannot be ignored. In fact, these boundary conditions typically influence the selection of seals and the support system design.

API 692 Annex 2D was developed to assist engineers in seal selection while Annex's 2A & 3A assist in the specification of these components and systems.

---

[3] API defines steady state as "Constant operating parameters."

[4] API defines standby as "Service state in which a piece of equipment is normally idle or idling and is capable of immediate automatic or manual start-up for continuous operation."

[5] API defines transient as "Operation outside the defined conditions or transition from one operating state to another such as normal start-up to running or running to shutdown."

[6] API defines slow roll speed as "Rotational speed (RPM) typically less than 5 % of the normal operating speed or the minimum speed permitted by speed control. Typical speed values for slow roll are about 500 RPM."

[7] API defines turning gear speed/ratcheting/barring as "Temporary operating condition during start-up and shutdown. Typically 50 RPM or less."

[8] API defines settle-out pressure as "Highest pressure that the machine experiences when the machine is not running and equilibrium has been reached." Determination of settle-out pressure requires consideration of the trapped volume of gas throughout the machine and its associated piping system.

Influences such as gas sources, ambient conditions, sealing conditions, compressor temperature limits, and compressor configuration can determine the minimum allowable temperature[9] and maximum allowable temperature[10] of the dry gas seal and dry gas seal support system. Most dry gas seals and dry gas seal support systems are rated to maximum allowable working pressure (MAWP)[11] of the compressor unit. In some cases the dry gas seal and dry gas seal support system may have different limits from the compressor unit.

Flows for the dry gas seal support system are determined by the needs of the dry gas seal, separation seal, and process side seal. When labyrinths are used in these areas, minimum velocities across these seals must be considered. It is important to associate a clearance value with these velocity values. In situation when the process gas cleanliness is a concern, these velocity values should be increased.

SEAL SELECTION

API 692 Annex 2D was developed to assist engineers in seal selection. For ethylene producers, the first decision is an easy one – there is likely no process that would allow application of a single seal. The next decision however is worth some consideration: high or low pressure? Any process with sealing pressures above 500 PSIG will eliminate double opposed seals as an option. There are multiple refrigeration processes whose sealing pressure operate near atmospheric, however double seals are not really suitable for closed cycle processes. A low or medium pressure body of a cracked gas compressor string may be good candidate however, assuming a reliable source of nitrogen seal gas is available.

If tandem seals are now your only choice then an intermediate labyrinth and secondary seal gas are essentially required to prevent process leakage to atmosphere. If air is going to be considered as a separation gas, then labyrinth type separation seals should be considered with enough flow to keep the secondary vent flow below the minimum explosive limit.

SUPPORT SYSTEM MODULE SELECTION

---

[9] API defines maximum allowable temperature as "Maximum continuous temperature for which the manufacturer has designed the equipment (or any part to which the term is referred)." The maximum allowable temperature is usually set by material considerations.

[10] API defines maximum allowable temperature as "Maximum continuous temperature for which the manufacturer has designed the equipment (or any part to which the term is referred)." The maximum allowable temperature is usually set by material considerations.

[11] API defines maximum allowable working pressure as "Maximum continuous pressure for which the manufacturer has designed the equipment (or any part to which the term is referred) when handling the specified gas at the specified maximum allowable temperature.

Module A, the Alternative Seal Gas Supply may be the least appreciated module of the support system. Its most basic purpose is to provide seal gas during start-up, shut-down, and settle-out conditions as the compressor discharge transitions in to, or out of a suitable supply pressure. Simple choices such as the type, location, or orientation of the check valves may affect how well this module performs during these critical conditions. Figure 3 illustrates how a swing type check valve can operate best in the vertical, flow up orientation, while a piston type check will work best in a horizontal orientation.

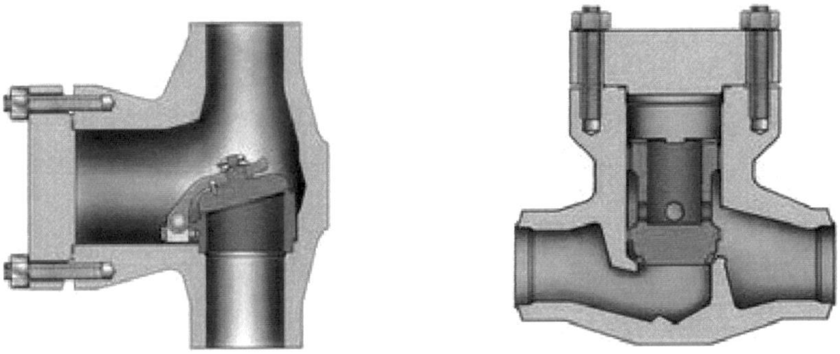

Figure 3: Check valve types & orientation

Special care must be taken at any point which two independent systems are merged. A classic example is that of a nitrogen back-up supply being contaminated by the process due to a leaking check valve, which can lead to disastrous consequences. Figure 4 is one example of how to safely isolate these two supplies when the back-up supply is at a lower pressure than the compressor discharge.

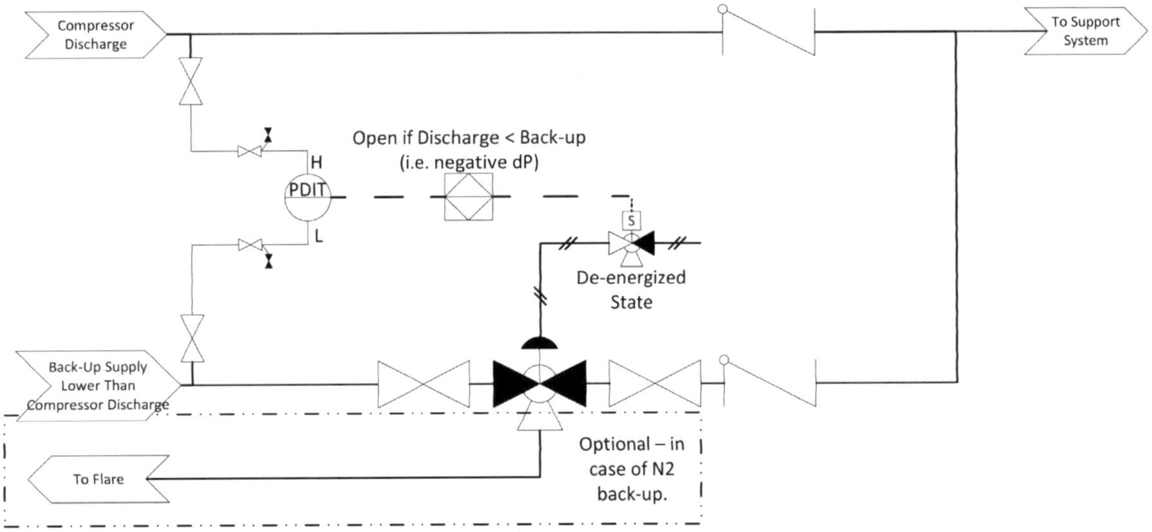

Figure 4: Alternative Seal Gas Supply
Back-up lower than compressor discharge pressure

What if the back-up supply is higher than the compressor normal discharge pressure? Check valves alone are certainly not the answer. Figure 4 is one example of how to modulate a higher pressure back-up supply, assuming the compressor normal discharge pressure is well above settle-out pressure to begin with.

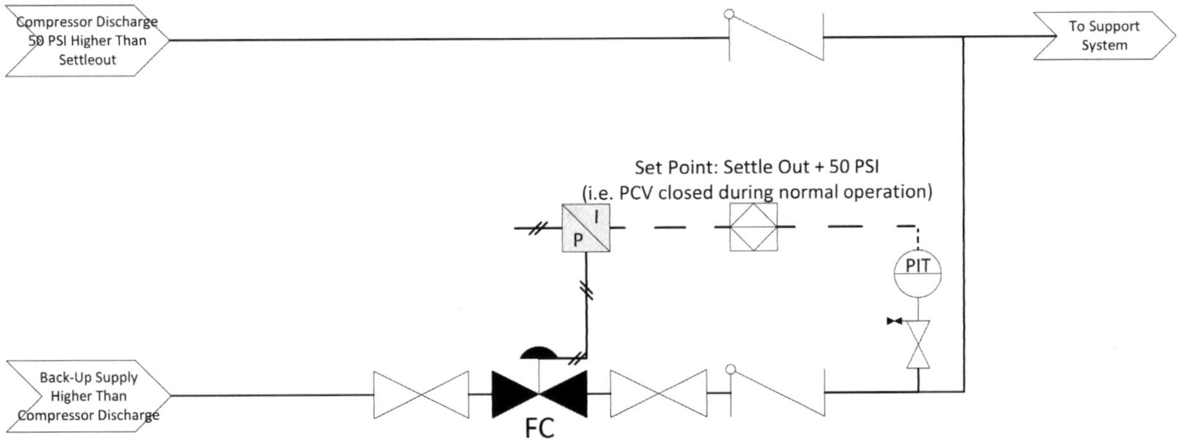

Figure 5: Alternative Seal Gas Supply
Back-up higher than compressor discharge pressure

The PID loop will modulate the valve until the compressor discharge exceeds settle-out pressure by a suitable margin (50 PSI). Then close off the backup supply while the compressor is in normal operation and the pressure at the PIT is above set point. These two simple examples further illustrate the importance of participation by all stake holders in the HAZOP.

One of the many additions in API 692, above and beyond what had previously been addressed in API 614, is the Module B - Gas Conditioning System (GCS) shown in Figure B.18 of API 692. Based on failures early in dry gas seal history, API 692 requires a GCS as a default scope of supply for any tandem seal support system. The intent of this was to insure reliability and prevent costly failures during commissioning and other alternate operating conditions in addition to normal operation. The intent is *not* to require every module described in API 692 as if they were code requirements and without consideration of the application. It is important for the purchaser and supplier to reach mutual understanding and agreement on how the GCS is to be implemented based on the application and previous experience. Clearly a cracked gas application has greater considerations with respect to seal gas conditioning than a refrigeration cycle.

Heat Exchanger (B1)

Although it may be the first component in the stream of the GCS, a heat exchanger should be regarded as a last resort. Clearly a water cooled heat exchanger increased complexity that could actually *decrease* system reliability if it is poorly selected. Somethings to consider before committing to the heat exchanger module:

- Seal gas supply source, interconnecting piping, and ambient conditions? How much variation will there be in gas inlet and outlet temperatures
- How much "approach" temperature exists between the plant cooling water supply and the desired process outlet temperature? Will say, a 100°F process outlet temperature be enough to achieve the desired results?
- What is the confidence and level of detail provided in the datasheet process conditions? Are there condensables that are missing or understated?
- Cooling fluid on the shell or tube side? Can tube leaks be detected and resolved?

If a heat exchanger is determined to be necessary, then a commitment to including a reliable separator and heater is also required.

Separator (B2)

Including a separator in the primary seal gas supply is inexpensive insurance against liquids that may have condensed in the seal gas supply piping that is likely uninsulated and lengthy. However for a separator selection to be meaningful, there must be thought given and agreement made not only to the condensate collection rates, but the specific gravity of the condensate.

Booster (B3)

First of all, let's consider the name "booster". Most people agree that this was a poor choice for a name that stuck early on. Renaming it "intensifier" has been attempted, but just hasn't caught on as hoped. The problem with these names is that they are misleading. Just like any pump or compressor, the "booster" creates *flow* – it is the piping system components that create the back *pressure*. These devices will not turn a "low" pressure supply into a "high" pressure supply regardless of it is a positive displacement type or centrifugal type. They simply create enough differential pressure to furnish the minimum seal/buffer gas flow.

In API 692, the minimum gas velocity for the process side seal is reduced during periods when the compressor is operating at speeds below slow roll (below 500 rpm).

When the seal/buffer gas can be supplied by a booster during compressor settle out or pressurized hold, the specified gas velocity for the process side seal is not expected to be maintained during these periods.

Heater (B4)

As stated in API 692, All seal gas and buffer gas streams (including alternate, backup, start-up sources) flowing into seal cavities shall be supplied at a temperature providing at least 20 °C (35 °F) margin from the dew point line in order to preclude the possibility of liquid entering or forming in the dry gas seal. Heater, Heat tracing, and insulation is to be provided to keep gas with limits.

Agreement on heat tracing and insulation of lines downstream of the separator must be addressed early on. Although heat trace and insulation can be applied in a shop environment by the seal support system manufacturer, it is should be noted that this work can be easily undone if the system is stored poorly or mishandled later at site. However the piping designer must still leave provision for insulation if it is to be applied later at site.

In some services gas is just above the dew line margin, insulation should start at the gas source as there are cases where the environment can rapidly cool the piping causing liquid to form in the lines. In some cases, rain water from a summer storm cooled piping was enough to have liquid form in the lines.

Assuming a tandem seal has been selected, the next big choice would be flow control verses differential pressure control. Consider performing a hazard and operability study (HAZOP) as early as possible in the project before fully committing to a flow control strategy. This can pay huge dividends later by preventing changes midway or even late in the project. (reference Figure 2) Participation by the rotating equipment and seal support system manufacturer is worth it.

If flow control is chosen, it is optimal to use independent primary supply flow control valves. Only using one FCV will essentially put the suction side of the compressor in differential pressure control – assuming the seal support system is following minimum velocity requirements.

Not following minimum velocity requirements may take a while to show up in the form of poor reliability, however reverse pressurization will identify itself immediately. Differential pressure control is a more robust method of control and with low signal select properly applied will prevent reverse pressurization of the primary tandem seal in services with low or near atmospheric suction such as refrigeration cycles.

API 614 has helped raise the reliability of oil systems requiring failsafe[12] design practices. Of course that experience and philosophy has been carried over to API 692. However with respect to alarm[13] and shutdown[14] points, API 692 is much more sophisticated. For instance, on oil systems the "Low-Low Oil Pressure" set point is universally understood to be the equipment "trip", which automatically shuts down the equipment as quickly as possible to minimize damage & repair. Since these signals were automatic two out of three (2oo3) voting is typically required to prevent spurious trips.

With early detection, experienced operators can now initiate an orderly, manual shut down of the equipment. This reduces plant upset, lost product, while increasing the information available for troubleshooting.

SUMMARY

The intent of API Standard 692 is to facilitate communication between the Owner, EPC, OEM, any other party involved in the design, manufacture, and commissioning of a dry gas sealing system.

Advanced planning and mutual understanding and agreement up front will prevent changes later in the project, keeping quality high, delivery on schedule, and costs under control.

Understanding the basis of the requirements in API 692 and "applying sound engineering judgement regarding when and where these publications should be utilized" will be key to a successful new equipment project or dry gas seal upgrade.

REFERENCES

1. American Petroleum Institute, "Dry Gas Sealing Systems for Axial, Centrifugal, Rotary Screw Compressors and Expanders," API 692, 1st Edition, 2018.
2. American Petroleum Institute, "Axial and Centrifugal Compressors and Expander-compressors," API 617, 8th Edition, 2014.
3. American Petroleum Institute, "Lubrication, Shaft-sealing and Oil-control Systems and Auxiliaries," API 614, 5th Edition, 2008.

---

[12] API defines failsafe as "System or component that will cause the equipment to revert to a permanently safe condition (shutdown and/or depressurized) in the event of a component failure or failure of the energy supply to the system."

[13] API defines alarm point as "Preset value of a measured parameter at which an alarm is activated to warn of a condition that requires corrective action."

[14] API defines shutdown set point as "Pre-set value of a measured parameter at which automatic or manual shutdown of the system or equipment is required."

AIChE Paper 35a

DEBUTANIZER BOTTOMS RECYCLE FOR ENHANCING GASOLINE/ WATER SEPARATION IN THE QUENCH SYSTEM

André Bernard
Process Design Engineer
NOVA Chemicals Corporation

Feridoun Fahiminia
Research Scientist
NOVA Chemicals Center for Applied Research

Prepared for Presentation at the 2019 Spring National Meeting
New Orleans, LA, March 31st – April 4th, 2019

AIChE Paper 35a

DEBUTANIZER BOTTOMS RECYCLE FOR ENHANCING GASOLINE/ WATER SEPARATION IN THE QUENCH SYSTEM

André Bernard
Process Design Engineer
NOVA Chemicals Corporation

Feridoun Fahiminia
Research Scientist
NOVA Chemicals Center for Applied Research

Abstract

Deficiencies in raw pyrolysis gasoline (RPG) can be experienced when processing larger volumes of light feedstocks. Beside challenging the process capability in maintaining RPG product specifications, it may affect the water/ hydrocarbons separation in the quench water settler. This could likely result in premature fouling of the dilution steam system. Poor separation can be cause by emulsion and/ or low RPG/ water ratio below the immiscibility compositions. While emulsion problems can be handled with emulsion breaker and proper pH control, operating below the liquid/ liquid immiscibility region can only be resolve by increasing the RPG/ water ratio. Recycling Debutanizer bottoms is one way of increasing the critical mass of gasoline in the settler within the RPG/ water immiscibility region. However, the stream contains a number of known fouling pre-cursors, which depending on the recycle location may increase fouling propensity in certain areas of the plant. This paper discusses the expected distribution of fouling precursors relative to the recycle location in the Quench system and the effectiveness of the recycle in increasing the RPG/ water ratio.

Introduction

The evaluation presented in this paper was initiated by interests expressed to the EPC Ethylene Feedstocks Sub-Committee regarding Debutanizer bottoms recycled to the Quench System for enhancing water/ hydrocarbon separation. Given the lack of operating data available for such practice, the evaluation largely depends on a review of literature, process simulations and available plant stream analysis. Numerous papers have been published on quench water treatment. From those papers a distinction in process requirements can be made between plants originally design for gas and those designed for naphtha feed. The gasoline characteristics and absence of a Primary Fractionator in a gas plant makes treatment of quench water more difficult than for a liquid cracker processing naphtha. As such, additional treatment steps such as, coalescing, filtration and/ or floatation will complement the oil/ water settler and process water stripper.

In recent years; many plants originally designed for naphtha cracking have been introducing or expanding their capability to process larger volumes of gas feed. Those conversions, largely driven by the availability of ethane from shale gas, are known to impact the effectiveness and reliability of the entire quench system.

A review of Quench system designs, feedstocks and associated problems is first presented. Key parameters influencing performance of individual quench system components are discussed and compared against Debutanizer bottom stream characteristics to assess the suitability of the stream in improving performance. Process simulation results are analyzed relative to the effectiveness of recycling Debutanizer bottoms and the impact on the distribution of fouling precursors throughout the olefins unit.

Quench System Overview

A typical naphtha cracker quench system is illustrated on Figure 1. Cracked gas exiting the furnaces transfer line exchangers is further cooled by quench oil in quench fittings prior to feed the Primary Fractionator. Quench oil recovered in the bottom of the tower is cooled by generating dilution steam; a slip stream of quench oil is further cooled by pre-heating naphtha feed and provides mid-reflux to the Primary Fractionator, while the bulk of the quench oil stream is recycled to the furnace quench fittings. Excess fuel oil is processed in a Fuel Oil Stripper where raw pyrolysis gasoline (RPG) and some light fuel oil are stripped back to the Primary Fractionator. The bottom section of the Primary Fractionator is often referred as a heat transfer section. The top section is refluxed with RPG recovered from the bottom of the Quench Water Tower. The reflux rate is set to meet a target end point in the RPG stream extracted from the Quench Water Tower bottom. Cracked gas from the overhead of the Primary Fractionator feeds the Quench Water Tower where further cooling is achieved by a quench water loop, which uses a number of process heat sinks and cooling water exchangers. The water/ RPG mixture is collected in a settler where RPG separates from water due to immiscibility of the fluids. The bulk of the water is recycled in the quench water loop, while the excess is treated in a Process Water Stripper and supplies water for dilution steam generation. RPG condensing across the Cracked Gas Compressor train is processed and recovered in a Gasoline Stripper. Lighter RPG moves forward downstream the Cracked Gas Compressor and is recovered in the Debutanizer bottom.

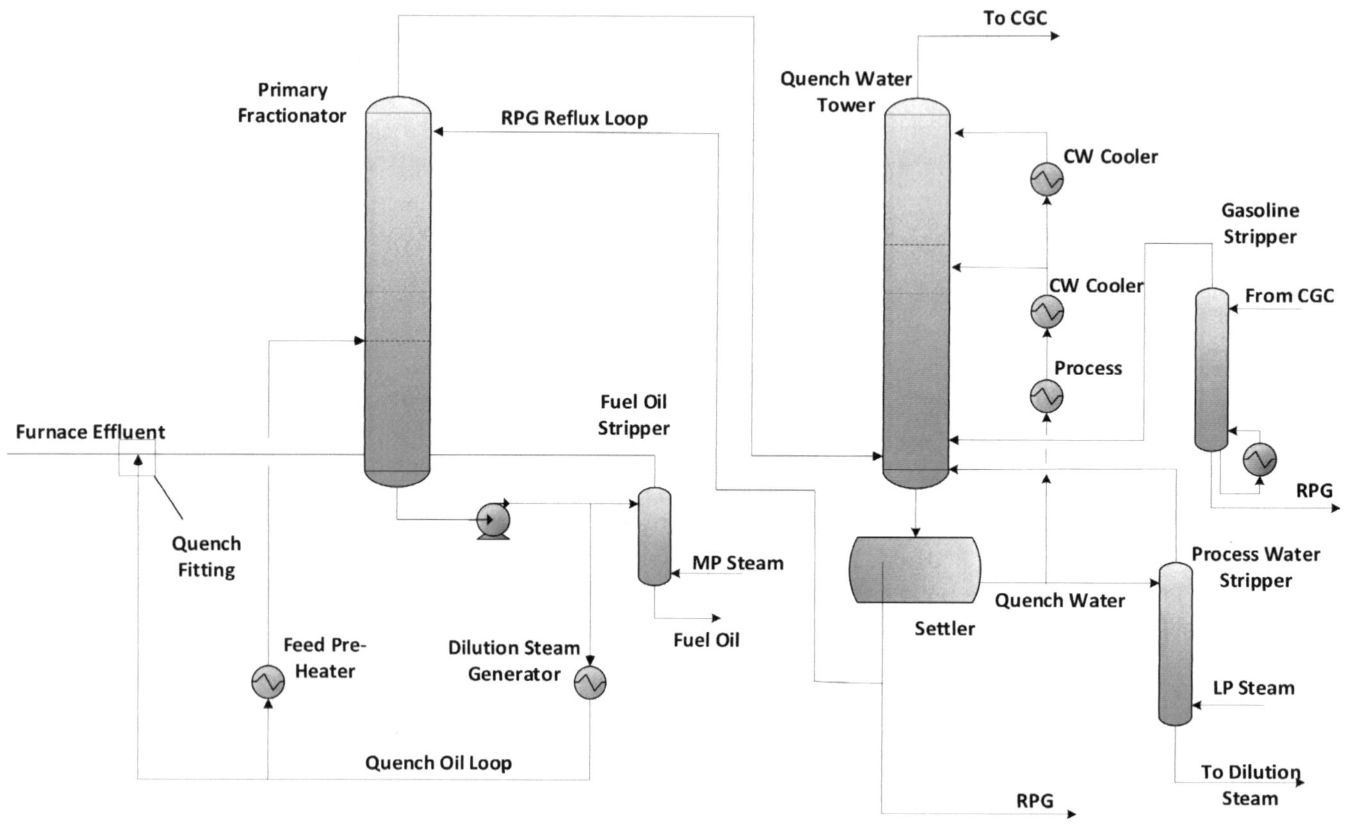

Figure 1: Naphtha Cracker Quench System

As shown in Figure 2, plants designed for gas feed are not equipped with a Primary Fractionator. Cracked gas from the Furnaces goes directly to the Quench Water Tower. As such, the hydrocarbon content in the settler includes tars, fuel oil and RPG. Due to the higher specific gravity than water, tars and heavy oil will settle in the bottom, water will form a middle layer, with light fuel oil and RPG floating on top. The settler is normally equipped with 3 compartments separated by baffles to ensure withdrawal of each of the phases. Similar to the Naphtha plant design, the bulk of the water is recycled back to the Quench Water Tower after being cooled by process users and cooling water. Excess water undergoes further treatment prior to feeding to the process water stripper. Some of the systems used for water treatment may include a coalescer, filtration, dispersed oil extractor (DOX) and/ or induced gas floatation. Pre-treatment upstream the Process Water Stripper is necessary due to oil and tar carryover in the water phase withdrawn from the settler. Separation difficulties in the settler due to emulsions is known to cause significant fouling problems in the quench water loop around the Quench Water Tower, the water treatment unit, Process Water Stripper and dilution steam system. The conventional approach to improve separation in the settler relies on good pH control (slightly acidic to neutral) and emulsion breakers. Cockshutt et al. [1] reported that the addition of sulfur compound such as DMDS in the furnaces contributes to the stability of the emulsion in the settler. Tighter control of DMDS dosage, pH control between 6 and 7 along with proper dosage of emulsion breakers proved effective. Overdosage of emulsion breaker can become ineffective in breaking the emulsion, as has been reported by Hammond [2].

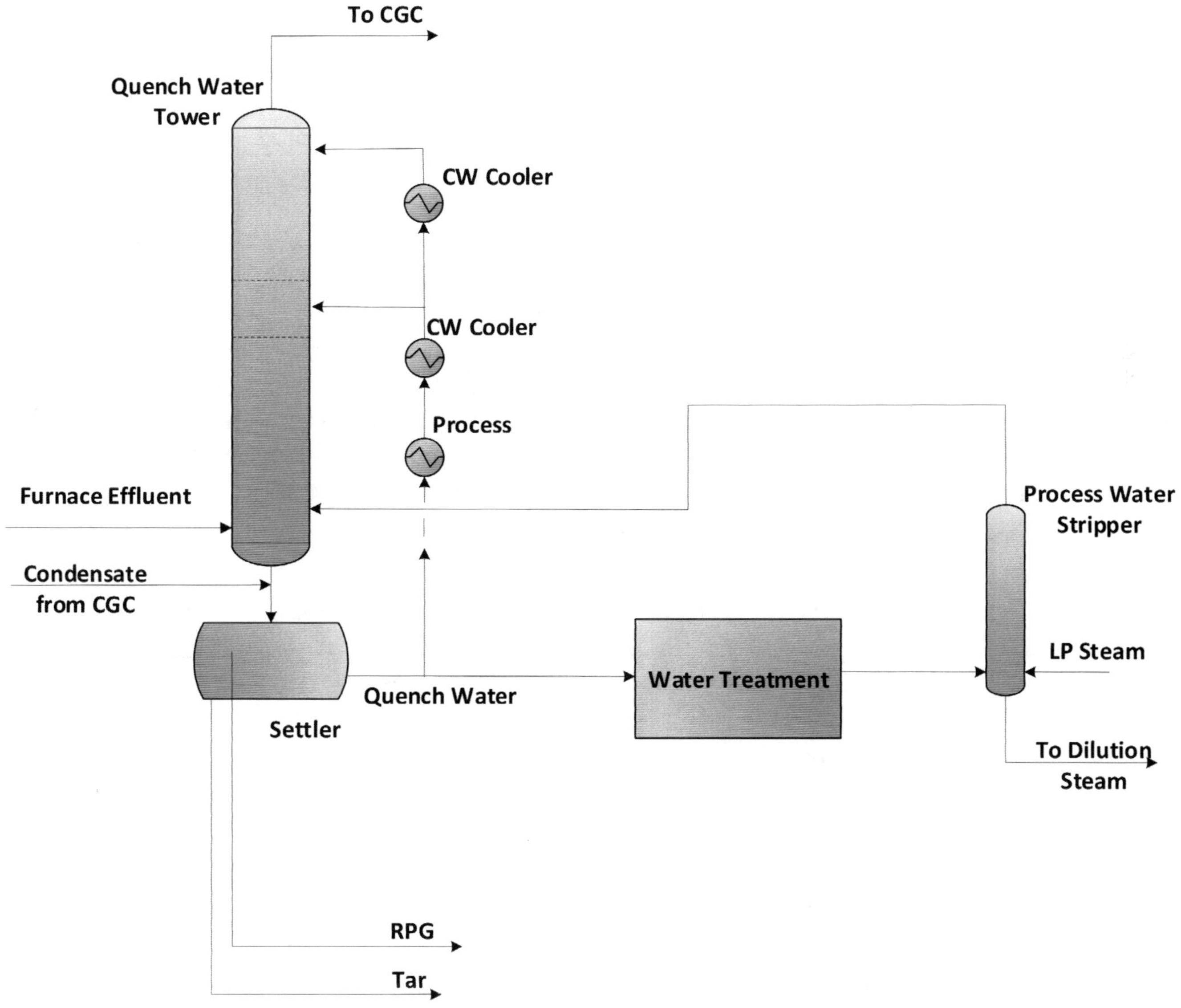

Figure 2: Gas Cracker Quench System

The use of Debutanizer recycles to different locations of the quench system has been reported by Cockshutt et al. [1] as an effective way to displace the tars and heavy oil, preventing them from accumulating. Kurukchi et al. [3] report that Debutanizer bottoms recycle used as wash increases the density difference between the water and hydrocarbon phase, hence enhancing the separation in the settler. However, he indicates that the presence of dienes and other fouling precursors contained in the Debutanizer bottom stream increase the risk of fouling in the Process Water Stripper and dilution steam system.

Impact of Feedstocks and Process Schemes

Figure 3 compares typical furnaces fuel oil and RPG make for different feedstocks. As shown, naphtha cracking generates significantly more fuel oil and RPG than natural gas liquid (NGL) feeds. Separation of tars and fuel oil in the Primary Fractionator prevents them from contaminating water recovered in the quench water settler. As such, Naphtha crackers do not need additional water treatment other than the Process Water Stripper. The characteristics of the RPG of naphtha cracking is also different than the one generated from NGL feeds. For instance, the RPG from naphtha cracking is easier to separate from water in the settler and contains components that provides higher solvency relative to fouling precursors such as dienes. Due to their solubility in water, components such as benzene will be present in the water exiting the settler but can easily be removed in the Process Water Stripper. Fouling problems in naphtha crackers are displaced to the Primary Fractionator and dilution steam exchangers but, unlike the gas cracker, the fouling manifests itself on the quench oil side of the dilution steam exchangers. Those problems have been discussed in numerous papers and are beyond the scope of this paper.

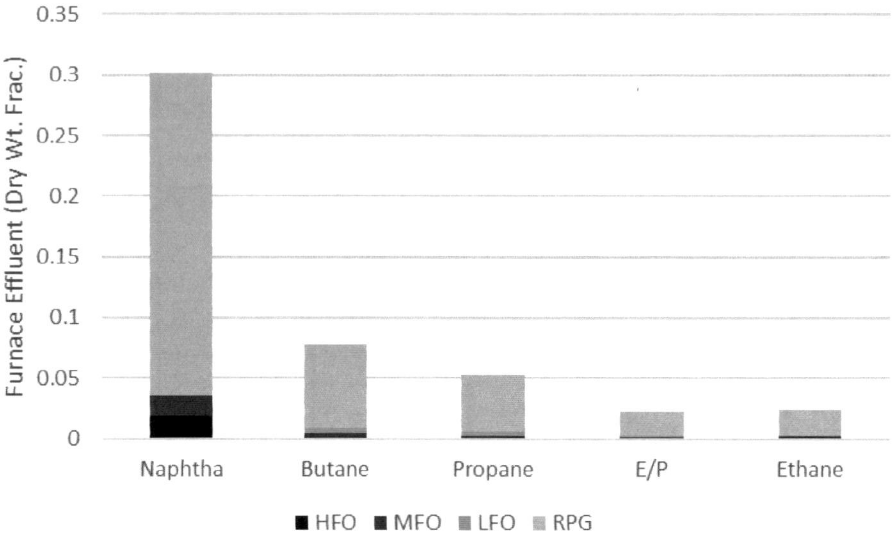

Figure 3: Furnace Effluent Gasoline & Fuel Oil Content vs Feedstocks

Butane cracking is known to generate RPG components with specific gravities close to water, making it hard to separate from water in the settler. However, a mix of butane and naphtha feed may minimize the problem. Providing an RPG stream to the settler that could increase the density differential between water and hydrocarbon could also be beneficial.

Ethane cracking generates significantly less RPG and fuel oil than naphtha, propane and butane. The RPG characteristics are also quite different then naphtha crackers as it has much less solubility for dienes. Dienes are therefore likely to contaminate the water and may cause fouling issues in the Process Water Stripper and dilution steam system. The low RPG/ water ratio in the settler can also make separation difficult.

Ethylene plants originally designed for naphtha cracking which expanded their ethane cracking capabilities while maintaining their flexibility to naphtha would likely experience difficulties due to the low fuel oil and RPG make. Numerous papers have discussed the use of flux oil to alleviate the deficiencies in fuel oil make. Emmert et al. [4] suggest the use of diesel and RPG import to overcome the above deficiencies when cracking ethane, while keeping the integrity of a naphtha cracker design and full feed flexibility.

Debutanizer Bottoms Recycle

Table 1 provides the Debutanizer bottom stream composition obtained by simulation and used as basis for the evaluation for ethane cracking. The composition is compared against analyses of actual ethane cracker Debutanizer bottom streams. The simulation differs from the actual stream analysis but is within a reasonable order of magnitude for evaluating the relative distribution of some fouling precursors throughout the olefins unit. Note that the stream analyses show a significant level of dicyclopentadiene (between 11 and 18 wt. %), which is not included in the list of components used by the simulation. The stream analyses also show about 3% of cyclopentene, which is lumped with pentene in the simulation. The components shown on Table 1 accounts for 94% of the simulated stream and about 87% of the stream analyses. Pseudo components characterized by a boiling range of 160 to 190 °C make up the remaining 6% of the simulated stream. Heavier gasoline components make up the remaining 13% of the stream analyses. Known fouling precursors make up 21% of the simulated stream. The lab analyses suggest a fouling precursors contribution of 23 to 30% when accounting for dicyclopentadiene. The stream has however some level of solvency due to the high benzene content. Toluene provides good solvency properties, especially with regards to dienes. However, it is present but at relatively low levels.

Based on the above discussion, consideration to Debutanizer recycle may sound beneficial relative to:

- Increasing the critical mass of RPG in the settler to enhance separation in gas cracker.
- Providing a wash stream in the quench system to prevent accumulation of tars and heavy oil in gas cracker.

➤ Supplementing RPG deficiencies in naphtha cracker design while processing ethane feed.

Table 1: Debutanizer Bottom Stream Composition

Components	Simulation (Wt. %)	Stream Analysis (Wt. %)
Cyclopentadiene[1]	13.96	10 – 11
Dicyclopentadiene[1]	Not Included	11 – 18
1,3-Pentadiene[1]	1.16	1.0 – 3.0
Isoprene[1]	0.07	3.0
Pentene	10.3	4.5
Benzene	53.03	43.4 – 43.8
Hexene	2.0	
Toluene	6.05	2.5 -2.7
Heptene	0.70	
Styrene[1]	5.77	0.34 – 0.44
Xylene	0.32	
Octene	0.03	0.02 – 0.03
Indene[1]	0.35	0.1 – 0.32

Note 1: Known fouling precursors

Critical Mass of RPG/ Liquid – Liquid separation (Settler)

Disregarding the potential for emulsions; effective water/ hydrocarbon separation in the settler is driven by a few fundamental principles; immiscibility, density differential, velocity of the continuous phase and residence time.

Literature [5] suggest that for good settling, Reynolds number based on the velocity of the continuous phase (water) shall be less then 5,000. Stokes law, which predicts the rise or fall of droplets, is normally used for sizing the settler. The size and geometry of the settler is to allow liquid droplets to either reach the surface when lighter or reach the bottom when heavier then water before exiting the vessel. As shown below, Stokes law relates the droplets vertical velocity as a function of density difference, droplet diameter and viscosity.

$$u = g_c\,(\rho_2 - \rho_1)\; d^2/18\,\mu$$

Assuming no change in liquid droplet sizes, recycling debutanizer bottoms would improve separation in the settler only if it could increase the density differential and/ or reduce the viscosity.

Simulation of an ethane cracker predicts a hydrocarbon/ water ratio of 0.0002 in the stream feeding the settler from the bottom of the Quench Water Tower. The predicted hydrocarbon phase contains about 38% of heavy gasoline and 62% heavy oil and tar. The hydrocarbon/ water ratio drops to 0.00008 for gasoline and 0.0001 for heavy oil when segregating the two expected phases. Densities of the heavy gasoline and heavy oil phase are estimated at 57.7 and 70.6 lb/ft^3 respectively. Water density at the settler temperature is 60.4 lb/ft^3. The low hydrocarbon ratio would suggest separation difficulties if components contained in each hydrocarbon phases have some degree of solubility in water. Given the limitations of simulation in addressing solubility, the heavier characteristics of the components involved ($C9^+$) and their representation by pseudo-components; total immiscibility with water was initially assumed. As such, the heavy oil was assumed to settle to the bottom and the gasoline was assumed to float over the water layer. Unlike the heavy oil layer for which a decent density differential exists with water; separation of the gasoline layer may be more difficult due to the density being relatively close to water. A Debutanizer bottom recycle stream to the settler was configured in the simulation, with recycling rates varying between 5 and 35% of the Debutanizer bottom rate without recycle. As shown on Figure 4, recycling Debutanizer bottoms to the settler proved effective in increasing the density differential of gasoline with water. However, negligible gains are observed when recycling more than 10% of the original Debutanizer bottoms flowrate. The gasoline/ water ratio increases linearly as it is driven by material balance. Using a 10% recycle rate increases the gasoline/ water ratio by a factor of 3.4. Although difficult to assess with a simulation, recycling 10% of the Debutanizer bottom is believed to be beneficial in terms of reducing the viscosity of the gasoline. Hence gasoline recycle contributes favorably to enhancing the settling performance by increasing the rising velocity of gasoline droplets, as suggested by Stokes law.

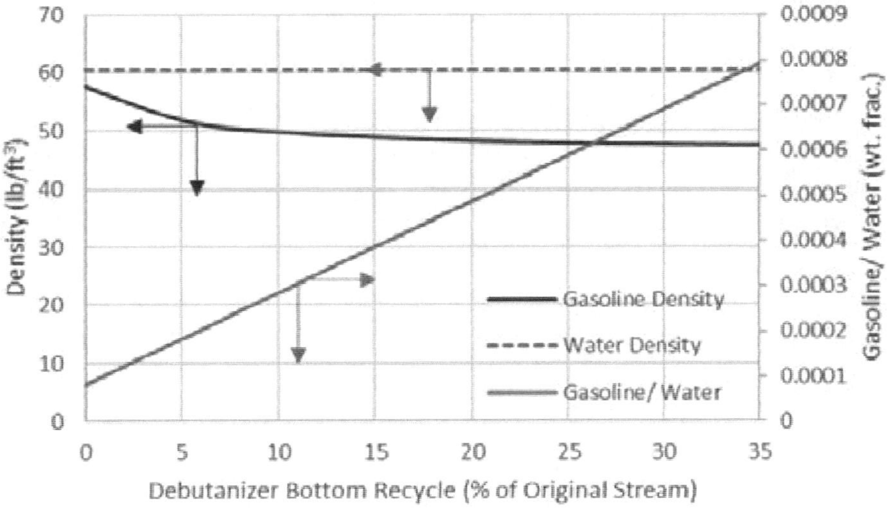

Figure 4: Impact of Debutanizer Recycle on the Settler

As indicated in the literature, recycling Debutanizer bottom is expected to increase the composition of fouling precursors. Table 2 compares the content of the fouling pre-cursors listed on Table 1 in the gasoline stream exiting the settler relative to 10 and 35% Debutanizer bottom recycle rates. Also included in Table 2 is the change in composition of benzene and toluene, which provides some solvency.

Table 2: Fouling Precursors Composition in the Quench Water Settler vs Debutanizer Recycle Rates

Components	No Recycle (wt. ppm)	10% Recycle (wt. ppm)	35% Recycle (wt. ppm)
Cyclopentadiene	0.106	28.4	99.3
1,3-Pentadiene	0.007	2.36	8.26
Isoprene	0.0004	0.14	0.49
Styrene	1.33	12.9	41.9
Benzene	1.2	108.7	378
Toluene	0.42	12.7	43

As mentioned earlier, the fouling pre-cursors listed in Table 2 have been reported soluble in water. Due to simulation limitations, solubility data in water for those individual components was gathered from literature and use as reference. Table 3 provides the solubility data of those components along with benzene and toluene in water at 25 °C. It is intuitively believed that the solubility is higher at the warmer operating temperature of the settler. However, comparing the data from Table 2 and Table 3, suggest that the entire mass of components recycled from the Debutanizer would dissolve in the water phase. To understand the migration of those fouling precursors a simulation of a Process Water Stripper was done. The feed to the Process Water Stripper assumed a hydrocarbon content in alignment with the solubility data of all the components shown on Table 3. Live low-pressure steam was assumed as the stripping medium. The simulation results suggest that 100% of the hydrocarbon content is stripped out of the tower and recycled to the Quench Water Tower. The simulation also suggests that the total mass of those components is stripped by cracked gas in the Quench Water Tower and is forwarded to the Cracked Gas Compressor.

Table 3: Specific Gravity & Solubility of Hydrocarbons in Water (at 25 °C) [6]

Component	Solubility (wt. ppm)	Specific Gravity	Boiling Point (°C)
Cyclopentadiene	660	0.805	41.65
1,3-Pentadiene	381	0.683	42.05
Isoprene	642	0.681	34.22
Styrene	312	0.903	145.31
Benzene	1,800	0.876	80.24
Toluene	520	0.865	110.73

The profile of fouling pre-cursors across the Cracked Gas Compressor when assuming 100% solubility of the components shown in Table 2 and 100% recovery from the Process Water Stripper overhead stream, is provided on Table 4 for the no-recycle and 35% Debutanizer bottom recycle cases. The simulation assumed a five-stage compressor with inter-coolers and knock out drums. The composition provided are gathered from the discharge of each stages.

Except for styrene, the profile across the compressor train is relatively flat for both cases. Styrene shows a small decrease on stage 4 and 5, which is more pronounced on the 35% recycle case. The decrease is attributed to composition reaching saturation, causing some of the styrene to condense in the inter-stage knock out drums. The higher composition in the 35% recycle case enhances condensation and this explains a relatively high drop in composition compared to the no recycle case. Comparison of both cases shows a consistent increase of 35% for cyclopentadiene, 1,3-pentadiene and isoprene. This suggests that if soluble in water, as per the above discussion, those components will form a closed-loop between the Debutanizer bottom and the settler. Styrene exhibits almost the same behavior but a fraction of the styrene, which condenses in the first three stages of compression is recycled to the Quench Water Tower settler and exits the process with the gasoline at that location.

Table 4: Fouling Pre-Cursors Composition Profile Across the Cracked Gas Compressor

No Debutanizer Bottom Recycle				
Compressor Stage	Cyclopentadiene (wt. ppm)	1,3-Pentadiene (wt. ppm)	Isoprene (wt. ppm)	Styrene (wt. ppm)
1	3,052	254	15.2	1,261
2	3,052	254	15.2	1,260
3	3,066	255	15.3	1,266
4	3,076	256	15.3	1,258
5	3,076	256	15.3	1,202
35% Debutanizer Bottom Recycle				
Compressor Stage	Cyclopentadiene (wt. ppm)	1,3-Pentadiene (wt. ppm)	Isoprene (wt. ppm)	Styrene (wt. ppm)
1	4,127	343	20.6	1,654
2	4,127	343	20.6	1,653
3	4,127	343	20.6	1,653
4	4,153	346	20.7	1,511
5	4,148	346	20.7	1,368

The results from the above evaluation must be interpreted with caution due to limitation of the simulation model used. Although potentially inaccurate on quantitative absolute values, the relative differences between the cases illustrate the sensitivity of the changes evaluated and allow the following conclusions.

- When disregarding solubility of the components; recycling Debutanizer bottoms to the settler suggests an increase in density differential with the water phase, which should improve separation. It also suggests, that beyond a certain recycle rate (10% per above simulation results) minimum gains are expected.

- The impact of solubility in water of many of the components present in the Debutanizer recycle stream is unclear relative to the ability of the stream to increase the density differential with the water phase, when supplied at rates below the solubility level. It is suspected that no change in density differential will be achieved if most of the stream dissolves in the water phase.

- Debutanizer components soluble in water are expected to be totally removed from the water stream in the Process Water Stripper and recycled back to the Quench Water Tower. An increase in fouling precursors would then be expected through the Process Water Stripper, Quench Water Tower and Cracked Gas Compressor train.

- Debutanizer bottom recycle rates beyond the solubility level of its components would be required for increasing the density differential when assuming solubility in water. The contribution of fouling precursors to the Cracked Gas Compressor would then be limited to the solubility of those components. Recycle material present above the solubility level would be purged out with the gasoline stream.

Wash Stream in the Quench System of Gas Cracker

As mentioned earlier, gas cracker designs do not typically include a Primary Fractionator. As such, heavy oil and tars are fed directly to the Quench Water Tower. Although expected to condense in the lower section of the tower, they are known to foul tower internals. If carried over with the water exiting the settler, they will flow with the quench water cooling loop and flow downward throughout the entire tower. They will also be a source of fouling for the Process Water Stripper and the downstream water treatment equipment. Cockshutt et al. [1] reported the use of a $C5^+$ wash stream injected in the top reflux and the bottom of the Quench Water Tower. The bottom injection point located below the cracked gas feed. The $C5^+$ stream was originally a mixture of Debutanizer bottoms and slop oil from an oily water system. Elimination of the slop oil from the recycle stream was reported to provide significant benefits relative to emulsion formation in the Quench Water Tower and significant reduction in oil and grease level in the circulating quench water. However, there are no data presented that allows a comparison between with and without Debutanizer bottom recycle. As stated by Cockshutt, the slop oil stream, which contained sulfur and solids was a contributor in stabilizing an emulsion.

To be effective, a wash stream would have to flow across equipment internals in the liquid phase and provide solvency to prevent fouling precursors from growing on solid surfaces. The Debutanizer bottom stream is relatively rich in benzene, which has good solvency. It also contains toluene, but in relatively low concentration. Simulations were conducted with recycle rates varying from 0 to 35% of the Debutanizer bottom rates. Injection in the top and mid-reflux section of the Quench Water Tower were evaluated. As shown on Figure 5a, recycling Debutanizer bottom to the top reflux location has relatively no impact on the benzene composition profile. Feeding in the mid reflux location provide more capability in increasing the profile as illustrated on Figure 5b, where the change in benzene composition varies between 0 in the very bottom of the tower and increase by about 35% above the mid-reflux location for the 35% recycle case.

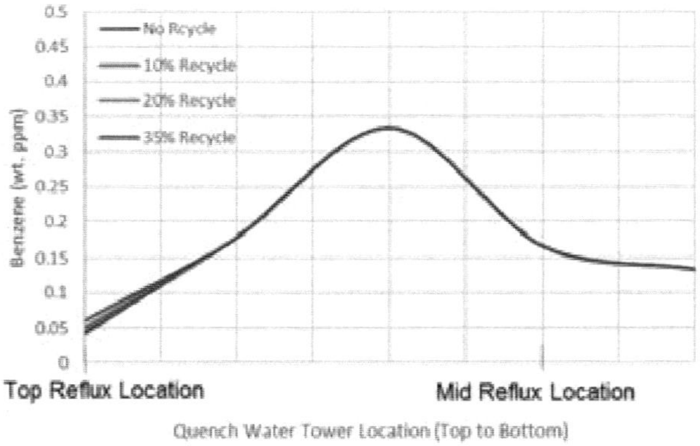

Figure 5a: Recycle to Top Reflux

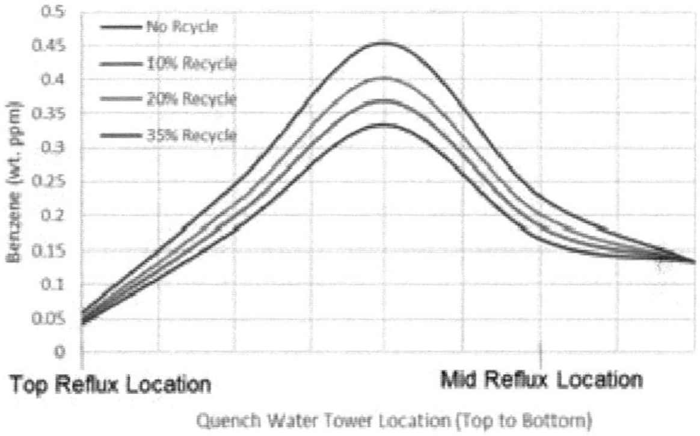

Figure 5b: Recycle to Mid Reflux

Figure 5: Benzene Liquid Composition Profile vs Debutanizer Recycle Rate to the Quench Water Tower

The simulation material balance suggests that 99% of the Debutanizer recycle stream exits the overhead of the Quench Water Tower and moves forward to the Cracked Gas Compressor train. The content of the stream is stripped off the tower by cracked gas. The mid-reflux seems more effective than the top reflux feed location. This behavior is believed to be caused by additional stages of separation above the feed location, which cause an internal circulation that allows benzene to build up. The negligible increase when feeding in the top is explained by the material being stripped right away, exiting the tower and not having the opportunity to build an internal circulation.

The content of fouling pre-cursors such as cyclopentadiene, 1,3-pentadiene, isoprene and styrene increase through the Cracked Gas Compressor train in a similar fashion as when injecting the Debutanizer stream in the settler. The composition profile across the Cracked Gas Compressor would then be consistent with the values reported on Table 4.

The above discussion suggest that the Debutanizer bottoms should be fed at location that minimizes contact with cracked gas. Locations such as the bottom of the Quench Water Tower below the cracked gas feed and the settler would prevent the stripping effect from contact with cracked gas. Feeding in the mid-reflux location, although better than the top reflux would not cover the full tower. However, regardless of feed location, the entire content of the recycle will most likely cycle between the quench system and the Debutanizer. The effectiveness of the wash is also questionable.

Supplementing RPG deficiencies in naphtha cracker design while processing ethane feed.

Plants originally design for naphtha cracking, which expanded their capability to crack ethane while maintaining their original configuration are different then gas crackers. The presence of a Primary Fractionator eliminates the problems associated with tars and heavy oil in the Quench Water Tower and downstream equipment. Fouling problems are contained within the Primary Fractionator and quench oil system. Also, as shown on Figure 1, the RPG reflux loop between the Quench Water Tower settler and the Primary Fractionator provides a much higher RPG/ water ratio in the settler. However, processing large volumes of ethane feed causes deficiency in fuel oil and gasoline make from the furnace. This affects the efficiency and reliability of the quench oil circuit and causes challenges in containing RPG within the Primary Fractionator reflux loop. Meeting the target RPG end point in the bottom of the Quench Water Tower is also a challenge. Those deficiencies can be addressed by adding external fuel oil and gasoline to the system when processing ethane. While the Debutanizer bottoms characteristics do not match the fuel oil deficiencies, it could be in alignment with the shortage of gasoline in the Quench Water Tower bottoms. As such, simulations were configured to evaluate the potential of recycling Debutanizer bottoms for improving the RPG maximum reflux rate to the Primary Fractionator and the resulting RPG 95% cut point.

The simulation is based on a naphtha cracker design consistent with the description on Figure 1. It assumes the use of flux oil (external light cycle oil) for alleviating the fuel oil deficiency but no external gasoline import. The unit has been simulated for 100% naphtha and ethane feeds to cover the book ends of the feed slate. The first difference observed is around the ability to contain the RPG within the Primary Fractionator reflux loop. For naphtha, the RPG reflux to dry feed ratio on the Primary Fractionator is about 0.7 with excess RPG withdrawn from the Quench Water Tower bottom. On ethane feed, this ratio drops to 0.6 without any excess material. In other words, 100% of the RPG recovered in the bottom of the Quench Water Tower is refluxed to the Primary Fractionator. The RPG 95% cut point (ASTM D2887) is estimated at 203 ^{o}C on naphtha vs 266 ^{o}C on ethane feed. Figure 6 illustrates the distribution of RPG across the unit as a weight percentage recovery from the feed. As shown, for a naphtha feed, 19% of the RPG is recovered in the bottom of the Quench Water Tower, 56% from the Gasoline Stripper and 25% from the Debutanizer bottom. On ethane feed, the simulation predicts 99.5 % recovery from the Debutanizer bottom and about 0.5% from the Gasoline Stripper.

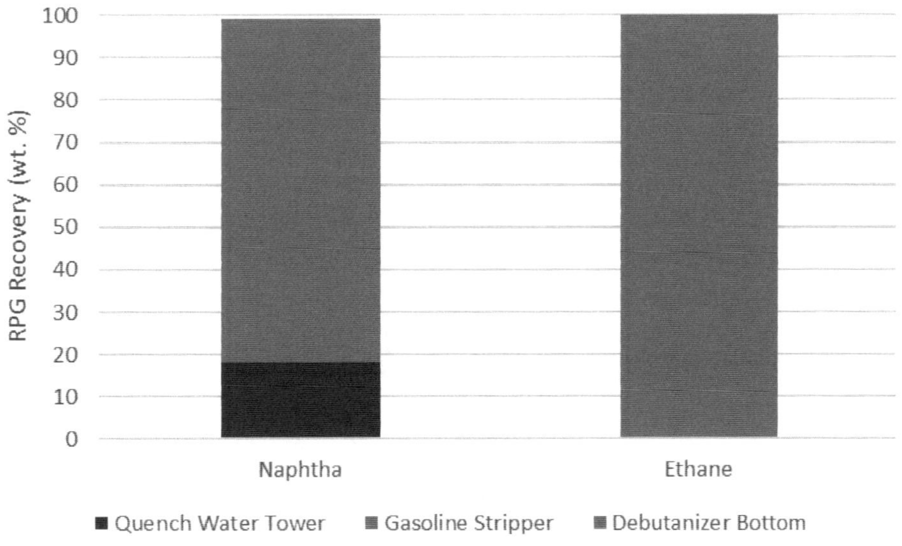

Figure 6: Raw Pyrolysis Recovery Distribution

A previous evaluation [7] on the impact of processing light feedstocks in a naphtha cracker suggested a significant increase in composition profile and residence time of indene in the Primary Fractionator. This increase is a result of gasoline deficiency in the feed, which requires concentrating the RPG in the reflux loop. The increase in RPG reflux rate eventually reaches an equilibrium where the gasoline components are pushed forward in the process. That equilibrium defines the maximum achievable RPG reflux rate to the Primary Fractionator.

Debutanizer bottom recycle rates between 10 and 80% of the rate without recycle were evaluated with respect to increasing the RPG reflux to the Primary Fractionator and impact on the RPG cut point. The distribution of fouling pre-cursors was also evaluated. The Debutanizer bottom recycle stream was fed to the Quench Water Tower settler in the simulation model. As shown on Figure 7, recycling Debutanizer bottoms provides no capability for allowing higher RPG reflux to the Primary Fractionator and does not affect the gasoline 95% cut point in the bottom of the Quench Water Tower.

The density differential between the hydrocarbon and water phase in the Quench Water Tower settler is not affected by the Debutanizer bottom recycle rate. The simulation predicts a hydrocarbon phase density of 50.5 lb/ft^3 on naphtha feed and 54.3 lb/ft^3 on ethane. The reduction in density differential may affect the performance of the settler. Since the settler of a naphtha cracker design is based on a larger density differential, the geometry of the vessel may not provide enough residence time to allow effective separation when processing ethane feed.

The Debutanizer recycle rate has little impact if any on the liquid composition profile in the Quench Water Tower. However, except for indene, the simulation predicts higher concentrations in the top of the Primary Fractionator, the relative difference falling close to the Debutanizer bottom recycle rate. Indene shows a reduction in concentration in the top section of the Primary Fractionator. This difference is explained by the distribution of indene in the process. 11% of the indene is extracted from the Gasoline Stripper and 84% from the Debutanizer bottom. The other fouling precursors tracked by the simulation are entirely recovered in the Debutanizer. As such, the reduction in indene content in the Debutanizer bottom stream provides some dilution as it mixes with gasoline in the settler prior to reflux the Primary Fractionator.

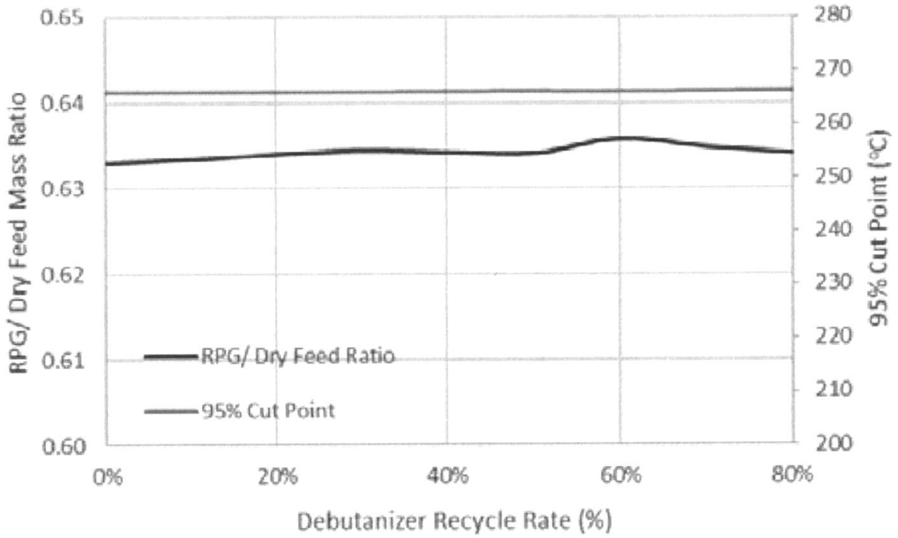

Figure 7: Impact of Debutanizer Bottom Recycle Rate on RPG Reflux Rate & 95% Cut Point

The profile of cyclopentadiene, 1,3-pentadiene, isoprene and styrene exhibit a relatively flat profile across the Cracked Gas Compressor train with relative composition increase in alignment with the Debutanizer bottom recycle rate.

The above behavior suggests that the Debutanizer bottom recycle stream enters the settler and mixes with the gasoline from the bottom of the Quench Water tower. As it refluxes the Primary Fractionator, the entire content of the recycle stream is stripped by cracked gas and remains in the vapor phase as it travels across the Quench Water Tower and through the downstream units until it is recovered in the Debutanizer. The recycle therefore creates a closed loop between the quench system and the Debutanizer. The stream provides no benefits relative to the RPG reflux loop, the density differential in the settler and the RPG cut point in the bottom of the Quench Water Tower. It does however increase the composition of fouling precursors in the top of the Primary Fractionator and across the Cracked Gas Compressor train.

Debutanizer Bottom Recycle Impacts on Fouling

Cracked Gas Compressor

The above cases evaluated have one point in common, they all increase the content of fouling pre-cursors across the Cracked Gas Compressor. Except for styrene, the relative increase is about the same as the Debutanizer bottom recycle fraction. Styrene also increases, but to a lesser magnitude, especially in the 4th and 5th stages.

Ethane cracking is known to cause more fouling problems in the Cracked Gas Compressor than naphtha feed. The difference is mainly explained by insufficient aromatic content of light gasoline flowing across the compressor train. The fouling mechanism involves components such as styrene and butadiene through free radical polymerization. Cyclopentadiene derivatives are reported to get involved through Diels Alder condensation. The fouling reactions are sensitive to the composition of fouling precursors, temperature, pressure and presence of oxygen impurities such as peroxides and iron oxide which act as initiators in free radical polymerization. Temperature and fouling precursors composition are key variables in the fouling reaction. The reaction rate constant typically follows the Arrhenius equation and increases exponentially with temperature. The rate of reaction is then directly a function of the product between the fouling precursors composition and the rate constant when assuming a 1st order reaction. Pressure is also believed to influence the rate constant as higher reaction rates are observed at higher pressure. However, the impact of pressure is not as significant as temperature.

Fouling in the Cracked Gas Compressor affects the compressor itself and the inter-stage coolers. The fouling rate of this system often dictates plant turnaround frequencies. The following fouling mitigation are often used on their own or in combination:

- Wash water injection in the compressor to reduce temperature.
- Wash oil injection in the compressor to prevent adherence of polymers and dissolve them.
- Coating on the compressor mechanical components in contact with the process to prevent adherence of polymers.
- Use of chemical inhibitors. Note that chemical inhibitors have been reported as ineffective relative to Diels Alder reactions [8].

Debutanizer

Increase of fouling precursor flowrate to the Debutanizer is another common feature of all the cases evaluated. The Debutanizer bottoms recycle rate affects the tower temperature and composition profile. Figure 8a illustrates the temperature and butadiene composition profile for a 0 and 35% Debutanizer bottom recycle rate. As shown, the temperature profile is slightly higher when recycling Debutanizer bottoms by a few degrees in the bottom part of the tower, while it remains unchanged in the very bottom and reboiler. As illustrated the recycle stream provides some dilution effects and cause a lower butadiene composition profile when compared to no recycle. Figure 8b shows the cyclopentadiene profile where a slight increase in composition is observed in the lower section of the tower. The composition becomes however equivalent to the no recycle case in the very bottom. The increase in cyclopentadiene composition occurs in an area of the tower where temperatures are relatively low relative to fouling reaction.

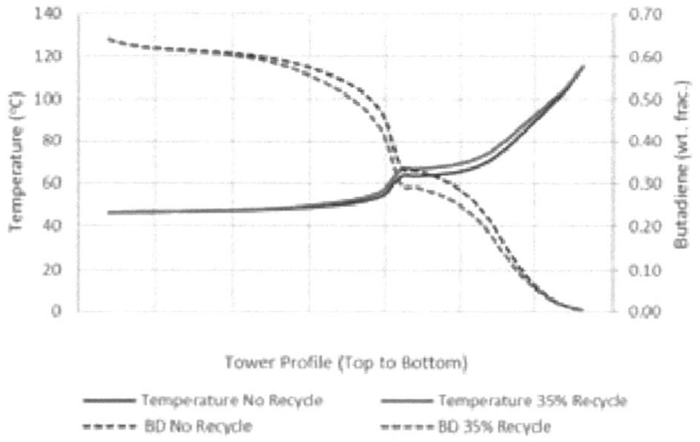

Figure 8a: Debutanizer Temperature and Butadiene Profile

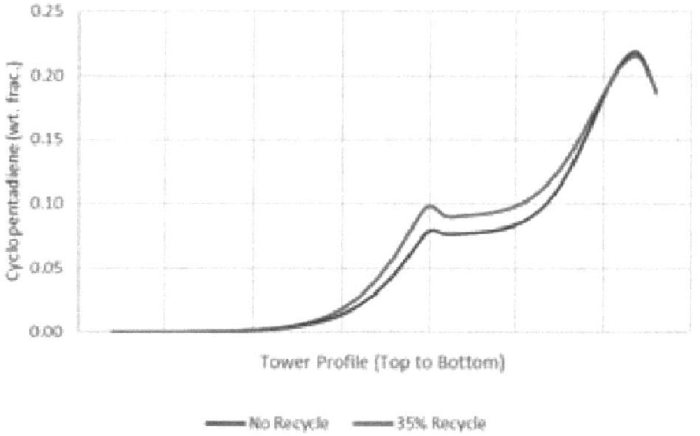

Figure 8b: Debutanizer Cyclopentadiene Profile

Quench System

The use of Debutanizer bottom recycle was found to have minor impacts in shifting the composition profile in the Quench Water Tower of an ethane cracker. The increase of fouling precursors to the Process Water Stripper is believed to be limited to the solubility of the components and would increase only if the water is not already saturated on a no recycle case. It will not improve the fouling propensity of this system. Although inconclusive from the simulation work, it could potentially increase fouling.

On a naphtha cracker configuration with ethane feed, increase of some of the fouling precursors would be expected in the top of the Primary Fractionator. However, without external addition of gasoline, this system would have a high risk of fouling due to high residence time of components such as indene and styrene. Recycling Debutanizer bottoms would not reduce the impact on the fouling propensity.

Dilution Steam System

Simulation of the Process Water Stripper suggests 100% removal of the assumed diene content in the water stream. The diene content is limited to components recycled from the Debutanizer bottom in this evaluation. However, literature [3,9] reports fouling due to dienes in the dilution steam system. This would suggest presence of dienes of higher boiling point and soluble in water in the heavy gasoline content found in the settler of ethane crackers. The simulation does not provide sufficient resolution for assessing the above due to the characterization of heavier gasoline components through pseudo-components. It is however suspected that the Debutanizer bottom recycle has a minor, if any, contribution to this.

Fouling – General Comments

The consequences of increasing the fouling propensity due to a recycle stream such as the Debutanizer bottoms would manifest itself over an extended period of time. While composition, temperature and pressure drive reaction rates, residence time will dictate how much polymer is generated. As an example, the fouling rate of a Debutanizer bottom stream has been evaluated in a batch reactor [10] and correlated to the formation of gum. Figure 9 shows the rate of gum formation as a function of time and temperature. As illustrated, the batch reaction occurs over hours with completion requiring about 200 hours. Cyclopentadiene was identified as key contributor to the fouling reaction in this experiment. The residence time of cyclopentadiene in that Debutanizer was estimated between 40 and 50 minutes depending on location in the tower. The rate of formation of the polymer would then be limited by the reaction rate and residence time. As such it may take a long time before the accumulation of polymer in the tower shows any symptoms of fouling. The residence time in the tower is believed higher than in flow-through type equipment such as compressors and exchangers due to internal re-circulation. The residence time of fouling pre-cursors in compressors and exchangers is believed to fall into an order of magnitude of seconds. Consideration for the use of external or recycle gasoline streams, which are known to contain contaminants must be done with care as the expected benefits may manifest themselves quickly while the consequences of fouling may not be detected until some months later. Fouling of systems such as the Cracked Gas Compressors, towers and/ or key exchangers will often result in pre-mature shut down of the unit.

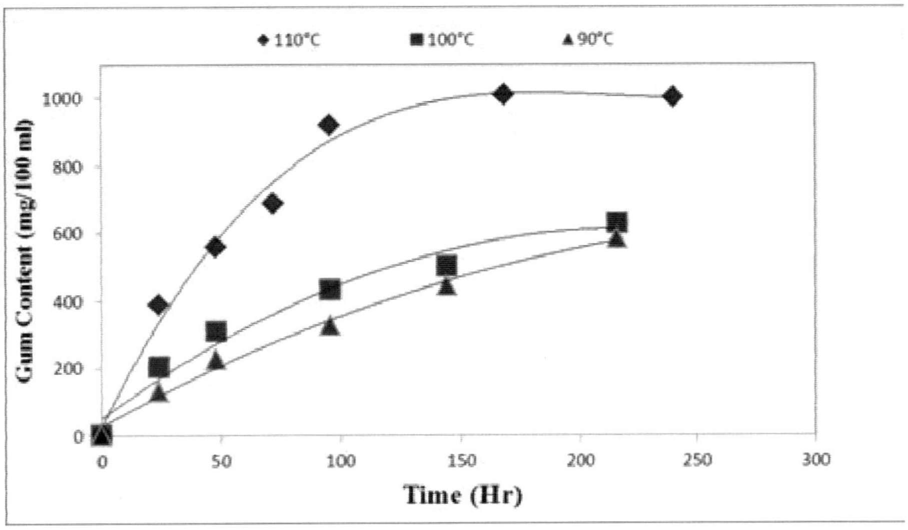

Figure 9: Debutanizer Bottom – Gum Formation in Batch Reactor [10]

Discussion on the Evaluation Methodology

The evaluation presented above relies on several assumptions and the use of process simulation. It would be difficult to perform such an evaluation on a live plant. Full control of process variables, response time of process moves and potential upsets resulting from such experiment makes it impractical. A simulation environment removes a lot of those barriers without causing any risk to the process. However, it also has some limitations, particularly in an area of the plant that involves a lot of empiricism. The heavy gasoline and fuel oil cuts are characterized by pseudo-components which may not provide a good fit with actual data. In practice, the distribution of those components would be calibrated with distillation curves and analyses of the plant streams. This analysis is at a high level and is intended to evaluate typical plant configurations for which no such data was available. This explains some numbers that might be questionable. For instance the RPG/ Water ratio of 0.0002 stated on ethane crackers is believed under estimated by a factor of 10 when comparing to published data [3]. As also discussed earlier, solubility and interactions between several components cannot be fully evaluated with the data available. However, it does provide an idea of the relative changes of a number of key variables.

Introducing external or recycle streams in an ethylene process can be risky and requires an approach that minimizes risk to the reliability of plant operation. The method used in this evaluation is believed to be appropriate, but a real-world evaluation should be complemented by several analyses and experiments at lab scale. Iteration between simulation and lab experiments would likely provide a more accurate result.

Pros and Cons of Recycling Debutanizer Bottoms

The following summarizes the pros and cons of recycling Debutanizer bottoms relative to the cases evaluated.

Pro

- The stream is available and can be easily recycled to the front end of the plant.
- The evaluation suggests that it could potentially increase the density differential in the settler of an ethane cracker.

Con

- Increase risk of fouling, specifically in the Cracked Gas Compressor train.
- Could potentially reduce the plant run length due to premature fouling in the Cracked Gas Compressor train (compressor and inter-coolers).
- This evaluation suggests that recycling Debutanizer bottom as wash to the Quench Water Tower of an ethane cracker is not effective.

- The Debutanizer bottoms recycle stream has no capabilities in supplementing RPG deficiency in a naphtha plant design when processing ethane feed.

Conclusion

This evaluation, although limited in terms of plant data, suggests few benefits to recycling Debutanizer bottoms to the quench system. Increasing the density differential in the settler of a gas cracker was the only potential benefit identified. However, the risk of fouling on key equipment is believed to outweigh the gain. Offsetting the increase of fouling propensity through the use of chemicals and higher wash oil rates to the Cracked Gas Compressor would increase operating cost and defeats the purpose of using a stream considered free.

Using Debutanizer bottoms as a wash stream in the Quench Water Tower of gas cracker is not effective. The content of the recycle stream gets stripped by cracked gas in the Quench Water Tower. Feeding the recycle in the mid-reflux location of the Quench Water Tower does provide some sensitivity for increasing the benzene composition profile. However, the increase is localized around the mid-reflux point and does not cover the entire tower. Increased concentration of fouling precursors throughout the Cracked Gas Compressor elevate the risk of fouling and could impact the plant turnaround frequencies and/ or cost of wash oil.

Alleviating gasoline deficiencies with Debutanizer bottoms recycle in a naphtha plant that processes ethane feed was not found to improve the RPG reflux to the Primary Fractionator and/ or improve the RPG cut-point in the bottom of the Quench Water Tower. The entire content of the recycle stream is stripped by cracked gas and becomes contained in a closed recycle loop between the Debutanizer and the quench system. As per other cases evaluated, it increases the fouling precursors composition profile across the Cracked Gas Compressor train. Gasoline deficiency in such a system should be addressed through addition of external gasoline streams with high aromatics content. Plants that retain naphtha cracking operation should consider storing RPG during naphtha operating campaigns for future use in supplementing the gasoline deficiency that occurs when processing large volumes of ethane.

Nomenclature

RPG: Raw Pyrolysis Gasoline
NGL: Natural Gas Liquid

References

1. T. Cockshutt, J. Dennehy, Successful Process Selection for Quench Water Cleanup, Ethylene Producer Conference, 1993 AIChE Spring Meeting, Houston, TX
2. C. E. Hammond, E. Walker Ethylene Quench System Fouling, Ethylene Producer Conference, 2006 AIChE Spring Meeting, Orlando, FL
3. S. Kurukchi, J. Gondolfe, Innovative Quench Water Pretreat Technology For Ethylene Plants, Ethylene Producer Conference, 2002 AIChE Spring Meeting, New Orleans, LA
4. T. Emmert, S. Antonas, G. Kracker, Cracker Feedstock Flexibility from Naphtha to Ethane: Design Criteria and Concepts for Quench Water and Quench Oil System, Ethylene Producer Conference, 2016 AIChE Spring Meeting, Houston, TX
5. J. R. Couper, "Chemical Process Equipment – Selection", 3rd Edition, Elsevier 2012.
6. I.M. Smallwood, "Handbook of Organic Solvent Properties", Elsevier 1996.
7. A. Bernard, F. Fahiminia, G. Kim, Challenge of Primary Fractionators on NGL Feed, Ethylene Producer Conference, 2013 AIChE Spring Meeting, San Antonio, TX
8. R. Metzler, H. Mo, Compressor Fouling Monitoring Methods Identify Changes in Fouling Mechanism to Direct Mitigation Performance Improvement, Ethylene Producer Conference, 2017 AIChE Spring Meeting, San Antonio, TX
9. S. De Haan, B. K. Sullivan, Overview of Cracked Gas Compressor Fouling: Theories and Practices, Ethylene Producer Conference, 2001 AIChE Spring Meeting, Houston, TX
10. F. Fahiminia, A. Bernard, P. Herrera, M. Stante, "Root Cause of Corunna Debutanizer Tower Operational Upsets, Ethylene Producer Conference, 2012 AIChE Spring Meeting, Houston, TX
11. J. L. Webb, Design Fundamentals Gas Cracking Quench Systems, Ethylene Producer Conference, 2002 AIChE Spring Meeting, New Orleans, LA

AIChE PAPER NUMBER: 35b

VISUAL CLUES INTO ETHYLENE SERVICE FOULING

Debby M. Rossana Ph.D.
Industry Technical Consultant
Nalco Champion

Jessica Burgin
Account Manager
Nalco Champion

Anahita Khanlari Ph.D.
Lead Chemist
Nalco Champion

Prepared for Presentation at the 2019

Spring National Meeting, New Orleans, LA (March 31 - April 4, 2019)

AIChE PAPER NUMBER: 35b

VISUAL CLUES INTO ETHYLENE SERVICE FOULING

Debby M. Rossana Ph.D.
Industry Technical Consultant
Nalco Champion

Jessica Burgin
Account Manager
Nalco Champion

Anahita Khanlari Ph.D.
Lead Chemist
Nalco Champion

Abstract

Fouling is an on-going issue in an ethylene plant. During an outage, dirty equipment may be taken out of service for cleaning as needed and deposit samples taken for chemical identification. Often, the chemical identity is already suspected; poly butadiene in the top of the debutanizer or polystyrene in the primary fractionator. To many plant operators, all foulants are similar and little consideration is given to color, texture, and location of the foulant. Questions such as: why is the foulant found in one location and not another? why is the deposit sticky or rubbery or crystalline? are not asked. One should consider that the foulant is located where it is and looks as it does because temperature, monomer concentration, oxygen, rust, feed rates, feed stocks, feed impurities, and design all impact the why, where, and how of foulant formation. Answering the who, what, where questions provide clues that help to better interpret what is happening in the process and how to tackle the issue. This paper will focus on using these questions to help elucidate foulant formation.

1- INTRODUCTION

Fouling and unwanted polymer formation is an on-going issue in an ethylene, butadiene or reactive monomer processing plant. Based on a study by HSB Soloman Associates in 2015 [ref 1], fouling generally represents one of the top 2 or 3 reasons why an ethylene plant may experience an unplanned shutdown. During an outage, dirty equipment may be taken out of service for cleaning as needed, inspected [ref 2, 3], and deposit samples taken for chemical identification. Often, the chemical identity is already suspected; poly butadiene in the top of the debutanizer or polystyrene in the primary fractionator. The chemical analysis of the deposit is often done to identify something unusual; for example, excess iron in the deposit.

To many plant operators, all foulants are similar and little consideration is given to color, texture, and location of the foulant. Questions such as: why is the foulant found in one location and not another? why is the deposit sticky or rubbery or crystalline? are not asked. One should consider that the foulant is located where it is and looks as it does because temperature, monomer concentration, oxygen, rust, feed rates, feed stocks, feed impurities, and design all impact the why, where, and how of foulant formation. Answering the who, what, where questions provide clues that help to better interpret what is happening in the process and how to tackle the issue.

The visual clues are particularly valuable in distillation towers, reboilers and baskets/strainers. This paper will focus primarily on these pieces of equipment but will touch on the other areas of the ethylene plant such as compressors and quench and dilution steam systems. Combining learnings from visual observations with analytical techniques will give the readers a comprehensive view of the system operation and ways to improve production and maintenance of the system.

2- Where Did the Foulant Come From?

2.1 Towers

In a distillation tower, fouling forms most often on the top surface of the trays at the lower section of the tower (**Figure 1**) [ref 4]. However, there are exceptions to this common observation (**Figure 2**). In these cases, the foulant formed in the vapor spaces such as undersides of trays and tower walls. In **Figure 2**, the foulant has peeled off the walls and tray undersides and has deposited on the tray below. When vapor space fouling is seen in a distillation column, a change to the inhibitor program might be necessary to address this occurrence.

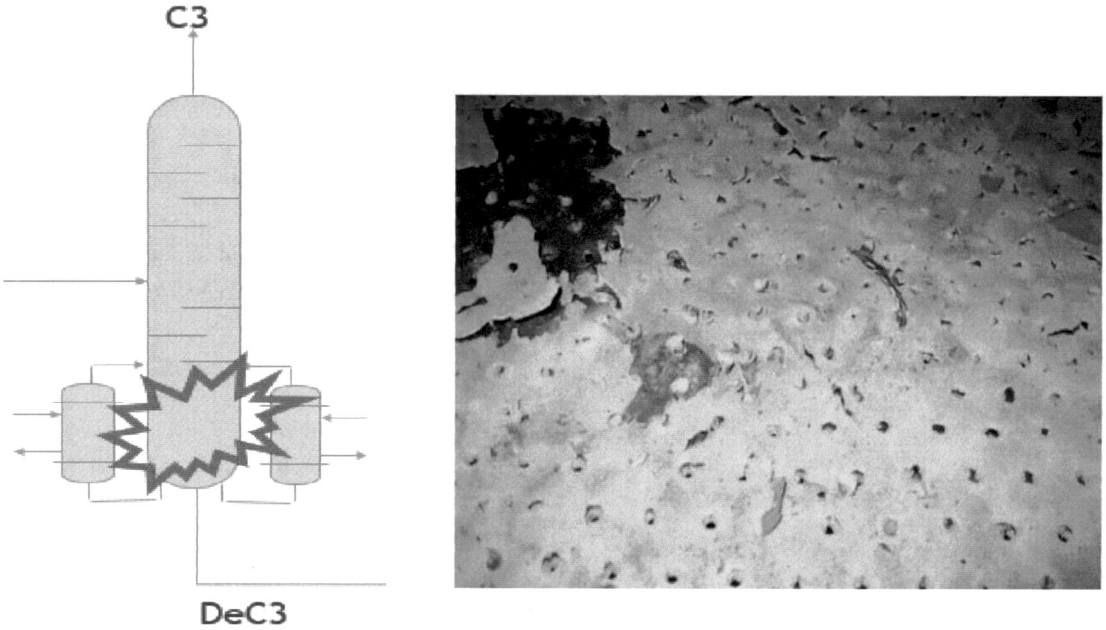

Figure 1: Typical location for fouling in a Depropanizer

Figure 2: Foulant on walls/under trays

Another deviation from normal fouling in a distillation tower is deposition on a tray at a feed point (**Figure 3**) or reflux return (**Figure 4a and b**). When fouling is at the feed point, it suggests the feed stream contains soluble polymers or gums that are losing solubility due to incompatibility of the feed stream with the part of the tower where it enters. This could be due to temperature or stream composition differences.

Figure 3: Fouling at feed tray

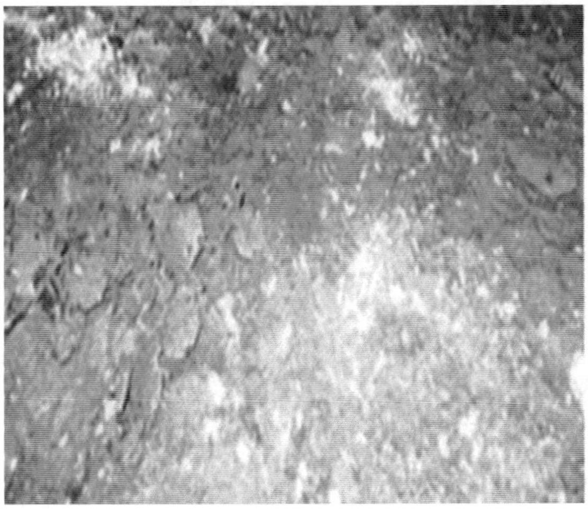

Figure 4a: Fouling at reflux tray

Figure 4b: Reflux tray after treatment

In one case, fouling was noticed 2 trays below the feed point of a light-ends tower. The tower was the plant bottleneck requiring shut-downs every 2.5 to 3 yrs. Addition of a dispersant in the upstream tower kept the gums in solution and fouling at the feed tray was eliminated. This tower now runs turnaround to turnaround (6+ yrs).

Fouling at the reflux tray is most often seen in a Primary Fractionator. The mechanism is similar to fouling at a feed tray but the soluble polymers that come out of solution are often from recycle streams that are sent to the oil/water separator, the oil side of which is reflux to the Primary Fractionator. As plants crack lighter feed-stock and pygas production is reduced, recycle streams make up a larger portion of the organics in the separators. In this case, more than one stream may have to be treated to control the fouling.

2.2 Reboiler Tubes

Sometimes, it is rather simple to pinpoint the location of foulant formation; this is usually the case for reboilers where process is on the tube side. **Figure 5 a and b** show two examples of foulant formed in exchanger tubes with the process on the tube side and the utility on the shell side. The physical shape of the foulants in these cases makes it easy to find the origin. These polymers typically form *in-situ* due to conditions such as high skin temperature or over-vaporization.

Figures 5 a,b: Foulants formed in reboiler tubes

2.3 Reboiler Bottom Heads and Bottom Baskets

When foulant is found in a tower bottom basket or the bottom head of a reboiler, it may be harder to determine where or how the foulant formed. Rarely does the foulant actually form in these locations although there are a few instances where it appears the foulant has come out of solution in the bottom basket. What is expected in a basket or reboiler head are usually pieces of polymer that are fairly small (**Figure 6**). Sheets of polymer larger than palm size may indicate fouling in the tower. However, the foulant's origin is less intuitive since baskets and reboiler heads are foulant collection points from several different places.

Figures **7 and 8** show foulant samples that are less uniform than **Figure 6** and are more likely a collection of foulant from the tower, reboiler or even further upstream. Their presence does not necessarily constitute a problem but is more of a challenge to interpret. In **Figure 7**, the material is fluffy and thin, suggesting the foulant came from the underside of a tray. There are also some longer, thinner rolled pieces that most likely came from the reboiler. In **Figure 8**, the dark rolled pieces look like foulant from the tube walls of the reboiler while the lighter chunks most likely came from the tower. However, the chunks are unusual and harder to decipher. One interpretation can be deposition of fine particles of foulant that have clumped together and rolled around to form hard little balls.

Figure 6: Normal basket foulant

Figure 7: Fluffy deposit possibly from under-side of trays and rolled deposit from reboiler.

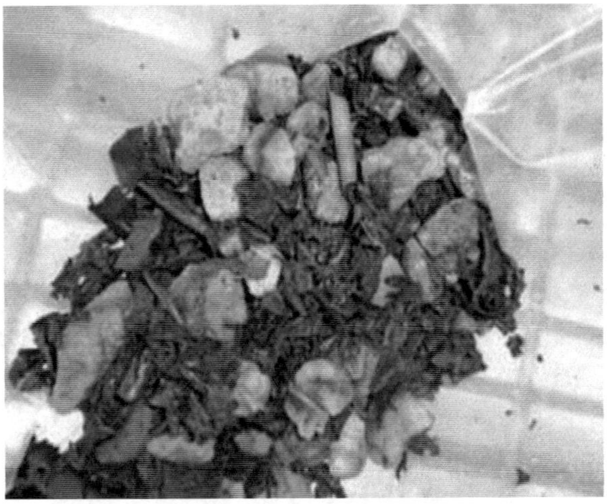

Figure 8: Rolled deposit from reboiler combined with light deposit from a tower.

2.4 Compressors

Compressor deposits do not lend themselves to the same "clues" that are found in distillation towers. Fouling in a process gas compressor is typically in-situ fouling; *i.e.*, the majority of fouling forms during the time cracked gas flow passes through the case. An exception to this fouling in the first stage due to pytar carry-over from the quench system. Depending on the formation mechanism (free radical polymerization, Diels-Alder condensation, and coking), presence of heavier polynuclear aromatics (PNAs), and run length, compressor deposits vary in texture from sticky to rock-like coke (**Figure 9 a and b**).

Depending on the treatment strategy, the location of fouling can also be moved from the case to the intercoolers. Wash oil is widely used to dissolve polymer and prevent it from sticking to the internal surfaces of the machine. Dispersant treatments can also move the foulant to the intercooler where, due to temperature drop and presence of physical barriers (distribution plate, tube bundle), the majority of solids come out of the gas stream and deposit inside the intercooler (**Figure 9c**). Chemical analysis can provide more insight to the nature of the foulant in a compressor while location, color and texture are less informative.

Figures 9: (a) sticky foulant from compressor wheel (b) hard, brittle foulant (c) (lower picture) foulant at the inlet of the inter-stage cooler

3- What Can You Learn From Color and Texture?

3.1. Foulant Color

While most foulants from an ethylene plant are brown to black, the color can vary considerably as can the texture. As a rule, light colored foulant in a service where foulant is most often dark, is an indicator that the fouling occurred in the vapor spaces/under the trays OR the foulant is fairly fresh, having not had time to discolor. However, the converse is not true. Dark foulant does not mean old foulant or liquid phase only foulant. "Curtain" polymer (due to its draping shape) foulant formed underneath of a tray is usually lighter in color and often very thin. Occasionally holes can be seen in the polymer formed by passage of liquid and vapor. A good example of underside tray fouling from a reactive monomer purification tower is shown in **Figure 10**.

Figure 10: Underneath tray fouling. The polymer is light and thin showing the tray holes.

In the case of a caustic tower, the color of the foulant is particularly indicative of the age of the polymer. As discussed in previous papers [ref. 5, 6], aldol polymer goes through several color changes as the molecular weight increases: yellow, orange, red, and black. These changes can be visually evaluated by taking samples at different points in time as shown in **Figure 11**. Here, vinyl Acetate is added to caustic and allowed to undergo an Aldol Condensation Reaction. Small oligomers condense to form larger and larger polymers and as they do so, the color changes. **Figure 12** shows **2** samples from different places in the same caustic tower.

15 min 30 min 2 hrs 5 hrs 24 hrs

Figure 11: Aldol polymer formation using vinyl acetate in caustic over time

Figure 12: Samples from the same caustic tower but different locations in the tower. The sample on the left is expected to have lower molecular weight of aldol polymer due to its yellow color.

One interesting type of caustic tower foulant is a jello like polymer (**Figure 13**). The deposit is the rag layer of an aqueous caustic in hydrocarbon emulsion, which is stabilized with aldol polymer [ref 7].

Figure 13: Rag layer from a caustic tower [ref 7]

3.2 Foulant Texture

As mentioned earlier, in the vapor space, some foulants are thin and soft and often seen on the undersides of trays. Another type of vapor space foulant is usually thicker, often darker and has the characteristic bumpy "cauliflower" surface. These types of foulants are often seen in the return lines of off-line reboilers and relief-valve stand-pipes. The bumpy surface is the clue to vapor space fouling since there are no physical restrictions in the vapor phase (**Figure 14a**). For comparison, foulant formed in the liquid phase is usually has a smoother surface (**Figure 14 b**).

Figure 14: (a) Polymer from vapor space (b) polymer from liquid phase

Figure 15 a and b are good examples of a very light, friable polymer seen most often in compressor discharge lines or equipment that is under higher pressure. When the line is opened, the polymer may fill half the line and it is hard to believe throughput wasn't affected. It is clear by the light and porous nature of the foulant that it was a thick liquid or syrup, taking up only a fraction of the space while on line. The system pressure is reduced, gas trapped in the thick liquid is released. At the same time, the deposit is cooling and solidifying, giving the foulant a porous look.

Figure 15 a and b: Light, friable polymers that puffed up as the gas was released

3.2.1 Sticky Polymers

The contrast to light, friable polymers are dense and sticky or gooey polymers. Different foulant locations have different interpretations for sticky polymers. Gooey polymers from towers with butadiene or isoprene such as those shown in **Figure 16** a and b usually represent low molecular weight polymers. However, foulants resembling grease and tar are prevalent in quench and dilution steam systems are also sticky but could be pytars as well as low molecular weight polymers such as polystyrene. (**Figure 17 a, and b**).

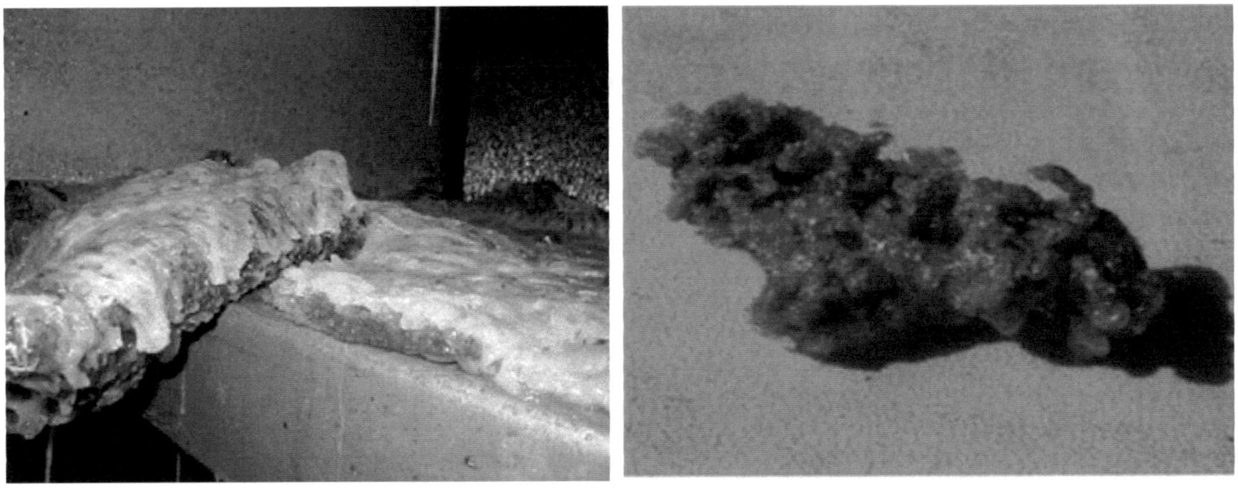

Figure 16 a and b: Low molecular weight foulants

Figure 17 a, b: Pytar type foulants from the dilution steam system

3.2.2 Crystalline Deposits

The best-known crystalline deposits are from "popcorn polymer" **Figure** 18 (a) and its cousins glass (**Figure 18 b**) and amber (**Figure 18 c**) polymers. These polymers have been discussed in earlier papers and the reader is directed to refer to these papers for more information [ref 8 and references therein].

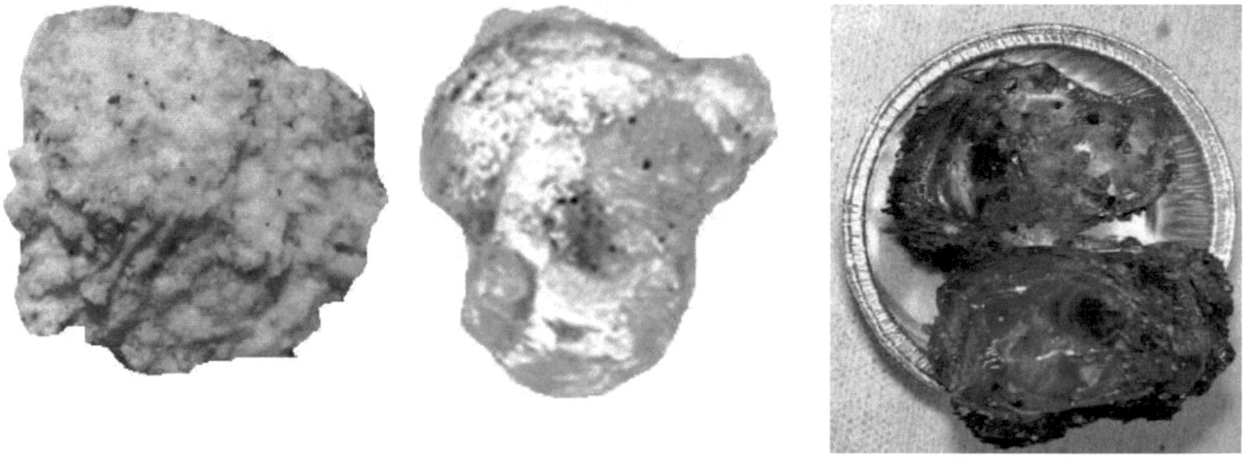

Figure 18: (a) Popcorn Polymer; (b) glass polymer; (c) amber polymer

An example of an inorganic crystalline foulant from a dilution steam generator is shown in **Figure 19**. As a rule, crystalline deposits in an ethylene unit is an indicator of trouble in the process. In such cases, further investigation and mitigation is necessary to prevent future mechanical failures in the system.

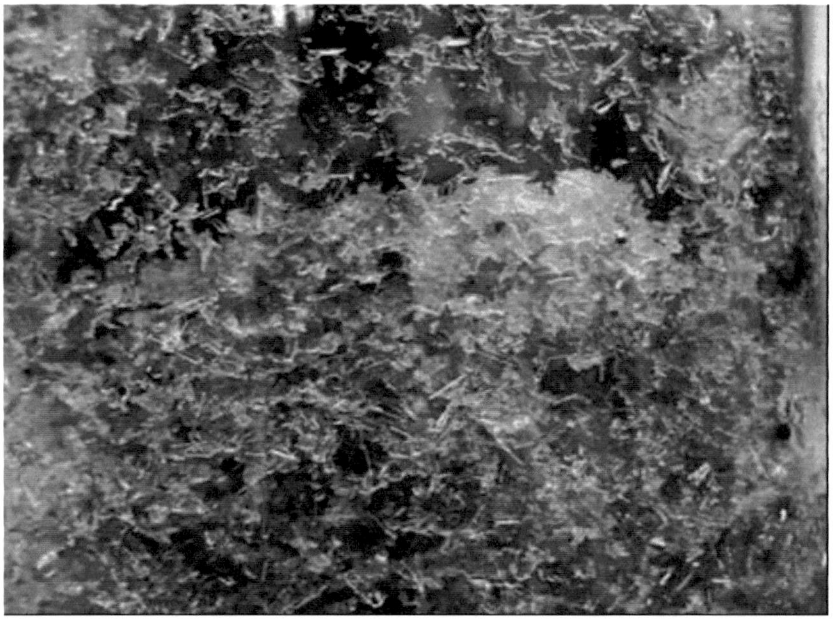

Figure 19: Inorganic crystalline foulant from a dilution steam generator

3.2.3 Hiding in Plain Sight

Because the mechanisms to form crystalline polymers are often more aggressive than those that are associated with non-crystalline foulants, crystalline deposits can often be hidden inside a non-crystalline deposit. When chunks of polymer are found, break it open to see if there is a core of a different looking polymer inside (**Figures 20 and 21**).

Figure 20: Popcorn polymer in rubber polymer

Figure 21: Amber polymer in rubber polymer

4- How Visual Inspection Can Be Used to Examine Fouling?

A recent example occurred when an existing ethylene plant added new light-ends reboilers. The reboilers began to foul every 30-60 days, whereas previous run-lengths from the old reboilers were more than one year. Analysis of the foulant showed no changes from normal polymer from this tower. However, visual inspection of the foulant identified "snake skins," suggesting the skin temperature was too hot. Later, it was discovered that the new reboilers no longer had de-superheated steam. The customer added condensate to de-superheat and reboiler run lengths improved significantly as shown in **Figure 22**.

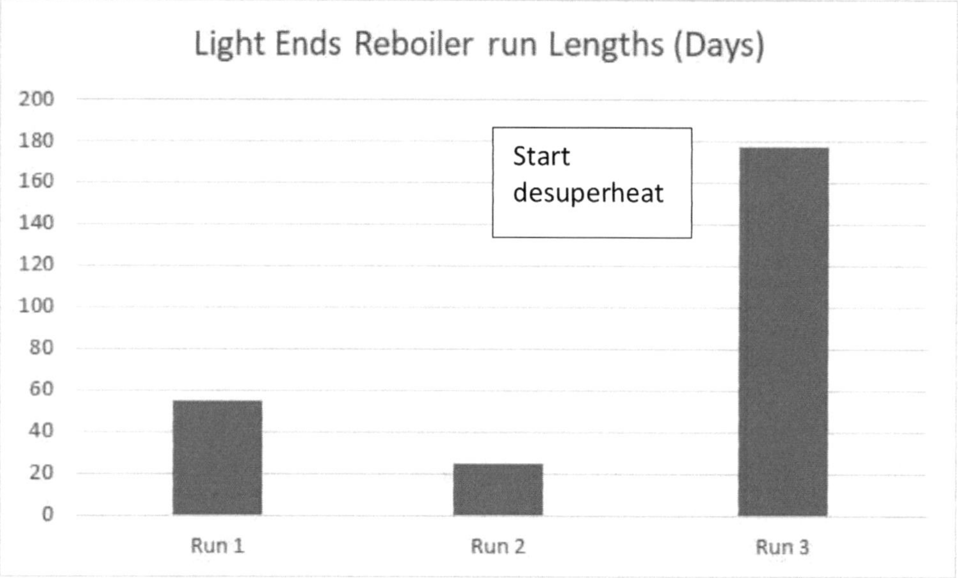

Figure 22: (a) "skin" foulant at the entrance of reboiler due to superheat temperatures (b) Run-length of reboiler before and after adding condensate.

The origin of fouling becomes harder to interpret when foulant forms somewhere else and ends up in the reboiler. An ethylene producer who historically experienced good reboiler runs on a light-ends tower (>300 days) began to see run-lengths drop to less than 90 days. Higher butadiene concentrations and bottoms temperatures were suspected as they moved to a lighter feed. Chemical analysis of the foulant showed no differences compared to previous foulants from this tower (poly BD). The foulant plugged the tube sheets only and not the tubes themselves (**Figure 23**). Visual analysis of the foulant suggested it formed in the vapor space (bumpy surface). Further inspection found a vent line plugged with polymer. Foulant from the line made its way to the sump and would plug the on-line reboiler tube sheet. After cleaning, reboiler run lengths returned to normal.

Figure 23: Polymer formed in vent line moved to the inlet of the reboiler plugging the tube sheet.

As ethylene plants have moved to lighter feedstocks, the incidence of fouling on the underside of trays in Primary Fractionators has increased. At one plant, prior to cracking lighter, the primary fractionator ran turnaround to turnaround. The performance dropped off when they went to lighter feeds. Inspection of the tower showed fouling on the underside of the trays. Although only lightly fouled, the material was heavy enough to cause some valves to stick shut from underneath. **Figure 24 a and b**. The chemical treatment program was changed to address underside fouling and the tower is now running with no limitations.

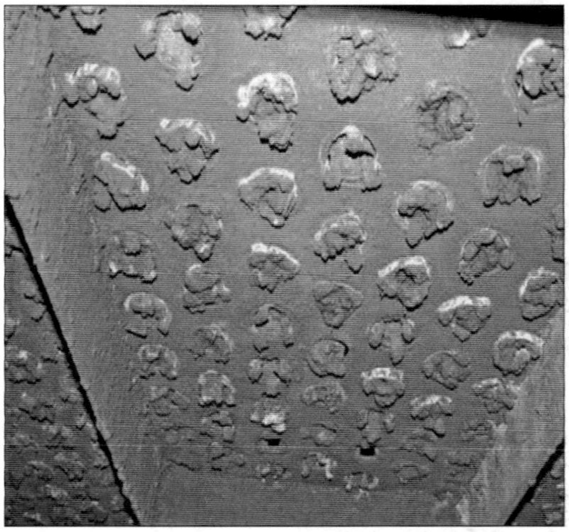

Figure 24 a: Valves stuck with polymer

Figure 24 b: Top of same tray with minimal fouling

5- Conclusion

Fouling is an unfortunate part of ethylene and reactive monomer production. Despite all precautions and treatments, some level of fouling is always expected. Usually little consideration is given to color, texture, and location of the foulant. This paper encourages those in the industry to not only look at the polymer deposit analysis but also its specific attributes and whereabouts. This will provide clues that help to better understand what is happening in the process and how to tackle the issue.

While it is understood that significant dollars can be lost in cleaning and lost production due to fouling, stepping back from the cost, allows for challenging the imagination, as shown in **Figures 25, 26, 27**.

Figure 25: "A very bad day" – a dump truck full of aldol polymer

Figure 26: Macaroni and cheese from a reboiler

Figure 27: A dip of chocolate ice cream

References:

1. Cagnolatti, C. L. HSB Solomon Associates, LLC., Hydrocarbon Processing April 2015, pg. 31-36.
2. Rossana, D. (2015), Nalco Champion "Tower Climbing Safety and Inspection" internal Nalco Champion document.
3. Bouck, D., Inspecting Distillation Towers Part 1: Turnarounds, CEP September 2018, pg. 26-35.
4. Hancock, J.M., Van Zijl, A.W., Robson, I., Vachon,J., Hashmi, A., "A chemist's perspective on organic fouling in ethylene operations", Hydrocarbon Processing June 2014.
5. Hammons, C., and Ham, V., "Caustic Tower Operation Considerations for Effective Performance"; 2009 Spring Meeting AIChE #81g – April 29th,
6. Sambandan, S., Alnasleh, B., Rossana, D., "Study of Aldol Polymerization in Ethylene Plant Caustic Towers", AICHE Paper 32c. 2018 Spring Meeting
7. Neilson, A., Nalco Champion internal report., TSR# HSHA-7TG2XA, 2009
8. Hancock, J., Rossana, D., Korf, S., "Popcorn In Light Ends", Hydrocarbon Engineering, Dec 2014.

AICHE PAPER NUMBER: 35c

RECYCLE CONTAMINANTS IN ETHYLENE PRODUCTION

Dr David Slim

Technology Advisor

Evonik Corporation

Dr Phillip James

Technology Director

Evonik Performance Materials GmbH

Prepared for Presentation at the 2019 Spring National Meeting

New Orleans, LA (March 31 - April 4, 2019)

AICHE PAPER NUMBER: 35c

RECYCLE CONTAMINANTS IN ETHYLENE PRODUCTION

Dr David Slim

Technology Advisor

Evonik Corporation

Dr Phillip James

Technology Director

Evonik Performance Materials GmbH

Abstract

To minimize waste, steam crackers tend to feature a range of recycle streams, either internally within the plant, or externally from neighboring units. These recycles can contain compounds that can adversely affect olefin operation. Examples of these factors include: a component returns to its place of origin and results in an increase in concentration e.g. MAPD contamination of recycled propane can increase MAPD from furnaces; a component is made and goes to a place upstream of where it is produced e.g. caustic red oil being routed to a primary fractionator a component is added for a specific function but can create a problem elsewhere; an additive in a recycle stream decomposes and produces a new contaminant upstream e.g. amine derivatives in light ends fractionators may produce NOx in furnaces which could be a cold box problem. With these factors in mind, an assessment has been made regarding the origin and impact of a number of common compounds. The effect of recycle streams will vary from site to site. However, there are some common themes and thus a mitigation strategy outline is included.

1 SUMMARY

Recycle streams can contain contaminants that can impact an olefin operation. An assessment has been made regarding the origin and impact of a number of contaminants. An outline is presented describing a mitigation strategy for managing contaminants in recycle streams

2 INTRODUCTION

This Subcommittee has studied deeply the issues created by contaminants that enter the steam cracker in the feedstock. Recent topics have included CO_2, CS_2 and oxygenates, and an industry survey. This year this presentation focuses on a slightly different topic, contaminants from recycle streams. This is a very broad subject encompassing many aspects of operating an ethylene plant. Some key topics have been identified, focussing on both internal recycle streams – where material is transferred from one location to another in the cracker, and external streams, where the olefins produced by the cracker have been through further processing such as polymerization, and the residual unreacted olefins are returned to the cracker for reprocessing.

3 DISCUSSION

Modern steam crackers are very efficient, and have a high degree of integration. As well as recovering heat, and reusing water, many streams that in the past were considered waste are now reprocessed. But there can be unintended consequences of recycling these streams

In this paper we will review both the origins of the contaminants and their impact as well as steps to minimise the impact

While no two steam crackers are exactly the same, there are two main categories. The first type process a gas feed, which is predominantly ethane based. Such crackers make relatively few product streams. Differences within this category are mainly a result in the design philosophy of contractors. The other category processes liquid feeds, ranging from a naphtha to a heavy gas oil. Such crackers involve the recovery of a larger number of product streams. Configurational differences are often a result of local economics. Below are generic flowplans of liquid and gas crackers showing typical recycle streams.

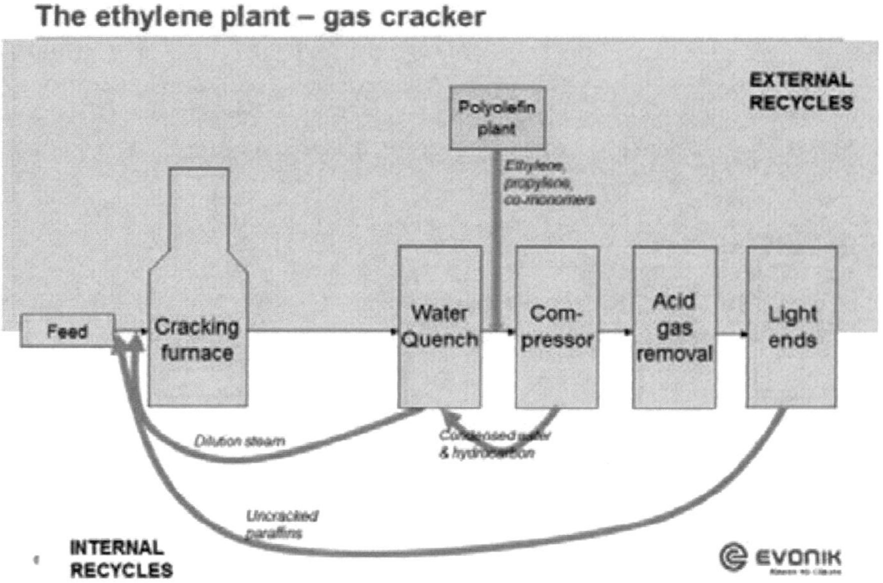

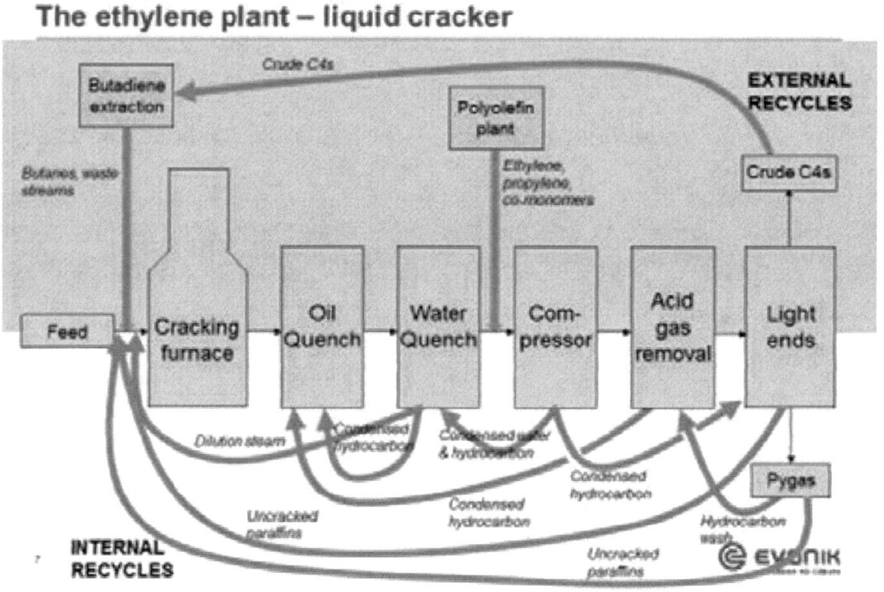

FIGURE 1 - SIMPLIFIED GENERIC FLOW PLANS OF GAS (TOP DRAWING) AND LIQUID CRACKERS (BOTTOM DRAWING) SHOWING EXAMPLES OF RECYCLE STREAMS

3.1 Impacts of problems resulting from processing of recycle streams

There are a number of possible impacts that contaminants in recycle streams can have. These include:

Existing removal system

Higher contaminant levels from recycle streams can overload an existing removal system. A removal system is designed to remove contaminants at a specific rate/concentration. As an example vinyl acetate to a caustic tower can overload an additive system to remove acetaldehyde

Catalyst Systems

Catalyst systems can be affected by contaminants. Some contaminants impacts are based on lb/hr (accumulation over time) and/or by concentration. These can adversely impact catalysts performance/life

Fouling management

Fouling is often managed by inhibitor addition or solvent washing. Recycle contaminants can require increased additive cost or increased solvent requirements

Safety

Some contaminants may impact safety e.g. NOx in cold boxes. This can lead to increased safety management. This topic has been discussed at previous AIChE Spring Meetings. In this paper we consider NOx in terms of cracker recycle streams, rather than as a feedstock contaminant.

Product quality

Contaminants can impact product quality leading to higher costs.

Given the above, a more detailed description of recycle contaminants is described below. We have categorized the recycle streams into two classes:

- internal - starting and ending inside the ethylene plant)

- external - starting in an derivatives unit downstream of the cracker where the unreacted olefins return to the cracker

3.2 Internal Recycle Streams

The following aspects of recycle streams will be considered: dilution steam, uncracked paraffins e.g. ethane from the C2 splitter, condensed liquids from the compression train and streams containing additives

Dilution steam

Generally the steam to the cracking furnaces is provided by a dilution steam generator. Invariably, the quench water is pretreated and then vaporized to produce the steam. Contamination following the pre-treatment step can reach the cracking furnaces. Such contaminants include potentially dilution steam can contain sodium compounds and ammonia/amines. Sodium compounds can damage furnace tubes, while ammonia can be a product quality issue (ammonia can be produced via cracking amines used for pH control, or deliberately added for the same reason).

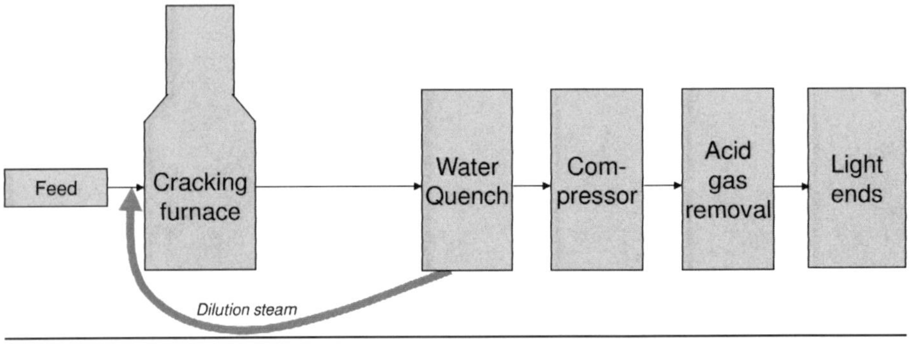

FIGURE 2 - RECYCLE OF WATER QUENCH

Uncracked Paraffins

In a number of cases, light ends streams can be routed to the furnaces. Examples include: C2 Splitter bottoms ethane, C3 splitter bottoms propane, Certain C4 streams (recycled from a butadiene extraction unit) and recycles from Pygas Units (C5 Recycles, C6/C7 Recycles etc). The above streams are generally paraffinic and give good yields of olefins. However they can contain contaminants that can impact furnace operation. Examples include:

C4 streams with high olefins may be routed to furnace feed which result in lower levels of high value products and make high levels of low value products (poor yields)

Propylene in recycle propane recycle results in poor ethylene yields. These streams may contain MAPD and/or MAPD Green Oil

Splitter bottoms can contain high ethylene which results in poor yields, while acetylene polymers (Green Oil) can result in fouling issues.

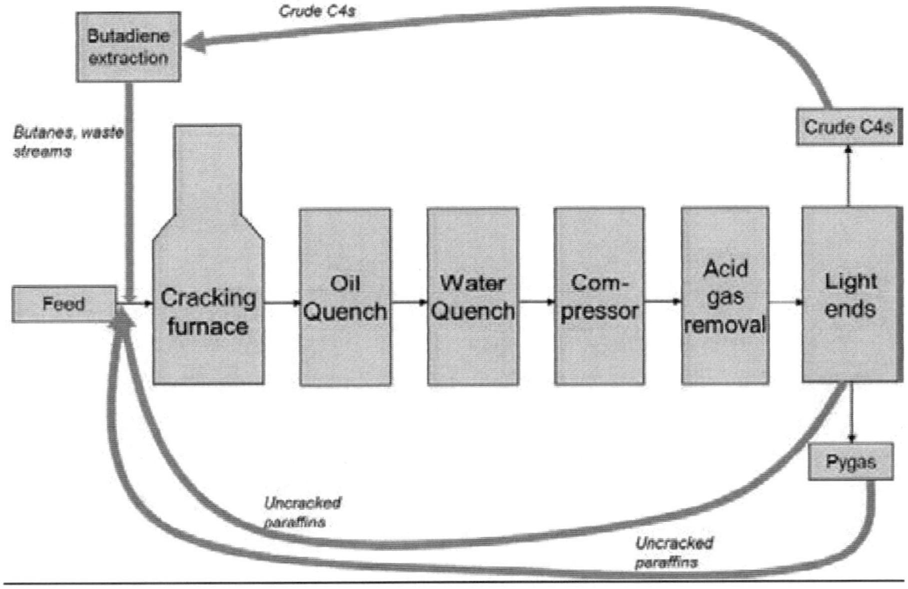

FIGURE 3 - RECYCLE OF UNCRACKED PARAFFINS

<u>Condensed Liquids from the Compressor Train</u>

After each stage of compression, condensed liquids are removed and recycled. Typically water streams recycle to the quench tower. Hydrocarbons either recycle to the quench tower or the light ends section. In the caustic tower, condensed hydrocarbons may also be recycled. The above streams may cause a range of issues. These include: distribution of process additives (see later) and can cause pH excursions in the quench water.

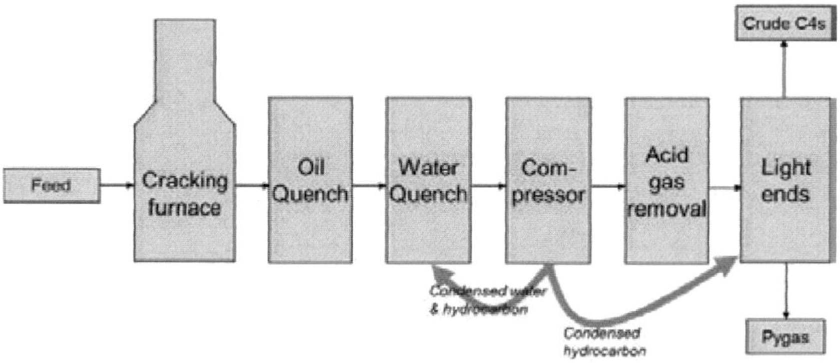

FIGURE 4 - RECYCLE OF COMPRESSOR TRAIN LIQUIDS

<u>Quench gasoline recycle loop</u>

In liquid crackers, condensed hydrocarbons in quench water section are often recycled to primary fractionator to provide reflux – "gasoline recycle". Certain hydrocarbons acts as low boilers in this tower and therefore return by distillation to the quench water section, thus creating a loop. Styrene can concentrate in this loop to reach >20%. At the top of the primary fractionator it is hot enough for styrene to polymerize. This can lead to fouling in the primary fractionator, either at the location of polymerization or lower down the tower if the polymers are washed out by the heavier liquid hydrocarbons. To manage these impacts it is recommended that the composition of the gasoline recycle be monitored and if necessary add an inhibitor to the loop to stabilize the styrene

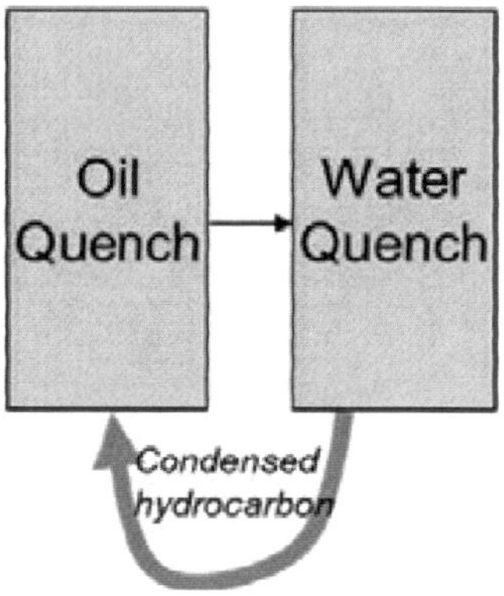

FIGURE 5 - QUENCH GASOLINE RECYCLE LOOP

Streams Containing Additives

Process additives are used throughout the steam cracker to control unwanted side-effects such as fouling, emulsions and foaming. If these are present in recycle streams, they may lead to unintended consequences. Examples include:

N-O compounds

Molecules containing NO groups can possibly result in the formation of NOx. Other nitrogen containing molecules tend to make ammonia or can remain in heavy aromatics. A number of light ends towers use N-O containing additives to inhibit fouling. These compounds may decompose to form nitric oxide in the furnace area. Nitric oxide is a precursor for higher oxides of nitrogen that may result in gum and salt formation in cold-boxes

Silicon compounds

Silicon compounds are used to inhibit foaming and emulsions in light ends systems and/or refinery streams that can be recycled back to the cracking furnaces. Polydimethylsiloxanes start to decompose at about 190°C to give lower molecular weight oligomers that can be found in pygas streams. These low molecular weight siloxanes can impact catalyst activity. Silica is also present in some antifoams; this can lay down on catalyst beds to block active sights.

Surface active agents

Dispersants that recycle to the quench water tower may upset water/oil separation

Alcohol Washing Products

Alcohol is often used to remove ice/hydrates in light ends systems. As a general rule, oxygenates decompose to produce secondary species but conversion is invariably incomplete. Methanol cracking has been discussed in several EPC meetings with respect to methanol contamination in propylene (unreacted methanol or contaminated propylene product). Methanol is produced from cracking streams containing MTBE and TAME. Methanol can be found in a variety of recycle streams but tends to boil with the C4's. Methanol in cracking furnaces tends to make CO

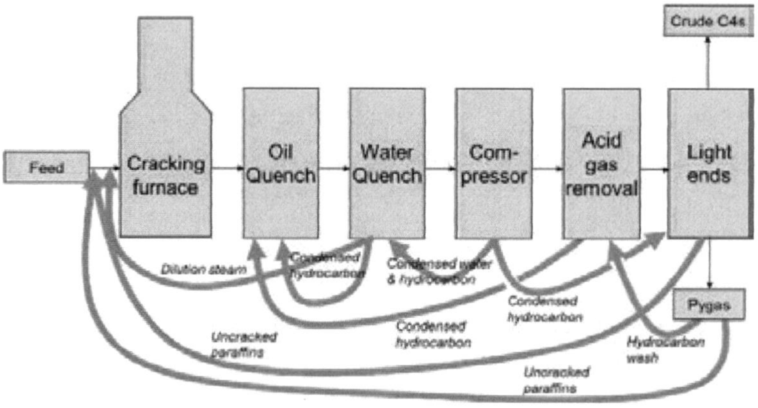

FIGURE 6 - ALCOHOL WASHING PRODUCTS

3.3 External Recycle Streams

External recycle streams can result from unreacted olefins from derivatives plants, including ethylene from PE units, oxygen and waste streams from butadiene extraction units

<u>Unreacted Olefins From polymers plants</u>

These compounds primarily include ethylene and propylene from polyolefins units. They may contain components that could cause issues including: catalyst particles, which could cause polymerisation or at least increased deposition. Polyolefins, which could increase deposition.

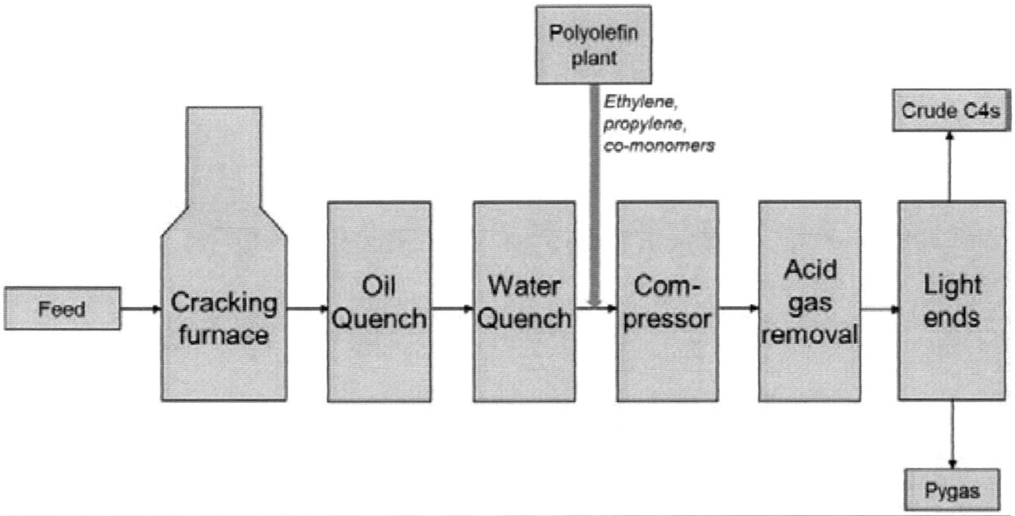

FIGURE 7 - UNREACTED OLEFINS FROM POLYMERS PLANTS

Vinyl Acetate

Vinyl acetate is a co-monomer in some polymer plants. On entering the caustic tower it is converted to acetaldehyde. Acetaldehyde is also naturally present in the cracker. It is believed to be formed in the furnace by hydration of acetylene (a known industrial process). Many plants monitor acetaldehyde in the caustic tower feed because it is known to form red oil in the column. This can be prevented by additives. Vinyl acetate will not be detected by the normal method, but increases acetaldehyde concentration, so more additive is needed

Oxygen

Oxygen can initiate polymerisation of reactive monomers – diolefins and vinyl aromatics

Butadiene extraction recycles

Steam crackers integrated with butadiene plants may have additional recycle streams. These include:

Solvent recovery purge stream

This can bring solvent residues to the cracker, including NMP, DMF, and ACN. If recycled to the furnace these may liberate ammonia, HCN and formic acid. If recycled after the furnace, the solvents can cause process upsets. Also present in this recycle stream may be process additives such as dispersants and inhibitors, as discussed earlier.

Butadiene product column bottoms stream

As well as containing reactive species which may cause fouling, certain monomers in this stream could concentrate as they pass through the cracker and back to the butadiene extraction unit

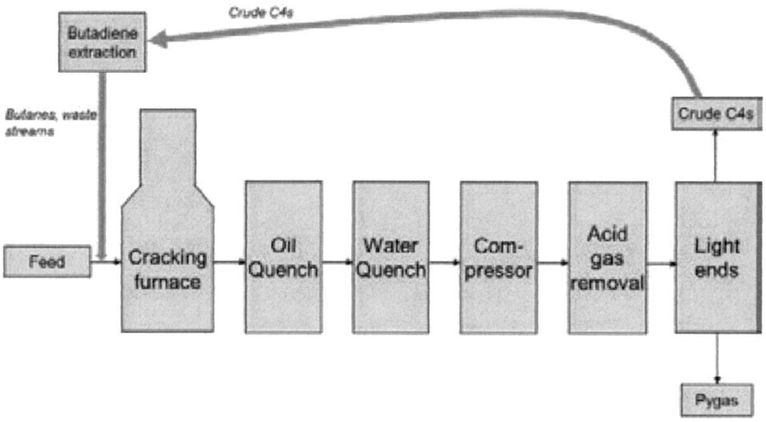

FIGURE 8 - C4 RECYCLE STREAMS

3.4 Need for mitigation strategy

Most locations have contaminant removal systems. Such systems will vary depending on contaminants expected, their quantity, desired removal efficiency and complexity. Recycle streams can increase a contaminant level, put a contaminant in an unexpected location or create new contaminants. There is no universal system. Several reasons include: plant configurations are different (liquid vs gas cracking), different process additives, different requirements etc. Given there cannot be a universal system, we have developed a generic two step approach for managing contaminants from recycle streams. This is described below.

Step 1 - Be aware of all recycle streams

Recycle streams may be part of the original design, and therefore should be fully documented on piping and instrumentation drawings. However, often they are added later for a number of reasons, such as a change in feedstock, construction of a new derivatives plant (e.g. polyolefins, C4s), or to deal with an issue that was not foreseen during construction. It is important to ensure all recycle streams are fully documented, so that present and future plant personnel are aware of the origin and destination of the recycle, the flowrate, its composition and specific hazards, and whether the stream is gaseous or liquid, and whether that liquid is one or two phases.

Step 2 - Understand the process chemistry

All contaminants influence the chemistry of the process in some manner. Thus it is important to have a thorough understanding of your own process. Examples of the areas where chemistry effects occur include: furnaces, acetylene and MAPD converters etc.. Following this it is important it is important to determine what is in recycle streams that are potential contaminants, including an understanding of vendor additive chemistry (how it works, what they can break down to under different environments and how this can affect the process etc.) and understand how process changes can impact the process chemistry e.g., the effect of moving the red oil wash from a quench water tower to the gas fractionator.

If this cannot be determined internally, advice can be sought from the plant licensor, additive vendors, or could be raised in an industry forum.

Step 3 - Manage the contaminant

For contaminants that have a major impact, several considerations must be taken into account. Below is a list of examples to actions that can be taken to manage contaminants. Ultimately the decision will be based on cost-benefit.

Possible Action Steps

Remove the contaminant-containing recycle or feed stream. This could be low or no investment but could be significant operating cost by the loss of an economic feed or recycle stream

Install a specific adsorbent for the contaminant. This could be high investment and medium operating because of the need to periodically regenerate and replace. Consider places where an existing structure can be used e.g., for mercury removal, there are products available that can be used as a separate layer in the charge gas driers.

Wash systems e.g. water washes to remove nitriles or caustic to remove acid gases. This is usually high investment cost and high operating cause because of the need to continually supply the wash or to regenerate it.

Find a purge point to get rid of the contaminant e.g. increased blow down off the dilution steam drum to remove sodium in the dilution steam or the reflux of the waste water stripper to purge ammonia.

If there is uncertainty, then we recommend checking the literature e.g. EPC Contaminants Tool, search EPC website (Top 5 Contaminants in Ethylene Plants)

4 CONCLUSION

Recycle streams can contain contaminants that can impact an olefin operation. An assessment has been made regarding the origin and impact of a number of contaminants. An outline is presented describing a mitigation strategy for managing contaminants in recycle streams

AIChE Paper Number:

Effect Of Contaminants On Coke Formation During Steam Cracking Of Propane

Ir. Manjunath Patil
Ghent University

Dr. ir. Marko R. Djokic
Ghent University

Prof. dr. ir. Guy B. Marin
Ghent University

Prof. dr. Marie-Françoise Reyniers
Ghent University

Prof. dr. ir. Kevin M. Van Geem
Ghent University

Prepared for Presentation at the 2019 Spring National Meeting New Orleans, LA,
April 1 – 4, 2019

Effect Of Contaminants On Coke Formation During Steam Cracking Of Propane

Manjunath Patil
Ghent University
Marko R. Djokic
Ghent University
Guy B. Marin
Ghent University
Marie-Françoise Reyniers
Ghent University
Kevin M. Van Geem
Ghent University

Abstract:
In conventional steam cracking feedstocks, contaminants such as sulfur, phosphine, and heavy metal components are generally present in ultra-trace (ppb) levels which have a direct effect on the coking in the reactor coils. It has been a normal practice to treat the feedstock to adjust the contaminant levels within close limits to achieve no negative results in steam cracking. The purpose of this study was to mimic industrial conditions for coke formation with contaminants such as sulfur and phosphorus compounds at ppm and even ppb level. These experiments are carried out in a thermogravimetric unit which consists of a plug flow reactor (PFR), which allows to evaluate coke formation in real time behavior on a Ni/Cr 35/25 and Al-containing reactor material with phosphine as a contaminant. The coupons were initially pretreated under the presence of steam from 300 °C to 900 °C with 50 °C/h to create a sufficient oxide barrier. The cracking experiments consist of two cracking cycles for each contaminant with four hours of propane cracking at 950°C followed by ramping-up to 1050°C for forty five-minutes per cycle, the latter resembling the end-of-run industrial conditions. The results indicate that the presence of small amounts of phosphine has a positive influence on coke formation by ~30% in mass of coke formed on the coupons. The presence of phosphine has a bigger positive impact on asymptotic coking rate than catalytic coking rate. The morphology of the coked reactor materials were studied using an off-line Scanning Electron Microscope with Energy Dispersive X-ray (SEM with EDX) images of coked coupons.

1. Introduction

Major amount of research is being done to reduce the coking rate during steam cracking of hydrocarbons, the dominant industrial process to produce light olefins. Formation of coke inside the reactor tubes is disadvantageous since it acts as a thermal insulator, increasing the required heat. Apart from that, it lowers the cross sectional area of the reactor tubes, increasing the pressure drop. This eventually leads to very expensive shutdowns of the steam cracker unit to decoke the unwanted carbonaceous deposits inside the reactor. A simple operational method to alter the environment inside the reactor, trying to reduce the coking rates, is to add specific compounds to the feed. Promising research has been done regarding the addition of organophosphorous compounds, results include a noticeable decrease of the coking rate and a different coke morphology and composition is obtained. Continuous addition of organophosphorous compounds is always applied, since it reduces the coke formation rate more compared to periodical addition.

Previously, several different organophosphorous compounds have been studied regarding this topic. The influence of triphenyl phosphine (TPP), tri-o-tolyl phosphine (TTP), and triphenyl phosphine oxide (TPPO) on the coking rate during steam cracking of naphtha has been studied by Niaei et al. in 2008 [1]. Conclusions were that all compounds inhibited the coking rate, proportional to the amount of organophosphorous compounds added. When more than 300 ppm of the latter compounds are added, the coking rate remains constant, at a value about 5 times lower compared to when no additional compounds are added. The explanation for this is that TPP, TTP and TPPO will decompose into radicals, which will chemisorb at the alloy surface, forming a phosphorous-metal complex. This complex acts as a protective film and deactivates the surface, which is known to catalyze coke formation. The more phosphor is added, the more extent this film will be, until the entire surface is covered at 300 ppm, explaining why no additional inhibition takes place beyond this concentration.

Similar results were obtained by Ghosh and Kunzru in 1988 [2] for the compound triethylphosphite (TEP) and by Kunzru and Chowdhury in 1993 [3] for benzyl diethyl phosphite (BDP). The coke formation reduction increased initially, until at higher concentrations the coking rate approached a constant value. The same conclusion was made, i.e. the phosphorous-containing additives passivate the alloy surface. When 200 ppm of BDP is added to the naphtha feed, the asymptotic coking rate is reduced by 87.7%.

Another beneficial effect of phosphorous addition is the altering of the morphology and metal content of the cokes being formed. Cokes formed in presence of phosphorous is typically finer and more porous compared to the reference case without inhibition. This is industrially preferred since the cokes can be removed more easily during decoking, reducing the decoking time. Lastly, it is proven by Niaei et al [1]. that the amount of metals in the cokes decrease when phosphorous is used as inhibitor. Since iron and nickel catalyze coke formation, this is very beneficial. This partially explains the inhibition effect of organophosphorous compounds, besides the effect of the protective film as explained earlier.

In this work, the effect of phosphine (PH3) on steam cracking of propane is being investigated. Its coke inhibition effect is studied for a classical steam cracking alloy (Ni-35, Cr-25 on Fe-base), and for an aluminum containing alloy. Apart from that, its influence on the formation and stability of the protective chromia and alumina scale layers is being investigated.

2. Experimental Setup

In this work, a plug flow reactor setup is being used. It consists of three parts: the feed section (with preheaters), reaction section and analysis section, see Figure 1. All feed streams are controlled with mass flow controllers to achieve the desired flow rates. The plug flow reactor is made out of quartz, within, a metal coupon hangs from the arm of an electrobalance. The latter measures the mass changes over time due to the coke deposition on this coupon, the used frequency is 1 Hz. After reaction, the reactor effluent is quenched with cooling oil to prevent further cracking. Its composition is measured with two GC's, both using nitrogen as internal standard, one measures components with less than 5 carbon atoms, the other measures C5+.

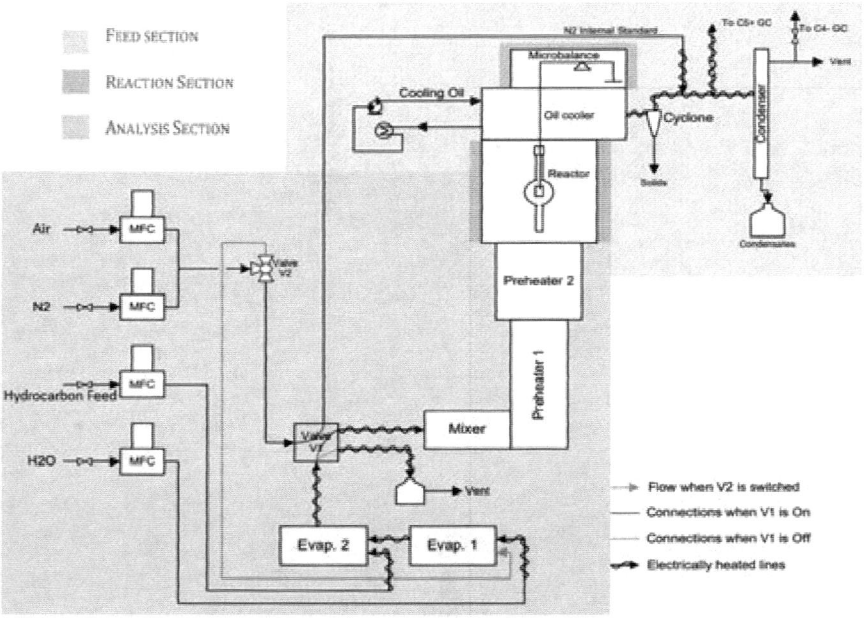

Figure 1: Flow diagram of Jet-stirred reactor setup [4]

The operating conditions for the experiments with phosphine, called 'industrial case', and without phosphine as 'reference case'. Both sequences of process steps are very similar, the only difference being the addition of 75 ppb phosphine diluted in nitrogen gas is fed to the reactor with similar flowrates compared to the reference case. Before the first cracking cycle a pretreatment with steam is performed, oxidizing the alloy surface. This is followed by presulfiding. All experiments consist of two cracking cycles, first 4 hours at 950 °C, then 45 min at 1050 °C to mimic the industrial metal temperatures corresponding to start-of-run and end-of-run industrial conditions, respectively. To prepare the coupon for the next cracking cycle, decoking of the coupon is performed for half an hour with air and steam mixture. The calculations of coking rates are illustrated in detail by S. Sarries [4].

Since it is impossible to make a reactor out of each alloy to be investigated, an inert quartz reactor is used, with a coupon of the respective alloys hung inside the reactor, representing the coking properties of that alloy. All coupons have dimensions of 10 mm x 8 mm x 1 mm. They are cut out of the inner reactor tube as can be seen in Figure 2. Since it is preferential that both sides of the coupon have the same composition and properties, i.e. those of the inner reactor tube, "back-to-back" welded coupons are used. These coupons are made out of two separately cut samples, welded back to back as can be seen in Figure 3Figure . In this work, two kind of alloys and thus coupons are used, one being made out of a standard steam cracking alloy; Fe-35Ni-25Cr, while the other contains aluminum as well.

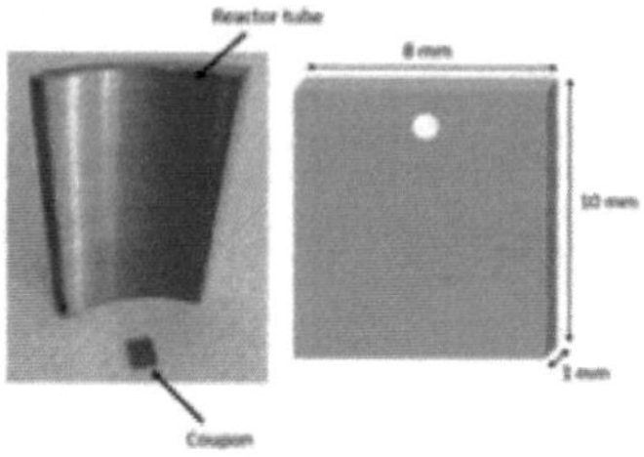

Figure 2: Reactor tube and coupon dimensions

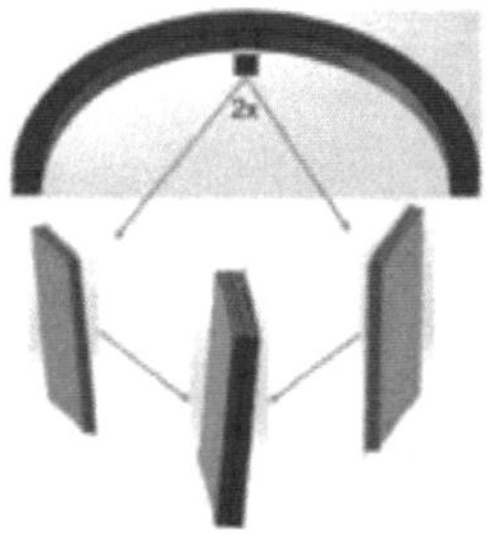

Figure 3: Design of a back-to-back welded coupon

3. Results & Discussion

i. Coke formation rates

The relative obtained amount of coke formed, with and without phosphine, is shown in Figure 4 for both alloys and both cracking cycles. Several conclusions can be made out of this figure. First of all, coke formation is more severe during 45 minutes at high temperature (1050 °C) compared to 4 hours at low temperature (950 °C), for both alloys, with and without phosphine. The coupons in steam cracking media at gas-phase temperature of 950 °C resembles start-of-run industrial conditions followed by increase in gas-phase temperature to 1050 °C resembling the end-of-run industrial conditions. The combination of both the steam cracking temperatures followed by a decoking procedure reflects aging of the material. Next, the aluminum content alloy performs better compared to Ni/Cr 35/25 for all configurations. Alumina scales are thus better coke inhibitors than chromia scales, which is in accordance with the literature.

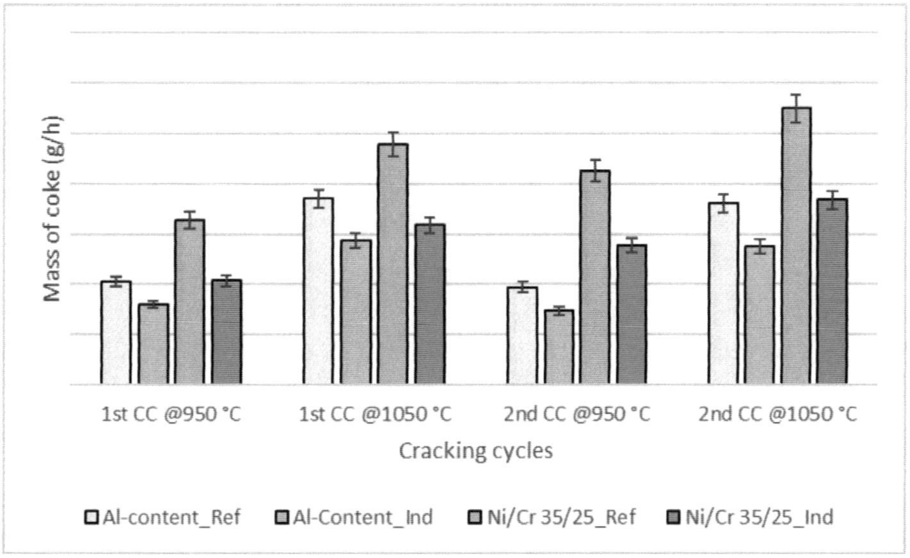

Figure 4: Relative amounts of coke formed per hour of cracking cycle

While there is a clear increase in amount of cokes formed in the second cracking cycle for Ni/Cr 35/25 compared to the first one due to the spallation of the protective Cr2O3 layer over time. Such a behavior is not observed for the aluminum containing alloy, since the Al2O3 layer is thermodynamically more stable at higher temperatures than Cr2O3 layer. The al-content material is more stable with aging of the coupon, therefore the coking rates for 1st and 2nd cracking remains stable in both the cases, reference and industrial.

The impact of phosphine cannot be overlooked. Both cracking cycles for both alloys showed a clear decrease in coke formation due to the addition of phosphine, which is in accordance with literature. The relative mass decreases are shown in Figure 5. Under all conditions, the mass decrease was more profound for the Ni/Cr 35/25 alloy, compared to the aluminum content alloy. This is due to the presence of aluminium oxide layer, which is more stable with aging of the material.

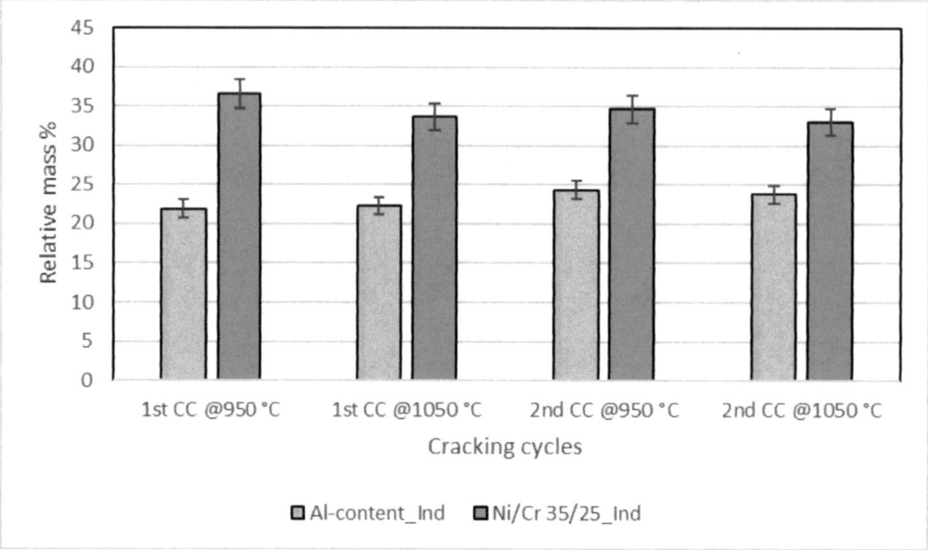

Figure 5: % reduction of coke in presence of phosphine under propane steam cracking conditions at a gas-phase temperature of 950 °C

Figure 6 and 7 show the coking rates for catalytic and asymptotic coking. During all sections, a reduced coking rate is observed due to phosphine addition; for both catalytic and asymptotic rates and for each cracking cycle and alloy type, although the influence of phosphine is more explicit during the asymptotic coking periods.

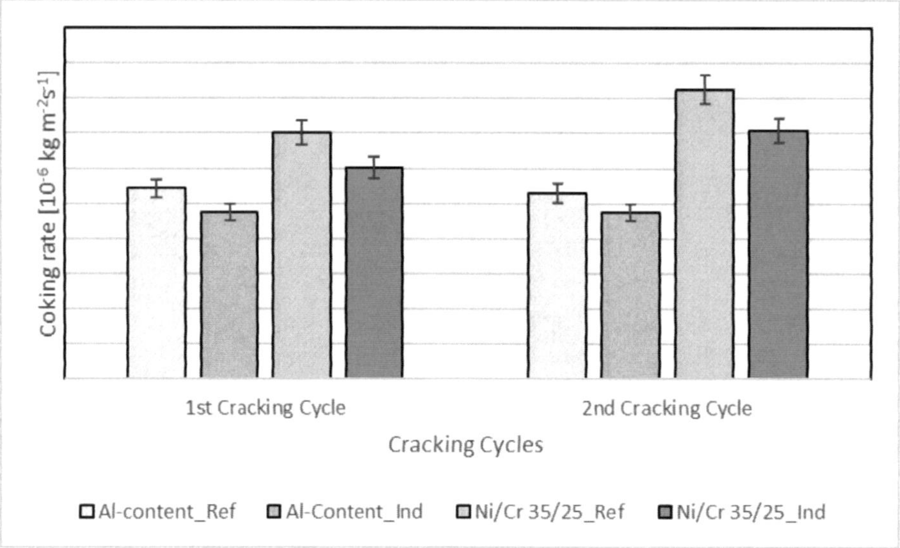

Figure 6: Catalytic coking rate of Al-content and Ni/Cr 35/25 alloys under propane steam cracking conditions at a gas-phase temperature of 950 °C

The pronounced positive effect of phosphine on asymptotic coking rates can be explained due to the absence of phosphor during the pre-sulfiding of the coupons. Phosphor will gradually chemisorb on the alloy surface while steam cracking, and thus its impact will gradually increase over a cracking cycle. During the catalytic coking rate, the amount of metal-phosphor chemisorption is lower, during this stage most of the coke formation reduction is due to the effect of the presulfiding the coupons. This presulfiding of the coupons comprises of DMDS addition of 500 ppmW S/H2O. When the cracking cycle proceeds, more phosphor will react with the surface, increasing its coke inhibiting properties, as can be seen by the reduced coking rates during asymptotic coking, can be observed in Figure 7.

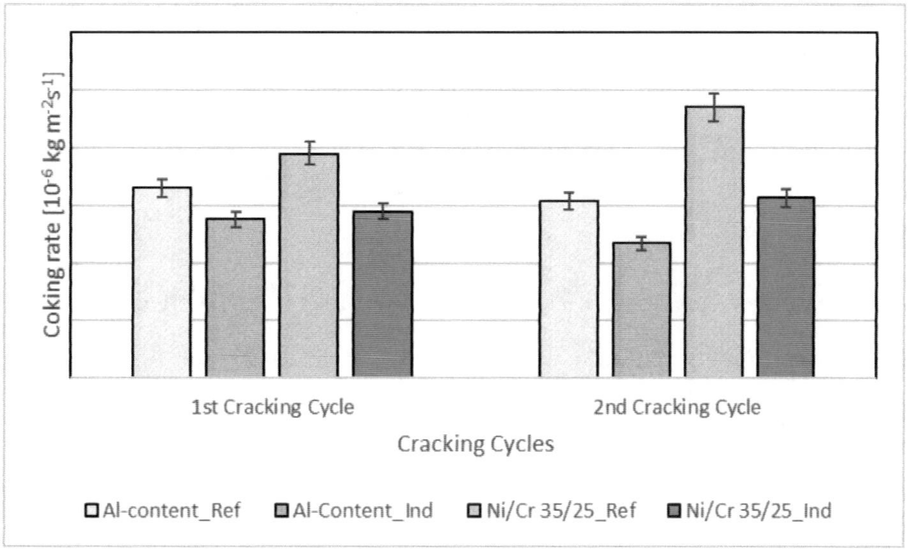

Figure 7: Asymptotic coking rate of Al-content and Ni/Cr 35/25 alloys under propane steam cracking conditions at a gas-phase temperature of 950 °C

ii. ***SEM with EDX discussion***

After the second cracking cycle, imaging of the coked coupons is executed using off-line Scanning Electron Microscope with Energy Dispersive X-ray. This is done in order to study the microstructural morphology of the oxide layers. The coked coupons from the experimental setup are embeded using Polyfast®, the coupons are etched electrolytically using 6V in 10% Nital. The oxide layer is not continuous and also not sharply defined, which would be preferential and due to spallation of oxide layers. Spallation of Cr2O3 has occurred on the edges of the image, as can be seen by the lack of Cr oxide at the surface of the alloy. Internal clusters of oxygen and carbon show that unwanted oxidation and carburization took place. The internal oxides are mainly SiO2, as can be seen by the clusters of Si at the same regions of oxygen.

4. Conclusions

The influence of phosphine addition to steam cracking of propane has been studied. Two alloy types were used, a conventional steam cracking alloy i.e. Ni/Cr 35/25, and an aluminum containing alloy. Electrogravimetric measurements showed that the mass of coke formed was reduced by about 35% for the former alloy, and about 20% for the latter. This is because of the presence of aluminum, forming a very stable and inert alumina layer, which has a strong influence on the coking rate regardless of the presence of phosphine. The coke inhibiting effect was found to be more prominent in the asymptotic coking regime. The reason for this is that the inhibiting phosphor-metal complexes takes time to be chemisorbed on the surface of the material, hence in the beginning of each cracking cycle no substantial inhibition of coke could be observed. SEM with EDX mapping resulted to the same conclusions. Phosphine can thus have in small amounts a beneficial effect on coke formation during steam cracking of propane.

Acknowledgements:

The authors acknowledge the financial support from the 'Long Term Structural Methusalem Funding by the Flemish Government' – grant number BOF09/01M00409.

5. References

[1] Niaei, A., Salari, D., Towfighi, J., Panahi, P., Nabavi, S. (2008). Effect of Organophosphorous Compounds as Coke Inhibitors on Coking Rate in the Pyrolysis of Naphtha. Petroleum Science and Technology. 26. 2170-2181.

[2] Ghosh, K., Kunzru, D. (1988). Reduction of coke formation during naphtha pyrolysis using triethyl phosphite. *Industrial & Engineering Chemistry Research 27* (4), 559-565

[3] Chowdhury, S. N., Kunzru, D. (1993). Benzyl diethyl phosphite as a coke inhibitor during naphtha pyrolysis. tubular reactor studies. Can. J. Chem. Eng., 71: 873-879.

[4] Sarris, S. (2018). Unravelling carbon formation on high temperature alloys and coatings.

[5] Verstaen, S. (2017). Investigation of cokes formation during steam cracking of hydrocarbons: design of a new experimental setup & validation, master thesis.

Leading Edge Technologies of Large Power Mechanical Drive Steam Turbine for 3.0 MMTA Ethylene Plants

Authors

Tomoyuki Nishikawa *is a Steam Turbine Engineer of Turbine Engineering & Design Section in Engineering & Design Division, Mitsubishi Heavy Industries Compressor Corporation, in Hiroshima, Japan. He has 12 years' experience in designing steam turbines. Mr. Nishikawa has BS and MS degrees in Mechanical Engineering from Hiroshima University.*

Norihito Fujimura is *Team Leader, Turbine Engineering & Design Section, Mitsubishi Heavy Industries, Compressor Corporation, in Hiroshima, Japan. He has 15 years' experience in designing steam turbines. Mr. Fujimura has B.S and M.S degrees in Mechanical Engineering from Oita University.*

Kyoichi Ikeno *is Manager, Turbine Engineering & Design Section, Mitsubishi Heavy Industries Compressor Corporation, Hiroshima, Japan. He has 20 years of R&D experience for mechanical drive steam turbines. Mr. Ikeno has B.S. and M.S. degrees in Mechanical Engineering, Miyazaki University and Kyushu University, respectively.*

ABSTRACT

According to the continuous trend of plant capacity increase from 1.0 MMTA to future 3.0 MMTA ethylene crackers, the required process flow for compressors and required power for steam turbine drivers increase. Mitsubishi Heavy Industries Compressor Corporation (MCO) has adapted this demand for the rotating equipment to assist the development of ethylene industry for long years and provided state of the art designs with high safety, high performance and high reliability to end users based on a lot of our experiences.

MCO has finished basic design of 140MW class mechanical drive steam turbine for future 3.0 MM TA capacity range plant with our experienced and proven components based on our site operating experience of 100MW class steam turbine. Following is the brief introduction of this large power variable speed steam turbine and what we think necessary as the design and manufacturing for such large power range. The authors' present ethylene plants trends, technology development history maps for design advancements, challenging technology, verification tests, application results in terms of transient fluid dynamics, thermodynamics, rotor dynamics, and blade vibration strength evaluation.

In addition, after commissioning turbomachinery performance and reliability tends to deteriorate during long-term operation. The typical damage and deterioration map for machinery is introduced, and the authors explain practical technologies (such as flow path surface treatment, effective on-line washing, anti-corrosion, erosion prevention and stage performance enhancement by partial component replacement.

INTRODUCTION

During this decade, the capacity of ethylene plants has tended to increase continuously. Consequently, flow rates in the high-pressure and low-pressure stages of the steam turbine driver have also increased. To meet this capacity increase, it has been necessary to apply larger sub-components including large flow governing valves, longer blades, and longer bearing span rotors. Compressor and turbine trains with capacities in excess of 1.8 MMTA have already been designed, manufactured, and are successfully operating. Therefore, in order to minimize power consumption and downtime losses, it is necessary to conduct machine monitoring for prevention and diagnosis of performance deterioration.

Turbomachinery equipment manufacturers have been moving forward to apply state of the art technologies for critical components in order to provide more efficient and reliable machines corresponding to the high power requirements for ethylene services. The authors introduce the technology development history map in the past and future for typical designs, verification tests, and application results in terms of transient fluid dynamics, thermodynamics, rotor dynamics, and blade vibration strength evaluation. In addition, after commissioning turbomachinery performance and reliability tends to deteriorate during long-term operation. The typical damage and deterioration map for machinery is introduced, and the authors explain practical technologies that are applied, such as flow path surface treatment, effective on-line washing, anti-corrosion, erosion prevention and stage performance enhancement by partial component replacement.

Mega Ethylene Plant Trend

In the past decade the capacity of ethylene plants has tended to increase continuously as shown in Figure-1. Consequently, flow rates in the high-pressure and low-pressure stages of the steam turbine driver have also increased. To meet this capacity increase, it is necessary to apply larger sub-components including large flow governing valves, longer blades, and longer bearing span rotors. Compressor and turbine trains with capacities in excess of 1.8 MMTA have already been designed, manufactured, and are successfully operating in existing plants.

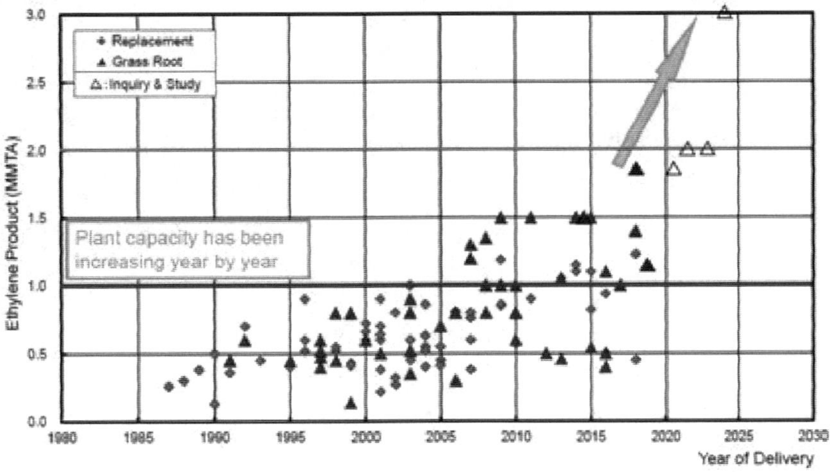

Figure.1 Ethylene Plant Capacity Increase Supported by Upgraded Equipment and Technology

Figure 2 shows the photo for recent Cracked Gas compressor drive steam turbine for Mega ethylene crackers. This steam turbine has more than 70MW output power. And, there are dual trip throttle valves for large HP steam flow.

Figure.2 Steam Turbine for Mega Crackers (1.3MMTA)

Every compressor's performance, including rotor dynamics, is measured and verified as a part of the manufacturing process. In addition, the steam turbine and compressors can be coupled and the complete train is tested to confirm combined mechanical performance, such as vibration and bearing condition, on a large test stand prior to shipment. Figure 3 shows a 1.5MMTA class mega ethylene CGC train in the same cooperation condition as the site. Through these processes the guarantee of high quality machines is provided to the clients.

Figure.3 Steam Turbine for Mega Crackers (1.5MMTA)

Challenging Technologies for Critical Components

To produce the increase power the required inlet steam flow (HP section) and exhaust steam flow (LP section) become larger and a larger component size for the flow path is necessary. Figure.4 shows key technologies to meet the requirement for large plant and machines which are used in steam turbine design. Optimized impulse blade is applied for the inlet and middle stage flow paths in order to increase efficiency and minimize the thrust load due to differential pressure across each blade inlet and outlet.

The last 3 stages in the LP section utilize a z-lock integral shroud and high rigidity blade profiles in the basis of un-tuned design are applied to reduce resonance vibration stresses in high variable speed and to increase stage efficiencies.

A critical design concept in modern rotor design is to increase rigidity as much as possible by optimizing blade stage number, bearing span, and HP/LP section efficiency at the required operating speed, in addition to applying high dynamic and static performance of larger sized journal bearing. A high rigidity rotor can reduce vibration response due to unbalance and increase vibration stability. Also, for large size journal bearings, self-aligning bearing pad to support rotor is applied to get the high dynamic and static performance.

Recently, operating intervals of over 8 to 10 years have been required to meet customer demand. For this requirement, several new technologies have been developed and applied to existing machines.

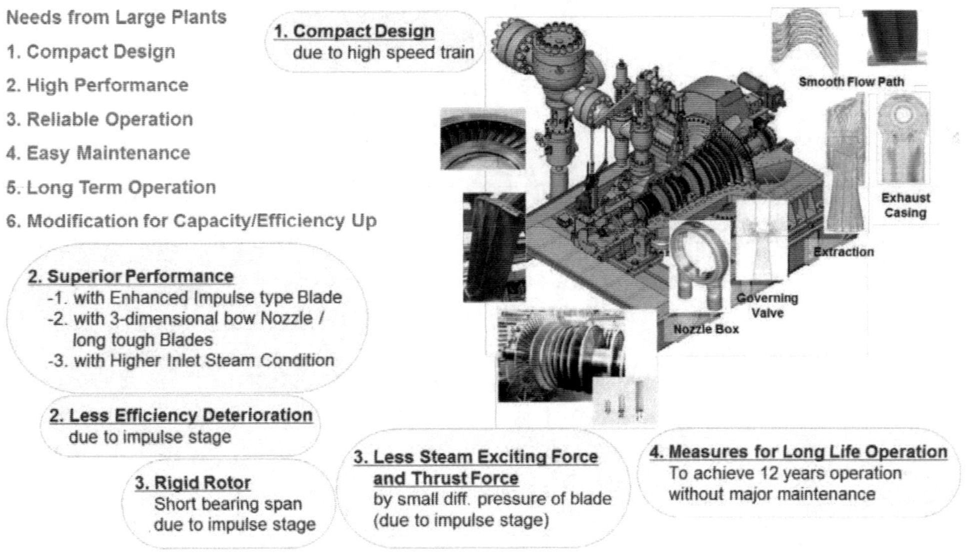

Figure.4 Steam Turbine Key Technologies

Figure 5 shows the design concept of 140MW class mechanical steam turbine for 3.0MMTA ethylene plants. In this paper, this concept of five critical components (HP section, MP&LP section, LP stage blade block, Blading (for intermediate stage), and Rotor, is introduced. For HP section, trip valve, governing valves etc., well proven application from large power generator steam turbine technology are applied for large HP flow, high-pressure and high-temperature. For MP and LP section, core parts such as extraction control valves, LP casing, etc., well proven application from mechanical drive steam turbine technology are applied for large LP flow and high exhaust vacuum. For LP stage blade block, un-tuned blades are applied in variable high speed which was already verified in validation test and field operation. For all stage blades, endless group structure with high efficiency profile is applied to achieve high reliability and high performance in long-term operation. For turbine rotor, advanced high temperature creep and high toughness material is applied for high steam temperature condition and keeping high rigid rotor structure. This rotor was successfully operated in many filed service. Except the above main components, there are various other turbine parts (Journal and Thrust Bearing, Turning device, etc.). The various other turbine parts for 140MW class mechanical steam turbine can be covered with the current leading edge technologies.

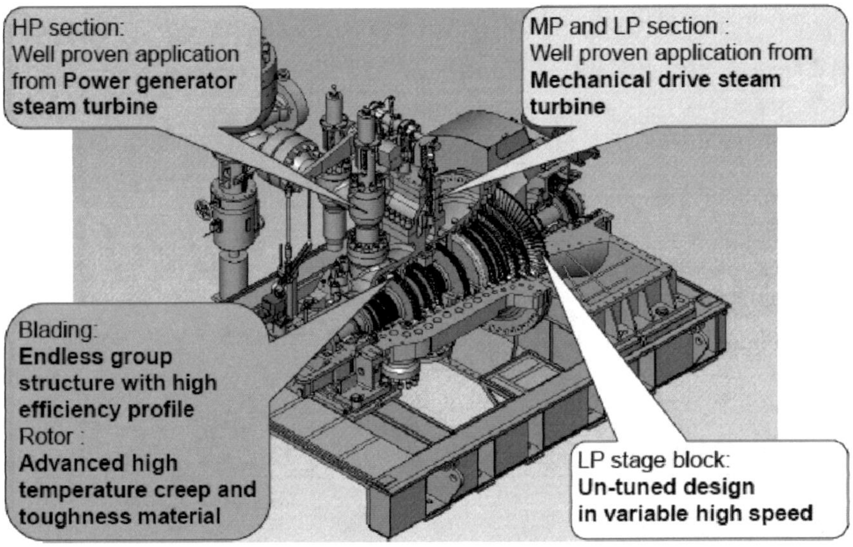

Figure.5 Design concept of 140MW Class Mechanical drive steam turbine

In promoting the design more detail, turbine operating condition for 140MW class mechanical drive steam turbine of cracked gas compressor is settled based on current experience as case study shown in Table 1. Actually, turbine output power and operating speed are required from compressor demand. On the other hand, steam balance for 3.0MMTA ethylene plants which process licensers configure in considering other trains (propylene ref. train and ethylene ref. train) should be considered. For cracked gas compressor drive steam turbine, the demand of extraction steam is highly depended on the whole arrangement of steam balance with core trains. If inlet steam pressure and temperature for HP section are higher, steam consumption may be relaxed. And, if exhaust pressure for LP section is higher vacuum, the thermodynamic efficiency will be increased. However, the condensate system design including the design of surface condenser should be considered for large LP flow and high vacuum. Then, the demand of maximum extraction flow is fixed by user utility in whole steam header system.

Table.1 Case Study - Mega CGC Steam Turbine Operating

Turbine Operating Condition	
• Turbine Max. Power	139000 kW
• Turbine Nor. Speed	3000 rpm
• Inlet Nor. Steam Press.	119 Barg
• Inlet Nor. Steam Temp.	520℃
• Max. Inlet Flow	1100 t/h
• Ext. Press.	40 BarA
• Ext. Max. Flow	900 t/h
• Exh. Nor. Press.	0.15 BarA
• Exh. Max. Flow	355 t/h

The rotor materials for critical components can be upgraded in high temperature applications. Figure 6 shows the upgraded materials of turbine rotor in high temperature applications as one example. This is an advanced rotor material which is called 10325MTE. This rotor material is made of Ni-2.25Cr-Mo-V forged steel. Maximum working temperature for this rotor in HP section is 510 Deg-C. And, Low temperature toughness in LP section is $FATT_{50}$ around 40 Deg-C. This advanced rotor performance is 30% up for high temperature creep strength and 10 time up for life time compared with conventional rotor.

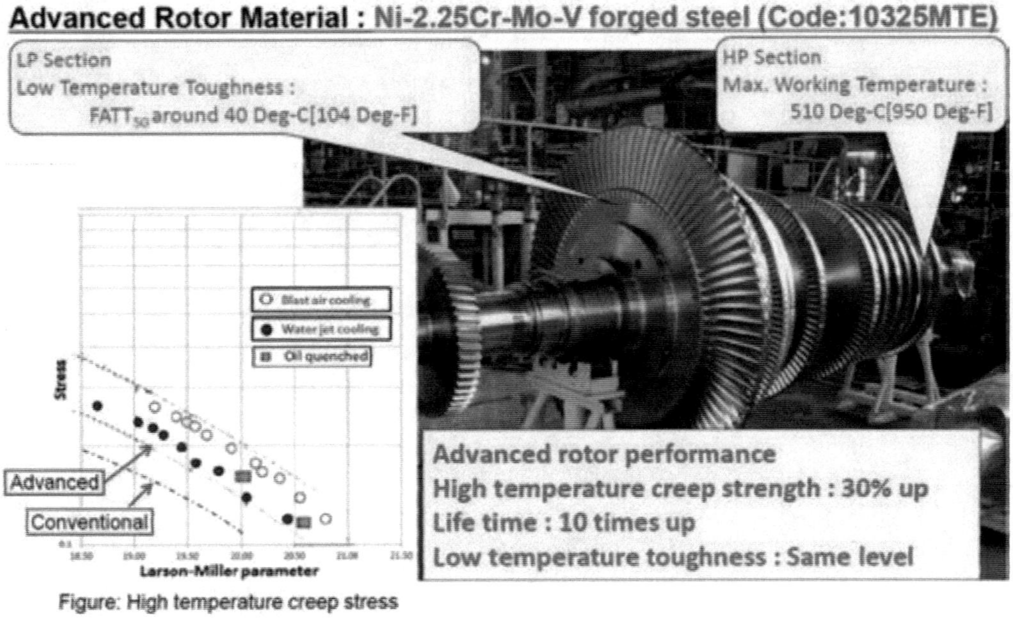

Figure.6 Large Steam Turbine Rotor Material Upgrades for High Temperature

Steam turbine casing is one of critical components, this casing thermal deformation with stress should be evaluated quantitatively in order to avoid any contact with rotor in operation including start-up. Also, the reliable seal performance is required to prevent any steam leakage in long-term operation. Finite Element Analysis (FEA) is one of design verification tools to check a soundness of steam turbine casing from the viewpoints of thermal deformation with stress and seal performance. Figure 7 shows the typical of full 3D model FEA of steam turbine casing with analysis results.

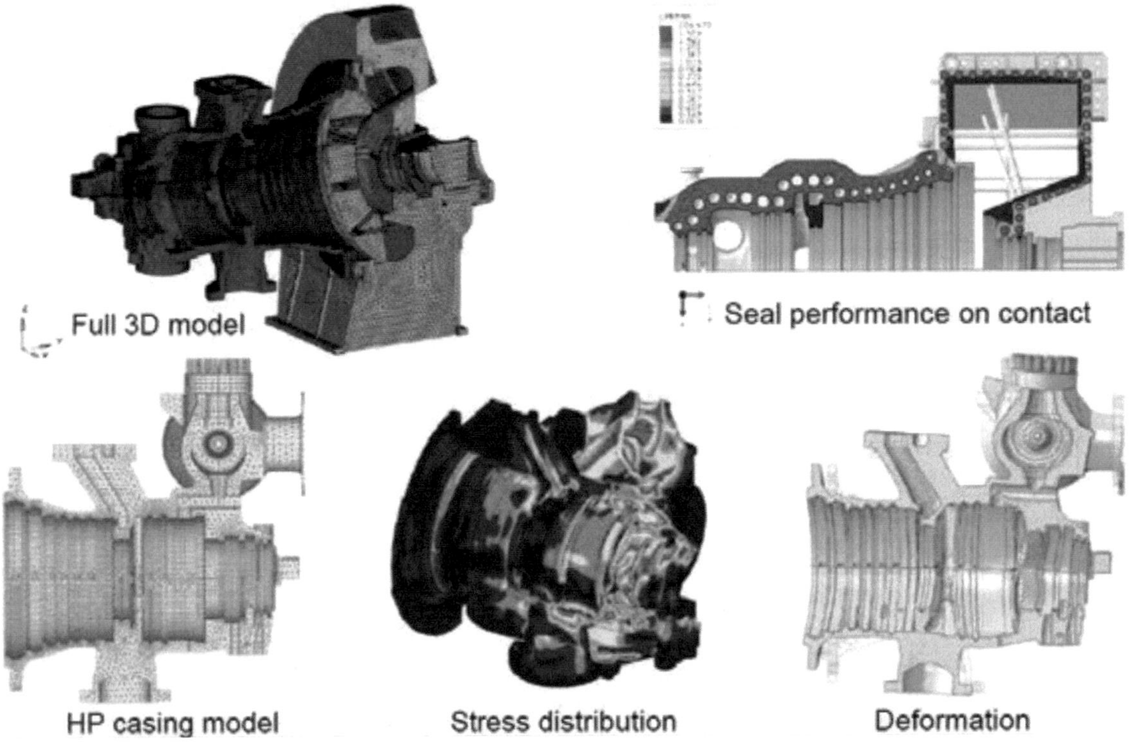

Figure.7 Full 3D FEA of Steam Turbine Casing with analysis results

The last stage blade for Mega cracked gas steam turbine is very important. Figure 8 shows last stage blade for large power steam turbine. The last stage blade height 500mm is required to deal with large exhaust flow. And, the last blade is in un-tuned design for wide operating ranges. The last stage blade is already applied for large power steam turbines and is successfully operated in existing plants.

Figure.8 Rotor and last stage blade for Mega CGC steam turbine

Figure 9 shows the applicable technologies for long term operation. Key technologies for steam turbine reliability improvement such as special blade coatings have been developed. In addition, applications of stainless steel materials to protect against drain erosion on the diaphragms, and the development of hybrid coatings of fluorine resin and ceramic have been utilized. One of the essential factors for long-term operation is coating technology. Regarding the deterioration phenomenon of internal parts, the two major causes are solid particle erosion and drain erosion. In the case of solid particle erosion, boronizing heat treatments are applied for high pressure stage nozzles, particularly the 1st stage nozzle. On the other hand, in case of drain erosion, Cr-TiN coatings and PTA welding are applied. In addition, On-line washing system is applied to remove fouling due to accumulation of foreign particle for low pressure stage.

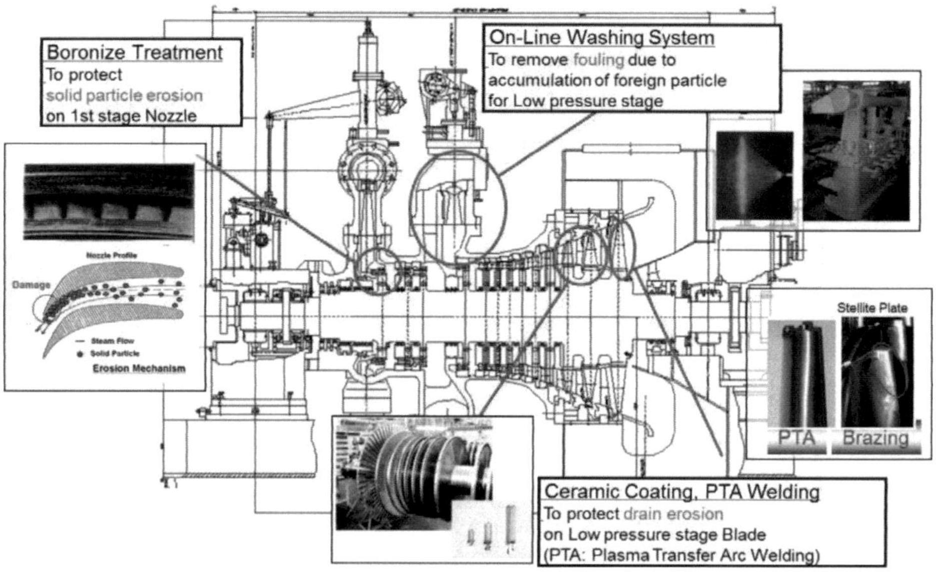

Figure.9 Applicable Technologies for Long Term Operation

Conclusion

1. MCO has finished the basic design of a 140MW class mechanical drive steam turbine based on our well experienced and proven components.

2. Large power variable speed turbine can be provided from synergy design with our power generation and mechanical drive steam turbine technologies.

3. For more efficient and reliable machines during long-term operation, the key technologies of advanced rotor material, on-line washing and anti-erosion prevention are introduced.

Reference

Satoshi Hata, et al, 2016, ATPS-Tutorial 9 _" Technical Challenges for Compressor and Steam Turbine for Efficient and Sustainable Operation in Mega Ethylene Plants"

AIChE Paper Number: T4 550803

Refrigeration Compressor Dry Gas Seal Panel Vent Header Design Consideration

Tim Dwyer
Machinery Engineer
LyondellBasell Industries

Andrew Herzig
Process Engineer
LyondellBasell Industries

Brian Kalfrin
Sr. Regional Engineering Manager
John Crane

Nick Hank
National Sales Director
Advanced Rupture Disk Technology, Inc.

Prepared for Presentation at the 2019 Spring National Meeting
New Orleans, Louisiana, April, 01, 2019

Refrigeration Compressor Dry Gas Seal Panel Vent Header Design Consideration

Tim Dwyer
Machinery Engineer
LyondellBasell Industries
Andrew Herzig
Process Engineer
LyondellBasell Industries
Brian Kalfrin
Sr. Regional Engineering Manager
John Crane
Nick Hank
National Sales Director
Advanced Rupture Disk Technology, Inc.

Abstract:

An ethylene plant recently underwent a debottleneck which included a new gas seal package for the refrigeration compressors. As is typical, the primary vent (normally >95% nitrogen) was routed to the flare header. Issues with potential back pressure during heavy flaring events were identified early in the design process, so the decision was made to route a bypass to atmosphere with a rupture disk used for containment. During the subsequent plant startup, high backpressure occurred but the rupture disk did not rupture per design. Inspection revealed a partially burst disk that held backpressure yet was unable to fully burst. Investigation and modeling revealed a unique design requirement for low pressure rupture disks. A modified rupture disk configuration, testing of the configuration and the final solution for the dry gas seal panel will be reviewed. We will also discuss the hazards associated with gas seal pressure reversal and methods of monitoring.

Refrigeration Compressor Dry Gas Seal Panel Vent Header Design Consideration

In December of 2016, an ethylene plant started up after undergoing a debottleneck that increased capacity by about 50%. To achieve this, the site installed two new refrigeration compressors with modern gas seals - provided by a well-known mechanical seal supplier. The design of these seals included a primary vent on the non-process side of the seal face, that for environmental and safety reasons was routed to a flare. Note: as long as the seal is healthy, the composition of the primary vent is >95% nitrogen at a very low flowrate. To maintain the seal closing pressure, the primary vent pressure needs to remain below the pressure in the seal chamber (process side of the seal face). Late in the design phase, a potential for minor flare header back pressure (during process upset conditions) was identified. Since the seal chamber pressure is low (driven by the compressor/process design), reverse pressure became a concern. To counter this, the seal supplier was asked to provide a relief system to prevent over pressure, thus preventing reverse pressurization of the seal. The system selected used 1" rupture disks under 1" relief valves, tied to an atmospheric vent (see Figure 1).

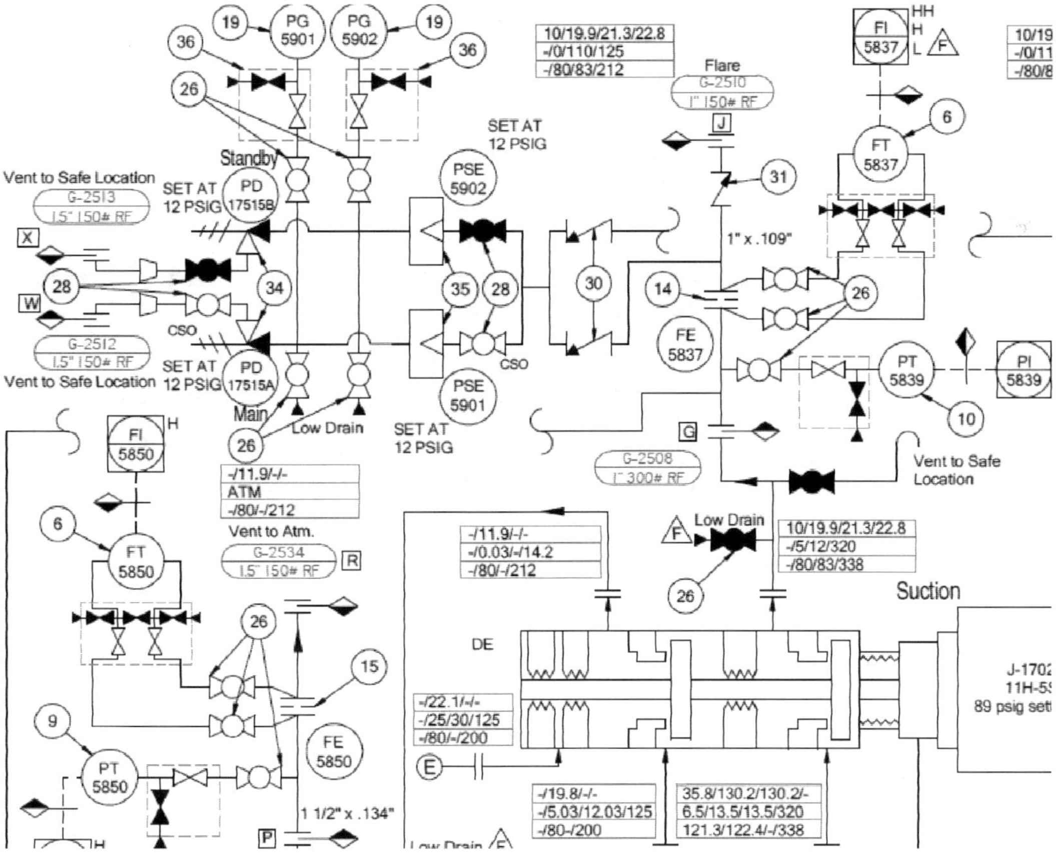

Figure 1: Seal Supplier P&ID

During the project execution, it was discovered that the relief valve inlet line pressure losses were excessive. If the relief valve had been internally designed and purchased, internal requirements would have identified this earlier in the project; but since it was provided as part of a package unit, this was not identified. There was limited PSV sizing documentation provided, so it was evaluated internally once it was discovered. The recommended PSV inlet line losses are 3%. This system, as installed, was over 20%. High inlet losses can cause rapid cycling (chattering) of the PSV, which can cause failure of the relief valve. If the relief valve fails in the closed position, it will not protect the seals.

Note that this application is outside of the scope covered by ASME Section VIII rules for at least two reasons: 1) A pressure vessel is not being protected, only the seals are. 2) The set pressure is below 15 psig. However, the design considerations for relief valve operating stability, internal company requirements, and other industry standards/recommendations still apply.

Based on this evaluation, the quickest and simplest option mid-construction was to remove the relief valves, and use the rupture discs alone for relief protection. Rupture discs are not negatively affected by high inlet line losses since they do not re-close. The drawback is that the rupture disc will continue to vent to atmosphere until it is replaced. In this process, this trade-off was deemed acceptable. In December of 2016, the relief system that went into service is shown in Figure 1 (minus the relief valve).

Reverse Pressure in Dry Gas Seals

Dry gas seals are much like conventional mechanical seals with some slight differences, primarily that they utilize special face features to create hydrodynamic lift. With these features (bidirectional grooves in this case), hydrodynamic pressure is self-generated where the pumping and compression action is created by the sealing gas entering the rotating grooves. The design of the bidirectional groove results in a hydrodynamic pressure increase, which in turn increases the operating film thickness and film stiffness of the seal interface gap. This hydrodynamic pressure maintains a non-contacting sealing interface by effectively balancing the remaining closing forces acting on the seal faces. The advantage of such a design is that it results in low heat generation, power consumption, and longer life without the need for external liquid lubrication. The pressure gradient created across the seal faces, along with positive sealing gas pressure, result in a separation between the seal faces of 0.0001" to 0.0002".

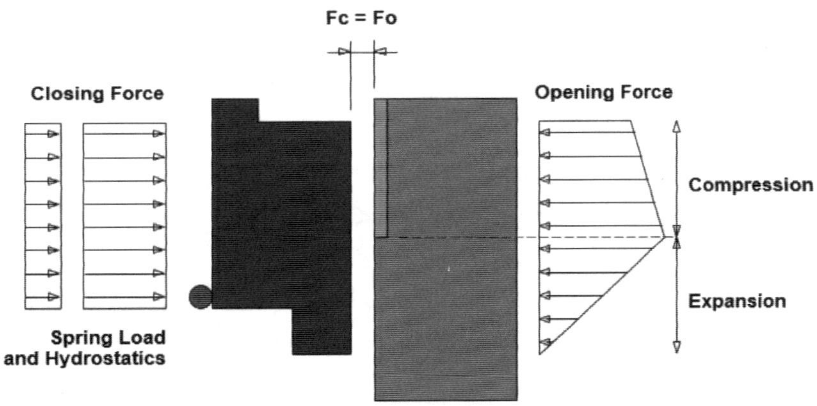

Figure 2: Dry Gas Seal Design Principle

Dry gas seals (much like most conventional mechanical seals), are designed strictly for OD pressure differentials (process pressure > vent pressure). Figure 3 below shows a pair of generic seal faces designed for OD pressure service, where PO > PI. The outer diameter of the seal faces are exposed to process pressure, or seal gas pressure in the case of a dry gas seal. This pressure exists along the outer diameters of the faces, down to the primary seal face's secondary sealing element at the shaft / sleeve diameter and up to the mating ring's secondary sealing element. Process pressure keeps the faces together and closing force can be tailored with the design of the seal and hydraulic face balance. In the example, since the mating ring's sealing diameter is greater than the primary ring sealing diameter, the net force keeps the mating ring against its support surface. The primary ring secondary sealing element is pushed against the forward edge of the ring counter-bore so that the hydraulic force reacts from the primary ring face outer diameter down to the shaft / sleeve diameter. The axial forces above the primary ring face OD are equal and opposite on the front and back surfaces of the ring.

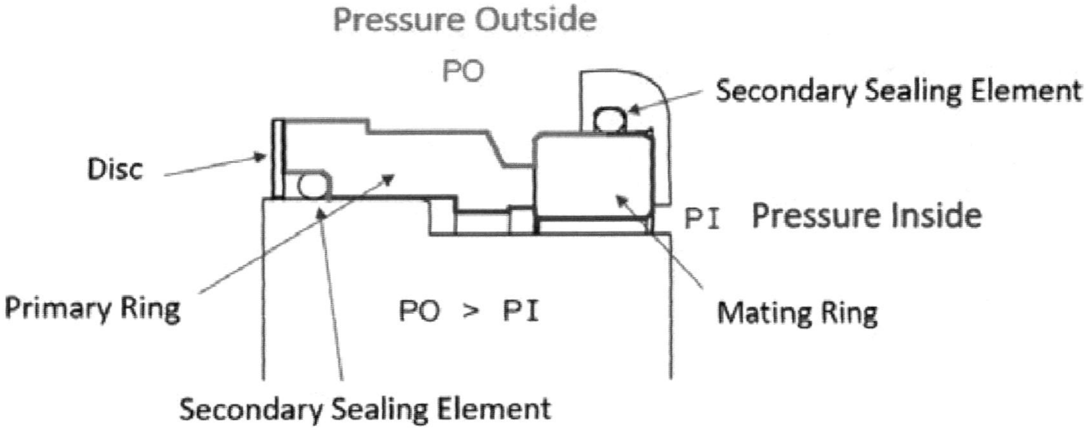

Figure 3: OD Pressure Distribution

In the event where a seal designed for OD pressure is subjected to an ID pressure differential, problems can arise. Figure 4 illustrates a generic seal designed for OD pressure in a scenario where ID pressure is increased, or PI > PO. The first issue is that since the mating ring secondary sealing element diameter is larger than the primary ring secondary sealing element diameter, the reaction force from the pressure behind the mating ring will tend to overcome the spring load and the mating ring secondary seal friction to eventually unseat the mating ring. The second issue is that with ID pressure, the primary ring balance diameter shifts to the secondary seal counter-bore diameter forcing secondary sealing element towards the disc and unloading the spring force. If the pressure is high enough it can force the secondary sealing element out of the counter-bore unless some form of retention is employed. Finally, the force ratio must be examined to check the primary ring's suitability to dynamic ID pressure conditions.

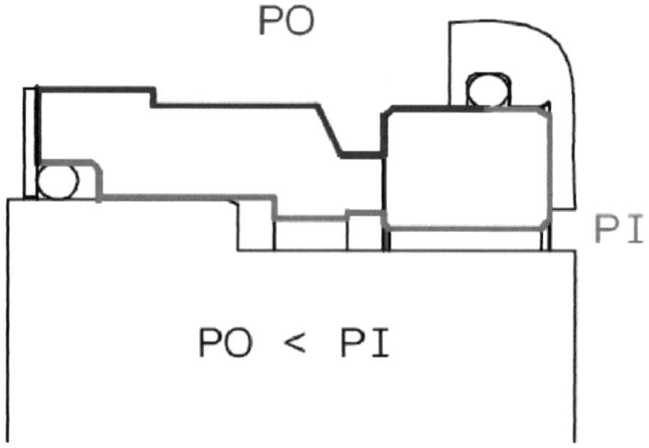

Figure 4: ID Pressure Distribution

The seal supplier has extensive experience with dry gas seals and understands the potential concerns with reverse pressure conditions applied to dry gas seals. As highlighted in the proceedings from the 25th Turbomachinery Symposium, one of the most challenging scenarios for a dry gas seal is operation during a prolonged, even small pressure reversal. In such a situation, lift off may not occur and the seal faces may begin to operate in a contacting mode, ultimately resulting in seal failure. If reverse pressure differentials are significant (typically >7psi but depends on the design of the seal), the seal faces can be forced open (Mayeux / Feltman, 1996). Dry gas seals have been successfully designed to accommodate minor pressure reversals through evaluation of bidirectional groove design using advanced computer modelling in addition to optimizing spring loads. In general, to achieve optimum reliability the support system and control logic should be designed and maintained to ensure the dry gas seal is not exposed to pressure reversal excursions.

Operating Issues and Investigation

After the debottlenecking outage, the end user began startup efforts which included significant flaring for clearing purposes. This flaring, as predicted, did result in back pressure in some of the smaller flare branch connections, including the one tied to the primary vent for the gas seals. Unfortunately, the rupture disk system failed to provide adequate overpressure protection, and the seal did indeed reverse pressurize. Due to the design of the seal and nature of the reversal, it was able to recover, but as discussed above reverse pressure is not a part of normal design.

Following the reverse pressure event, the rupture disks were pulled for inspection which showed incomplete rupture (see Figure 5). It appears that the disk attempted to rupture but did not have sufficient volume to completely tear the disk. The result was a small perforation that allowed the pressure upstream of the disk to continue to rise while not completely rupturing the disk due to the now present downstream pressure.

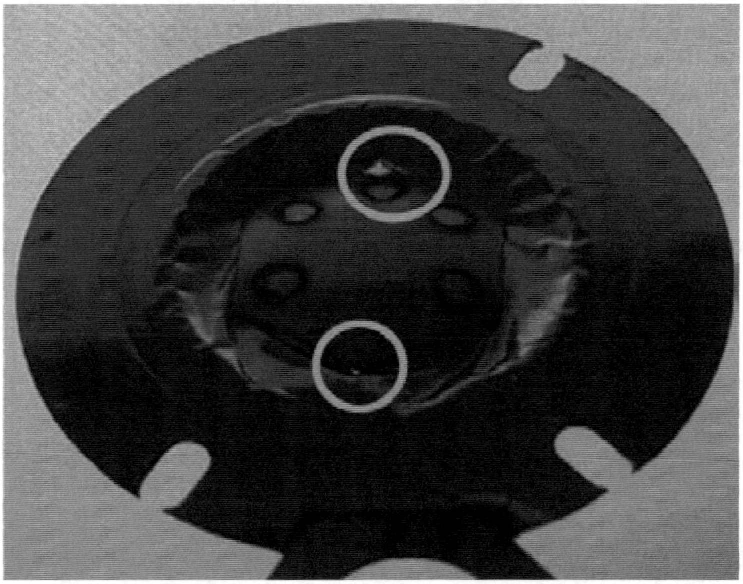

Figure 5: Original service rupture disk with pinholes but incomplete rupture

The disks employed for this application were precision, low pressure, reverse buckling style rupture disks with an operating ratio of 90%. They are designed to operate with extremely low burst pressures while maintaining an all-metal design. Initial testing of the 1" disks showed that they functioned accurately under factory test parameters.

While simple in concept (thin membrane that breaks at a designed point) rupture disks have specific application conditions that vary across a variety of disk designs. In our case, the disk needed to operate below 90% of set pressure and needed reserve volume to violently tear the disk in order to completely open, and that is how the factory acceptance test is designed. If ruptured completely, the pressure drop across the disk will be relatively low. Our application, in contrast, had very little reserve volume coupled with a slow pressure rise and restricted outlet. Since it was believed that the primary reason for the incomplete rupture (and subsequent back pressure) was the lack of reserve volume, the end user worked with the rupture disk supplier to test the initially installed design (and a proposed solution) under test stand conditions that simulated the actual service conditions.

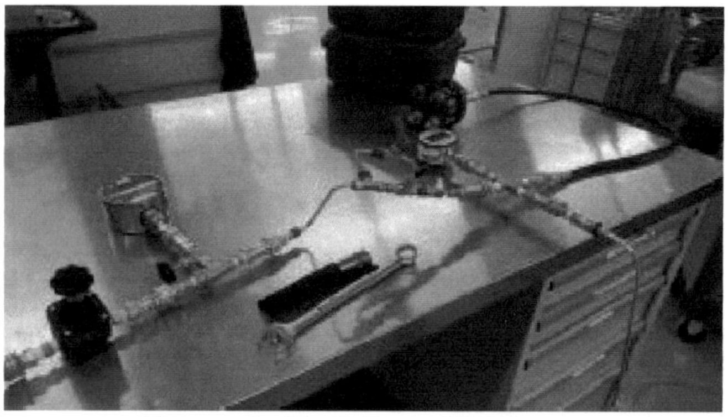

Figure 6: Test Stand Apparatus

Figure 7: Test Stand Disk Results From Initial Design Configuration

These tests used a bench top apparatus to simulate pressure rise and reserve volume (See Figure 6). This testing was able to replicate the field results (See Figure 7) and provided a method of evaluating future solutions using rupture disks.

Once the rupture disk supplier established the test parameters they worked with the end user to design a solution that would meet the unique application. Minimal information was available on how large the volumes should be, so a prototype was constructed and tested. It was agreed that a 6" disk and accumulation chamber with a volume ratio of 2 to 1 would likely increase the opening consistency of a compression loaded rupture disk in what is a quite difficult rupture disk application: low volume+ low set pressure+ slow buildup of pressure+ restricted outlet.

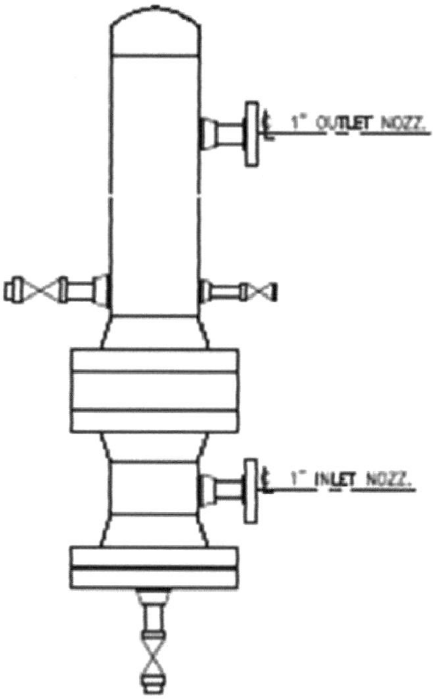

Figure 8 Rupture disk volume chamber

The end user then fabricated a sample of the proposed volume chamber (See Figure 8) and sent it to the rupture disk supplier for testing.

Testing at rupture disk suppliers manufacturing facility was performed using the fabricated accumulation chamber, a pressure rise procedure simulating past relief scenarios (rise rate of ~ .25 psig/min), and two models of 6" disks: one was the original all metal compression loaded or reverse acting disk, and one was a tension loaded or forward acting composite disk. Both designs were set at 12.75 psig @ 72 Deg. F. During the test timeframe, the final outlet piping configuration had not been fully designed. To prove performance under worst-case piping configuration conditions, the testing was performed with the outlet of the rupture disc chamber blocked. This would confirm that the outlet volume provided was sufficient, regardless of downstream restrictions.

The results (Figures 9) eliminated concerns of post rupture pressure rise and proved out the end user's design concept. Ultimately, the original all metal reverse buckling design (benefiting from the increased size and volume) had more consistent burst characteristics, and was selected for final installation.

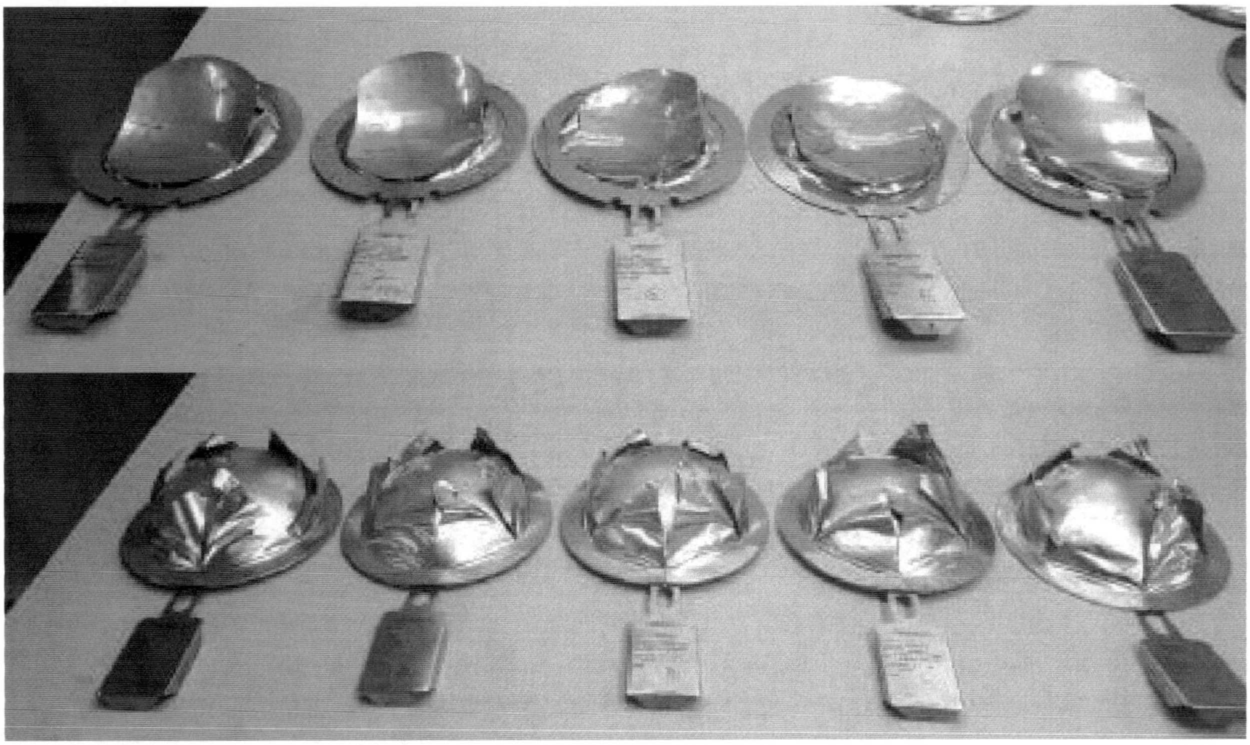

Figures 9: Tested disks, Top- reverse buckling, Bottom- tension loaded

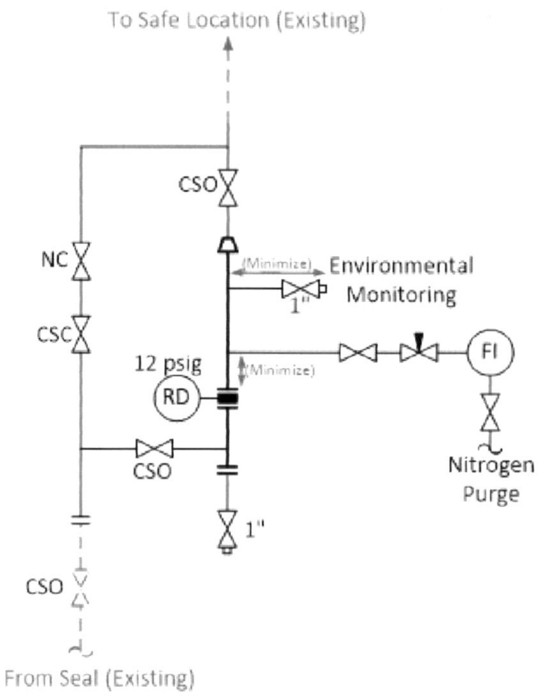

Figure 10: System Design

System Design and Results

The final rupture disk system was constructed as shown in Figure 10. The only available tie points for adding overpressure protection were the existing manual vents to atmosphere. These vent lines do not have check valves, eliminating the concern that they could fail closed and violate the purpose of the rupture disk. A bypass line was included around each rupture disc chamber with two block valves in series accessible from grade or existing platforms. This preserves the existing ability to manually vent these streams, if necessary. Rupture disc chambers were constructed out of 6" diameter pipe to allow for the use of the larger diameter rupture discs. The chamber also provides sufficient upstream and downstream volume (confirmed by the testing described above) to ensure a complete burst when required. Downstream volume was considered in this case because the outlet of the rupture disc chamber reduced to 3/4" and 1" piping, matching the initial vent sizes. Utilizing these small diameter lines allowed for the existing vent piping to be used which exhausts to a safe location above the roofline of the compressor building. A 1" environmental monitoring connection was included downstream of the rupture disc for required periodic monitoring to confirm that the rupture disc is not leaking. A small nitrogen purge was included downstream of the rupture disc. This reduces the risk of backflow of oxygen into the flare system after the rupture disc bursts and prevents false environmental monitoring results by purging the stagnant area downstream of the rupture disc. A full-size flange was provided at the base of the upstream section of the rupture disc chamber to allow for quick removal of rupture disc debris, if any. The chosen rupture disc is non-fragmenting under normal conditions. All block valves upstream and downstream of the rupture disc chamber are car sealed open (CSO) or closed (CSC) and controlled by operating procedure. This is to avoid valve misalignment which could isolate or defeat the relief protection.

While the rupture disc system as described above is permanent, one item being considered for the next maintenance turnaround is to move the normal vent to a less restricted flare location. The flare backpressure can be minimized by connecting the vent to flare directly into the main header, instead of the small-diameter liquid drain lateral. This would reduce (or perhaps eliminate) instances where the rupture discs activate.

Shortly after installation of the above described system, a process upset resulted in high vent back pressure. The new rupture disk system function exactly per design and prevented a seal pressure reversal.

Figure 11: Post Installation Rupture

References:

1. Mayeux, T.P., and Feltman, P., 1996, "Design Improvements Enhance Dry Gas Seal's Ability to Handle Reverse Pressurization," *Proceedings of the Twenty-Fifth*

Turbomachinery Symposium, Turbomachinery Laboratory, Texas A&M University, College Station, Texas

Figures 1-4 courtesy of John Crane.

*** Formatting Issue - Best Available Paper/Graphic ***

AIChE Paper Number: T4 550804

Ethylene Product Pump Mechanical Seal Upgrade Lessons Learned

Tim Dwyer
Machinery Engineer
LyondellBasell Industries

John Choma
Machinery Engineer
LyondellBasell Industries

Brian Kalfrin
Sr. Regional Engineering Manager
John Crane

Prepared for Presentation at the 2019 Spring National Meeting
New Orleans, Louisiana, April, 01, 2019

*** Formatting Issue - Best Available Paper/Graphic ***

Ethylene Product Pump Mechanical Seal Upgrade Lessons Learned

Tim Dwyer
Machinery Engineer
LyondellBasell Industries
John Choma
Machinery Engineer
LyondellBasell Industries
Brian Kalfrin
Sr. Regional Engineering Manager
John Crane

Abstract:

An ethylene plant experienced an increased number of seal failures during an API Plan 11/72/76 seal upgrade (originally an API Plan 11/52) on their ethylene product pump. These failures were originally attributed to contamination in the seal flush that built up inside the seal. Ultimately two sources of contamination were identified: 1) iron oxide that resulted from a lengthy outage with equipment open to atmosphere, and 2) internally generated polymer from local seal face hot spots promoting the ethylene reaction. Additionally, failures were also experience in the secondary seal associated with gradual loss of back pressure and ultimately process leakage. Several modifications were made on an expedited basis with the help of the seal supplier. The modifications that were made to solve the failures will be discussed such that the end results can be used as a template or shortcut for future applications of a similar nature.

***** Formatting Issue - Best Available Paper/Graphic *****

Ethylene Product Pump Mechanical Seal Upgrade – Lessons Learned

The end user is a world scale producer of ethylene, with a peak production of approximately 300M lbs./hr. To provide ethylene to market, the facility currently operates a 2 out of 3 parallel pumping arrangement, using 13 stage Pacific (now Flowserve) IJ pipeline pumps. These pumps take liquid ethylene at -17 degrees F. from 330 psi to over 2500 psi. References to these three pumps will be made using the ID numbers J-1732 (A/B/C). The J-1732 pump drivers are 1,500 HP motors that turn at 1,800 RPM. These motors are coupled to speed increasing gearboxes with a ratio of 2.9 to 1. The final output speed is a nominal 5,200 rpm. Due primarily to the rotational speed, there were no other pump applications within the end users enterprise that would be considered technically equivalent applications for comparison.

Figure 1: J-1732 C Multistage Barrel Pump

Application History

The IJ line of multistage barrel pumps (see Figure 1) were a common API configuration in the 70s & 80s. They used a slender shaft arrangement that relies on the hydrodynamic stabilization at the inter-stage shaft bushings and impeller wear rings. Modern designs (like the Flowserve WIK) will typically employ a stiff shaft design, where the clearance in the inter-stage bushings/wear rings is less critical. The J-1732 pumps were upgraded with composite material at the wear ring and shaft bushings in 2008. This allowed the running clearances to be reduced, improving the shaft stability and pump

Figure 2: API Plan 11/52 Compnent Seal

***** Formatting Issue - Best Available Paper/Graphic *****

Figure 3: Black Residue

efficiency.

Originally the J-1732 pumps were equipped with non-cartridge component seals, and API 11/52 piping plan (unpressurized dual seal) using methanol as a barrier fluid (Figure 2). These seals required frequent manual intervention (refilling the methanol barrier fluid reservoirs), suffered from high infant mortality, and had an MTBR that was less than optimal (3 years - driven by secondary seal leakage). Additionally, the seals required a long break-in period (1-2 weeks) where the leak rate (measured by the secondary seal pressure on the vent to flare) would cycle up and down. Data for this event is limited as the only instrumentation on this seal design was a pressure switch on the secondary seal vent that alarmed at 25psi and rising. Failure analysis of these seals showed small quantities of what was believed to be residue (black particles) from the seal carbon ring (Figure 3). It was thought that this residue resulted from the break in period, and then remained suspended in the methanol barrier fluid and accumulated in the seal pot.

A cartridge API Plan 11/72/76 was selected as the preferred upgrade (Figure 4) to eliminate the issues associated with API Plan 11/52. This new seal arrangement uses a similar process fluid API Plan 11 primary seal flush, and a gas (lift off) secondary seal with nitrogen used as a purge gas. This arrangement has been used successfully across industry where the pumped liquid is vapor at ambient conditions, and in very clean service. Since this is the ethylene product stream, it was presumed to be very clean, and would also be in the vapor state at ambient conditions.

Several improvements were added as part of this upgrade: 1) moving the primary ring and springs to the stationary portion of the seal, 2) adding composite

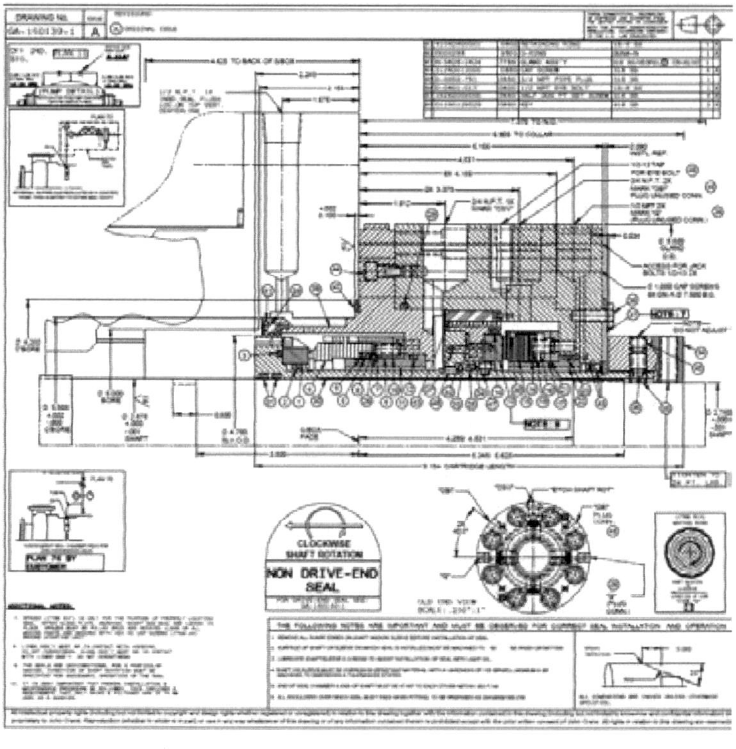

Figure 4: Original API Plan 11/72/76 Cartridge Seal

material in the close clearance bushing areas, and 3) additional instrumentation - which allows monitoring both the upstream and downstream sides of the purge gas. These improvements coupled (with the inherent benefits of a cartridge seal being supplier assembled and bench tested) made the expectations high that these new seals would provide consistent performance upon installation and an increased service life. After installation, (and contrary to expectations) these seals experienced a high failure rate (3-6 month MTBR).

Service history

The first mechanical seal failure occurred several months after the initial upgrade, and a total of seven seal repairs have been performed to date (since December 2016) for these three pumps. The damage found during these repairs had three main symptoms: 1) heavy contamination, 2) high heat generation, and 3) low/unstable containment pressure profile.

The most immediately visible symptom was solids packed around both the primary and secondary seals. It was determined that these solids, found around the dynamic o-ring position, were causing hang-up of the mechanical seal faces. Elemental analysis of the solids confirmed a high amount of organic components, with a significant amount of iron oxide also present. While that analysis suggested some ethylene polymerizing, ethylene polymerization requires heat/pressure which the bulk conditions (-17Deg F & 330psi) would not support. Looking at the seal at the micro level, it is possible for very localized areas on the seal face to develop temperatures that could support ethylene reaction. Over time, solids from this reaction could build-up in the areas around the seal that were considered stagnant (such as

Figure 5: Solids on Primary Seal Ring (Vent Side)

Figure 6: Fouling Under the Primary Seal O-Ring

the inner or atmospheric side of the primary seal).

The second symptom was heat related damage to the primary seal faces. The indications of this were straightforward, and included grooving, pitting, deposits, and both ID and OD contact. One of the mating rings also experienced thermal shock damage (which often indicates trapped vapor pockets around the seal faces).

The third primary symptom was low/unstable pressure at the secondary (containment) seal. The containment seal is a lift off (gas) type seal. These seal types are very effective at high pressure, but have diminishing dynamic response as the closing pressure approaches zero. The initial design of the secondary seal provided effective response to primary seal failures, but experienced difficulty managing normal leakage where pressures were very low (<5psi). Failures in this mode presented as fugitive emissions, with the primary seal functioning well (low/normal vent pressure).

While the path forward was iterative, generally speaking the corrective actions followed the three primary symptoms: 1) improve the ability of the seal to manage contamination, 2) improve the thermal properties of the primary seal, and 3) provide the containment seal with sufficient normal closing force without adversely affecting its ability to deal with primary seal failures.

Figure 7: Primary Seal Face Thermal

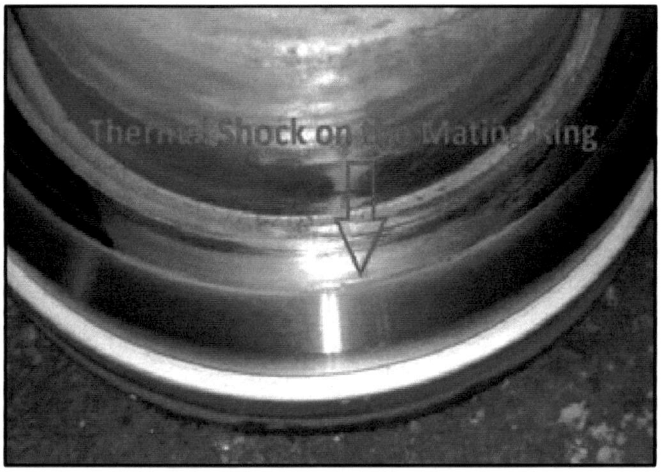

Figure 8 : Primary Seal Mating Ring Thermal Shock

Improved Inner Seal Geometry

Several initial failures of the revised seal design highlighted concerns in the areas of seal face lubrication and heat generation. Typical primary seal face wear patterns were indicative of high wear and insufficient cooling. Damage consisted of deep grooving of both the antimony carbon stationary seal face, and a fine phonographic finish on the tungsten carbide mating ring surface.

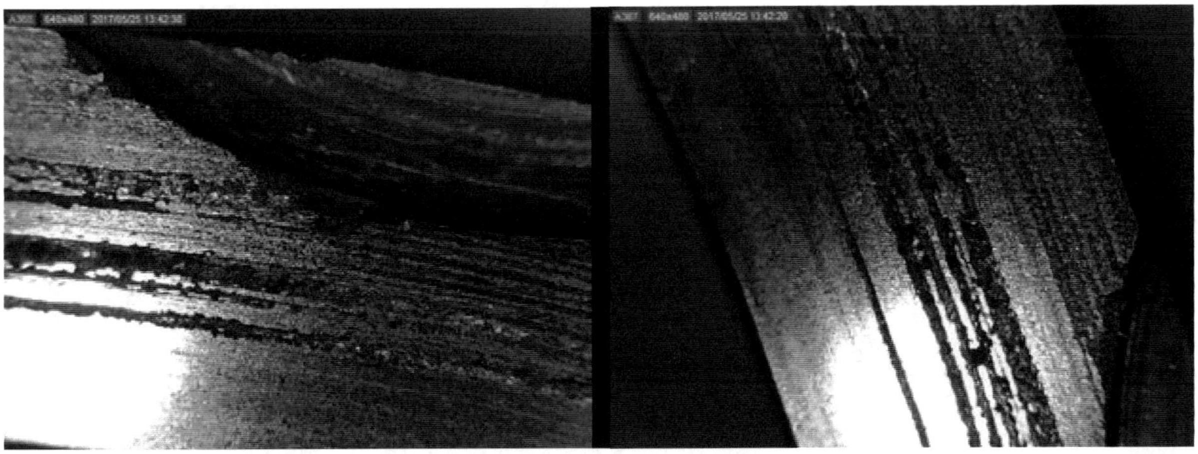

Figure 9: Carbon Seal Face Damage

The inner seal face inspection highlighted that there was excessive face contact, more heavily concentrated at the seal face ID, but also prominent in the OD regions of the face as well. This damage indicated that there was inadequate fluid film formation in the sealing gap (between the two faces). A sealing interface is subject to a closing force, which is the sum of the hydraulic pressure and the spring force. This is resisted by an arithmetically equivalent opening force, which is the sum of the hydrostatic pressure (of fluid in the interface), hydrodynamic pressure (caused by rotation), and contact pressure (usually just at the asperities). If the fluid film is not formed adequately, then the contact pressure must increase to provide equilibrium - and this leads to a rapid increase in heat generation, vaporization and material degradation.

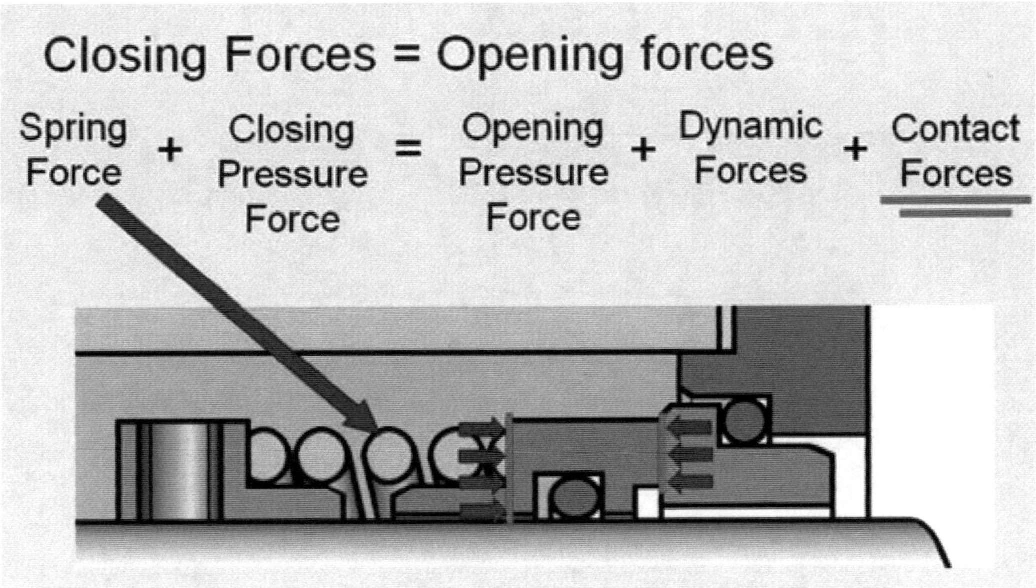

Figure 10: Seal Equilibrium Equation

It is not uncommon for a mechanical seal operating in this service to experience face wear of some degree, based on the limited amount of hydrodynamic load support generated by the low viscosity process fluid. However, the severity and concentration of the wear patterns was significant enough to warrant additional review. A useful tool in evaluating mechanical seal performance post failure is to replicate field operating conditions and simulate performance through the use of computer-aided engineering (CAE) tools, specifically advanced Finite Element Analysis (FEA) software. The mechanical seal supplier has extensive experience with modelling and evaluating high-duty applications such as this, and a comparison model replicating the seal performance was constructed to provide insight into the face wear patterns.

The FEA program takes into account combined effects of pressure and temperature distortions, fluid properties, interface friction and operating conditions. The mathematical and numerical models that predict the seal face performance are based on design principles and empirical data. The software utilized by the mechanical seal supplier in this case incorporates many features supporting analysis of mechanical seals that are not typically found in commercial FEA programs. To evaluate the performance of the mechanical seal in this application, several inputs were required including component geometry, shaft speed, material properties, fluid properties, pressure, temperature and boundary conditions to define the seal environment. From these inputs, calculated performance values such as power loss, interface temperature, pressure distortion, thermal distortion, interface gap, leakage, and contact pressure can be obtained. The results are typically shown as either graphical interpretations or numerical values (Tirunagari/Letizia, 2018).

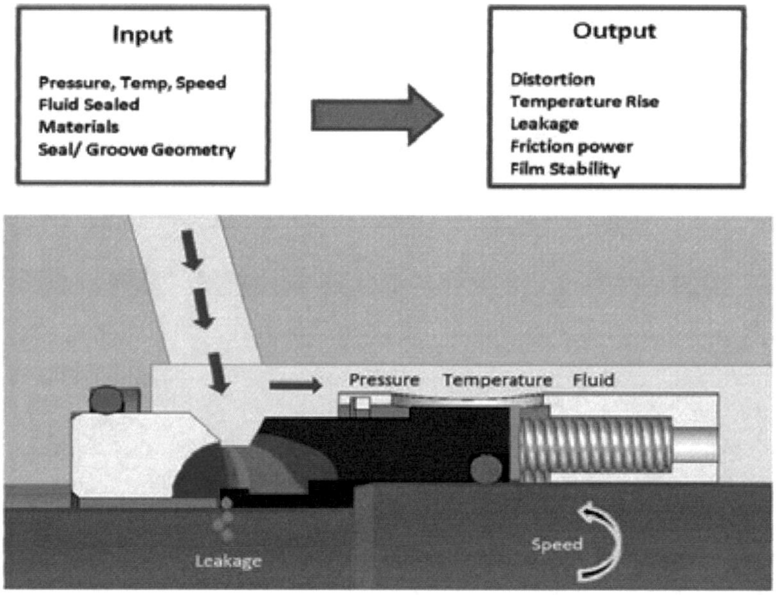

Figure 11: FEA Analytical Tool Representation

Several iterations were evaluated using FEA software to illustrate mechanical seal performance at the specified operating conditions and through varying seal chamber conditions to simulate potential transients that the seal might experience. As a baseline reference, the original mechanical seal design was evaluated initially to verify if there was a substantial improvement in performance with the existing design. The evaluation of the original API Plan 11/52 component seal yielded the worst results in terms of the highest seal face frictional heat values, highest average temperature rise in the surrounding fluid, highest interface temperature and contact pressure. Additionally, the average pressure rotation value across the simulations was beyond what would be acceptable - considering the hydrodynamic surface features being utilized on the seal face. What this data indicated to was potentially high leakage rates and undesirable contact at the face OD.

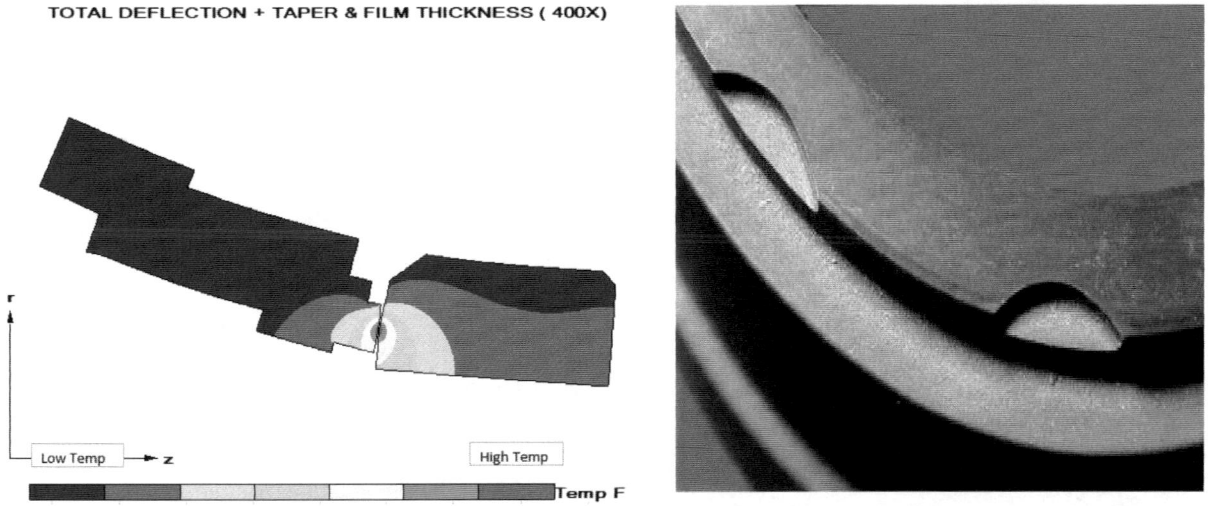

Figure 12: Original Seal Simulation (left); Hydrodynamic Surface Structures (right)

The next simulation focused on performance of the existing design configuration. The results highlight substantial improvement in the areas of face frictional heat generated, the interface temperature and contact pressure being on average 27% less than the original design. There was also a noted improvement in average pressure rotations being 42% less than the original value. The significance of the results were that they did indeed validate there were improvements gained from the original design to the existing configuration. However, review of the results indicated that there was still potential for regions of the seal interface to experience high contact pressures to satisfy the equilibrium equation (i.e., no hydrodynamic load support). This was consistent with the wear patterns visible in the worn seal faces upon inspection.

***** Formatting Issue - Best Available Paper/Graphic *****

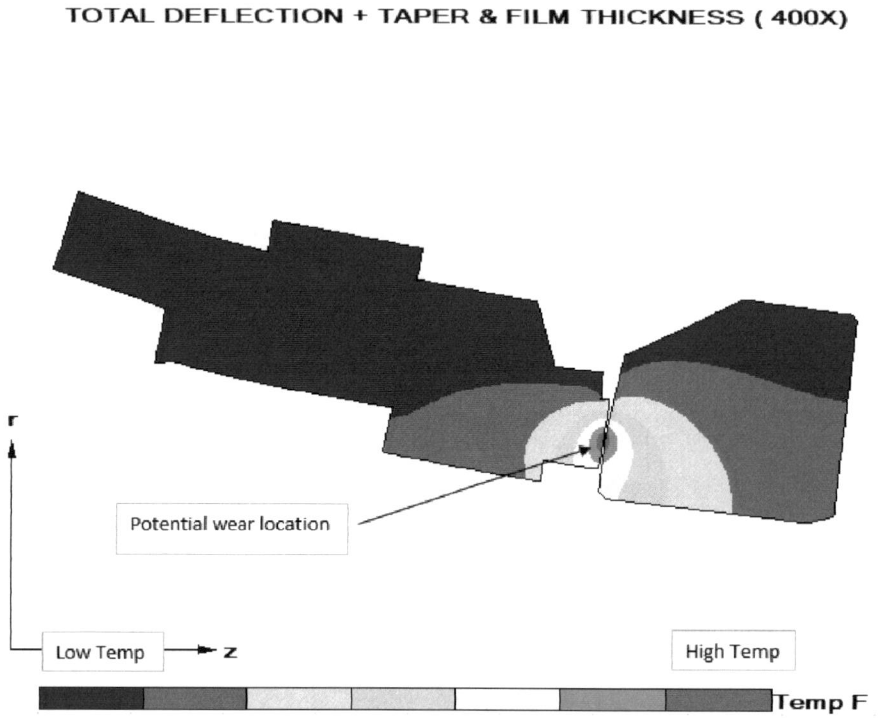

Figure 13: Existing Seal Simulation

The next step was to utilize the capabilities of the FEA software to determine what potential improvements could be made with minimal impact to the overall design of the seal. As these mechanical seals were cartridge assemblies, creating completely new hardware components would not be cost effective or practical, considering the high criticality and visibility of the application. Bearing this in mind, modifications to the existing seal configuration with minimal impact to the remainder of the seal cartridge were explored and evaluated utilizing the computer modeling. A desirable solution was reached that incorporated significant adjustments to the primary ring geometry while still maintaining minimal adaptive hardware impacts. The primary seal face geometry was modified to further mitigate pressure distortion by removing the dynamic o-ring from the ring counter-bore, and utilizing a thrust disc assembly. The placement of the o-ring in the thrust disc also allowed for pressure-balancing of the hydraulic pressure down to the secondary sealing element on the backside of the face, producing no "levering effect" of the carbon.

Additionally, there were modifications to the hydraulic balance of the face, while still incorporating the same hydrodynamic surface features. The coupled effects of the face balance and thrust disc design yielded significant reductions in overall seal face frictional heat (26% less than the existing design), average face temperature (4% less than the existing design), and contact pressure (10% less than the existing design). Further, the modifications yielded an additional 4% improvement in reducing overall pressure distortion values while increasing predicted film thickness by 5%. While the numerical values show improvement, graphically the predictions show that even marginal distortion improvements allow for increased hydrodynamic load support to satisfy the face equilibrium equation and reduce the potential for wear.

*** Formatting Issue - Best Available Paper/Graphic ***

The proposed design, when compared against the original seal, would have substantial benefits in overall seal face frictional heat (46% less), average face temperature (50% less), contact pressure (54% less), and overall pressure distortion values (48% less). Even with these geometry adjustments, the process seal head would be the only significant modification with the existing sub-sleeve, mating ring and remaining hardware re-used.

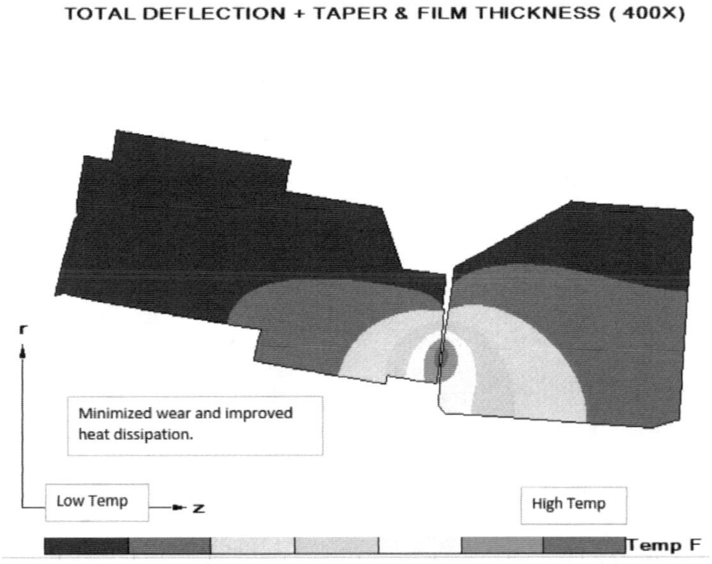

Figure 14 Simulation of Proposed Modifications

Improved Seal Flush Distribution

In addition to reviewing the seal face geometry, a review of the flush hardware was conducted to further maximize performance of the mechanical seal. Findings from previous seal observation activities found that there were apparent pockets of what appeared to be sufficient seal flush lubrication, but other areas of the process seal cavity remained stagnant - and subsequently visible debris accumulation was noted in the hardware. Figure 15 outlines both the primary seal head assembly and the ID of the internal gland cavity with the problem areas.

Figure 15: Adaptive Hardware Findings

The existing seal design utilized an API Plan 11 flush that was supplied from the first stage discharge pressure region on the pump's case, and injected into the inner seal cavity. The injection port for this flush was through an existing casing connection integral to the pump. Based on the orientation of this flush connection in relation to the mechanical seal faces, an integral liner assembly was purposely built into the seal cartridge to aid in distributing the flush flow and directing it to the seal faces. The liner assembly also included an integral throat bushing to help create additional back-pressure in the seal face area, and improve upon vapor pressure margin.

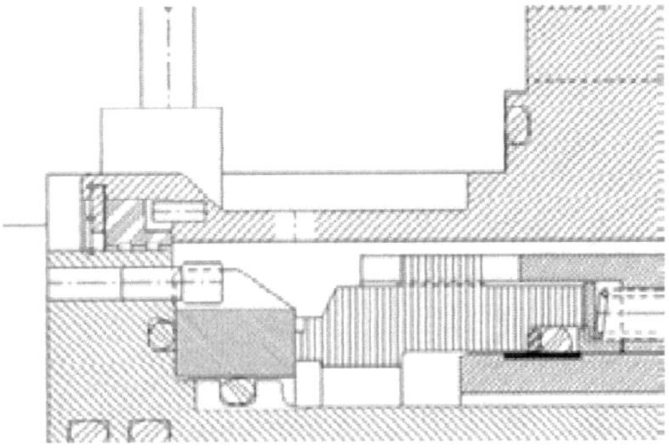

Figure 16: Existing Inner Seal Flow Guide

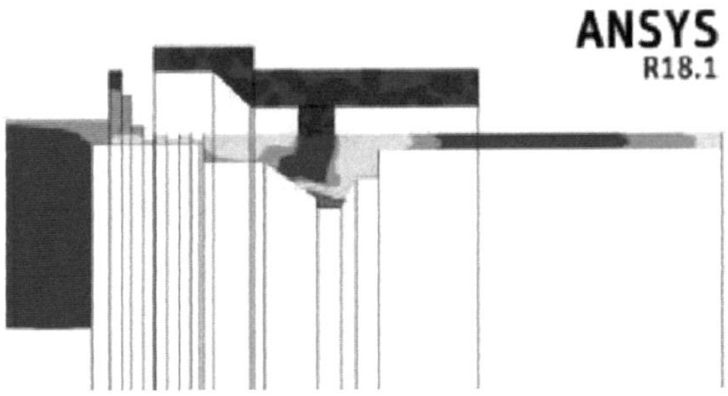

Figure 17 Existing Seal Flush CFD

The condition of the internal seal components indicated that there were potential deficiencies with the configuration of the flush flow guide, namely:

- Current orientation promoted stagnant areas in the seal chamber, allowing debris and particulate to potentially settle out in critical areas of the seal (dynamic o-ring location and springs).
- Fit of the gland adapter to the pump seal chamber bore, above the throat bushing location, was a clearance fit. It was theorized that some of the anticipated flush flow rate intended to lubricate the seal faces was bypassing the injection holes, moving directly toward suction pressure regions in the pump.
- Placement of the injection holes over the seal faces may be detrimental to seal face lubrication in this particular application for reasons that will be explained below.

To address these potential deficiencies, several modifications were incorporated to the inner seal gland adapter and flow guide. To create more flow through stagnant locations of the seal, the flush fluid was routed through multiple drill holes at the rear of the inner seal, through an annular space, then towards the seal face region. This modification ensured that the spring pocket and dynamic o-ring locations would receive a constant stream of flush flow to deter settling out of debris and particulate. As the flush flow passed over the seal faces, a diverter baffle was integrated into the liner piece to force fluid flow down to the seal face region and create slight turbulence to aid in heat dissipation. The injection of the flush flow from the rear of the seal also served to overcome potential centrifugal effects induced by one seal face rotating at the 5200 RPM shaft speed. Through previous studies of high-speed shaft seals using Computational Fluid Dynamics (CFD), the seal supplier was able illustrate that the pressure differential required to force a constant flow rate can be reduced by moving an inlet port to one side.

*** Formatting Issue - Best Available Paper/Graphic ***

Relocation of the port to the side of the rotating face overcomes centrifugal-force-induced vortices that are created at the OD of the face. These vortices force the flush flow towards the throat bushing, along the liner and away from the faces, therefore making cooling difficult. An additional vortex on the retainer side traps the debris and promotes accumulation. In offsetting the flush injection, the required pressure differential to force a constant flow rate is reduced. The diverter baffle would further ensure that the flush flow would overcome fluid stratification effects, compared to the warmer seal faces (therefore removing heat and providing lubrication). To avoid vapor pocket formation adjacent to the seal faces (particularly at startup) a slot was milled axially through the length of the diverter baffle to facilitate venting of trapped vapor.

Finally, the liner component above the throat bushing location was modified to include an o-ring. The purpose of the o-ring was only to provide sufficient restriction so all incoming flush flow was routed through the liner holes and annulus as intended. Along with the hardware modifications, pressure drop and flow through the API Plan 11 flow circuit was verified to ensure that the seal was receiving the required flow rate for cooling. The restriction orifice (RO) in the piping leg to each mechanical seal was re-sized to maximize the flow rate. This was done not only to remove heat from the seal faces but also to force vapor bubbles that may have formed on adjacent surfaces out of the seal face region.

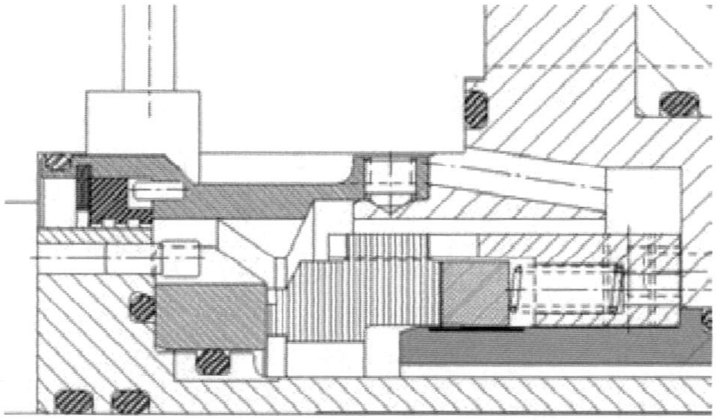

Figure 18: Revised Inner Seal Flow Guide

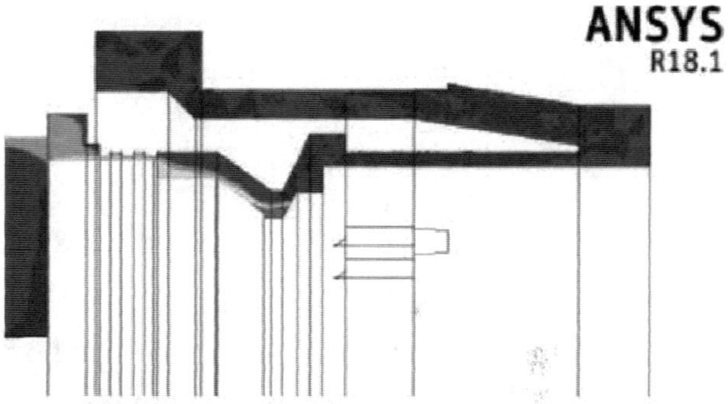

Figure 19 Revised Seal Flush CFD

Containment Seal Modifications

As part of the overall design review, some additional concerns with the dry-running, non-contacting containment seal being utilized were presented. The non-contacting containment seal is designed to divert normal vaporous process seal leakage to the plant flare system and contain more significant leakage amounts should the process seal fail, allowing for an orderly shutdown. The containment seal cavity (cavity between the process seal and containment seal) is connected to the flare system and susceptible to exposure from flare side contaminants. To mitigate these effects, a low-pressure nitrogen purge or API Plan 72 is injected into the cavity. The subsequent nitrogen flow and vapor leakage from the process seal is vented to the flare system, through a local monitoring panel (API Plan 76). The API Plan 76 panel is outfitted with instrumentation to provide an indication of inner seal condition.

*** Formatting Issue - Best Available Paper/Graphic ***

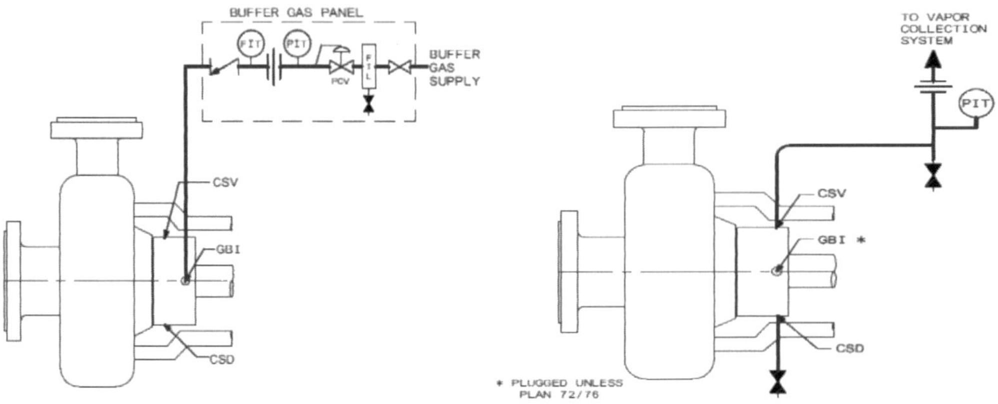

Figure 20: API Plan 72 (left), API Plan 76 (right) diagrams

The API Plan 72 control logic is essentially based on a flow logic to provide a sufficient purge rate to blanket the containment seal cavity and minimize emissions. The pressure required to generate this flow rate is characteristically very low. This presented some challenges with regards to the non-contacting containment seal being used in this particular application. In some early failures of the initial upgraded seal, the end user reported high API Plan 72 flow readings and low pressure, indicators that the containment seal was not functioning as intended. Non-contacting containment seals, by design, utilize special face features to generate hydrodynamic lift and run on film of gas during normal operation. In conjunction with the hydrodynamic face features, typical spring loads on these designs are very low to promote face separation. As with any mechanical seal, the spring force is a small component of the face pressure, and subsequently, spring load is effectively irrelevant in mechanical seals, unless there is an absence of hydraulic pressure.

In this instance, the optimized spring load utilized by the containment seal was potentially a detriment in terms of allowing the seal to contain the low-pressure buffer gas. To overcome this potential concern, a simple adjustment of the spring load was proposed. The modification was proposed after the non-contacting seal design was evaluated using FEA modelling to ensure no detrimental impacts to the seal performance would be realized. To further provide some additional slight back-pressure on the containment seal, a modification to the API Plan 76 panel was also implemented. As outlined in a tutorial from the 31st International Pump Users Symposium, utilizing a 5 or 10 psi (0.3 or 0.7 bar) cracking pressure check valve as an alternative in the API Plan 76 piping over the typically low cracking pressure check valve can be advantageous. The higher cracking pressure valve maintains back-pressure on the containment seal cavity, and better ensures positive retention in a reverse flow scenario (as the valve is not continuously fully open). A controlled buffer gas flow set at 1.5 SCFH is be supplemented to the outer seal cavity, providing a small amount of flow to slightly open the check valve and keep the flare piping dry (Kalfrin, 2015). This higher cracking pressure valve also provides sufficient back-pressure on the containment seal cavity to ensure consistent seal performance.

Results

***** Formatting Issue - Best Available Paper/Graphic *****

As the API plan 11/72/76 cartridge seals were upgraded in iterative fashion, they have enjoyed progressively improved performance. Review of the purge gas pressure/flow trends (IP 21-See Figures 21 & 22) for starts on the most current design show a much improved profile, with little to no break in. In addition to the seal modifications, some of the improvement can be attributed to the natural reduction in iron oxide contamination as the system flushes out after an extended outage. The end user has renewed optimism in the performance of this application going forward.

Figure 21: J-1732 A- Post maintenance start from May 2017, Unimproved API 11/72/76 Showing Rapid Pressure Cycling

Figure 22: J-1732 A- Post maintenance start from Nov 2018, Improved API 11/72/76 Showing Stable Pressure Profile

References:

1. Kalfrin, B., and Gonzalez, L., 2015, "API 682 Arrangement 2 Configurations – Considerations for Outer Seal and Support System Design," *Proceedings of the Thirty First International Pump Users Symposium,* Turbomachinery Laboratory, Texas A&M University, College Station, Texas

2. Kalfrin, B., 2016, "Mechanical Seal and Support System Considerations for Negative Temperature Hydrocarbon Services: NGL Processing and Ethylene Production Focus," *Proceedings of the Thirty Second International Pump Users Symposium,* Turbomachinery Laboratory, Texas A&M University, College Station, Texas

*** **Formatting Issue - Best Available Paper/Graphic** ***

3. Tirunagari, P. and Letizia, R., 2018 "Advanced Computational Technologies for Mechanical Seal Analysis", *Pumps & Systems Magazine*, Cahaba Media Group

Figures 2,4, and 9-20 courtesy of John Crane

High Cycle Fatigue in Turbine Blades

by

Samuel Snively

Steam Turbine Engineering Manager

Elliott Group

Jeannette, Pennsylvania

Shadi Gilke

INEOS

Olefins & Polymers USA

Sam Snively is the Steam Turbine Engineering Development Manager for Research and Development at Elliott Group in Jeannette, Pennsylvania. Responsibilities within the turbine development group include principle aerodynamic engineering, structural and modal analysis, and general product development. Mr. Snively received his B.S. degree (Mechanical Engineering) from the University of Pittsburgh in Pittsburgh Pennsylvania and M.B.A. from Seton Hill University in Greensburg Pennsylvania.

Abstract

The steam turbine blade is widely considered as the single most important component of today's mechanical drive steam turbine. Responsible for the efficient conversion of thermal energy to mechanical work the blade has experienced revolutionary changes since Sir Charles Parsons' modern turbine invention in 1884.

Despite all of the advances in computational engineering tools such as computational fluid dynamics and finite element analysis the turbine blade is not immune to failure during operation. This paper will explore the design, evaluation and operation of a modern steam turbine blade with a focus on high cycle fatigue and how the phenomenon is evaluated and mitigated.

Introduction

A typical ethylene crack gas 100MW impulse steam turbine will have approximately 1,000 blades on the rotor operating 24 hours a day, 7 days a week for upwards of 7 years before a turnaround and inspection for a total of 61 million blade operating hours. Statistically speaking, it can be said that modern steam turbines are essentially operating without failure. That means nothing to the ethylene plant manager whose turbine has tripped after 6 years and 6 months of operation. His concern is the lost production in the hours and days that follow from a plant shutdown caused by a steam turbine blade failure. Steam turbine blade designers spend all of their time developing the next generation of airfoils that are more efficient, can operate at faster speeds and above all else, operate without failure. That last objective, to run without failure, carries with it challenges that are difficult to quantify, time dependent, random in nature such as erosion and corrosion. Turbine blade design is a complex a methodical process requiring contributions from multiple disciplines such as aerodynamics, material engineering, stress and vibration analysis.

Blade Failure Mechanisms

Blade failures can happen for a number of reasons:

- Foreign object damage
- Tensile failure due to over-speed
- High cycle fatigue due to alternating stresses
- Environmental conditions – erosion and corrosion

Foreign object damage occurs when the rotating blade is impacted by some object carried into or dislodged from inside of the turbine. These failures are impossible to predict and design for since the mass and speed of the impacting object will never be known. Impact can range from scale and rust from upstream piping, water slugs in the steam supply line to dislodged vanes from upstream stators or blades.

Tensile failure due to over-speed can occur if the turbine speed exceeds the terminal velocity of the turbine blade. Generally this happens after a coupling break when the turbine is instantaneously shed of applied load resulting in a rapid speed increase. Even though modern trip systems react in less than 250 milliseconds a large turbine can reach terminal speeds of 160% of the rated speed in less than a second. Turbines are designed to withstand even those extreme speeds but systems do not always work as planned, most commonly because of trip valve closing lag. A 50 millisecond delay of closing time can mean the difference between a safe trip and rotor damage.

High cycle fatigue (HCF) failures happen when small elastic strains under a high number of load cycles leads to failure. The physics of this phenomenon are easily quantified with fracture mechanics and mitigated on paper. In reality the turbine blade is subjected to extreme and harsh conditions that can greatly affect the material's resistance to applied stresses. This paper will focus on the design, mitigation and operational conditions that lead to HCF failures.

Designing for High Cycle Fatigue Mitigation

First consider what constitutes a "cycle" for a blade. Upstream of every row of blades is a stationary stator, consisting of vanes that create a converging flow path which increases the velocity of the steam as shown in Figure 1. This jet velocity must be faster than the relative velocity of the blade, typically on the order of two times the speed of the rotor to efficiently produce power. A typical stator in the low

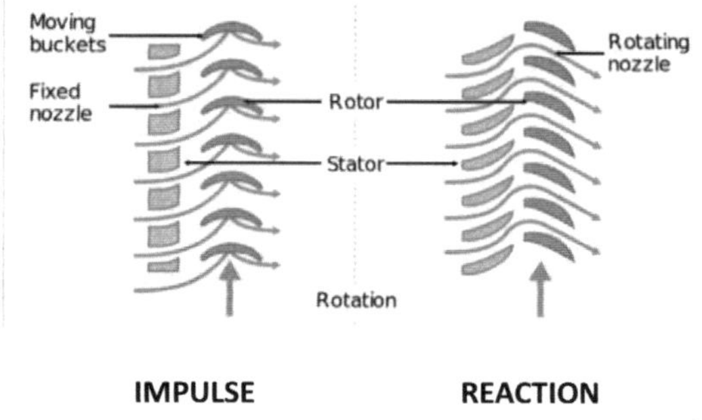

Figure 1 - Nozzle and blade interaction

pressure region of the turbine will have 50 nozzles, therefore the blade will experience 50 cycles every revolution.

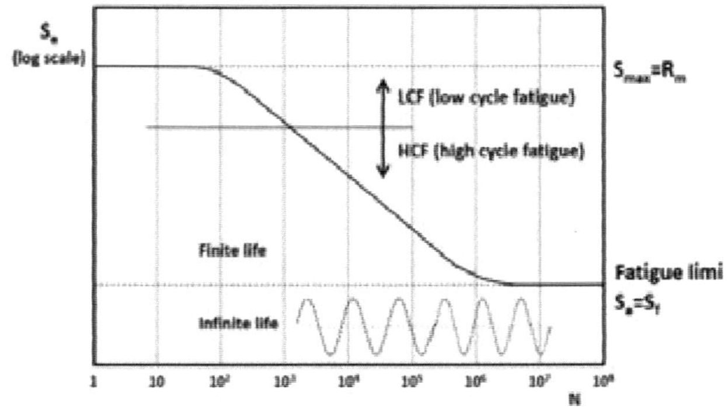

Figure 2: Fatigue Graph

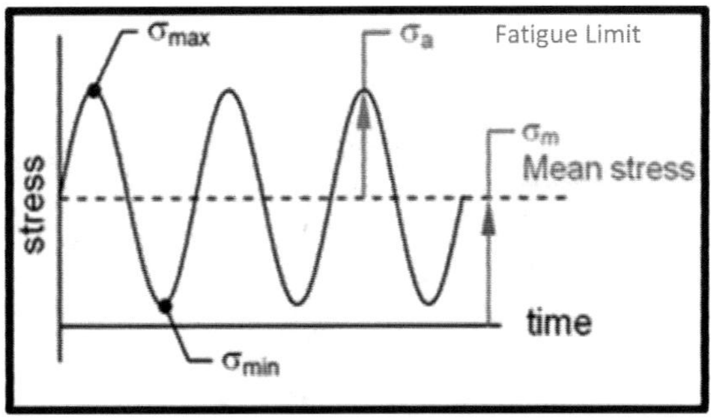

Figure 3 - Detailed view of alternating stress and fatigue relationship

Next consider how many cycles are considered as the minimum threshold for high cycle fatigue. The LCF-HCF curve, as shown in Figure 2, is not a step change so the transition is not a specific value. In general for ductile steels, such as those used for steam turbine blades a minimum of 10^6 cycles is used for the HCF evaluation. For the low pressure turbine blade with 50 nozzles up stream and operating at 4,000 rpm the blade will see 200,000 cycles per minutes and cross the HCF cycle threshold in a very short amount of time. Number of cycles alone does not mean a failure will occur. Alternating stress failure will only occur if the stimulus stress exceeds the fatigue limit stress for infinite life. A delayed failure is likely caused by the introduction of additional stimulus interaction and or material degradation attributed to environmental effects such as corrosion and erosion.

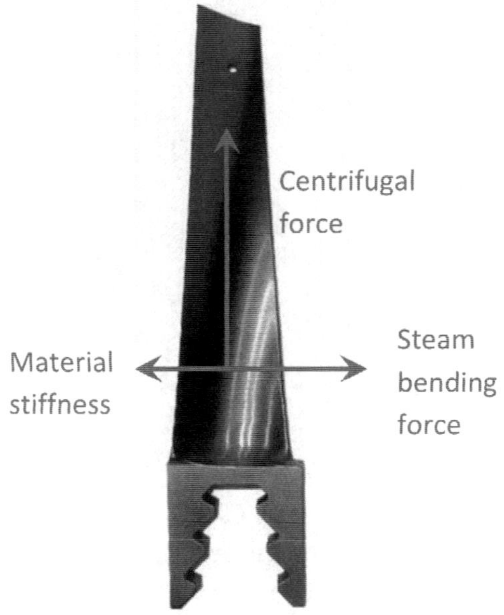

Figure 4 - Blade Forces

The primary forces, shown in Figure 4, applied to the blade are centrifugal from the rotation of the shaft and tangential from the impact force of the jet of steam. The centrifugal force is radial and constant with constant speed. The tangential force alternates as the blade passes through the jet of steam and is pushed down stream, then as the bucket passes behind the nozzle the material stiffness springs the blade back until the next nozzle again pushes the blade downstream. This operation can be considered as a steady state force at a fixed speed. This however is more of a text book representation and not actual operation. In reality mechanical drive steam turbine blades are operated at varying speeds and powers, with varying steam conditions (pressure, temperature, moisture, impurities, etc.) and fluctuating stimulus from inconsistencies in the upstream nozzles.

Geometry Considerations

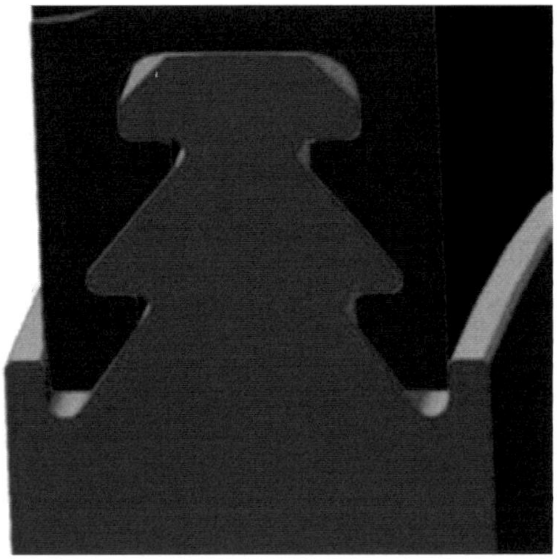

Figure 5 - Typical bucket root geometry

The number of blades required for each row on the rotor is a compromise between aerodynamic and structural limits. For aerodynamic efficiency a thin bladed cross section which allows for maximum area with minimum height and minimal secondary profile losses is ideal. Structurally this is not acceptable since the section modulus would be inadequate to handle the power generated by each passing nozzle. Historically as turbine power requirements increased there were two options: add more rows of blades or add more blades per row.

Adding more blades per row will reduce the absorbed power per blade and keep the loading within the material allowable stress for the least amount of cost. Adding additional rows of blades would require a longer shaft, a larger casing, additional blades and additional stator(s) and significant overall cost to the turbine. To "nest" more blades into a row the blade roots need to be slanted (Figure 9 & Figure 10) to fit each individual airfoil onto the platform. This solution corrects the airfoil problem but inadvertently creates a stress riser in the root section. Slanting the root

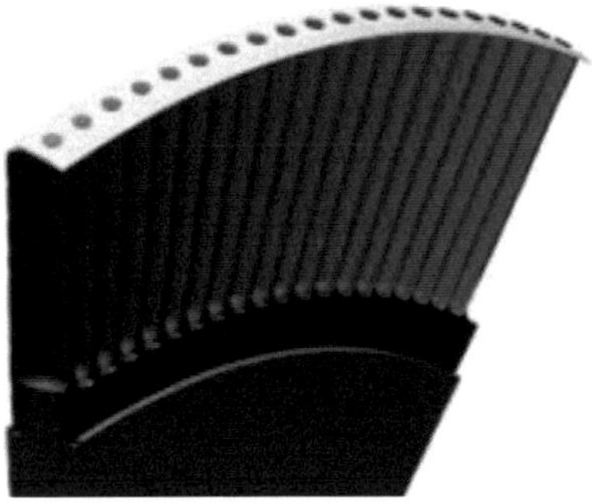

Figure 7 - Partial row of blades

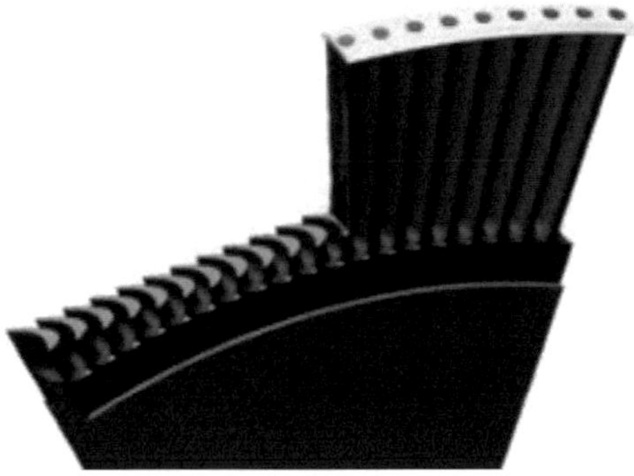

Figure 6 - Partial section showing blade nesting

creates a point on each of the loading lands and creates geometry that promotes high stress loading and crack initiation. Before the refinement of FEA the impact of this geometry was not completely understood. While it's easy to understand that the point geometry creates a focal point for stress, quantifying how much loading the root could handle was difficult to quantify without the assistance of plastic deformation FEA analysis. Even though the slanted root geometry is not ideal it should be noted that millions of turbine blades are currently operating around the world with slanted roots.

Figure 9 - Slanted root configuration

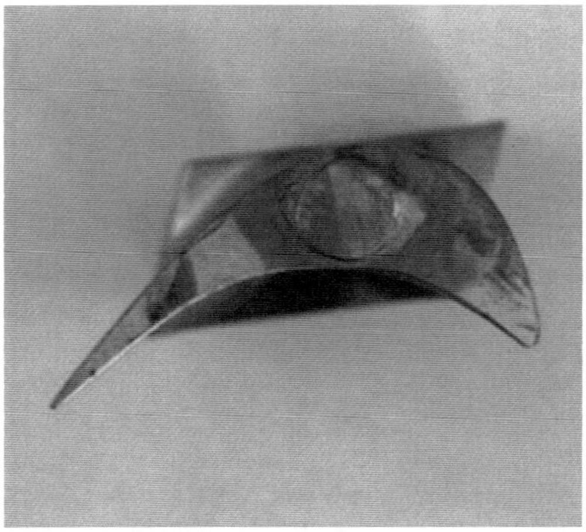

Figure 10 - Slanted root blade

As blade designs have evolved various design modifications were added to improve the reliability of the turbine:

- Pressure land roots were slanted to more evenly distribute the stress through the cross-section (Figure 14).
- Additional fillets and rounds were added to remove the stress risers at the point of the pressure land (Figure 12).
- Root geometry was modified to add more pressure lands to help more evenly distribute the load.
- Slanted roots were straightened to remove the stress risers which also required that the trailing edge of the airfoils be allowed to be unsupported (Figure 13).

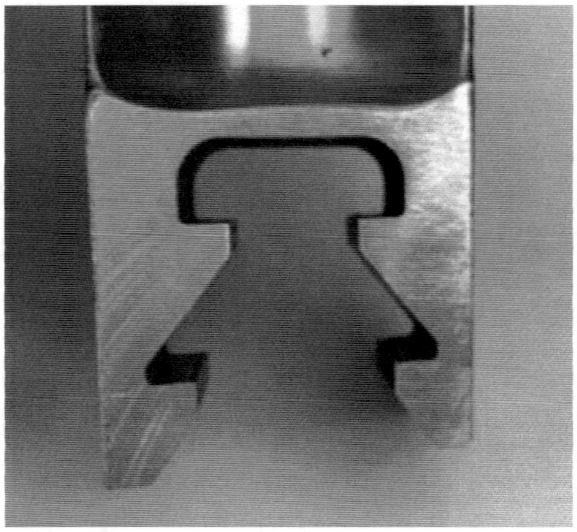

Figure 8 - Two land slanted root with straight pressure lands

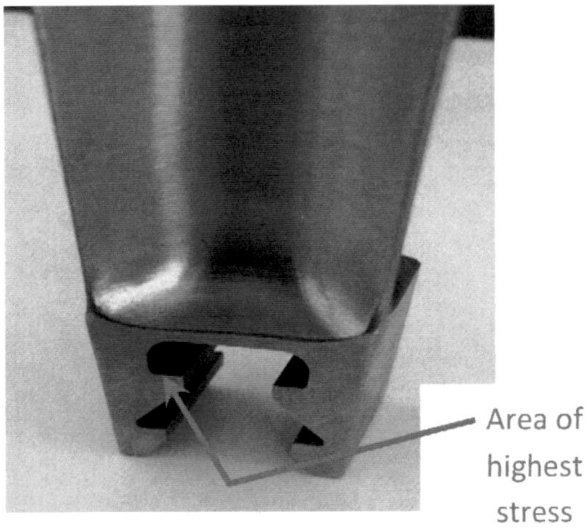

Figure 11 - Two land root (before shot peening)

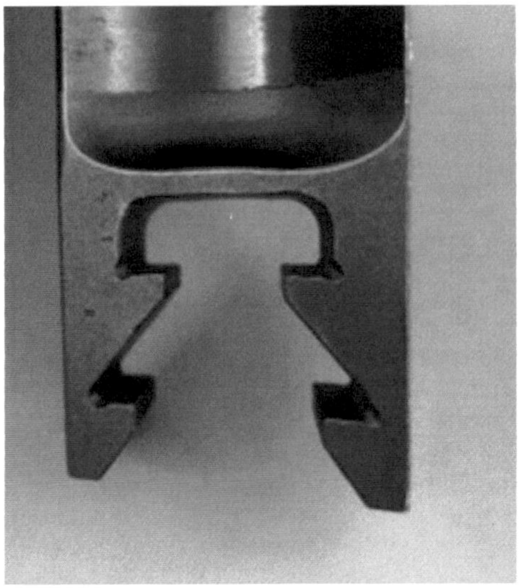

Figure 12 - Added radii to reduce stress risers

Fundamental Frequency Identification

The science and understanding of vibrations and stimulus amplification is complex and can encompass countless papers and text books. The goal of the blade designer is simple; ensure the alternating stress stays below the endurance limit and both of those things, applied stress and material endurance will vary from the design calculation sheet through operation.

Figure 13 - Straight root with overhung trailing edge

Thirty years ago modal analysis of blades was limited to simple rap testing to identify fundamental axial and tangential frequencies. Through time, experimentation, experience and advances in FEA technology, blade designers are now able to identify fundamental as well as higher order frequencies as well and the frequencies attributed to the interaction of adjacent blades and the shaft disk. This allows engineers to better understand the frequencies that reside within the speed operating map for operation. The frequencies of blades and their relationship to the operating speed are shown in a Campbell diagram (Figure 15).

Amplification and attenuation can be accounted for and the blades designed with sufficient strength to withstand the predicted alternating stress.

Stimulus is hard to quantify because it varies with nozzle throat openings, fouling, erosion, pressure differential fluctuations, etc. The endurance limit will always degrade with time due to erosion, corrosion and reduction in notch sensitivity.

Figure 14 - Three land root with slanted pressure lands

Blade airfoils and roots are not designed for each turbine but rather for a range of conditions such as speed and power. When the appropriate blades have been selected for a particular application each blade will be analyzed for the complete range of operating conditions to confirm the design integrity. To ensure safe and reliable

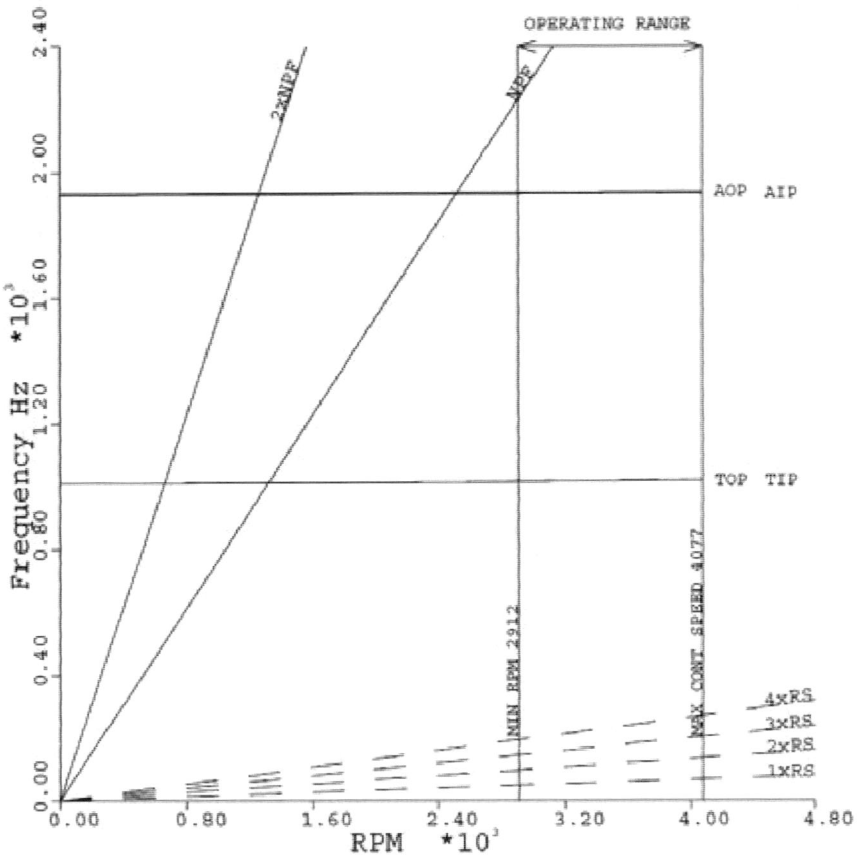

Figure 15 - Typical Campbell diagram

operation turbine manufactures define safety margins from the modified endurance limit. This is documented and recorded with the Goodman diagram.

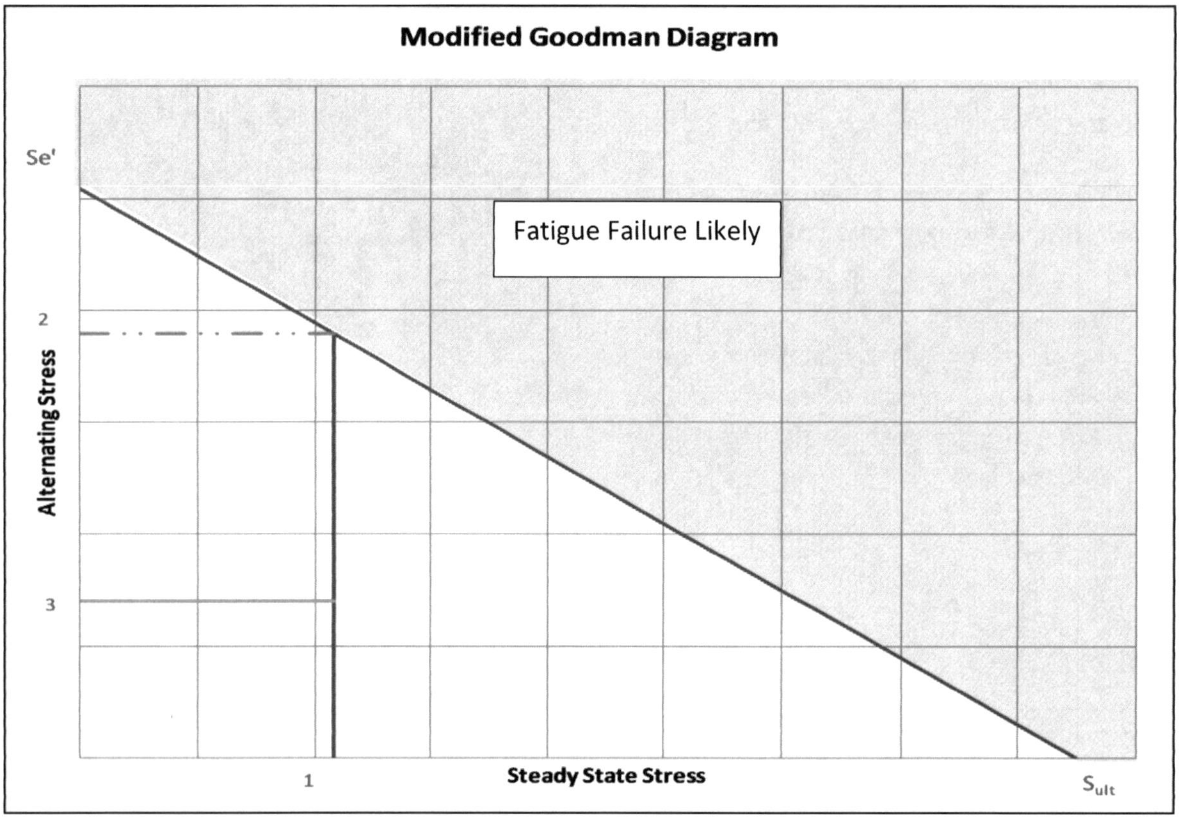

Figure 16 - Goodman Diagram

The Goodman factor, shown in Figure 16, is the most common visual representation of the fatigue life and the interaction of mean and alternating stresses applied to a turbine blade. Sult is the ultimate tensile strength of the material. Se' is the endurance limit, corrected for temperature, fracture mechanics and environmental factors and the line connecting the two is called the Goodman line. Any point below the Goodman line will not fail due to fatigue and any point above the line will eventually fail from fatigue. This approach does not predict cycles to failure but rather the magnitude of the stresses required for a fatigue failure. The following is an explanation of the lines shown in Figure 16: The red line shown in is the mean stress which is a combination of the centrifugal and steam bending stress. The green line #2 is the maximum allowable alternating stress. Line #3 is the actual alternating stress. The ratio of line #2 to line #3 is what is commonly known as the Goodman Factor. As the actual alternating stress, line #3, is reduced the Goodman factor will increase, indicating a design with a higher separation from the failure line. Failure will occur when the alternating stress increases (Figure 18) and/or the endurance limit of the material is reduced (Figure 17).

Causes of High Cycle Fatigue Failures

As mentioned previously a failure resulting from HCF results from either an increase in alternating stress or a reduction of endurance strength in the material. Increasing alternating stress can result from but is not limited to:

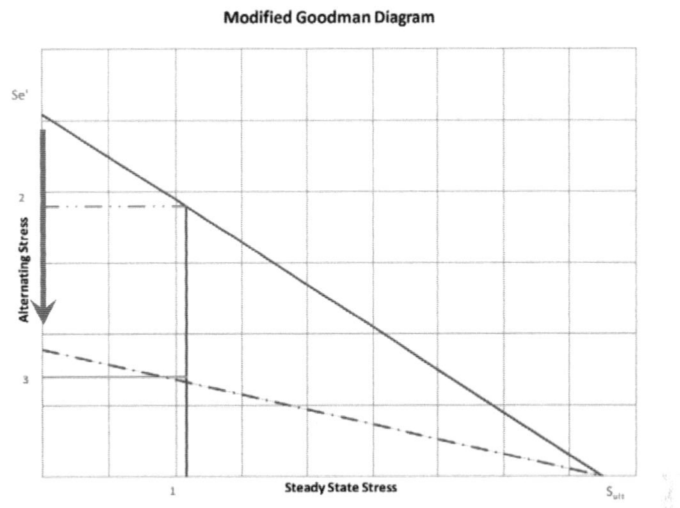

Figure 17 - Goodman factor showing reduced endurance limit

- Change in rotor speed that places the blade into a resonant condition.
- Damage to one or more nozzles upstream of the blades will cause an excitation that can generate a per revolution excitation.
- Rapid decrease in exhaust vacuur volumetric steam flow through the la
- Crack initiation within the bucket.
- Excitation from water droplet impin

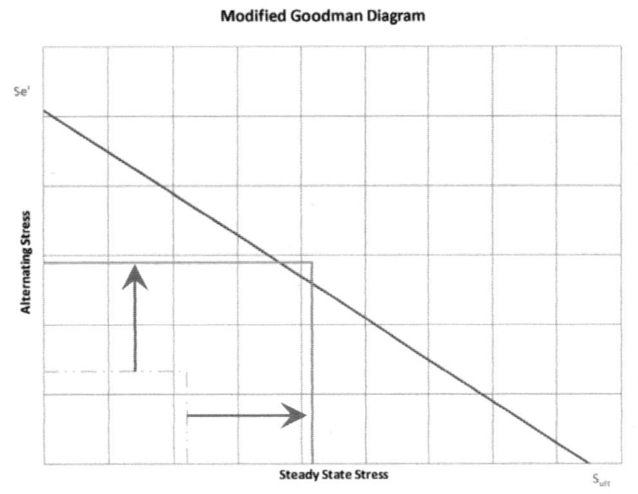

Figure 18 - Goodman diagram showing increased steady and alternating stresses

As a designer it would be best to develop a rotating blades. This is possible for fixed resonance can typically be avoided by chan modifying damping characteristics or varyin speed machine in a mechanical drive ap frequencies and avoiding all of them for a impossible. For rows that have unavoidabl blade strength to attenuate the alternating s

When one or more upstream nozzles are resulting stimulus can cause HCF. Unstea which occur at harmonics of the rotor speed due to deviation from cyclic symmetry of the stator geometry. Such deviations are primarily due to nozzle spacing variation, but may also include out-of-roundness, split-line interruption, and misalignment. Split-line interruptions are the most common deviation because the nozzles that are cut at the split-line are difficult to match to the adjacent nozzles.

These deviations 180 degrees apart generate a two per revolution harmonic that can match certain natural frequencies of the blades or row of blades.

The effect of exhaust vacuum reductions should not be understated. Typical exhaust pressures are generally 4" mercury absolute for normal operation. Reducing the exhaust pressure to 3" mercury (a delta of 0.5 psi) results in a steam volumetric increase of 30%. This increased volumetric flow can quickly overwhelm the rotating row of buckets and create unstable aerodynamic flow with increased and unpredictable alternating stresses. Discharge circumferential pressure variation, on the last row of blades, which also occur at harmonics of the rotor speed and due to uneven flow distribution at the discharge of the stage are typically related to the exhaust as a function of the exhaust velocity.

New blades go through rigorous inspection to ensure they are fit for service before they are installed on the rotor. Highly stresses areas such as the base of vane and root area are shot peened to induce compressive stresses to help resist any crack initiation. Any crack that develops during operation is likely to initiate at the area of highest stress. In most cases this is the top land root of the bucket. Once a crack has initiated the loss of tensile area will be reduced, thus altering the frequencies of the blade and possibly increasing the alternating stresses which will further reduced the tensile area and the cycle will continue to repeat. As this dynamic system evolves everything within the system is changing and eventually the blade will fail from fast fracture. Unfortunately once the blades have been installed on the rotor it is impossible to inspect the areas of highest stress, such as the top land root.

Endurance limit factors

A decrease in the endurance limit of the material is generally attributed to changes in:

- Notch Factor
- Surface finish
- Environmental factors

Notch Factor

Nothing is more detrimental to the reliability of a turbine blade than a surface crack in a highly stressed area. The ability of the blade to withstand defects is measured by the notch factor of the material, the geometry of the bucket and the stress concentration factor. No material is 100% homogenous and defects and inclusions will always be present throughout the entirety of the blade. The most stressed area of a turbine is typically the top land root of the bucket. A top land root crack is typically initiated by an intergranular defect from corrosion and/or excessive stress at the outer most fibers of the section where the stress and strain are the highest.

Surface Finish

The surface finish of a material has considerable effect on the fatigue life and is just one of the factors that account for the adjusted endurance limit. Fatigue failures almost always begin at the surface of a material. The reasons are that (a) the most highly-stressed fibers are located at the surface (bending fatigue) and (b) the intergranular flaws which precipitate tension failure are more frequently found at the surface (EPI, Inc., 2012).

Blade surface finish can degrade during operation as noted in Table 1 (based on EPRI, 2001). In this case, a specific turbine was examined that appeared to have deposit buildup on the rotating blades and erosion of stationary vanes.

Table 1 - Surface Finish for non-Elliott Turbine Blades During Overhaul (EPRI, 2001)

Blade Location and Condition	Surface Finish (μin rms)
HP Turbine Rotating Blade Finish as Manufactured	35
HP Turbine Rotating Blade Worst Finish After Operation (Stages 1-6, Concave Side)	145
HP Turbine Rotating Blade Worst Finish After Operation (Stages 1-6, Convex Side)	170
HP Turbine Stationary Blade Worst Finish as a Result of Operation (Stages 1-6)	60-100
HP Turbine Stationary Blade Worst Finish as a Result of Operation (Stage 7)	300

The cause of rotating blade surface finish degradation was not fully described in the EPRI report, whether due to deposits or due to erosion/corrosion. The erosion of turbine blades depends upon many factors including droplet size and relative velocity, but these factors are ignored for now. Corrosion is more than just a degradation of surface roughness such that the surface finish factors and environment factor can be treated as compounding each other. A rough surface finish on fossil and nuclear turbine blades has been noted to result in higher initial deposition rates (EPRI, 2001). And conversely, the experience of at least one turbine vendor shows that good surface finish combined with residual compressive stresses can almost eliminate corrosion cracking in LP turbines.

Surfaces are assumed to degrade in accordance with Table 2 during operation:

Table 2 - Surface Finish Degradation

Steam Condition	Vane Surface	Base of Vane Surface	Root Surface
Dry Steam	+135 μin rms	+135 μin rms	+67 μin rms
Wet Steam	+220 μin rms	+220 μin rms	+110 μin rms

Environmental Factors

The allowable alternating stress is based on an adjusted endurance limit as previously noted. The environmental modification factor accounts for operation of steam turbine blades in the moisture transition zone where harsh elements are present in their highest concentrations.

The first step to determine a modification factor for the phase transition zone is to determine what environmental conditions are present. There has been a number of researchers looking to determine the environment, including several EPRI sponsored projects and work done using the Moscow Electric Institute Research Turbine. The basic findings are summarized below.

Stainless Steels in the Phase Transition Zone

Many researchers begin looking at the problem by analyzing the Mollier Diagram. An example is shown below in Figure 19 that shows regions of deposition and condensation and their effect on corrosion. As can be seen in the diagram, above the saturation line is the salt zone that predicts a ~30% NaCl solution.

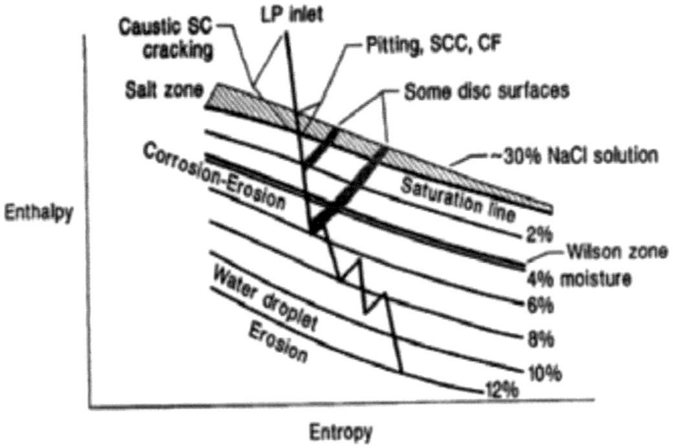

Figure 19 - Mollier diagram

Given that this very concentrated solution is predicted and NaCl is notorious for causing corrosion problems in stainless steels, it is very easy to jump to testing in a similar concentrated solution and for simplicity, using uncontrolled amounts of oxygen or purposely aerating the solution. Testing in this type of solution gives results similar to those presented by Atrens (1984) in Figure 20, which shows results of testing smooth 403 stainless steel specimens in a 22% NaCl solution that was aerated at 95°C. For this data, the factor that would go into a Goodman Factor calculation would be in the 0.1-0.125 range (~50 MPa/~400 MPa). Atrens states that this was selected as a worst case scenario.

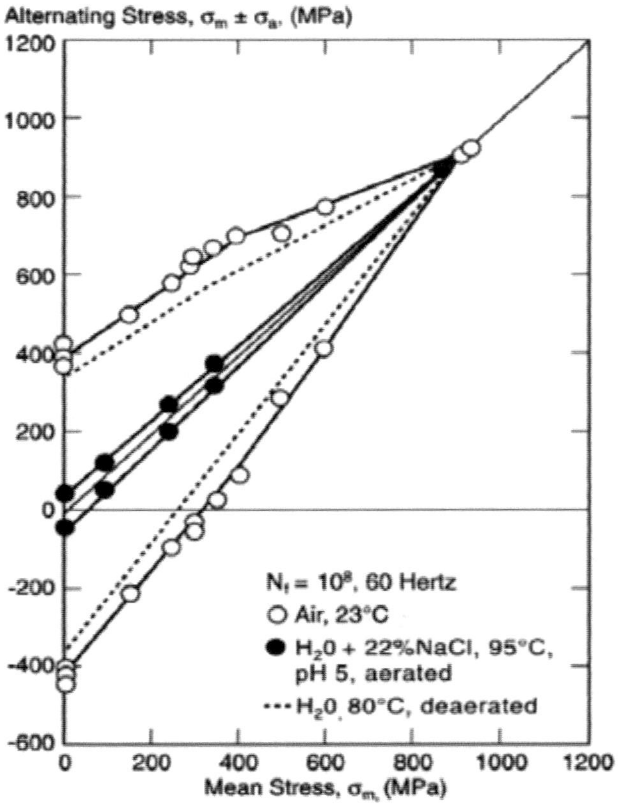

Figure 20 - Smith diagram

Due to the fact that pitting is the precursor to corrosion fatigue and stress corrosion cracking, the problem is truly a fracture mechanics problem. Both of these have threshold stress intensity values that must be exceeded prior to crack growth. The stress intensity can be related to a threshold stress value and to a critical pit size. It may be possible to relate the pit size to pitting kinetics, which in turn could eventually be related to the incoming steam conditions.

While it is difficult to completely eliminate pitting during a shutdown, it can be mitigated to some extent by using shutdown procedures that limit exposure of wet blade surfaces to oxygen. Being that recommendation is towards less conservative modification factors, it seems prudent to develop recommended procedures for customers to follow that would minimize pitting during shutdowns.

Recommendations for Turbine Operation

Turbine blades operate in harsh, dynamic and every changing environmental conditions making failure predictions impossible. Corrosion and erosion can have significant and detrimental impacts on the blade life. Operators should monitor their steam purity to minimize the corrosive attack of chlorides. Steam conditions must also be carefully controlled since the turbine inlet pressure and temperature along with the exhaust pressure will control the location of the phase-transition zone. Allowing the steam conditions to fluctuate will promote the nucleation and development of deposits on multiple rows of blades surfaces, reducing efficiency and creating a larger corrosive environment.

Critical service component inspection and replacement

Critical turbine installations are typically managed with main and spare rotors for faster turnaround during planned outages and to also quickly respond to unplanned outages due to turbine failure. Any rotor that has been recently removed from service after an extended amount of run time should be carefully inspected to determine the condition and fitness for future operation.

There is no fail-safe inspection procedure that can identify future problems so any repair or replace decision is purely based on a risk assessment of the past and future operation. Operators and OEMs should work together to determine the best time to replace the blades on a rotor. Re-blading a turbine rotor is by no means an inexpensive repair option although if it can prevent an unplanned outage it will pay for itself many times over.

Conclusions

The engineering of turbine blades and knowledge of material science has made significant advancements in the past twenty years. With a solid understanding of the applied forces, vibration attenuation and material integrity steam turbine manufactures are more capable of getting the most power and run time from the machines. However, despite the best efforts of man and computer there is simply no way to accurately predict the precise lifecycle of a turbine blade to ensure all 100% of the useable life has been consumed. Turbine blades operate in a dynamic and ever changing environment that has many variables and unknowns. As producers push to increase the mean time between maintenance it is imperative that sensible repair/replacement program is considered for the steam turbine rotor since the cost associated with an unplanned outage will greatly outweigh any cost for replacing the rotor during a planned maintenance cycle.

References

[1] EPI, Inc., "Metal Fatigue and the Factors That Influence Fatigue", http://www.epi-eng.com/mechanical_engineering_basics/fatigue_in_metals.htm , Copyright, 2012.

[2] Steam Turbine Efficiency and Corrosion: Effects of Surface Finish, Deposits, and Moisture, EPRI, Palo Alto, CA: 2001. 1003997

[3] Atrens, A., Meyer, H., Faber, G., and Schnieder, K., "Steam Turbine Blades," *Corrosion in Power Generating Equipment*, Speidel, M. O, & Atrens, A., Ed, New York: Plenum Press (1984).

AIChE technical paper reference number: 547357

Title: Challenges with Compressor Start-up/Dry out runs with dissimilar MW gases than the process gases.

Authors:

Kalyan Malnedi, is an application engineer for Elliott Group in Houston, Texas. He has more than 15 years of experience in centrifugal compressor research and development and selecting turbomachinery for various markets/applications including olefins markets. He holds MS in Aerospace engineering from Indian Institute of Technology -Bombay, Mumbai, India.

Shane Harvery, P.Eng., is the Manager of Americas Application Engineering for Elliott Group, a leader in supply of compressor and turbines for ethylene plants. He has more than 12 years of experience in sizing and selecting turbomachinery for the olefins market. He holds a BSc in mechanical engineering from university of Calgary and is a licensed Professional Engineer in the province of Alberta. He is the chair of the Rotating equipment subcommittee of the AIChE Ethylene Producers' Conference.

Abstract:

This paper discusses the challenges with compressor start-up/dry out runs using heavier or lighter Mole Weight (MW) gases than the actual process gas. It is very common for the end user to specify a compressor dry out case using Nitrogen, Hydrogen, air or a gas mix. Sometimes, it is challenging to accommodate these dry out cases due to significant volume mismatch between the process gases and dry out gas. These dry out runs also poses challenges with respect to maximum allowable working temperature (MAWT), control system design, ease of use by the operator in the field and the driven equipment and driver configurations.

Introduction:

Compressor type referenced and discussed in this paper is centrifugal compressor type. It is critical to optimize the compressor selection based on the application, service and all operating conditions including start-up/dry out conditions or run in on air capabilities. It is important to not to significantly alter the design of the compressor in accommodating the start-up/dryout conditions. In this paper, the authors discuss the theory behind the volume mismatch and strategies to implement these dry out cases. The authors also discuss various compressor configurations such as back-to-back, extraction, speed changes, decoupling parts of string and independent control loops etc., which may provide an economically viable solution to solve the dry out run problems. The paper also covers the considerations for dry gas seal mechanical design, control system design and ease of operator to run the start-up case in filed. Finally, the authors present a case study where one these above strategies was successfully implemented.

How A Centrifugal Compressor Works:

Centrifugal compressors are the dynamic type; compression is accomplished through the conversion of kinetic energy to static energy. A simplified cross-section of centrifugal compressor is shown below;

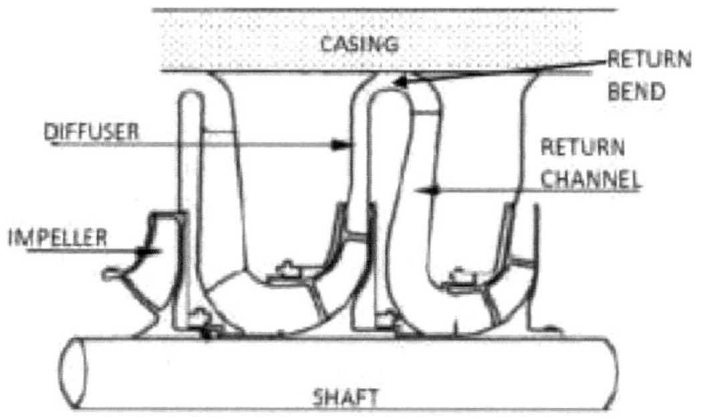

Figure 1: A simplified cross section of a centrifugal compressor

The two main characteristics of centrifugal compressors are volumetric flow through the unit and the head generated. Head generated inside a compressor is a function of the inlet gas MW. For the same suction and discharge pressure, a lighter MW gas requires a higher head than the heavier MW gas. Similarly, for a given mass flow a lighter MW gas will have a higher volume flow.

Below listed two equations, explains the relation between specific volume and the head generated as the gas go through each stage of compression.

$$v = \frac{ZRT}{144P}$$

$$v \; \alpha \; \frac{ZT}{(MW)P}$$

v	Specific volume [ft3/lb]
Z	Compressibility of gas
R	Temperature [°R] = 1545 / MW
T	Gas Constant [ft-lb/lb°F]
P	Pressure [psia]
MW	Molecular Weight [lb $_{mass}$ / lb $_{mole}$]

Equation-1

$$H = ZRT\left(\frac{n}{n-1}\right)\left[\left(\frac{P_2}{P_1}\right)^{\left(\frac{n-1}{n}\right)} - 1\right]$$

$$\left(\frac{n}{n-1}\right) = \left(\frac{K}{K-1}\right)\eta_p$$

H	Specific volume [ft3/lb]
Z	Compressibility of gas
R	Temperature [°R] = 1545 / MW
T	Gas Constant [ft-lb/lb°F]
P	Pressure [psia]
MW	Molecular Weight
$(n/n-1)$	Polytropic Exponent
K	Ratio of Specific Heats (C_p/C_v)
η_p	Polytropic Efficiency

Equation-2

As can be seen from the specific volume equation (Equation-1), for higher MW gas the volume reduction increases significantly through each stage of compression. If the MW between process gas and the start-up gas is significantly different, then volume reduction also significantly different between these gases through each stage of compression, which causes volume mismatch.

Volume mismatch between the process gas and dry out gases:

Below is a sample illustration of what happens inside the compressor between lighter process gas vs. heavier start-up gas and heavier process gas vs. lighter start-up gas.

Heavier process gas vs. lighter startup gas:

Surge

					Start-Up Case			Normal Case		
WHNO	SCALE	EFF	UEQ	FLC		Range			Range	
Impeller-1	1	0.81	739	0.106	9	U	79	36	S	25
Impeller-2	1	0.857	701	0.0872	17	U	71	25	S	31
Impeller-3	1	0.862	695	0.0633	17	U	53	25	S	31
Impeller-4	1	0.866	664	0.0529	32	S	43	33	S	41
Impeller-5	1	0.858	639	0.0451	33	S	43	24	S	63
Impeller-6	1	0.798	591	0.0439	46	O	20	25	S	61
Impeller-7	1	0.723	575	0.0398	53	O	5	24	S	65

Choke

Figure-2: Compressor characteristics – Heavier Process gas vs. Lighter Startup gas

The above figure shows the each impeller range in percentage from surge and choke. "U" stands for impeller in under-loaded region and "O" stands for impeller in overloaded region, "S" stands for impeller in the select region of the each impeller design curve.

As can be seen from the above Figure-2, the lighter start-up gas volume is not reduced as much as the process gas, front end of the compressor tends towards surge and back end of the compressor is approaching choke (or stonewall). For reference, please see the normal case impeller range in percentage from surge and choke.

Lighter process gas vs. heavier startup gas:

Choke

					Start-Up Case			Normal Case		
WHNO	SCALE	EFF	UEQ	FLC		Range			Range	
Impeller-1	1.058	0.794	564	0.0592	42	O	10	30	S	44
Impeller-2	1.058	0.794	547	0.0532	40	O	16	30	S	45
Impeller-3	1.058	0.797	532	0.0481	38	O	21	30	S	45
Impeller-4	1.058	0.803	517	0.0435	36	S	26	31	S	44
Impeller-5	1.058	0.807	504	0.0396	30	S	39	27	S	53
Impeller-6	1.058	0.781	491	0.036	10	U	43	28	S	51

Surge

Figure-3: Compressor characteristics – Lighter Process gas vs. Heavier Startup gas

As can be seen from above Figure-3, the heavier start-up gas volume is reduced more than the actual process gas, front end of the compressor is approaching choke and back end of the compressor is approaching surge. For reference, please see the normal case impeller range in percentage from surge and choke.

Strategies to implement the dry-out or air-run in cases:

In addition to volume mismatch, another issue that may be often encountered in running the start-up cases is the discharge temperature. Sometimes, these startup cases results in a high discharge temperature which may not be acceptable from compressor design, seal design or process limitations. At times, it is as easy as a change the compressor/O-ring materials to accommodate this higher discharge temperature.

Below are some of the strategies that can be considered to accommodate the start-up/dryout cases with volume mismatch. This is only a partial list of strategies and by no means, not a full list.

a. Back to back design: Modern compressors use Dry gas seals as casing end seals. Back to back configuration puts the higher final discharge temperature gas in the middle of the compressor away from the seals.
b. Speed reduction: When the process control is variable speed, it is often possible to run the compressor at the minimum speed or at the maximum continuous speed to accommodate the dry out case. A heavier start-up MW case than the process gas require to run at the minimum speed, whereas a lighter start-up MW case compared to the process gas may need to be run at the max continuous speed in order to overcome the volume mismatch.
c. Independent control loops: In a multiple sections/multiple body machines it will be very beneficial to have a separate control loop for each section. Having this will help to control the volumetric flow of the each section independently to overcome the volume mismatch.
d. De-coupling part of the strings: In a multiple body, it may be possible to de-couple the 2nd body and only run the start-up/dryout case on the first body.
e. Gear ratio change: Though this is little more complex, when a gear is present between the driver and driven equipment, change in gear ratio is another option to lower or increase the speed and accommodate the dryout case.
f. Extraction: When the start-up gas is lighter than the process gas, as discussed earlier, the front end of the compressor tends towards surge and back end of compressor approaching choke (stonewall). Having an extraction inside the compressor will help to overcome this situation and successfully run the dryout case. This also helps to lower the compressor discharge temperature, which helps with compressor design and dry gas seal design. The next section in this paper discusses a case study, which used this extraction strategy to overcome both volume mismatch and higher compressor discharge temperature.
g. Pole change design: For fixed speed motor driven compressors, a pole change option can also be considered to accommodate the start-up/dryout cases.

h. Mix start-up gas: Where process allows, a combination of lighter and heavier gas mixture may be considered for start-up/dryout runs, examples- Nitrogen and Hydrogen combination or Nitrogen and Helium combination.

In addition to the above when running on hot Air or Nitrogen into a system that may contain hydrocarbons, auto ignition temperature concerns also need to be considered, Reference: Bureau of Mines Circular 8187.

Case Study, Extraction:

Below compressor case study is a real example of an Elliott Group compressor design. In this particular example, the process gas MW is higher than the start-up/dryout gas. Also, the start-up is available at a suction temperature of 105 deg F, which resulted in a very high discharge temperature, greater than 500°F, which the end user is not comfortable. This is a seven stage straight thru compressor used in olefins process.

Elliott Group designed an extraction port after 3rd stage of compressor, which not only helped to keep the backend of the compressor out of choke but also helped to lower the final discharge temperature.

Section-1:

WHNO	SCALE	EFF	UEQ	FLC	Range		
Impeller-1	1	0.824	665	0.1551	15	U	57
Impeller-2	1	0.835	658	0.1203	34	S	38
Impeller-3	1	0.84	613	0.1135	37	O	19

Section-2:

WHNO	SCALE	EFF	UEQ	FLC	Range		
Impeller-4	1	0.881	595	0.0662	23	S	58
Impeller-5	1	0.89	576	0.0577	37	S	42
Impeller-6	1	0.863	560	0.051	46	S	22
Impeller-7	1	0.725	564	0.0419	52	O	10

Figure-4: Compressor characteristics – with extraction only for the Start-up/Dryout case.

The above figure shows the compressor characteristics with extraction after 3rd stage of the impeller. Please refer to the figure-2, where with no extraction, the front end of the compressor tends towards surge and back end of the compressor is approaching choke (or stonewall). The extraction also helped to keep the final compressor discharge temperature <425 deg F.

Below picture shows Elliott Group's compressor design showing the extraction for start-up only

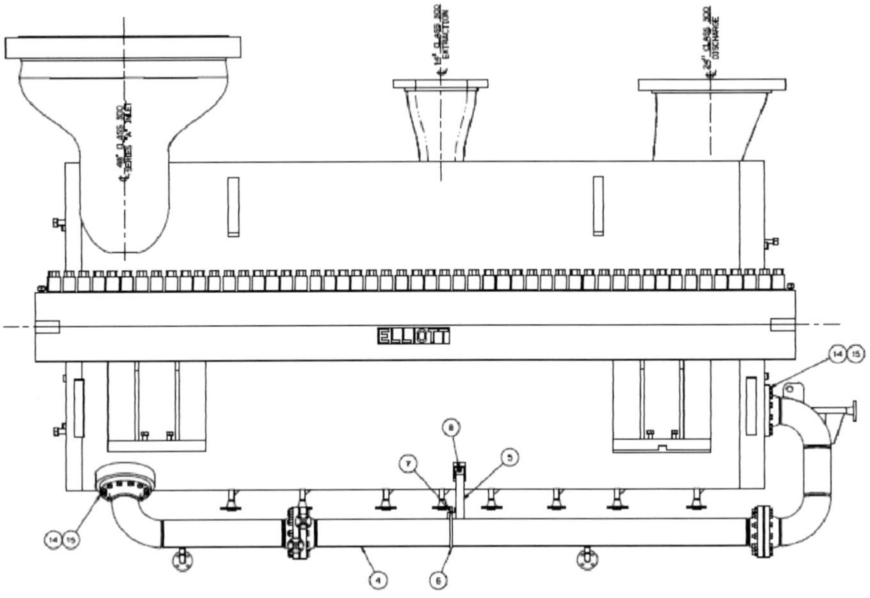

Figure 5: Elliott Group compressor casing showing the extraction line

The sketch shown below is a representative P&ID showing the extraction.

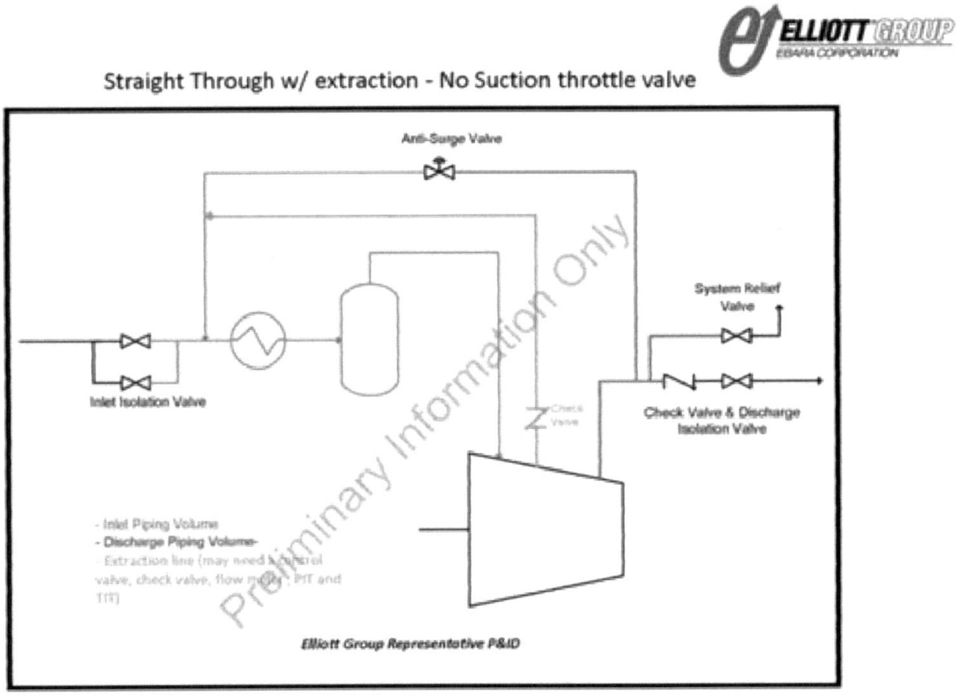

The above schematic shows the extraction line returning upstream of the cooler and suction drum. This is one of several options one could consider. Other options are taking the extraction to vent or a safe place or way upstream of compressor cooler/suction drum. Extraction line may need additional instruments/valves.

Figure 6: A schematic P&ID showing the extraction line

Having an extraction only for start-up case also produces unique challenges with respect to the control system and anti-surge and performance. Elliott group diligently worked with the controls vendor in designing an anti-surge and performance controls to accommodate this extraction.

Also, based on end user requests, Elliott Group added functionality for the operator to run the dry out case and switch over to process gas for normal start-up scenario. Please below screen shots taken from the actual control system HMI showing the functions of the dryout case. The operator has the flexibility to switch over from dryout to process gas through touch screen HMI and this also allows the operator to either start the unit on process gas or dryout gas. There is also trip on compressor discharge at 450 deg F.

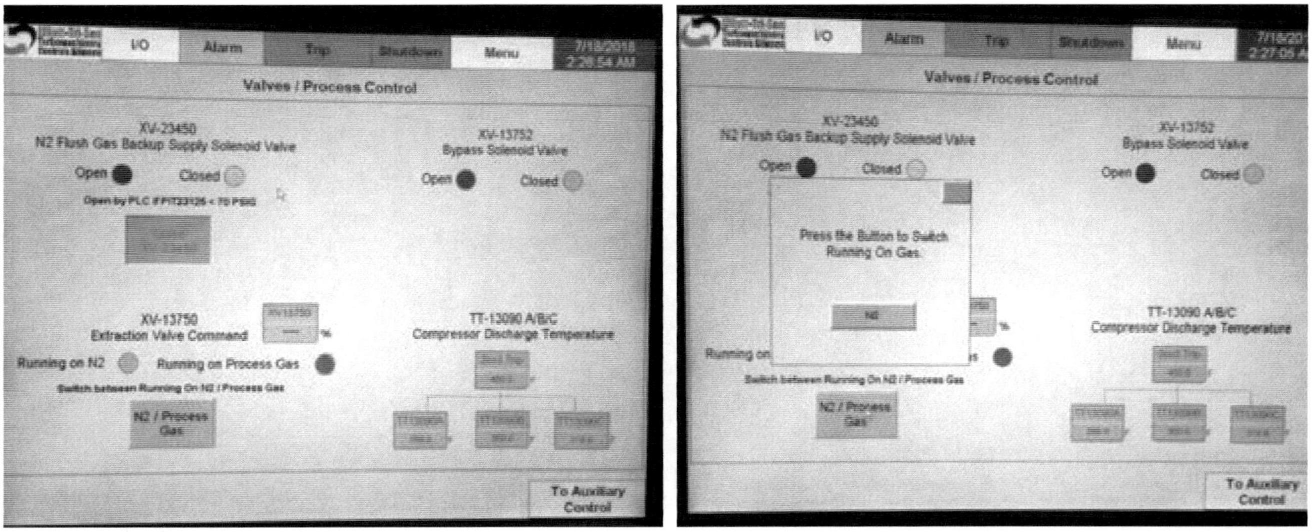

Figure 7: Control system HMI with Extraction for N2 dryout case only.

Conclusions:

At times, it is very challenging to design the compressor suitable for the start-up/dryout case while delivering an optimized performance for the main process gas/conditions. The authors discussed the challenges with using heavier or lighter MW gases than the actual process gas for startup/dryout and listed various strategies to accommodate these start-up scenarios. The authors also discussed the additional design considerations including dry gas seal design and control system design in accommodating these start-up scenarios. In conclusion, working closely with the compressor OEM, process licensor, EPC, end user, controls and seals gas vendors it is possible to come-up with an economically viable design to accommodate the start-up/dryout needs.

Acknowledgements:

The authors thank Elliott Group for allowing them to publish this paper. The authors also thank their colleagues who contributed to the work cited in this paper.

550880 Active Magnetic Bearing for Turboexpanders in Ethylene Plants: Technology and Advantages

Active Magnetic Bearing (AMB) turboexpanders are becoming an industrial standard in ethylene production plants. End customers are choosing this technology over oil bearing system for better reliability and efficiency, smallerfootprint and lower life cycle costs. This article briefly describes the AMB technology and its evolution since the first patents followed by the comparison between the AMB and oil bearing technologies. The analysis includes technical and economic factors.

550589 Digital Ethylene Plant: The Convergence of Process and Operational

Digital Ethylene Plant: The Convergence of Process and Operational

"Refiners are poised to adopt new technologies at a faster pace after years of incremental progress; this effort could save the industry billions of dollars." (Wood Mackenzie)

Given this important convergence of process and operational worlds in the Ethylene industry, this paper looks at how operators can integrate asset performance management (APM) solutions, with process simulation models and data analytics. This convergence allows for continuous interrogation of plant data against the process model, thus providing the ability to predict and quantify operational decisions in near real-time to manage risk and boost profitability. It will reference use cases from customers who operate in Ethylene and illustrate workflow and the outcomes achieved.

AIChE Paper Number: 77B

Harnessing Ethylene Plant Manufacturing Operations Reliability using Data Driven Analytics

Dr. Pratap Nair

President & CEO

Ingenero

Kaushal Sampat

Senior Manager, Analytics & Modeling

Ingenero

Prepared for Presentation at the 2019 Spring National Meeting

New Orleans, Louisiana, April 1 - 4, 2019

Harnessing Ethylene Plant Manufacturing Operations Reliability using Data Driven Analytics

Availability of the Ethylene plant asset is an important factor that determines the Overall Manufacturing Asset Effectiveness. It is not uncommon for advantages realized from optimization and advanced controls, to get wiped out by an unforeseen plant shut down. Retro analyses of such Reliability issues have often shown that the root cause of the issue almost always exhibits symptoms way before the event. However, capturing such symptoms before it is too late and proactively addressing the issues are critical for Asset Reliability Improvement.

The symptomatic indicators reside in the large volumes of a variety of data that is measured on a continuous basis from the manufacturing operation and archived. However, identifying such symptoms is difficult and time consuming for the unaided human mind, unless it shows up obviously on a commonly monitored parameter. Such identification would entail sifting through thousands of a wide variety of data points, collected at minute wise or higher frequencies. Several of these parameters measured are non-linearly related to one another, further complicated by different types of errors in the collected data.

Big Data Analytics techniques can be utilized to help effectively analyze thousands of minute wise tag data for an Ethylene plant, in real time, to help "find problems before they find you". This paper describes how Machine Learning Techniques have been applied to Ethylene Manufacturing to provide useful, non-intuitive insights that are usually hidden in the large volumes of data collected. When carefully integrated with fundamental process models and combined with domain knowledge, it helps identify symptoms well in advance and allows for Root Cause Analysis to be done before it precipitates as a Reliability issue. The integrated approach is essential to minimize alerts on "false positives", which is important for the sustained utilization of such an application.

The multivariate nature of these techniques makes this approach different and more effective than the traditional Statistical Process Control (SPC). Sixteen Unit operations have been identified as common areas prone to unforeseen reliability issues in the Ethylene Plant. These techniques have been effectively hooked to live plant data for effective Reliability monitoring and automatic outlier identification, with focus on the sixteen identified areas. Also, coupling this with vendor equipment performance curves and fundamental predictive models, allows for live comparison of the equipment performance with its design. Commonly occurring failures can be predicted early based on pattern recognition using failure maps and signatures. Not so commonly occurring failures are also identified with indicators that allow basic engineering analysis to predict possible reliability issues.

Case examples in Ethylene plants where these techniques have enabled early identification of possible anomalies; causes of possible failures and increased equipment effectiveness are described in this paper.

The Significance of Reliability in Process Manufacturing

Starting up and sustaining production levels consistently from an ethylene plant does not always match expectations.

The raison d' être for the ethylene plant asset is to get best possible production of ethylene, without excessive costs and without incurring safety and compliance incidents. The Overall Asset Effectiveness (OAE) is a useful metric to quantitatively indicate how the Asset is performing. Details of the OAE and its calculation are covered in an earlier paper (2). The OAE provides a measure of how well the asset is doing what it is supposed to do – produce product. The key parameters impacting the OAE are Throughput (total feed to the plant), Yield (kgs product produced/kgs feed) and Availability (on-stream time of the plant).

Throughput (availability of feed and the ability of the plant to use the available feed) and Yield (the effective conversion of feed to product and its effective separation) are important to ensure performance and several initiatives focus on improving these aspects. However, it is not unusual to see the unavailability of the asset to produce offsetting these throughput and yield improvements. The Availability is largely determined by the Reliability of equipment operation.

Figure 1 shows a case in an Ethylene plant, where the plant is operating well above design on both the Throughput and Yield. Yet its total annual production falls below what the plant was designed to produce on account of below expected Availability of the asset on account of unforeseen Reliabilty issues, because of which the plant had to take unplanned shutdowns.

Figure 1

Case Example where Reliability is shown impacting the Overall Plant Performance

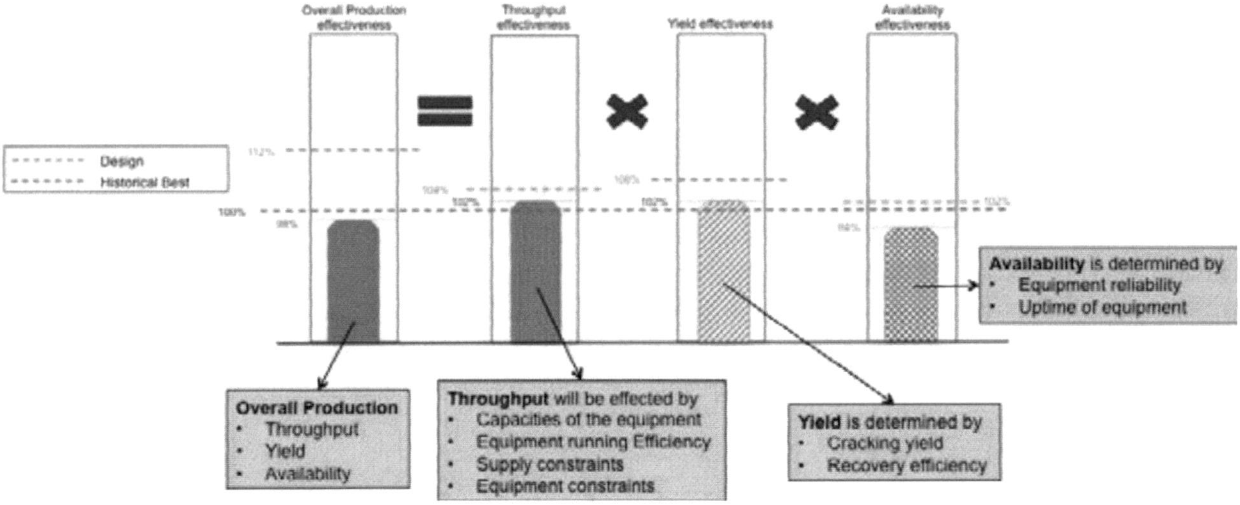

Key Challenges to be addressed

Most of the Reliability issues in a plant that requires an unplanned shutdown on account of an unforeseen issue with an equipment, upon retro analysis shows that signs of deterioration of an equipment on account of which, the plant has to be taken offline or the product quality deteriorates, was visible but hidden in the data collected from the plant and archived. However, these indicators may not be seen in the parameters the operations team normally focuses on. There are only so many parameters the human mind can observe and make sense of, at a time.

There is no dearth of data collected in most Ethylene plants, which clearly qualifies as Big Data (1,3). However, a large part of the data is just archived and not even looked at, on account of human limitations to process all the data and lack of tools to utililze the data (Figure 2).

Figure 2

<u>Data availability not a constraint but utilization is</u>

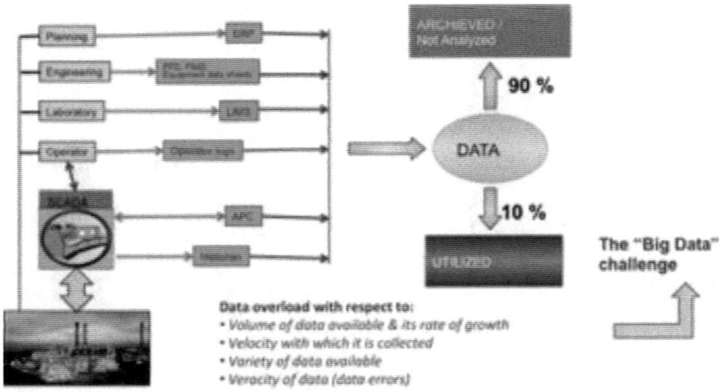

One of the key challenges that make it difficult to use the measured and stored data from the manufacturing process, as is, are <u>Errors in data</u>. Recent surveys have consistently shown this to be the top challenge (1), organizations face when trying to use the collected and archived data, as is. The data errors results in non-closure of the material balance around the plant. Understanding the types of errors (Figure 3) and being able to identify and weed them out of any analysis is key.

Figure 3

<u>There are various types of errors that can skew metrics and conclusions from an analysis</u>

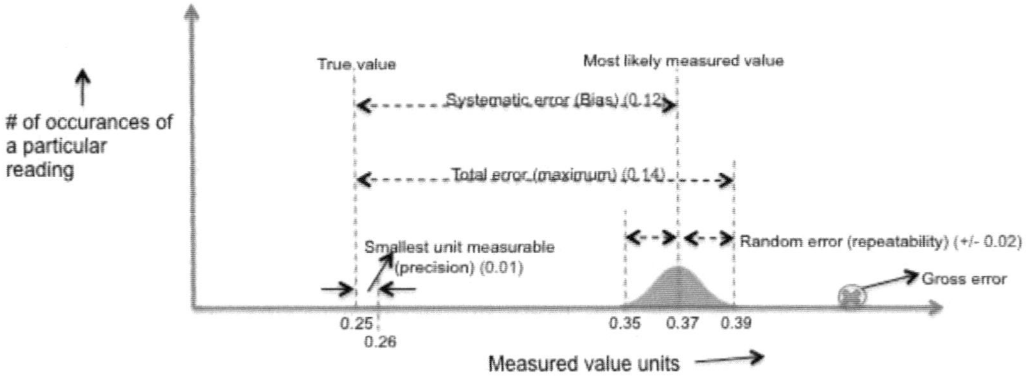

With Digitalization becoming an initiative being undertaken indsutry wide, leveraging the connectivity and affordable computing costs at feasible speeds, the veracity of the information being presented is important, since this would drive decisions. The connectivity and the viewer based dashboards from Digitalization initiatives addresses the lack of human bandwidth to process the big data. However, making sure that the information is not skewed by errors, resulting in erroneous conclusions, requires the data to be processed by analytics tools (Figure 4).

Figure 4

A basket of analytics tools and models essential for use of data when Digitalizing

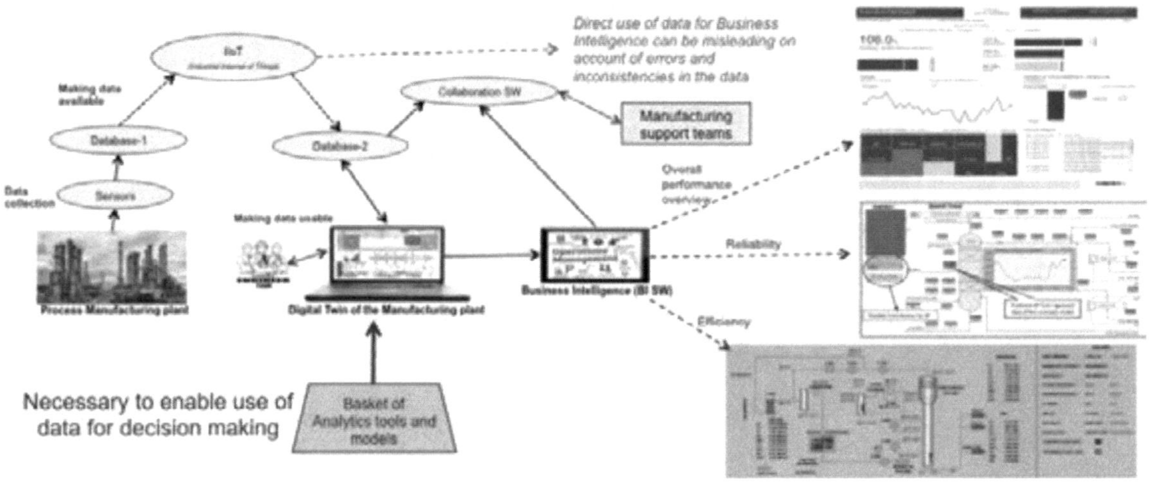

For the chemical process industry, fundamental models provide the intelligence necessary for such processing. However, use of fundamental models alone will entail a time lag between the collection of data and the conclusion. Machine Learning models, infused with intelligence (obtained from fundamental models) allows such analysis to happen near real time.

Data Analytics to improve Reliability

Reliability issues arise from equipment malfunction or quality excursions on account of changes to the health of the equipment. The key to improve Reliability, is to get an indication of the issue well in advance, so that there is sufficient time to address the potential problem or delay its ramifications till the next planned shut down, without suffering the losses on account of unplanned outages or quality excursions.

Predictive Analytics using auto adaptive Machine Learning models helps find problems before they find you with sufficient time to identify root causes and fix before an unforeseen shutdown or quality excursion (Figure 5).

A pure Data Science approach to use data from Process Manufacturing is likely to result in several false positives (false alarms). Alert fatigue from a purely data driven approach can be alarming. Studies have shown less than 5% of anomalous alerts flagged by a pure data driven Machine Learning algorithm, were actual threats – rest were false positives.

Figure 5

Early identification for improved Reliability

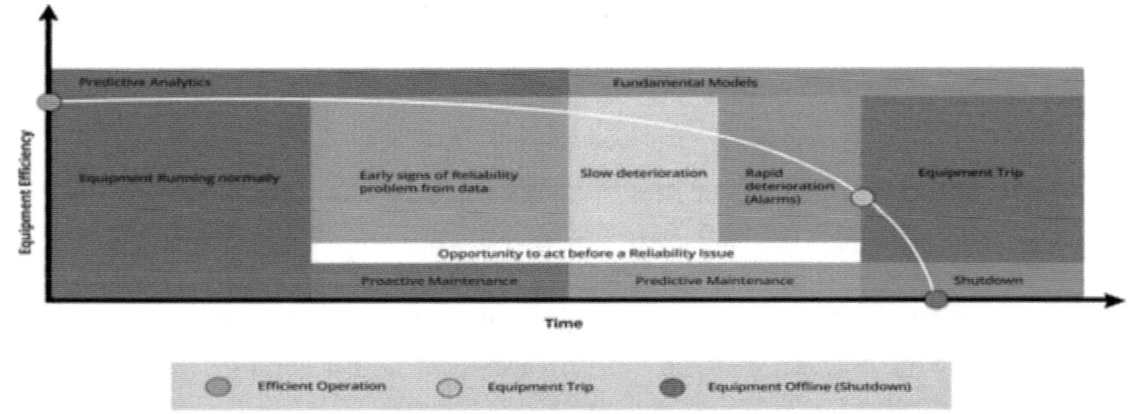

Figure 6 shows an application of utilizing Machine Learning to track equipment and flag early warning alerts, minimizing false positives.

Figure 6

Integrated solution for early identification of Reliability issues

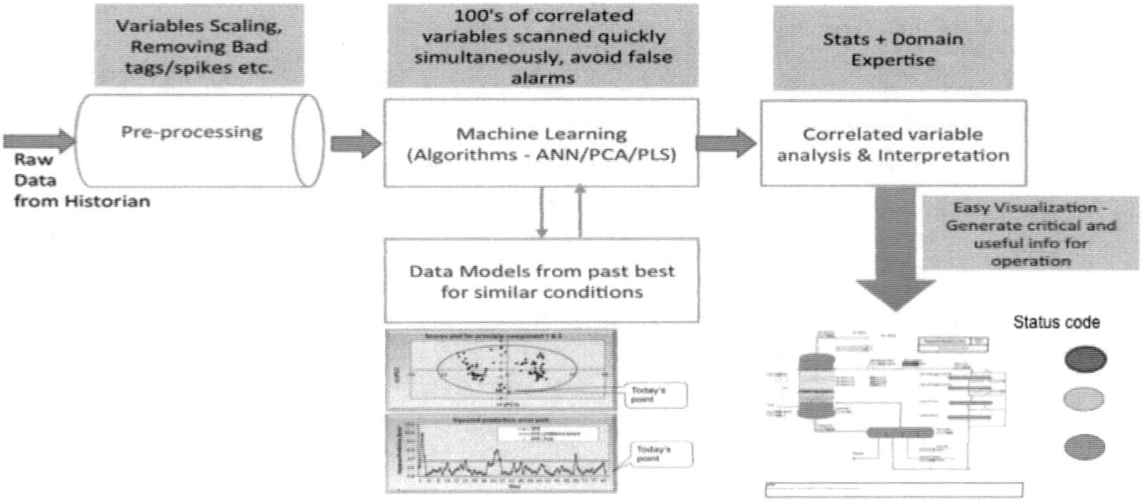

Insights we have gained through experience indicate several factors need to be considered when building Data Driven Machine Learning Models for Process Manufacturing

applications:

- ➢ Objective for which the model is being built should be clear
  - ✓ This will determine the choice of Training and Testing data used to build and validate the model
    - *For example, fouling detection models must be trained on clean condition data*
- ➢ Choice of Inputs and Outputs requires domain knowledge
  - ✓ Outputs determined by what we want to predict
  - ✓ Identifying variables that have maximum impact on predictions and whether their coefficients make directional sense from process perspective
  - ✓ The Degrees of Freedom and the control logic of every equipment have to be thought through before finalizing the input-output variables
  - ✓ Use of inferred variables as inputs instead of just measured variables greatly improves the prediction quality
    - *For example L/G, ΔT, SHC, P/E versus just the raw measurements*
- ➢ Proper filtering of training data
  - ✓ Based on purpose for which the model being built
  - ✓ Effectiveness of the Training data to enhance its applicability and reliability
    - *For example, Signal variation to noise ratio high enough*
- ➢ Combining Classification Techniques with Regression techniques (CART) can simplify the model prediction effectiveness and efficiency
  - ✓ Domain expertise aids the process in addition to classification based purely on the variable tree hierarchy
    - *For example, choose Feed type versus age of coil for furnace run length prediction*
    - *For example rather than building a non-linear model that covers all scenarios, categorize (classify) data by feed slates i.e. different models for different feed slates (9E-2P, 8E-3P, 7E-4P) versus combining all feed slate models in one*
- ➢ Provision in the model to learn on its own as equipment health changes and the model can be re-calibrated
  - ✓ Required to prevent false positives or recurring flagging of anomalies, even after it has been recorded

Machine Learning based Early Warning Tracker

An early warning tracker for Ethylene plants was developed and is being used as part of Monitoring and Tracking initiatives across multiple facilities. For the Ethylene facility sixteen key equipment areas were identified, based on historically occuring areas where Reliabilty issues are commonly observed, and monitored by the tracker:

- Furnaces
- Quench Tower
- CGC Machine
- Turbines *(CGC, PRC, ERC)*
- Cold Box
- QW loop
- Caustic Tower
- CG Dryer
- Saturator
- DOX Box
- MAPD Bed
- Condensate Stripper
- Deethanizer
- Debutanizer
- C2 Splitter
- C3 Splitter

The tracker is connected to the Historian and tracks in real time and detects anomalies early, just as it begins to see variations, not just on one specific variable but considering several related variables. Some of the features of this tracker:

- It sends automatic emails upon Outlier identification
- False Positives reduced by following the integrated solution approach as shown in Figure 6
- First Principles Models/ Vendor Performance Curves/ Column Loading Charts incorporated
- The analysis is multivariate hence different from SPC (which is typically uni dimensional)
- Identifies early signs of failure or Equipment performance degrade
- Possible causes of the outliers are identified and alerted
- Allows trending of outliers
- Easy deployment at site and linking to Historian with minimal maintenance

A snapshot of the typical output from the tracker is shown in Figure 7.

Figure 7
Early Warning Tracker output

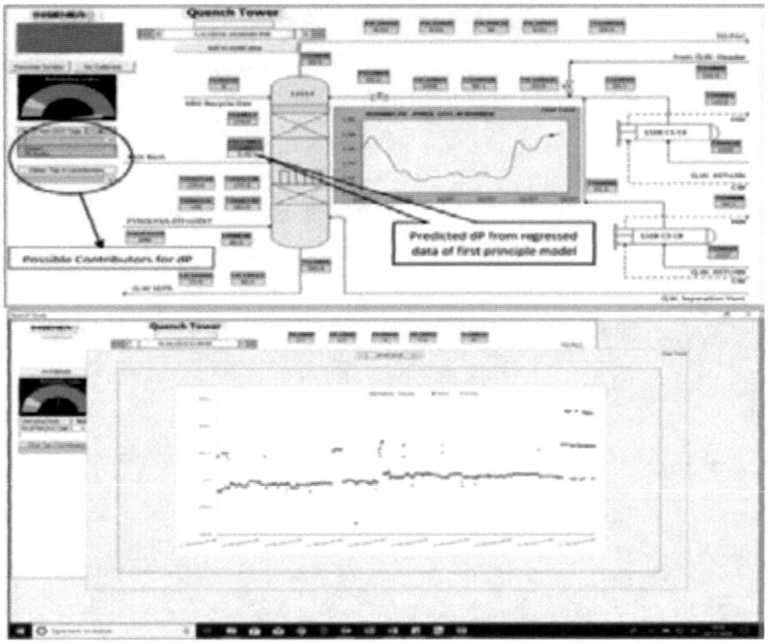

Scanning thousands of correlated variables in a typical Ethylene heat-integrated plant is not possible for the naked eye. Finding problems early helps avoid unplanned production losses and also helps in narrowing down variables for Root Cause Analysis. The automatic analysis of several correlations is only possible through machine learning models and takes and order of magnitude higher effort if it were to be done manually. The intelligence incorporated from independent runs of fundamental models and domain expertise to select appropriate inferred variables as model inputs, is essential to make the analysis results "reliable".

Equipment Reliability improvement Case Examples using the Tracker

Three case studies have been selected, to illustrate the utility of the Reliability Tracker:

1. Live Fouling detection for the C3 Splitter Reboiler
2. Live Machine Learning based operator guidance for C2 Splitter
3. Retro analysis using the Reliability Tracker for a Quench tower

Live Fouling detection for the C3 Splitter Reboiler

The C3 Splitter reboiler is heated by Quench water on the utility side. The quench water circuit is prone to fouling and this is a common occurrence on this reboiler, which then starts impacting the performance of this column making it a plant constraint.

A purely UA based calculation, which is the typical paradigm, is too simplistic and leads to many false positives. The parameters around this system are strongly correlated to each other, making simple multivariate regression unviable as a solution.

Once the Machine Learning model was well posed, it was found to successfully identify early signs of fouling in the reboiler, almost a month in advance of it starting to impact the process performance. The model is trained on clean reboiler conditions, which then serves as a multidimensional reference for "true" outlier detection, that indicates early fouling. The Quench water outlet temperature was found to be a key indicator, reflecting this comparison, to be shown on a 2-D plot, for easy visualization of deviation (Figure 8).

Figure 8

Early indication of fouling more than a month before process impact

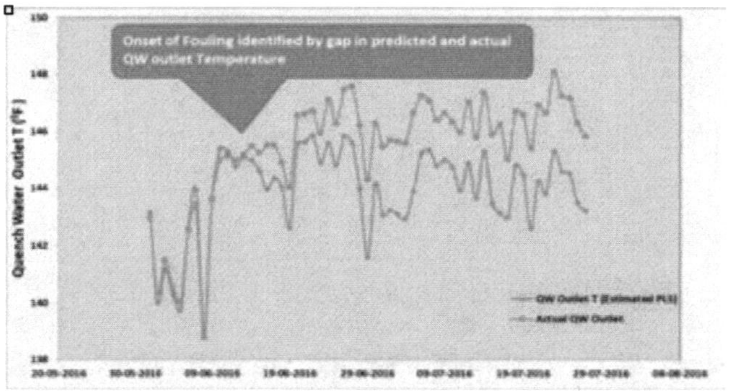

Choice of inputs and outputs, with appropriate data filters to capture the input/output variability adequately, to weed out noise and calculate "true" correlation coefficients, is required and critical (Figure 9).

Figure 9

Machine Learning Model I/O for a C3 Splitter

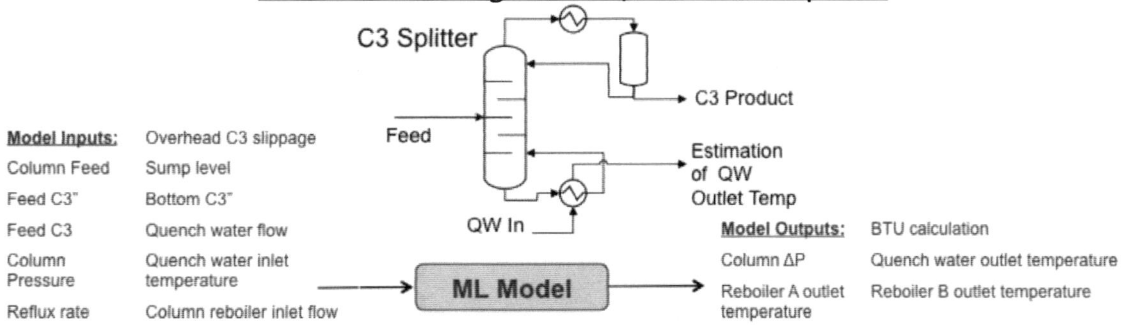

Live Machine Learning based operator guidance for C2 Splitter

The C2 Splitter Controls are often placed on manual (even with an APC) and Setpoints often end up being based on operator intuition. The Ethylene product often goes off-spec on Ethane in the Ethylene product, especially when feed rates go up.

The Reliability Tracker, by virtue of the underlying Machine Learning based model is able to provide live guidance to operators, beyond just alerting on deviations while tracking, to help avoid quality excursions, which is another issue related to asset operations Reliability in delivering not just volumes of product but also the required quality of product. Figure 10 shows an example of live recommendations of Setpoints provided to ensure on-spec product.

Figure 10

Operator Guidance for C2 Splitter from the Tracker

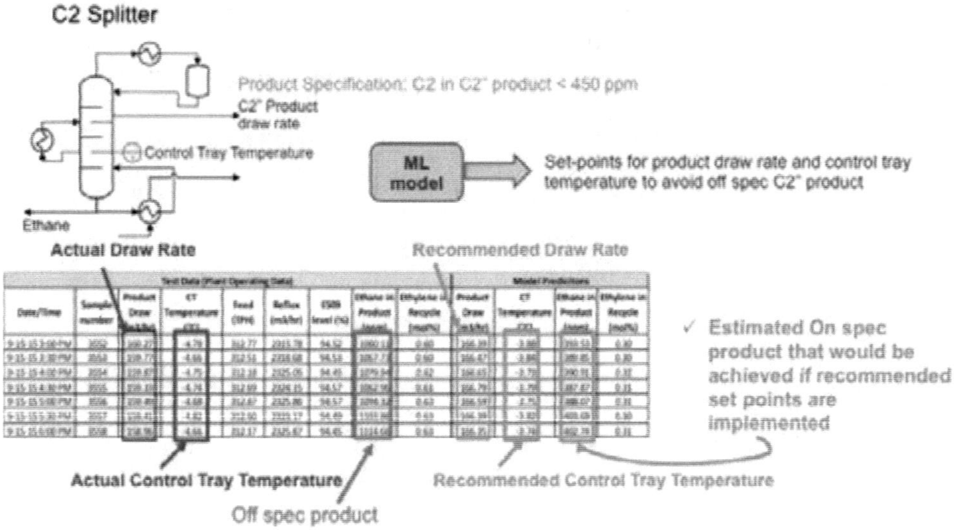

Retro Analysis using the Reliability Tracker

The Reliability Tracker was used to perform a Retro analysis of a past failure of a Quench Tower, based on historical data. The tracker was configured and then made to run through the historical data, like it was doing a live tracking of it, in a fast forward mode.

The Tracker started indicating variations several months before the failure occurred and even indicated possible root causes of the failure, as part of its flagging potential causes for the variations. The interpretation of the indicators flagged by the Tracker helped the process monitoring team easily identify fouling of the Quench tower, on account of consistent operation of the furnaces at high Propane to Ethane (P/E) ratios, without increasing the Steam to Hydrocarbon ratio, to reduce the formation of greater quantities of tar, on account of the higher heavies content in the feed. More over, the removal frequency/rate of tar from the Quench bottom was also not increased from when the plant was run at lower P/E ratios of feed.

The diagnosis would seem obvious to a process engineer monitoring the plant, once the key parameters are seen in conjunction, which may be easily missed if it is not presented or flagged by exception, as done by the Tracker. Had the Tracker been operational before the failure, the plant process monitoring team would have picked up on this quickly and fixed the root cause, thereby preventing the fouling to escalate to a stage of equipment failure and an unplanned shutdown to change the damaged column and loss of about one month of production.

Manual independent trend analysis of the historical data, post obtaining the insight from the Tracker analysis, as a cross verification, indicated that the furnaces were indeed run at higher Propane to Ethane feed ratios and lower Steam to Hydrocarbon ratios, instead of higher SHC ratios as would have been prudent (Figure 11).

Figure 11

Manual trend analysis of historical data

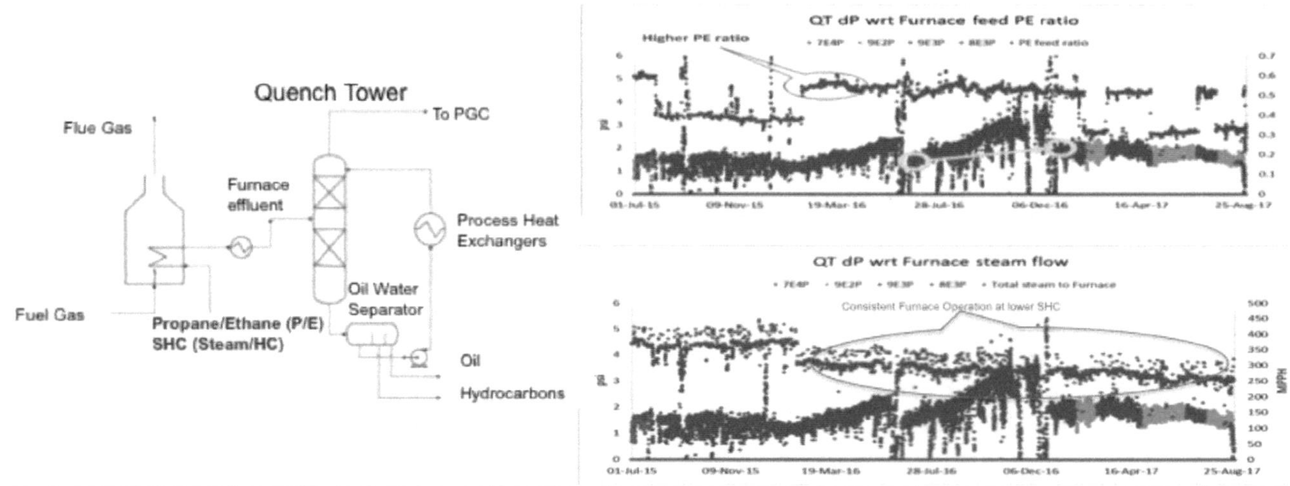

In the Quench model, predicted pressure drops from regressed data of a first principles model, was used a reference to prevent it being erroneously flagged as an outlier. This is an example of "adding intelligence from a first principles model" in the Machine Learning model (also shown in Figure 7).

Significance of domain expertise in the Reliability Tracking analysis using Machine Learning

Domain expertise is a major enabler for successful application of Machine Learning algorithms to Process Manufacturing applications. The impact of inferred variable selection on the results that can be inferred is normally significant, which requires domain knowledge:

- Machine Learning algorithms are more effective if critical "Inferred" variables used to build data driven models
- This approach gives true outliers and helps doing a quick root cause analysis versus just doing the analytics on pure data from tags leaves the interpretation to the end user
- Chances of flagging false positives are high if only raw tags are used to build data driven models
- Knowing equipment's typical pain point and inferring tags around that highlights non-intuitive insights

In the Quench Tower example above, the following selection of inferred variables made a difference versus relying on measured tags alone:

- Temperature Tags across the mid-section of the Quench Tower were replaced by Delta Temperatures across the bed
  - ✓ Uniform distribution of furnace effluent inside the packed bed should indicate minimal T gradients
  - ✓ Channeling or fouling in the bed would flag DT as outlier
  - ✓ Raw T tag may not flag as outliers
- To track relative Vapor-Liquid traffic inside the quench tower, tracking L/G ratio vs only L and G separately has multiple advantages in correlating with column ΔP
- Ratio of Total Propane to Ethane feed to the furnaces and the total overall SHC of all furnaces together is a critical inferred parameter which impacts fouling in Quench Tower

Value added analytics to the Machine Learning based Reliability Tracker

Associated analytics using trends and Historical data from the Reliability tracker adds value. Two such examples are, failure prediction and live benchmarking, which will be briefly covered as part of this paper.

<u>Failure prediction from Degradation analysis</u>

Failure prediction requires past data from the time of a similar failure. This is then used as a fingerprint for recognizing the same pattern in future amongst the thousands of data points. Often times failure data may not be available. In this case another possibility is to do a probabilistic model analysis based on tends from the Reliability Tracker. Such an analysis may be valuable, as it gives a sense on the possible time to failure and the time that one has to fix the problem or prepare for it.

This was applied successfully to a highly mechanical system for a process with eighty Coal Gasifiers (Figure 12).

Figure 12

Degradation Tracker analysis for Time to Failure

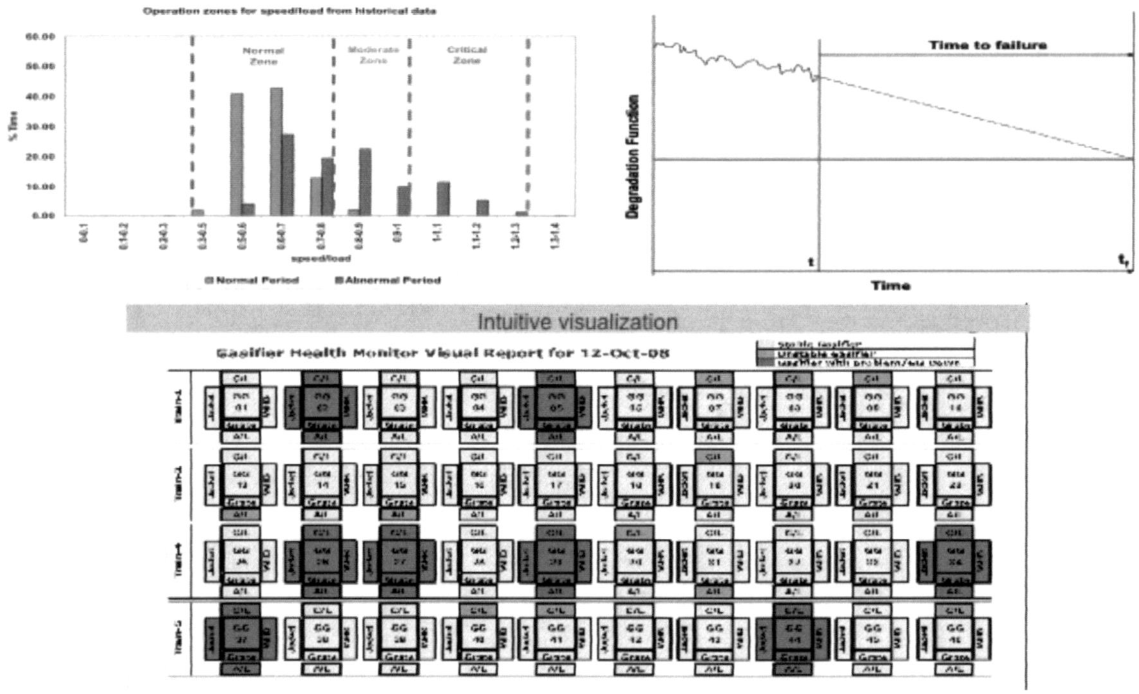

The Tracking coupled with the degradation analysis

- Enabled proactive maintenance and minimized unforeseen production capacity losses
- 7% improvement is asset availability
- 30 days alert versus just 4 days to take proactive action

This application is applicable to several scenarios in the Ethylene plant and other process manufacturing facilities:

- Component/equipment health monitoring
- Maintenance scheduling. Prioritizing maintenance within a fleet of equipment
- RCA and corrective action in mechanical and operational aspect in case of abnormal deterioration pattern
- Generalisation to any system/process
  - ✓ Reactors/Catalyst
  - ✓ Furnaces affected by coking in tubes

- ✓ Heat exchangers, condensers, after coolers
- ✓ Pumps, compressors, expanders

Live Benchmarking

Comparison with a yardstick is required, if we are to conclude on performance. Benchmarking with previous best operation and with design are two readily available yardsticks that can be used with available data, without having to run separate predictive models. Comparison to an ideal benchmark requires tuned fundamental models, which is more complex and time consuming and cannot be done in real time.

Machine Learning techniques are being used to do automatic live benchmarking of key metrics, like Yield or Production with previous Historical best and Design, for comparable conditions, from available historical data. The quick benefits that are gained from this automatic analysis would usually take several man-hours of trending and data analysis even for experienced domain experts. It saves the manual effort of interpretation derived from trending several variables together on a historian.

This analysis enables, the automatic identification of the best condition, based on a defined Performance metric that the plant has operated at, from the historical data made available to it and compare it with the current operation, for any defined equipment or section of the plant. The equipment/unit operation constraint and any possible operating change that could be readily made to close the gap, if any, between the current operation and the historical best, is identified and indicated.

The defined section of the plant could be, individual equipment, a Unit operation, the entire hot or cold side or the entire plant. The Performance metric would be the one that drives the comparison, while the result would be an explanation and/or an action to close the gap. The tool based on this analysis, ensures that the historical operating point selected is an "apples-to-apples" comparison, by matching key operating parameters of the plant from the current operation.

Such automatic benchmarking analysis when coupled with the live Reliability Tracker offers various benefits. It enables:

- ➢ Automatic Identification of Heat Exchanger Fouling
- ➢ Tracking Coil degradation and how much it is costing in terms of Yield
- ➢ Identifying faulty instrumentation automatically
- ➢ Hydrate formation identification in Cold Section
- ➢ On-the-fly comparison of Summer-Winter, Day-Night bottlenecks and how much it costs
- ➢ Early identification of Compressor degradation
- ➢ Identifying APC model drifts and tracking efficacy of APC
- ➢ Live Comparison of APC limits and Setpoints from historical data
- ➢ Tracking Furnace Differential Cracking, Furnace Efficiency and Burner Health, Decoking efficiency etc.

- Live Hydraulic tracking of Distillation Columns like Demethanizer, Deethanizer, Depropanizer, Debutanizer, C2Splitter and C3 splitter
- Automatic ERC, PRC comparison and its overall impact on C2 losses
- Tool can be configured at equipment level, section level and plant level
- Helpful for tracking plant Overall Asset Effectiveness, Equipment Availability and helps in decision making for Capex/Revamp projects
- Identifying Aging/non-performing equipment and its impact on Margin.

Figure 13 shows an example of the Deethanizer as the defined section, with the bottom ethane spec as the Performance metric. It automatically identifies and shows previous best Bottom Ethane slippage from the available history, for conditions equivalent to the current operating conditions. In this example, the gap is partly caused by C3" in Ovhd which is the constraint, but the gap can be closed by increasing Control Tray Temperature Set point.

Figure 13

Live Benchmarking for a Deethanizer

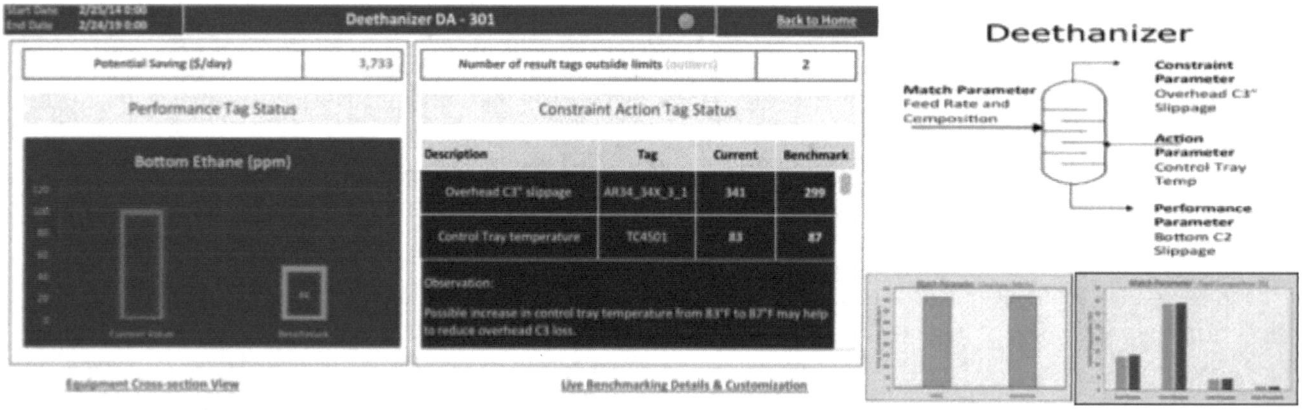

Conclusions

- Equipment Reliability and hence availability of the plant asset is one of the key parameters that impacts the overall effectiveness of the asset
- There is sufficient data available to be able to detect anomalies and potential reliability issues, albeit not readily usable in the raw form available in the historians
- Utilization of Machine Learning techniques enables us to closely track and draw out the early warning indicators, allowing us to find reliability problems before they find you and address
- Machine Learning algorithms allow closer to real time information
- Machine Learning models used by itself could result in several false positives
- Machine Learning models used in conjunction with distilled intelligence from fundamental models and domain expertise provide more robust and usable results
- Other Machine Learning techniques for failure prediction and live benchmarking offer enhanced value from a reliability tracker

References:

1) P. Nair, (March 7, 2017) "Tailoring Big Data Analytics for Optimizing Ethylene Plant Asset Performance" 2017 Spring National AICHE Meeting, Paper 2B (Ethylene Producers Conference).
2) P. Nair, (April 23, 2018), "Big Data Analytics for Improved Overall Asset Effectiveness of Ethylene plants" 2018 Spring National AICHE Meeting, Paper 169B (Ethylene Producers Conference).
3) L.F.Colegrove, M.B. Seasholtz, C. Khare., (March 2016) "Big Data – Getting Started on the Journey", CEP, pp. 41-45.

AIChE Paper Number 77c

Enabling Transfer Line Exchanger Utilization Optimization based on Advanced Data Analytics

Dr. Jörg Weidenfeller
Director RnD / Technology
SCHMIDTSCHE SCHACK | ARVOS GmbH:

Prepared for Presentation at the 2018 Spring National Meeting
New Orleans, Louisiana, April 01-04 2019

AIChE Paper Number 77c

Enabling Transfer Line Exchanger Utilization Optimization based on Advanced Data Analytics

Dr. Jörg Weidenfeller
Director RnD / Technology
SCHMIDTSCHE SCHACK | ARVOS GmbH:

Abstract:

Operation and reliability of transfer line exchangers are critical in the production of Ethylene and the economics of the plant.

The focus of the operator is on the process itself, for which standard sensor readings are monitored. Those readings up- and downstream the TLE nowadays are used to derive the condition of the heat exchanger. The interpretation of this data depends on the knowledge and experience of the individual operator. In addition regular inspections are conducted which only provide a snap-shot with limited confidence for future condition.

To support the needs of Ethylene plant operators and maintenance team the TLE functionality was further improved by applying Industrial Internet of Things technology (IIoT). In this case the TLE is equipped with additional sensors, installed fixed at measurement locations of interest for the TLE but not for the process. From the sensor readings meaningful values and equipment Key Performance Indicators (KPI's) are derived and monitored. Additionally the data are used for models which are based on OEM's domain knowledge with respect to specific thermal and mechanical design experience. In addition artificial intelligence (AI) is used to discover abnormalities in operation, to forecast the reliability and performance of the TLE and to provide diagnostics and future recommendations.

Introduction

SCHMIDTSCHE SCHACK | ARVOS GmbH, formerly known as ALSTOM Power Energy Recovery is designer and manufacturer (OEM) of transfer line exchangers (TLE) in Ethylene steam crackers and waste heat recovery systems in a number of chemical, petrochemical and metallurgical processes. The company was founded more than 100 years ago and was the first who introduced and established steam super-heaters for steam locomotives. Until then steam locomotives operated on saturated steam with an inappropriate poor efficiency. The founder of SCHMIDTSCHE SCHACK had the idea to improve this product by implementing a new technology in the existing envelope of the steam locomotive which significantly increased the efficiency of the steam process.

This company philosophy and motivation have not changed since then, but the products and applications did. More than 50 years ago, SCHMIDTSCHE SCHACK developed the transfer line exchanger (TLE) and established it in the Ethylene industry to replace the quench mode. Today, further enhancements to the TLE are being introduced continuously while keeping the existing envelope.

The operation and reliability of the transfer line exchangers are critical issues in the Ethylene process with impact on plant economics. In the transfer line exchangers the sensible heat in the cracked gas of an ethylene furnace is converted into high pressures steam (HPS). HPS is used to drive turbines, compressors and pumps in the separation train of an ethylene plant and is most important in the crackers energy system. Any loss of steam production by failure or decreased performance due to fouling of the TLE has to be compensated by HPS import from boilers requiring additional fuel gas or fuel oil with a negative effect on plant economics. High reliability and performance of the TLE's are most important, as utilization of the TLE's have a significant impact on the plant economics and require a high quality of the TLE, for maximum utilization and performance.

The status quo in Ethylene plants is that sensors and instrumentation are installed to monitor and control the production process and the process flows. From those measurement locations upstream and downstream the transfer line exchanger the status and condition of the hardware might be derived. The interpretation of these data strongly depends on the knowledge and experience of the individual operator. The real status and condition of the transfer line exchanger is obtained only during regular shut-down of the production line and inspection of the TLE. This indirect approach has uncertainties due to

- data are not collected directly at the TLE and do not describe TLE condition
- inspection during shut-down only provides a snap-shot with limited confidence for future condition

In general availability and performance are independent of each other, as a heavily fouled TLE will operate at a low performance level. Maximizing the availability and performance of the TLEs by online monitoring and predictive maintenance increases the asset utilization and will improve the plant economics significantly.

Enabling The Smart TLE

To support the needs of Ethylene plant operators SCHMIDTSCHE SCHACK started to further improve TLE's functionality by applying Industrial Internet of Things (IIoT) technology while keeping TLE's envelope. This requires a stepwise approach as shown in Figure 1. At the base level the TLE is equipped with additional sensors which will be installed during the manufacturing process in the SCHMIDTSCHE SCHACK workshop. At site the sensors will be installed after erection of the TLE. The measurement locations are at the points of interest for the TLE but not for the process. Those sensor readings are input for further evaluation and assessment of the status and condition of the TLE. The data are acquired and stored in an independent data acquisition system (DAQ) which will be installed nearby the TLE. The DAQ consists of Input/Output controllers connected to an edge-device which sends the live data via high-speed wireless network to the SCHMIDTSCHE SCHACK data storage. The complete DAQ system is enclosed in an explosive protected (ATEX) control box.

This approach ensures that the Data Control System (DCS) of the plant will not be impacted or disturbed. Optional operational data can be integrated by connecting to the DCS if the operator requests this. The SCHMIDTSCHE SCHACK data storage manages the storage and data handling. Live data are evaluated and visualized, historical data are stored, meaningful values and equipment Key Performance Indicators (KPI's) are derived and monitored.

The data are used for further models which are based on OEM's domain knowledge with respect to thermal and mechanical design. This enables diagnostic of the current TLE status. In addition artificial intelligence (AI) is used to discover abnormalities in operation, to forecast the reliability and performance of the TLE and to provide diagnostics and future recommendation service.

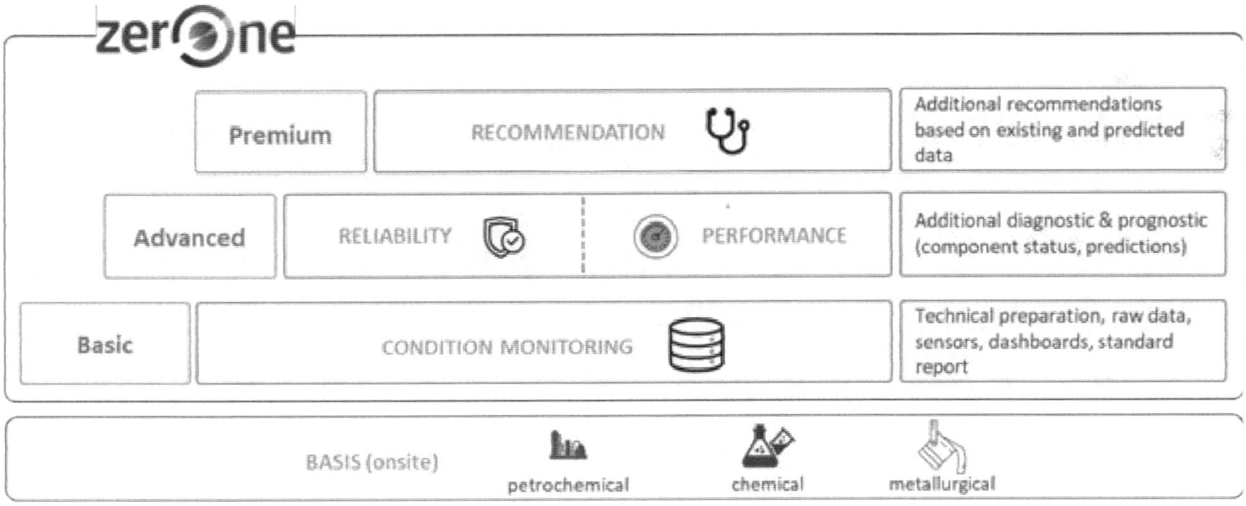

Figure 1: Stepwise approach to increase asset utilization

In summary this describes the new SCHMIDTSCHE SCHACK Zero.One platform for collection and analysis of data from TLEs. Zero.One is a cloud-based platform as a service and enables asset performance management and analytics by connecting equipment, data and people.

Smart Information Create Value

A typical view of a DCS monitor is shown in Figure 2. The data are displayed in time series and significant sensor values are marked. Although this presentation of data is well known it is difficult for the operator to identify the current status of the TLE as well as the future condition.

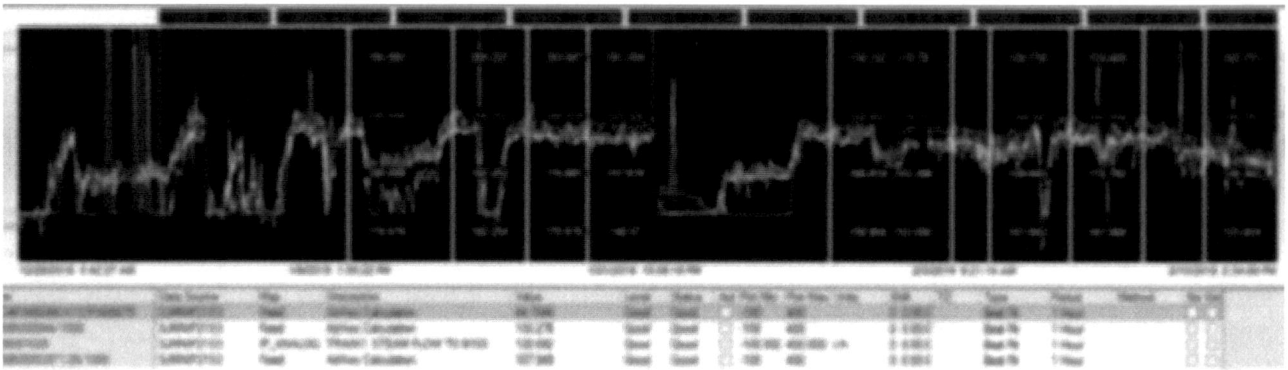

Figure 2: Typical DCS monitor view (blurred)

In contrast to that, Figure 3 shows a monthly KPI dashboard for the TLE. Several KPI dashboards for each TLE installed at the plant are aggregated to a single plant wide KPI dashboard. This information ensures a quick overview about the current and future condition of all TLEs by accessing the web-based platform mentioned above.

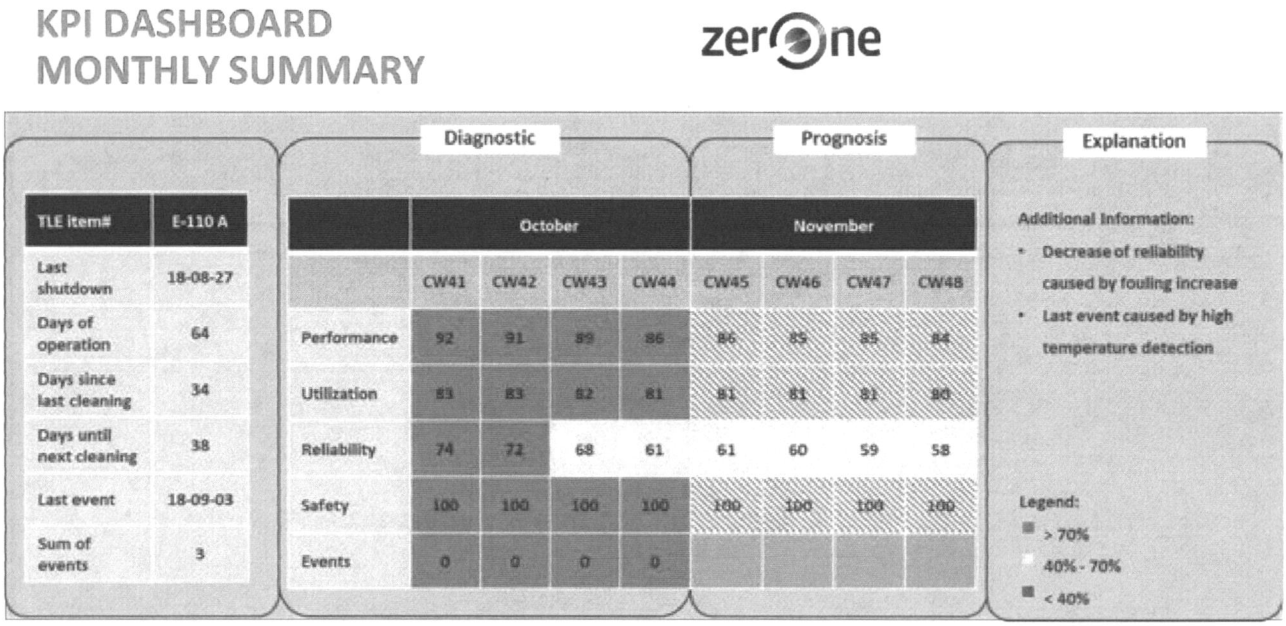

Figure 3: KPI Dashboards for single TLE

This dashboard looks clear and simple, however to derive those KPI's from sensor readings the application of OEM's design knowledge and design algorithms as well as AI is required. The user can drill down the dashboard to get more detailed information about thermal and mechanical conditions. In the following several use cases are presented to provide a deeper insight into the capability to monitor the current and future status of the TLE.

Use Case "High Inlet Temperature"

During operation the operator of a non-Ethylene plant recognized an increase of syngas inlet temperature. For a period more than two hours the temperature could not be controlled and the actual temperature reading was 90°K higher compared to the value which was used to design the heat exchanger. The operator was concerned about a potential impact on his equipment. Since Zero.One was not installed he contacted SCHMIDTSCHE SCHACK for support and to assess this event. Several discussions with the operator were conducted to clarify the situation, operational data were requested and the data were evaluated. Overall this took some days to get the right information.

In contrast to that procedure, with Zero.One the operator could have access to the web based system and could look at the condition dashboard as shown in Figure 4. This dashboard provides more details of the current TLE status. All graphs show the minimum, maximum and average value over a defined period of time as bar chart. These values are compared to upper limits like design values (shown as horizontal lines).

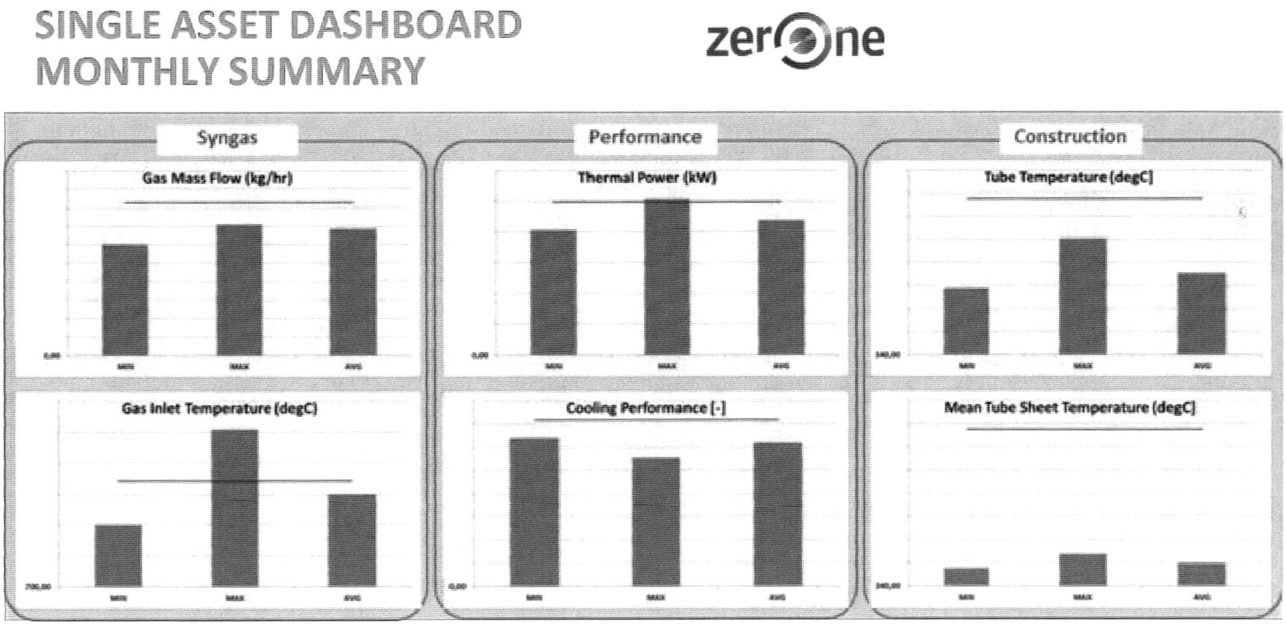

Figure 4: TLE condition dashboard for use case "High temperature event"

In the left column the values are shown which are important for the heat input into the TLE. It can clearly be seen that the syngas mass flow is always below the design value, whereas the high temperature event leads to an exceedance of the syngas temperature used for the design. Although the syngas mass flow is low, the high inlet temperature leads to a thermal power close to design value as shown in the middle column. Nevertheless the cooling capability is still excellent since all values are below the limiting value. Consequently the metal temperatures of the process tubes as well as the tube sheet are still below the design values as shown in the right column. Hence this high temperature event did not have any impact to the mechanical integrity of the equipment. This information can be obtained quickly by accessing the web based system. In addition, in case of a critical situation the operator will be informed automatically by email.

Although the information in Figure 4 looks simple, the procedure to provide thermal performance and mechanical integrity information requires the application of OEM's algorithms to calculate flow, heat transfer, hydraulic water circulation and mechanical. It is worth to mention that those algorithms are not based on commercial software but on in-house developed calculations which originally are used to design the heat exchangers. Those calculations are verified since decades by the TLE fleet operating at customers` Ethylene plants.

Use Case "Decreasing Water Side Pressure"

The second use case happened in the past, where Zero.One was also not installed. An operator informed about a damage of the exchanger and was asking for the root cause of failure. After several discussions the customer incidentally reported that after a regular maintenance shut-down the water side pressure level was decreased for further operation (the motivation to do that was not explained). A lower water pressure in steam drum and TLE results in a lower saturation temperature which indicates a better cooling capability. Although this is in general correct, it can be very critical to draw this conclusion for existing equipment which was designed for different conditions.

Assuming that Zero.One would have been installed the appropriate algorithms would have been applied based on operation data. This procedure transforms the raw sensor readings to smart data with valuable information for the operator. The results are displayed in the TLE condition dashboard as shown in Figure 5. The heat load to the TLE by the crack gas is very stable since minimum, maximum and average values are very similar. The maximum gas mass flow exceeds slightly the design value whereas all other crack gas values are below (see left column). The maximum thermal power also exceeds slightly the design value whereas the average value is also below. However, these KPI's are no cause for concern. In contrast to that the cooling capability always exceeds the limiting value (see middle column). This already indicates that the TLE cooling at water side is not further secured. The impact on the mechanical integrity is shown in the right column. The metal temperatures of the tubes as well as the tube sheet are always above the design value which directly indicates the negative impact on the mechanical integrity.

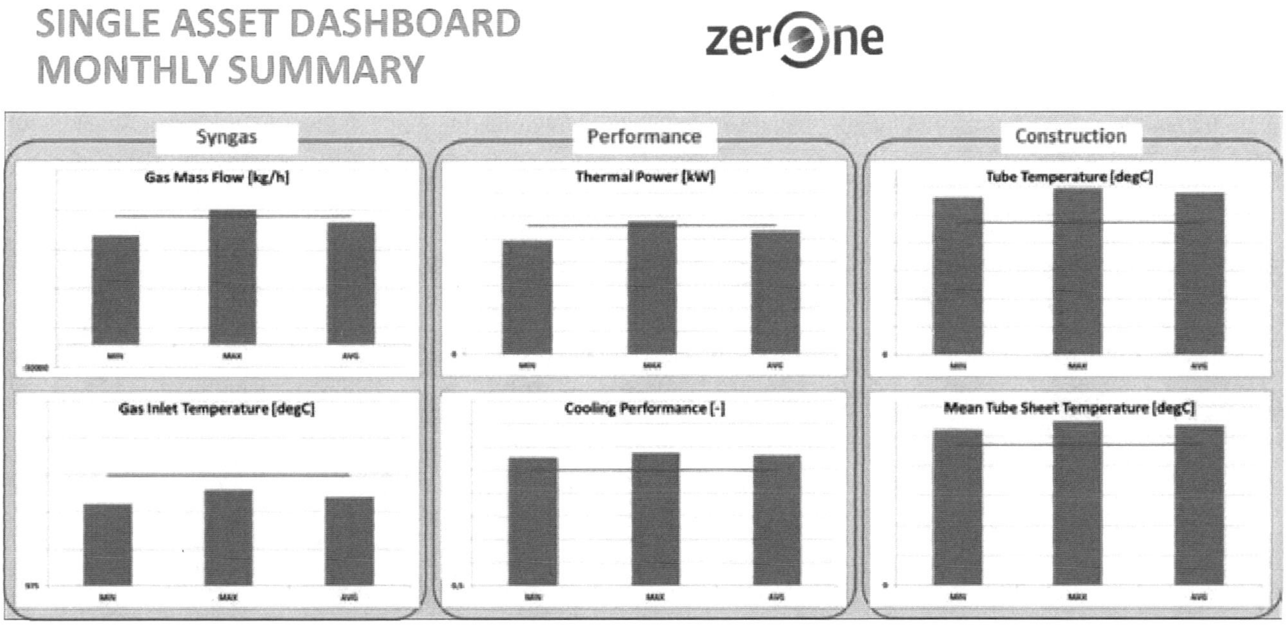

Figure 5: TLE condition dashboard for use case "Decreasing water side pressure"

To identify the root cause of the reduced cooling capability the operator can drill down to the cooling performance dashboard as shown in Figure 6. On the right side the measured sensor data for water/steam pressure and temperature are displayed as minimum, maximum and average value. The black bars describe the situation before the pressure level at water/steam side was reduced because those values are close to the design values (marked by the horizontal line). The status after reducing the pressure level is shown by blue bars. On the left the cooling performance map of the TLE is shown. The black dots that correspond to the black bars on the right are time dependent operation points. All points are placed along the expected cooling operating line that is located in the area for secured cooling capability. In contrast to that the blue dots are located in an area marked by red color for which proper cooling is not ensured. The reason for this is the decrease of water/steam pressure as shown at the right side. With this dashboard the operator directly can see the effect of lowering the pressure level. An active Zero.One system would have sent an alert message to the operator when the actual cooling performance would have reached the unsecured area. With that information the operator has the chance to apply countermeasures that bring back the performance to the valid operation range. In consequence the risk of equipment damage is significantly reduced.

The physical explanation of that use case is that due to the pressure decrease and the lower saturation temperature at water side with nearly unchanged heat load from gas side consequently the heat flux increases. In parallel, the critical heat flux as threshold for departure from nucleate boiling decreases as well as the steam fraction increases. This leads to the situation of local film boiling that dramatically reduces the cooling effect. In consequence the material temperatures increase. All calculations which describe this behavior and assess the current operation are included in the Zero.One algorithms to create the cooling performance map.

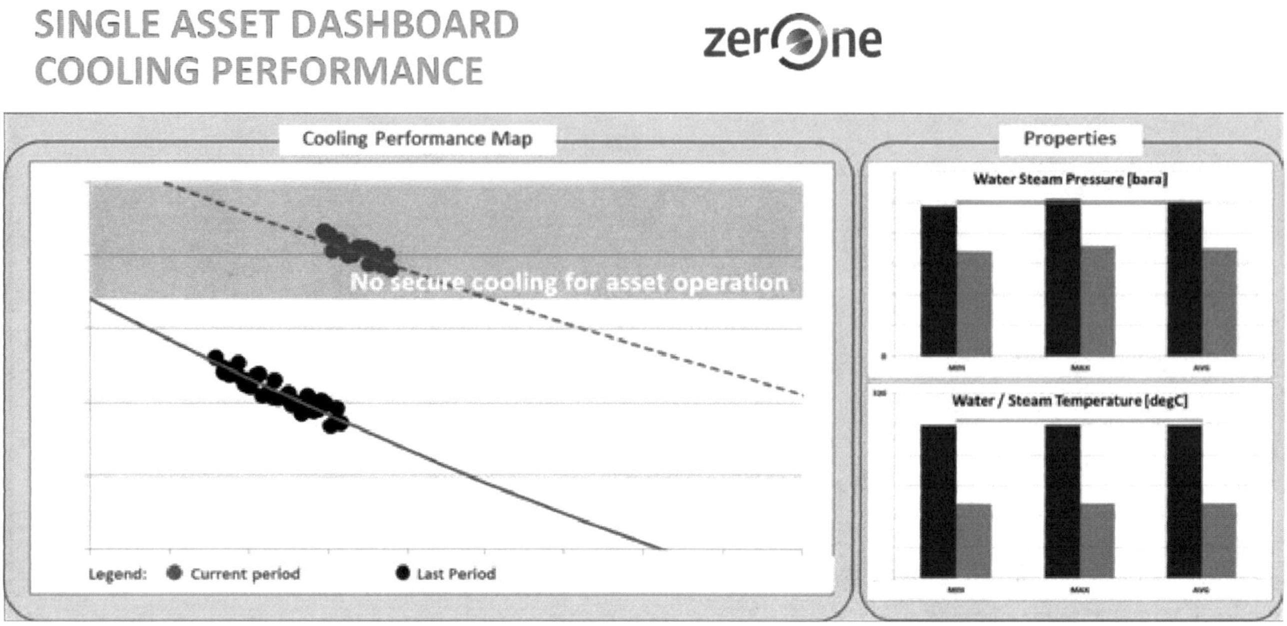

Figure 6: TLE Cooling Performance dashboard

Machine Learning

The KPI dashboard shown in Figure 3 provides also information about event or abnormality detection, prognosis of future KIP's describing the TLE condition as well as expected time period until the next de-coking. This requires the application of statistical methods, AI or the combination of AI with physical models, depending on the application.

The challenge when applying AI is to identify the right approach and algorithm. After this is obtained the model is going to be trained in a way that the error between output from the model and the known target value is minimal. When training a model, further verification is of the trained model is required to verify

- how the trained model reacts to new and unknown data
- if the training did not specialize the model in that way that accurate predictions are only possible with the training data

One use case for machine learning is to predict the time period until next de-coking of the TLE. The signal of gas outlet temperature was used as training input. The TLE is operated with different feeds, first with heavy feed then the feed is changed to light feed until de-coking. After that the operation sequence again starts with heavy feed and again switches to light feed.

The training was performed only for three operation cycles as mentioned above. After the training the algorithm was applied to unknown test data to predict the remaining time until de-coking. In addition, several unknown boundary conditions were implemented (see Figure 7):

1. heating surface reduced by 15%, 100% load, reduced operation time
   (blockage of tubes reduces heating surface, for 100% load the gas outlet temperature increases faster, in consequence operation time is reduced)

2. heating surface reduced by 15%, 90% load, normal operation time (operator reduces load to 90% to achieve the normal operation time)
3. no reduction of heating surface , 90% load, extended operation time (operator reduces load to extend operation time)
4. no reduction of heating surface , 100% load, normal operation time

The algorithm predicts the time until de-coking with an average error of +2/-4 days for the unknown boundary conditions of reduced heating surface, reduced load, extended or reduced operation time. In case that the training would consist also these events, a further accuracy improvement for the prediction is expected. However, this use case clearly shows that the chosen model is feasible for this prediction.

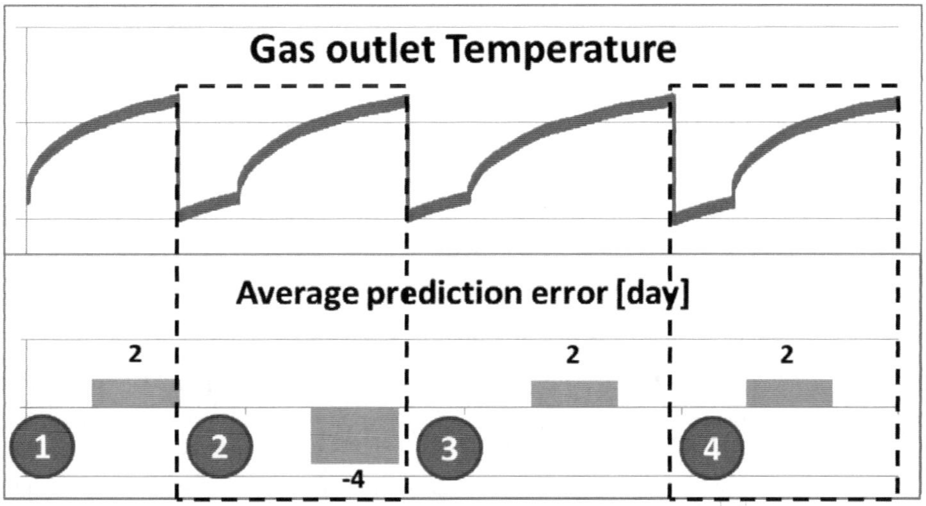

Figure 7: Test data, error between time until de-coking and prediction

Conclusion

This paper illustrates the next consequent step in further improving the functionality of SCHMIDTSCHE® Transfer Line Exchanger by applying new technologies. Zero.One comprises additional sensors installed at the equipment and applying a cloud based platform to analyze and assess the TLE. This service combines OEM design knowledge and artificial intelligence to describe the current status of the TLE, to detect abnormalities during operation and to predict future events. This creates transparency in the TLE condition, helps to optimize availability and productivity and improves planning reliability for the operator and the maintenance team.

The next possible integration steps are:

- online erosion / corrosion damage detection
- coil coking behavior for cleaning prediction
- further mechanical models for service assessments.

549518 Mitigating Risk: A Commercializing New Catalyst Technology in a World Scale Ethylene Plant

Bringing a new ethylene plant on stream and to full capacity quickly and smoothly is a tremendous undertaking. This is especially true in ethylene plants which employ "Front End" acetylene hydrogenation reactors which have traditionally been prone to run away at transient conditions common during streaming, such as:

- Low (or no) forward Cracked Gas feed flow, with catalyst temperatures at or near reaction initiation level
- Low inlet CO concentrations (e.g. 50-150 ppmv prevalent with new furnace designs)
- CO fluctuations
- Inlet temperature fluctuations

With ethylene plant capacities hitting record highs, the cost of an off-spec incident during startup is substantially higher than even five years ago. As such, the market began to demand an ultra-stable front-end selective hydrogenation catalyst.

Utilizing their Innovation Process, Clariant has identified a new catalyst formulation and applied High Throughput Experimentation to optimize a new, multi-component Front End catalyst. OleMax®260 shows greatly reduced reactivity of ethylene relative to that of acetylene. It is also highly stable at very low and fluctuating CO levels. Along with achieving maximum selectivity, its unique performance characteristics deliver significant operational risk mitigation.

The Dow Chemical Company partnered with Clariant toward the new catalyst's commercialization throughout several years of extensive laboratory and pilot testing. Despite what might be considered significant risk in selecting an unproven catalyst for use in a world-scale ethylene plant, Dow elected to proceed as the first adopter of this exciting catalyst technology at their new world scale cracker in Freeport, TX.

OleMax®260 was successfully commercialized in September 2017. The catalyst was very quickly and smoothly brought on specification despite low inlet CO levels and process variability which could have easily resulted in a runaway. Performance data to date confirm both its high activity and exceptional selectivity to ethylene.

AIChE Paper Number: 78b

Root Cause Analysis of a Premature Flooded Debutanizer Using Gamma Scanning, Engineering and Installation of a Bypass Line by Hot Tapping

André Perschmann
Linde Engineering

David Bruder
Linde Engineering

Thomas Walter
Linde Engineering

Fabian Schroeder
Ruhr Oel GmbH BP Gelsenkirchen

Prepared for Presentation at the 2019 AIChE Spring Meeting
New Orleans, LA, March 31 – April 4, 2019

ROOT CAUSE ANALYSIS OF A PREMATURE FLOODED DEBUTANIZER USING GAMMA SCANNING, ENGINEERING AND INSTALLATION OF A BYPASS LINE BY HOT TAPPING

André Perschmann
Linde Engineering
David Bruder
Linde Engineering
Thomas Walter
Linde Engineering
Fabian Schroeder
Ruhr Oel GmbH BP Gelsenkirchen

Abstract:

A Debutanizer in a steam cracker unit in Germany flooded prematurely. The column never caused problems during the typical turnaround cycles and was at that time in operation for two and a half years since the last turnaround. A sudden increase of the pressure drop was detected. At the same time the product purity deteriorated. Consequently, the column became a plant capacity bottleneck, limiting the ethylene production significantly.

A theory was developed that local fouling was causing the trays to flood. Gamma scans were performed at different operating conditions and proved the validity of the theory. Following this, a process simulation was conducted and demonstrated that the desired product specification can still be achieved when bypassing a part of the liquid. Thus, a bypass line was engineered and installed by hot tapping.

Bypassing a part of the liquid solved the capacity limitation and avoided a costly shutdown of the steam cracker unit. Fouling was found in the bottom section after the column was opened in a later shutdown.

1. Introduction and Basic Information

Flooding is a non-steady state operating condition in which liquid accumulates in a column. It is the most common capacity limit of a column [1]. The build-up of liquid in a column can be initiated by a variety of mechanism, e.g. jet flooding, downcomer choking and backup flooding. Flooding is usually detected by a sudden increase of the column differential pressure, a reduced bottom flow rate and liquid carry over. Several reasons for column malfunctions were reported in the past. Many column malfunctions are caused by plugging and coking [2].

Here the authors present a successful troubleshooting of a premature flooded Debutanizer in a steam cracker unit in Germany. The column has been operating for many years and never caused problems during the typical turnaround cycle. The column was in operation for two and a half years since the last shutdown when the flooding was detected the first time.

A rapid rise in pressure drop was measured by the plant owner in October 2017. Shortly after that, the bottom product purity followed by the overhead product purity deteriorated and the column operation became instable. The feed rate to the Debutanizer was reduced to achieve steady-state operation again. Due to this, the column became a plant capacity bottleneck limiting the ethylene production to approximately 90 % and, later, to approximately 80 %. Thus, it was urgently necessary to identify the root cause and to develop a solution to overcome the capacity bottleneck.

As seen in Figure 1, the investigated Debutanizer is equipped with 20 two-pass standard valve trays in the bottom section below the lower feed inlet, 11 one-pass standard valve trays in the middle section below the upper feed inlet and 17 one-pass standard valve trays in the top section. Each section has straight downcomers at a tray spacing of 600 mm (~ 24 in). All trays have 51 mm (2 in) high outlet weirs and a downcomer clearance of 38 mm (1 – 1/2 in). The trays are not equipped with regular inlet weirs, but with 13 mm (1/2 in) high barrier weirs to reduce the liquid momentum and avoid weeping at the inlet areas of the trays.

A C4+ fraction from an upstream Pre-Separation Column is fed to the column's lower feed inlet and a C4+ fraction from an upstream Depropanizer is introduced to the column's upper feed inlet. The feeds are separated in a C4 overhead product and a C5+ bottom product. The column is normally operated at an overhead pressure of 4.75 bara (68.9 psia) and a bottom temperature of 105 °C (221 °F). Polymerization inhibitor is fed to the upstream fractionators to avoid polymer fouling in the column.

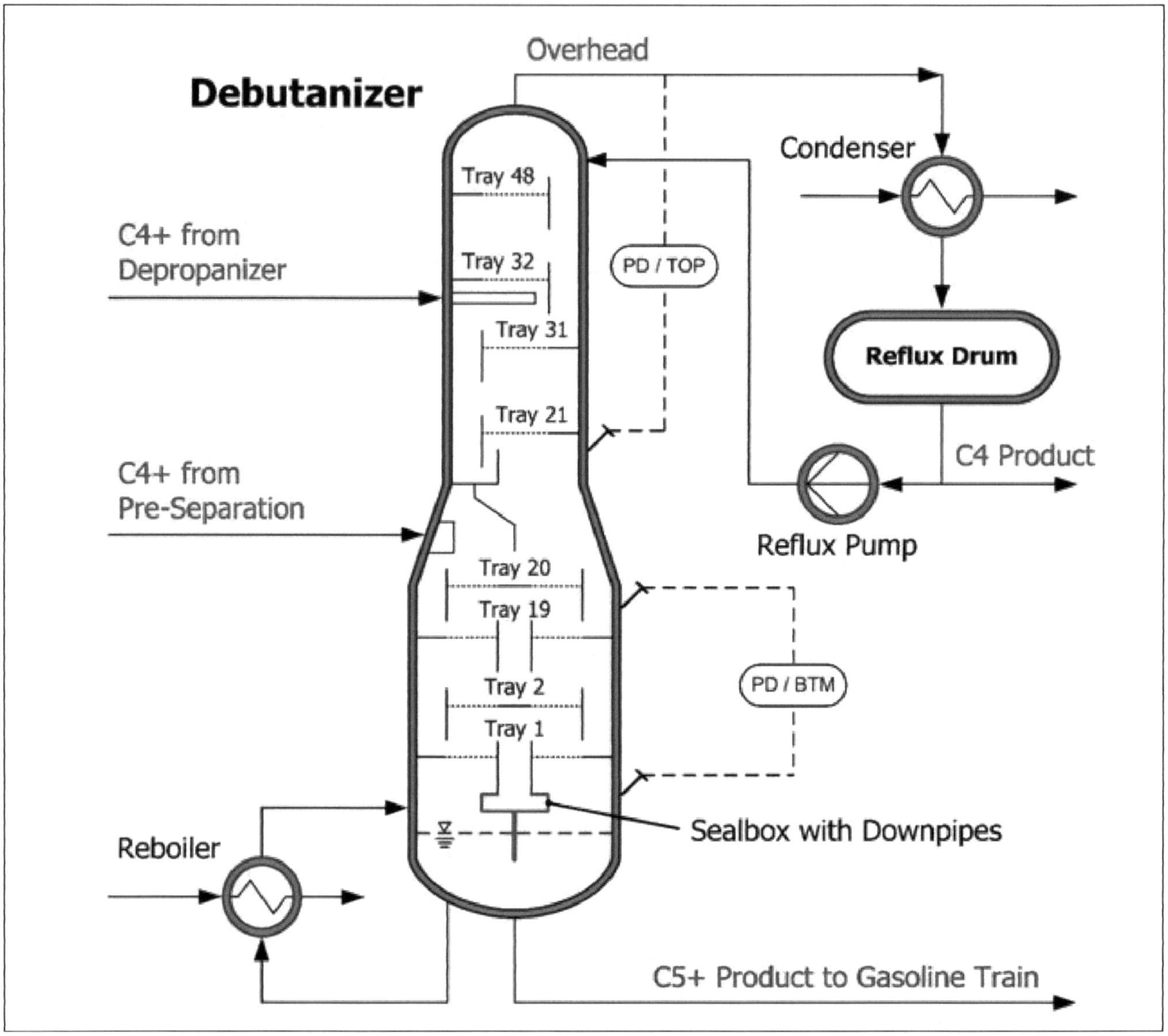

Figure 1: Simplified process scheme of Debutanizer.

2. Root Cause Analysis

First, field data provided by the plant owner was analyzed to gain a better understanding of the flooding phenomena. Second, a hydraulic rating of the trays was performed using Linde's in-house column design software. Based on that, Linde developed a theory which then was proved by additional gamma scans.

2.1. Investigation of Field Data

The differential pressure across trays 1 thru 19 as well as across trays 21 thru 48 is continuously monitored. A sudden increase of the differential pressure in the bottom section was noticed in October 2017. Shortly after the bottom pressure drop had established a plateau, the pressure drop across the middle and upper section started to increase significantly. As shown in Figure 2, the pressure drop in the bottom section went from 180 mbar to 350 mbar (72 in H_2O to 141 in H_2O), followed by the pressure drop in the middle and upper section that rose from 160 mbar to 380 mbar (64 in H_2O to 153 in H_2O). After the feed rate to the column was reduced, the liquid build up on the trays was assumed to have immediately stopped and then declined. The pressure drop in the middle and upper section decreased until the initial value was reached again. Shortly afterwards, a reduction of the pressure drop in the bottom section was detected. Steady-state operation and a normal pressure drop was achieved again after all the accumulated liquid was drawn from the trays.

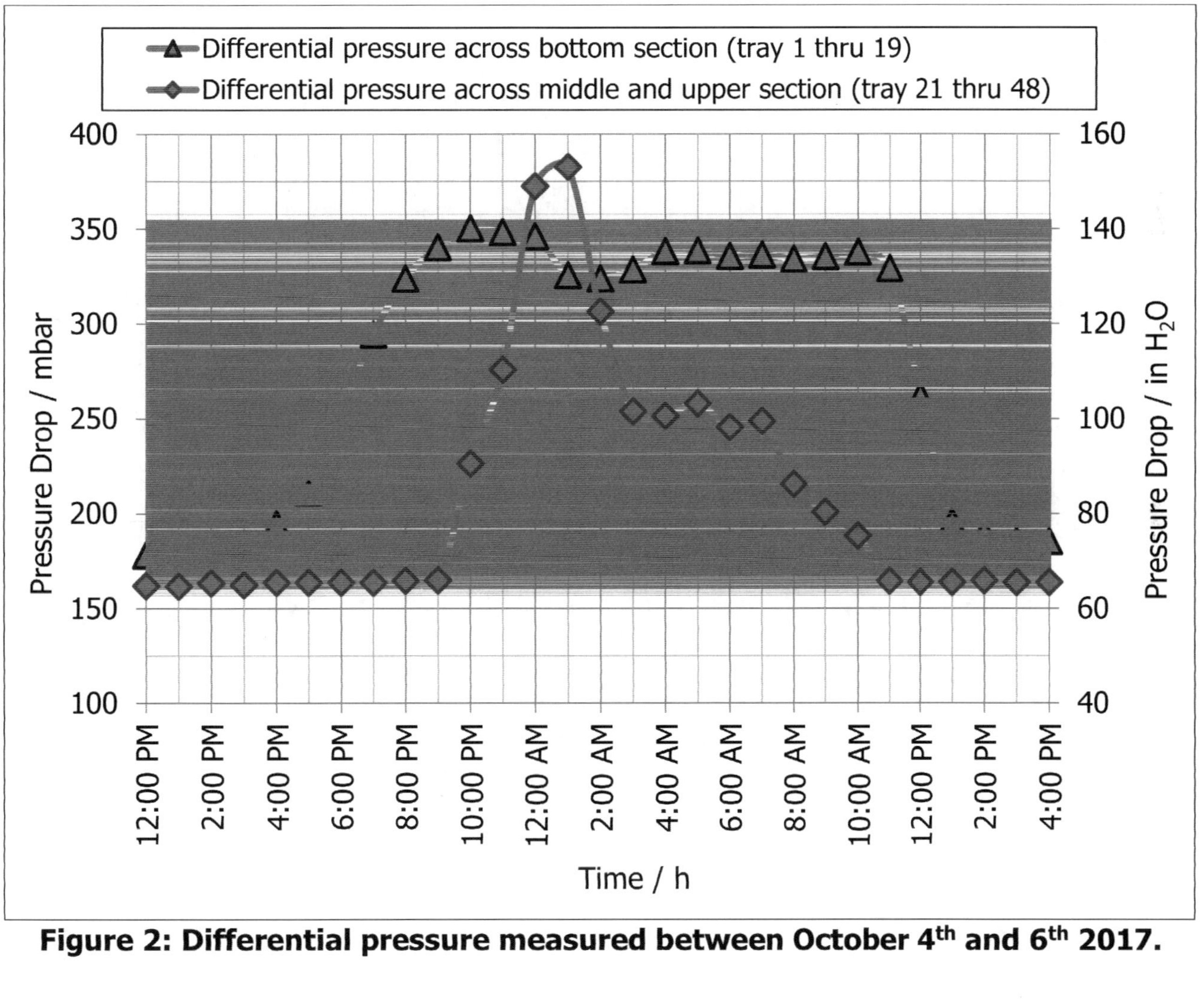

Figure 2: Differential pressure measured between October 4th and 6th 2017.

The investigation of the pressure drop data was a first indication that the root cause of the premature flooding can be found in the bottom section of the column. The reason being, that liquid accumulation and hence an increased pressure drop always takes places above the restriction in the liquid flow.

To conduct a thorough analysis of the operational problems, it was necessary to further investigate the history of the Debutanizer. Therefore, the differential pressure of the last years was studied. The pressure drop across the bottom section was constant 90 mbar (36 in H_2O) until end-2015. As seen in Figure 3, afterwards the differential pressure continuously rose up to 180 mbar (72 in H_2O) in the following two and a half years. A short shutdown took place in mid-2015. The average pressure drop across the middle and upper section was approximately 170 mbar (68 in H_2O). A constant increase of the pressure drop in the middle and upper section was not measured.

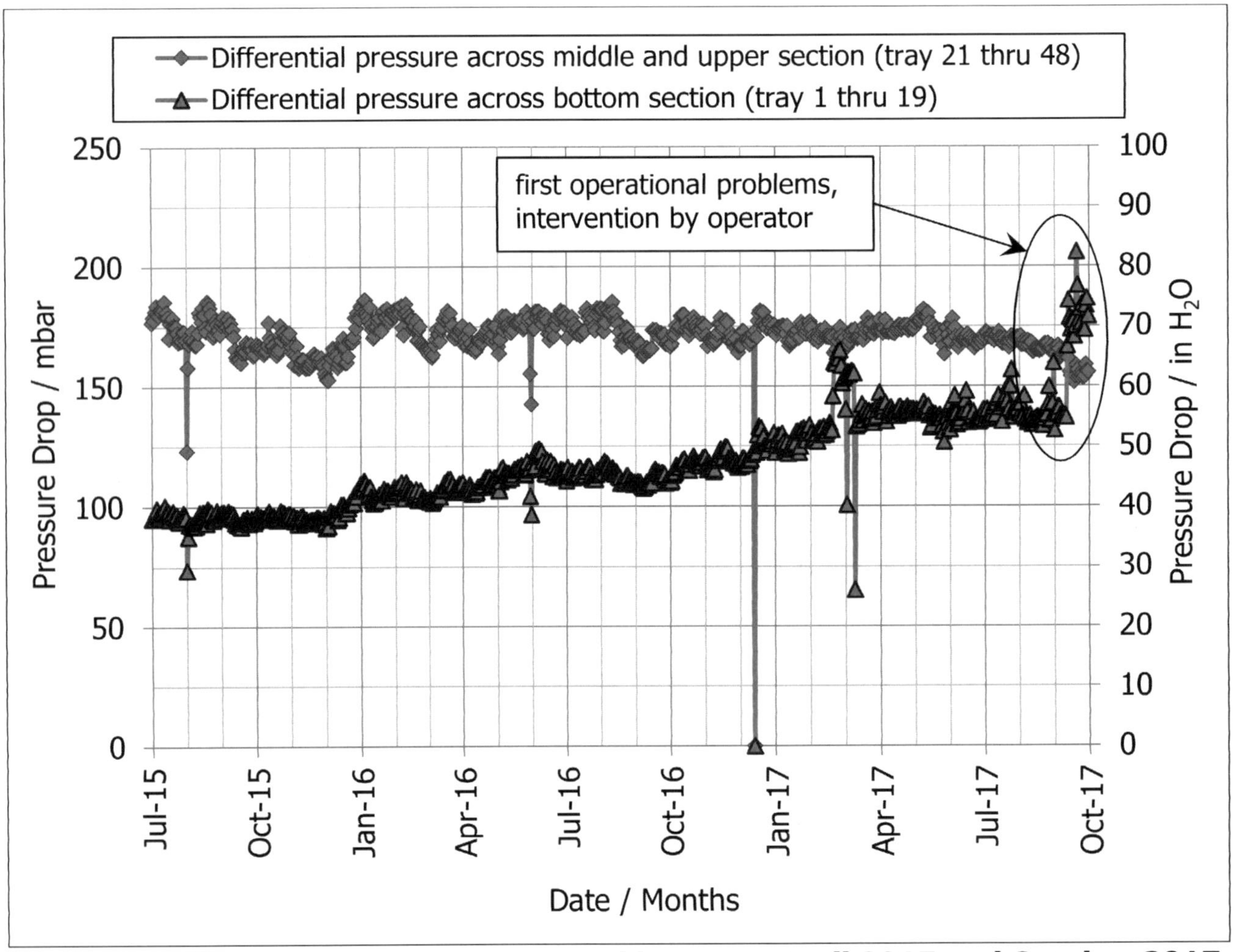

Figure 3: Differential pressure measured between April 2015 and October 2017.

The Debutanizer is well known for butadiene polymerization fouling. Hence, this observation was a strong indication that polymer fouling in the bottom section could have caused blockage of the tray active area and plugging in the downcomers. Nevertheless, several aspects of the tray design were studied before the entire investigation was focused on flooding caused by fouling.

2.2. *Tray Hydraulics Rating*

A process simulation was performed using plant data from a past high load operation of the column. The hydraulic performance of the column was rated for this case using Linde's column design in-house software which has been widely used for the design of Debutanizers. Table 1 gives an overview on the results of the tray hydraulic rating. A Linde proprietary correlation was used for the calculation of the downcomer backup flooding. The pressure drop for valve trays, the downcomer choke flooding and the percent jet flooding was determined with the equation from the Glitsch Design Manual [3].

Table 1: Tray rating results for high load operation.

	Top section (trays 32 - 48)	Middle section (trays 21 - 31)	Bottom section (trays 1 - 20)
Percent jet flood	41 %	36 %	32 %
Percent downcomer choke flood	24 %	34 %	29 %
Percent downcomer backup flood	33 %	32 %	30 %
Downcomer backup (froth)	214 mm (8 – 3/8 in)	210 mm (8 – 1/4 in)	197 mm (7 – 3/4 in)
Total tray pressure drop	4.5 mbar (1.8 in H_2O)	4.9 mbar (2.0 in H_2O)	4.9 mbar (2.0 in H_2O)

The results proved that the valve trays have sufficient capacity to accommodate the normal flow rates when no plugging occurs. The calculated overall pressure drop including the static head and the pressure drop of the overhead nozzle is 280 mbar (112 in H_2O). The calculated pressure drop well matches the measured pressure drop of 290 mbar (116 in H_2O) for the high load operation and hence proved the validity of the simulation. However, the high overall pressure drop at normal operating conditions that was measured just before the flooding occurred could not be predicted by the pressure drop correlation.

It could be concluded that the valve tray design is not limiting the capacity of the column. However, there was still the possibility that the trays were collapsed, which could only be verified by gamma scanning.

2.3. *Gamma Scans*

Gamma scanning is a non-intrusive inspection technique to diagnose malfunctions of pressure equipment while it is in operation. During the measurement a radioactive source which is emitting gamma-ray along with a radiation detector is lowered down the outside of the column. The radiation beam passes though the pressure equipment and its intensity is measured on the opposite side by the radiation detector in terms of a count rate. Interaction of the gamma-ray beam with pressure shell, internals and the process fluid causes attenuation of the gamma-ray. The detected count rate correlates to the average material density.

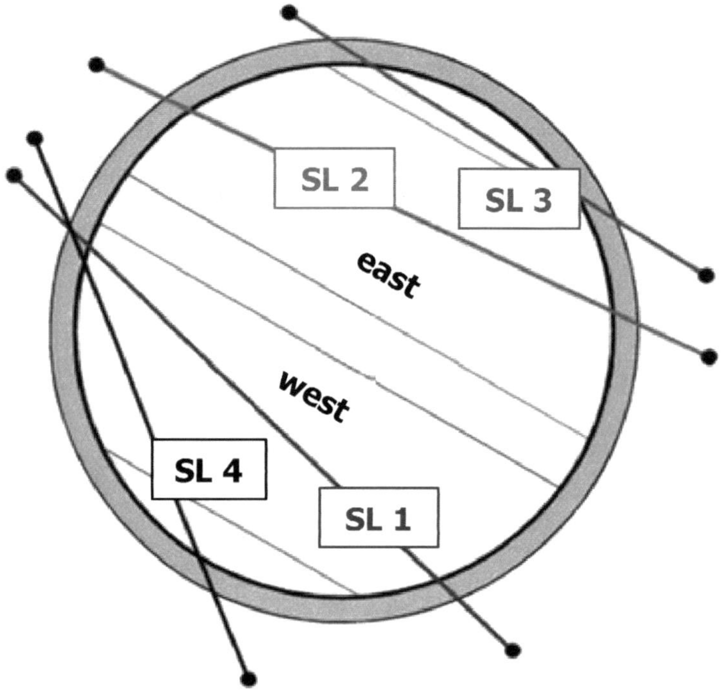

Figure 4: Scanline orientations (by courtesy of Tracerco).

A baseline scan was already conducted by Tracerco on behalf of the plant owner under normal operating conditions in August 2017, three months before the premature flood occurred for the first time. The gamma scan report was provided to authors at the time Linde started with the root cause analysis. The orientation of radiation beam (SL 1, blue) is depicted in Figure 4. The active areas of all trays were scanned with a Cs-137 source. For the two-pass trays in the bottom section only the western active areas were scanned. As shown in Figure 5 (dashed, blue), all trays were holding liquid, hence good mechanical condition of the trays was confirmed.

Another observation caught the attention of the authors. An increased froth height was detected on tray 2 while at the same time a normal froth height was measured on tray 1. Such scan profiles are often taken as measurement errors caused by obstructions in the scanline. However, the same observation was made by one of the authors in a previous troubleshooting, knowing that the increased froth height on tray 2 is a typical indicator for a bottleneck. In the earlier investigation, the cause was a physical blockage of the clearance under the side downcomers of tray 2 due to fouling which caused the liquid to back up onto the tray active area. A normal froth height was detected on tray 1 due to a specific sealing of the center downcomer below. A seal box was attached to the center downcomer of tray 1 instead of a commonly used seal pan (see Figure 1). Downpipes bolted to the seal box and submerged below the liquid level routed the process fluid into the column sump. Unlike a seal pan a seal box does not have a clearance under the downcomer. Accumulation of fouling material in the downcomer exits of tray 1 was not possible since all the solid material was washed away via the downpipes. Therefore, downcomer backup flooding can start from tray 2 or the above trays only.

The Debutanizer of the current investigation was also equipped with a seal box leading to the theory that plugging of the downcomer clearance due to butadiene polymerization caused a rise of the downcomer liquid level of tray 2 and flooding of the entire column eventually. Further gamma scans were performed in December 2017 to prove the proposed theory. The scanlines for these measurements are also given in Figure 4. First, the August measurement of the entire column was repeated (SL 1, blue) under flooded conditions as given in Figure 5 (solid, blue with yellow highlighter). The scan revealed that the entire column was flooded starting from tray 2. Only tray 1 showed a normal froth height.

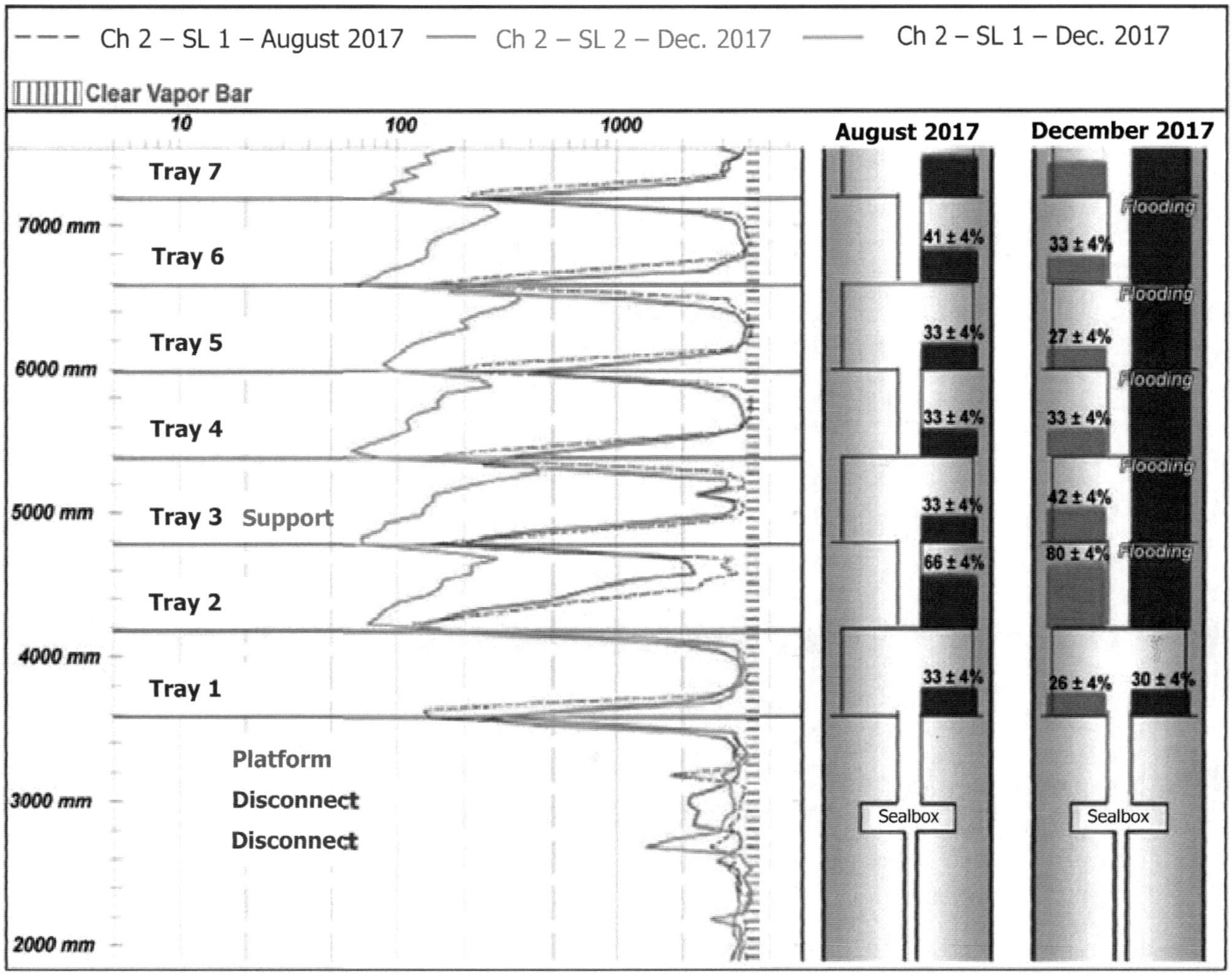

Figure 5: August 2017, baseline scan of western active area at normal operating conditions. December 2017, scan of eastern active area at normal operating conditions and scan of western active area at flooded conditions. Top of column not shown (by courtesy of Tracerco).

Another gamma scan of the eastern active area (SL 2, red) was performed at normal operating conditions to finally confirm the mechanical integrity of all trays, as shown in Figure 5 (solid, red). Tray damage was not causing the column to flood. As observed in the baseline scan from August 2017, an increased froth height was detected again on the eastern active area of tray 2. Furthermore, scans at normal operating conditions were shot through the side downcomers in the bottom section (SL 3, green and SL 4, black). As depicted in Figure 6 (solid, green and solid, black), the downcomer scans showed that the side downcomers of the bottom trays were close to flood even at normal operating conditions.

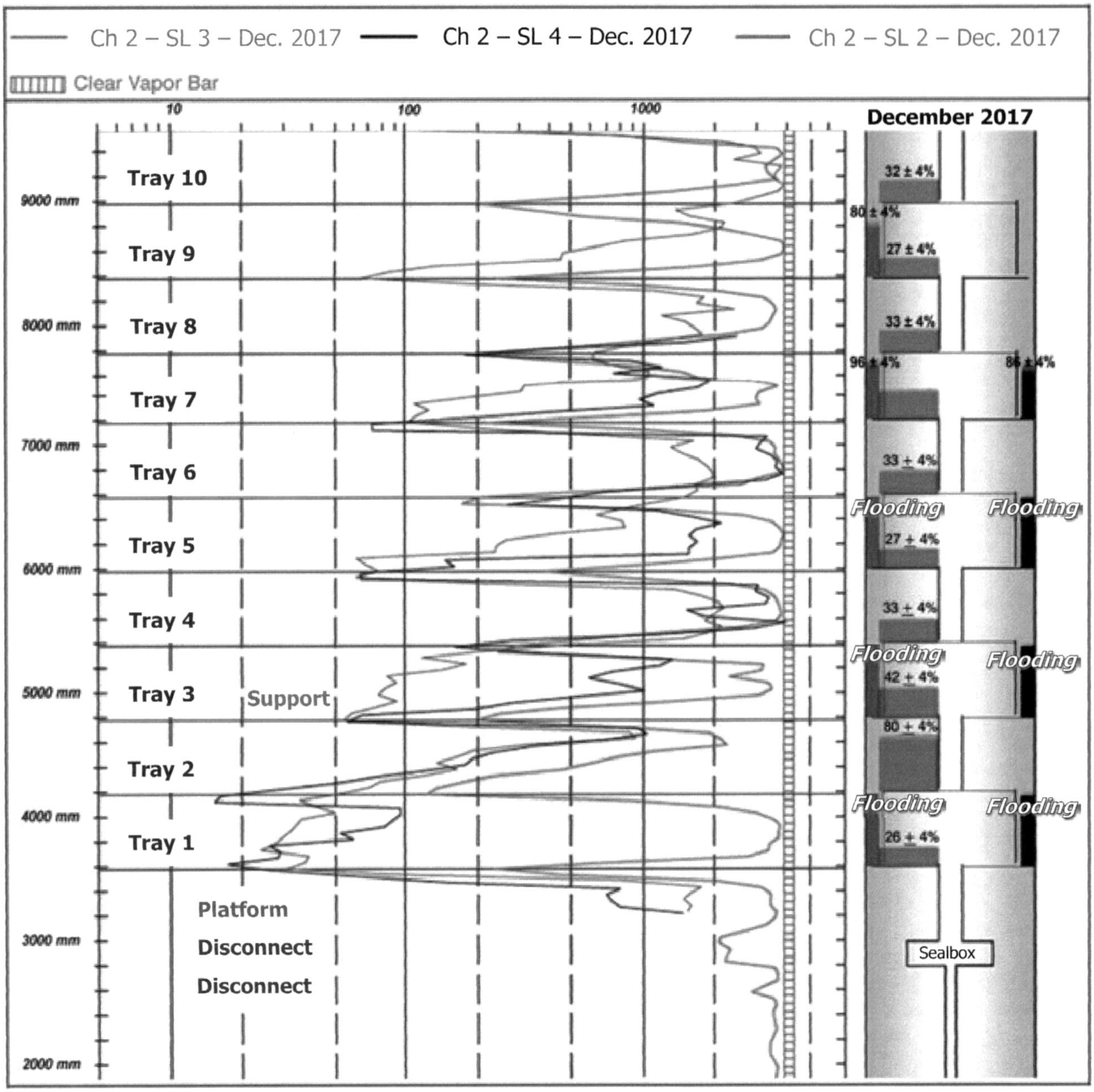

Figure 6: December 2017, scan of eastern active area, scan of eastern downcomer and scan of western downcomer at normal operating conditions. Top of column not shown (by courtesy of Tracerco).

The observations described above provided strong evidence for plugging of the trays and downcomers caused by polymerization fouling. However, cleaning the trays and downcomers was not considered an option since this would have required a costly shutdown of the entire steam cracker unit. Therefore, the load in the bottom section needed to be reduced. The authors decided to bypass a part of the liquid rather than a part of the ascending vapor for the following reasons.

The relative froth density, the height of the liquid on the tray deck, the pressure drop due to liquid flow under the downcomer and the total tray pressure drop contribute to the aerated liquid backup in the downcomer [4]. The pressure drop of 180 mbar (72 in H_2O) across the bottom trays, that was measured at incipient flooding, was simulated with Linde's in-house rating tool by reducing the valve count and hence the open area. A downcomer backup of 326 mm (12 – 7/8 in) was calculated for the aerated liquid. However, the gamma scans showed that the downcomers were flooded or close to flood under these loading conditions. In other words, the aerated liquid backup in the downcomers was approaching the 600 mm (~ 24 in) tray spacing. This means that the increased tray pressure drop contributed only partially to the overall liquid backup in the downcomers. Hence, the remaining liquid backup was mainly caused by plugging of the clearance under the downcomer. Due to a partly blockage of the downcomer exits, the effective clearance under the downcomer was smaller than the design value. This led to a significantly greater downcomer exit velocity which results in an increased backup of the liquid in the downcomer. Moreover, the barrier weir installed close to the downcomer exits may promote the accumulation of solid material and restrict the liquid flow under the downcomer apron further. Therefore, reducing the liquid load contributes largely to a reduction of the downcomer backup. Under consideration of the lessons learned from a past debottlenecking investigations, it was decided to install a liquid bypass line in operation.

A process simulation was conducted to demonstrate that the desired bottom as well as overhead product specification can still be achieved, while bypassing a part of the liquid into the sump.

3. Solving the Flooding Problem

3.1. Process Simulation

The column should be capable of working at plant high load conditions after installation of the bypass. Therefore, two different actual operating cases of the column were simulated. Case one represents the past high load operation of the column that was already described in section 2.2. Case two simulates the normal operating conditions of the column at the time of the downcomer gamma scans, just below the flood point. As typical for a distillation column, the process simulation showed that the highest gas and liquid load occur in the bottom section, especially on tray 1 thru 5. For case one, a maximum liquid load of 99.2 m^3/h (437 gpm) was simulated, while for case two, a maximum liquid load of 78.3 m^3/h (345 gpm) was simulated. The difference between both cases is 20.9 m^3/h (92 gpm). This load could not be processed by the column due to plugging of the clearance under the downcomer. It was necessary to overcome the column bottleneck by bypassing at least this amount of liquid around the affected trays.

As shown in Figure 6 (solid, green) at normal operating conditions the eastern side downcomers of the tray 2 thru 8 are flooded or close to flood. The side downcomers of tray 10 and the above trays were operated at approximately 80% downcomer backup or less. This means that the liquid load on tray 1 thru 8 needed to be reduced. A hot tap in the center downcomers was not feasible due to poor accessibility and the narrow downcomer width. Therefore, two bypass lines were installed, one in each side downcomer of tray 8.

A further process simulation proved that the desired bottom and overhead product specification could still be achieved at high load operation when bypassing 20.9 m^3/h (92 gpm) liquid. Only a minor change of the reboiler and condenser duty of less than one percent was predicted by the process simulation. The reason being that the old column has a relative big margin in the number of trays and theoretical stages. The slightly increased reflux and reboiler rate could be accommodated by the existing trays.

3.2. *Basic Engineering and Installation of the Bypass Line by Hot Tapping*

The bypass lines had to be installed at the bottom of the side downcomers to ensure degassing of the two-phase mixture. However, due to turbulences in the downcomers it was likely that there was no clear liquid at the bottom of the side downcomers. Therefore, sizing of the bypass lines was done considering Simpson's self-venting velocity [5]. Below the self-venting velocity, entrained gas can ascend against the downflowing liquid in a pipe. The calculation showed that each new bypass line must have a nominal pipe size of six inch to comply to the self-venting correlation. An additional safety margin was considered during the design of the bypass lines in case more liquid than calculated needs to be drawn off from the column. The bypass lines needed to be tied in below the column sump's liquid level to prevent vapor bypass. It was necessary to provide a hand valve in each line to control the amount of liquid drawn from the column. The detail engineering including a risk assessment and the installation of the bypass lines by hot tapping was done by the plant owner.

Figure 7: Active are of tray 1 showing plugging of the valve units.

Hot tapping is a technique to make a connection to a pressure vessel while it remains in service. A further gamma scan was performed on tray 8 in preparation for the hot tapping to determine the exact orientation of the downcomer bolting bars and the precise elevation of the support ring. In total four hot taps were installed on the pressure shell, two in the side downcomers of tray 8 and two below the normal liquid level in the column sump.

4. Results of the Implementation of the Bypass Line

The bypass lines were put in operation in mid-2018. The capacity of the steam cracker unit could be increased up to 97 % ethylene production. On-spec overhead and bottom products were produced while the bypass was in operation. The Debutanizer was opened in a later shutdown and severe polymerization fouling was found on the tray active areas and in the downcomers. (Figure 7).

In summary, it can be stated that a thorough analysis of field data and tray hydraulics together with gamma scanning are key factors for a time- and cost-efficient troubleshooting. The close collaboration between plant owner, scan experts and engineering company made the troubleshooting discussed here a success.

5. References

1. Perry, R. H., & Green, D. W. (2008). *Perry's chemical engineers' handbook.* New York, McGraw-Hill.
2. Kister, H. Z. (2003), *What Caused Tower Malfunctions in the Last 50 Years,* Trans. Inst. Chem. Eng., Vol. 81, pp. 5-26.
3. Glitsch, Inc., (1980) Bulletin H-4900, 3rd ed., *Ballast Tray Design Manual,* Dallas, TX.
4. Lockett, M. J. (1986), *Distillation Tray Fundamentals,* Cambridge University Press, Cambridge.
5. Simpson, L. L. (1986), *Sizing Piping for Process Plants,* Chemical Engineering, June 17, pp. 192-214.

547548 Liquid Heater Upgrade for Improved Operation with Ethane Feed

Shale gas development has led to the processing of large volumes of ethane feed in Crackers that were designed for heavier feeds. This can result in some significant operational challenges for cracking furnaces that were originally designed to process significant quantities of liquid feedstocks.

Four Naphtha cracking furnaces had previously been upgraded to increase capacity and to allow for the cracking of NGL feeds. The basis for this conversion included a design case for ethane feed but processing of that feedstock was not required until recently. The furnaces have experienced short run times on ethane feed and this is primarily driven by pressure drop in the cracking tubes. The end of run constraint is governed by maintaining appropriate pressure ratios across the critical flow devices that establish equal flows to each tube. These furnaces were found to have a very narrow operational window for pressure when cracking ethane feed due to the focus of their design optimization being towards other feedstocks.

This paper will discuss the upgrading of a Naphtha/Butane/Propane furnace to increase the operational window with ethane feed and to ensure considerations are included to properly integrate the operation with future Cracker debottlenecking plans.

548782 Decoke Tutorial

When a cracking furnace reaches its end of run (EOR) condition, an online steam/air decoke will be conducted to gain back the performance of the radiant coils and primary quench exchangers. Decoke is a short-term operation and it has some unique features compared to normal cracking operation. During decoke, steam, air and heat are introduced into the system to burn off or carry away coke. Depending on the disposal method for the decoking effluent (decoke to firebox or to decoke drum), particulate matter (PM), NOx and CO emissions will be quite different. In this tutorial, procedures unique to the decoking operation and their impact on furnace equipment and components will be reviewed.

548511 Tutorial on Adsorption Unit Troubleshooting

The cracked gas driers are a reliable unit operation in the ethylene plant. Because issues with adsorption units are rare, the process engineer may not be able rely on past experiences to detect these issues early on or may not know what corrective action should be taken to mitigate the issue. This tutorial will introduce you to the art of trouble shooting your adsorption unit. Honeywell UOP, is not only the largest adsorbent producer worldwide, but also has decades of experience in the petrochemical industry serving a wide variety of customers. This allows UOP engineers to come across many of these uncommon issues. Grouping all these issues allowed to develop a certain troubleshooting approach. During this tutorial, we will look at a couple of these groups: bed support failures, analyzer anomalies and liquid carry over. In troubleshooting an adsorption unit, it is also important to analyze the unit's performance, this is done with a software solution, UOP ARMGuard™ System (**ARM** - **A**dsorbent **R**emote **M**onitoring). Increasing your knowledge of the adsorption unit, using a software tool to analyze your unit's performance, and working together with adsorbent experts will increase the speed at which a resolution can be found to an adsorbent issue and potentially avoid a multimillion-dollar emergency shut-down.

AIChE Paper Number 120a

CASE STUDY OF STEAM HAMMER EVENT IN AN ETHYLENE PLANT

Tatiana Randall
Fixed Equipment Reliability Lead
Flint Hills Resources

Michael Nugent
Principal Engineer
The Equity Engineering Group

***James R. Risko**, CEM, PEM*
President
TLV Corporation

Prepared for Presentation at the 2019 Ethylene Spring Meeting
New Orleans, Louisiana, March 31 - April 4, 2019

CASE STUDY OF STEAM HAMMER EVENT IN AN ETHYLENE PLANT

Tatiana Randall
Fixed Equipment Reliability Lead
Flint Hills Resources
Michael Nugent
Principal Engineer
The Equity Engineering Group
***James R. Risko**, CEM, PEM*
President
TLV Corporation

Abstract:

A steam release occurred on a 600 psig high pressure extraction steam line from the cracked gas turbine in a Gulf Coast Ethylene Production Facility. A detailed review of the third-party failure analysis, along with a site visit and field walk down of the equipment determined that the cause of the failure was a condensation-induced water hammer. This is a rapid condensation event, sometimes known as a "rapid steam bubble collapse". Under certain conditions, a steam pocket can become totally entrapped in subcooled condensate. It is why it is important to keep steam lines drained from pooled condensate through condensate discharge locations (CDL) - those locations which include steam traps and associated piping and valves. When the steam gives up its heat to the adjacent condensate and pipe walls, steam changes from a vapor to a liquid state. As a liquid, the volume formerly occupied by the steam shrinks by a factor of several hundred to over a thousand times depending upon the pressure, and the rapid collapse results in high condensate acceleration to fill the void, generating extremely high force. A review of one such incident will be discussed with attention on the root cause and preventative measures.

Introduction and Incident

The cracked gas compressor is a major component for ethylene production. The cracked gas is compressed in four stages by centrifugal compressors driven by the 1300 psig steam extraction/condensing turbine. It condenses heavier gasoline and water by increasing the pressure of the cracked gas. The discharge pressure permits separation of hydrogen and methane from C2 and heavier hydrocarbons by fractionation and chilling train system.

In May 2015, a through-wall crack was identified on the 600# steam line downstream of the steam driven cracked gas compressor in the Light Olefins Unit. The same location had previously cracked through-wall in June of 2014. The June 2014 failure was weld-repaired following only grind-out of the crack (i.e., no metallurgical analysis was performed). A photo of the failure location is shown in Figure 1.

Figure 1: Failure location at reducing tee 12x12x10

Inspection and repair

Visual inspection found the crack parallel to the weld and located both in the HAZ (Heat Affected Zone) and directly adjacent to the weld toe on the side of the weld closest the NPS 10 PRV line.

Visually, the crack was approximately 3 inches on the OD and 8-9 inches on the ID of the pipe. Picture of the crack is shown in Figure 2.

The crack has been repaired by removing a section of the NPS 10 pipe and reducing tee, and replacing with an in-kind pipe segment.

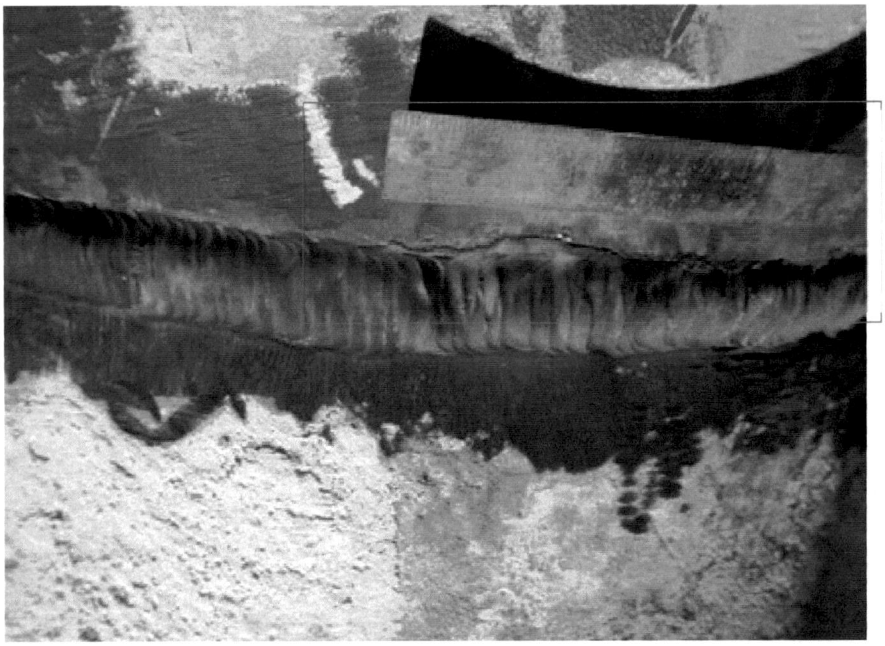

Figure 2: Crack location

Failure analysis

The Site Owner contracted a Materials/Mechanical Engineering Consultant to help perform failure analysis of this and many similar events at different locations. In addition, a Steam System Specialist organization was later involved for providing guidance and best practice recommendations for the EPC Conference presentation.

Initial objective was to perform piping stress analysis of the 600# steam line from the NPS 12 flange of the compressor to the 600# steam header. The goals of the piping analysis included the following:

- Evaluate primary and secondary stresses at the failure location.
- Provide recommendations to optimize the existing piping support to minimize stresses at the failure location.
- Verify B31.3 code compliance for all modeled location.

The 600# steam line was modeled starting at the steam turbine outlet shown on Figure 3. The model included the three pressure relief valves on the dead leg off the cracked tee, the 12-inch check valve line, and the 12-inch gate valve line.

The resulting stress analysis concluded the following:

1. The preceding fatigue analysis indicated the fatigue failure from the measured vibrations is highly unlikely.
2. The piping segment between the 600# steam compressor outlet and the steam header meets B31.3 code compliance.
3. The piping analysis does not indicate severe sustained weight and thermal expansion stresses at the failure location; therefore, further piping support modifications are not required.
4. Results suggest the measured vibrations do not represent the vibrations responsible for the identified damage.
5. The crack measured significantly longer on the ID than the OD, indicating internal shock force pulled the pipe apart from the inside of the weld.

Further process and design data review led to the apparent root cause of the internal shock as "condensation-induced" water hammer.

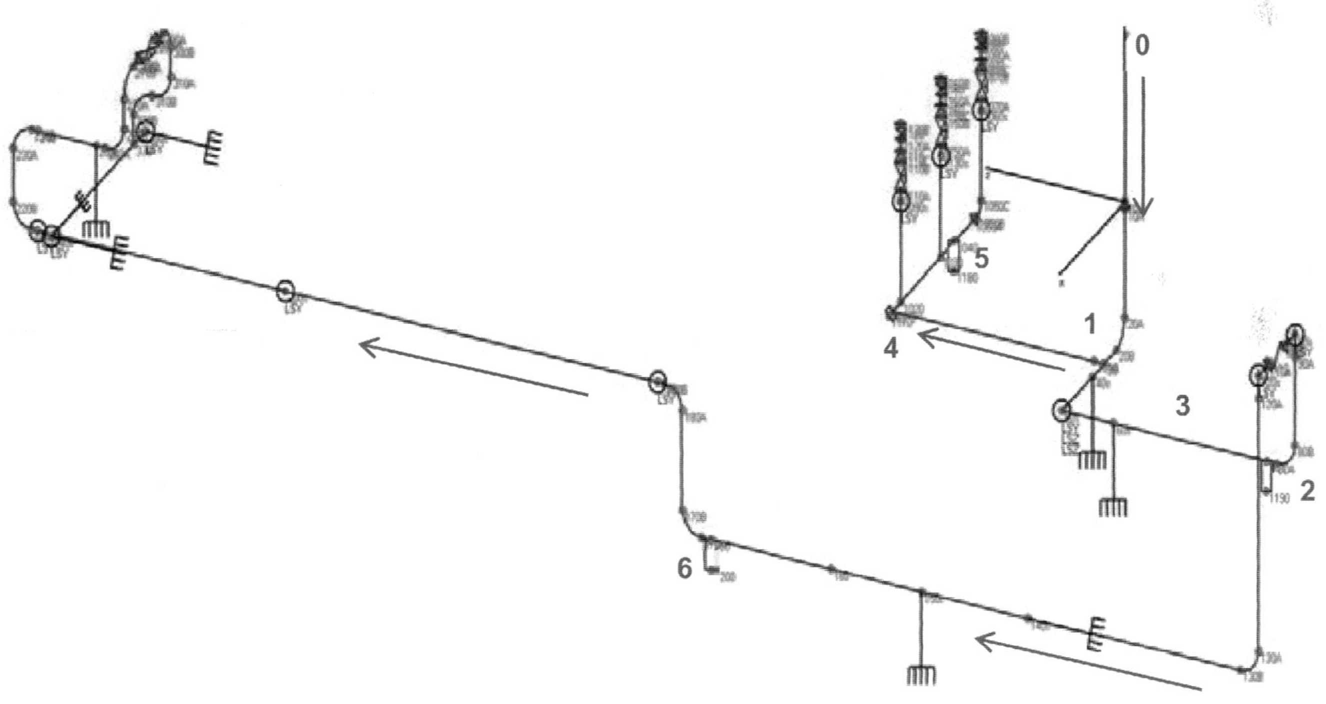

Figure 3: Piping model

Model Analysis, Condensation-induced Water Hammer

Before explaining how "condensation-induced" water hammer (CIWH) occurred, first the phenomenon of more common, "condensate-induced" water hammer (CWH) is explained to show the difference (see "Terminology" at end of this paper).

CWH is known to most steam users and can occur in pipes when the flow of condensate, propelled by steam velocity, is suddenly stopped by a riser pipe or the rapid closing of a valve. High velocity condensate is relatively incompressible, and the wall of the pipe is perfectly rigid, so from the laws of mechanics, the more rapid the deceleration, the greater the force generated by the water hammer. So, how does the fast-moving condensate decelerate and dissipate its force? Typically, this occurs with a shock wave reverberation or a steam line/valve packing/flange gasket blowout. The prescription to avoid CWH is relatively understood by site users: keep lines drained of disentrained condensate so there is no pool of condensate to accelerate. However, CWH is not felt to be the probable cause of the Ethylene plant incident analyzed, but rather CIWH instead.

CIWH results from the unusually rapid condensation of steam in a steam (not condensate) pipeline. Normally this requires a large transfer of heat energy from the steam to cooler water, either pooled in the steam line or from a rising water table on the exterior of the pipe. The greater the difference between the steam temperature and that of the condensing water, the more rapid the heat transfer and resulting shock (Figure 4).

CIWH in a Dead-end Steam Line

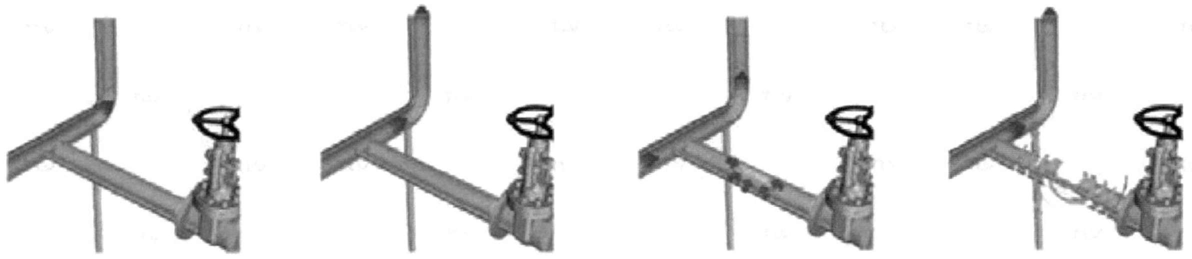

Figure 4: Steam pocket entrance and collapse in a dead-end steam line creates violent shock

The water hammer incident that occurred in the Ethylene plant was caused by CIWH, brought on by distributed steam entering a pool of sub-cooled condensate, collapsing the steam when it became surrounded - most likely in the dead-end section of low-lying horizontal piping. The resulting shock wave caused pipe movement pulling away from the tee, thereby stressing the tee at its weld. A deeper review of Figure 3 provides further explanation.

The 600 psig extraction steam flows vertically downward from point "0" on the diagram. Steam in extraction lines from turbines can flow at extremely high velocity, but obstruction such as pooled condensate or dead-end pipe can quickly bring velocity to a standstill. The tee is located at point "1" and it is notable that there is a substantial length of 12-inch piping on the same horizontal plane, and the 10-inch dead-ended pipe's free volume is located within the same plane (points 1-5). As a related factor, 10-inch pipe can hold water weight of 31 lbs/ft and 12-inch pipe of 41 lbs/ft. Long length can correlate to significant water mass to serve as a heat sink, drawing heat energy from entering steam, thereby causing a rapid collapse.

Examining points 1-5, it seems likely that steam collapse occurred in the piping section from points 1-4. A collapse of a steam pocket in the stagnated line filled with pooled condensate, could create a force vector toward the elbow at point 4, thereby pushing the pipe in that direction and creating pull tension at the tee weld.

There is a CDL containing a steam trap at point 2. This drainage location, 2 (and at points 5, 6), should be designated by the site as either "safety critical" or "critical" so that the low-lying horizontal sections of this pipeline are always kept free from pooled condensate. It is essential that condensate is removed quickly, and there are at least several ways identified that the section of pipe from points 1-6 can pool condensate. The most perilous time may be right after start-up from a turnaround.

1. Steam trap or some part of the CDL line is blocked, i.e. "cold".
2. Return line pressure is too high at the first entry of steam to enable the condensate discharge through the trap (negative differential pressure).
3. Steam trap has cyclical operation (not instantly discharging).
4. Trap is undersized for load.
5. If the trap does not discharge to grade, it is not readily recognized when it requires service.
6. The dead-ended line at points 4-5 pools condensate from sagging or wrong inclination.

It should be noted that although CIWH is felt to be the cause of this incident, such an event could also occur from CWH if for example, there was pooled condensate in an undrained, low-lying loop prior to point 0. To mitigate against CIWH from internally pooled condensate and CWH from undrained condensate, it is essential to sustain active management of the steam trap population paying attention to blocked/cold traps. To help understand the potential importance of mitigation efforts, some historical water hammer incidents are provided below.

Historical Steam/Water Hammer Incidents

Steam pipe water hammer caused explosions to occur in petrochemical facilities, pipelines, utility facilities. Possible CWH and CIWH incidents photos are showed below in Figure 5,6 (CWH),7 (CIWH):

Figure 5: Failure of a steam distribution line at a combined refinery and petrochemical complex. UK Health and Safety Executive (HSF), 2003 Major incident investigation report BP Grangemouth, Scotland: 29th May-10th June 2000.

Figure 6: Internet Images from Energy Resources Conservation Board. MEGEnergySteamLineFailure.pps

Figure 7: Image by Jason DeCrow/Associated Press. New York Times, July 18, 2017

Terminology

Within the steam industry, there may be some difference of opinion as to what to call such water hammer phenomena. Water hammer is caused by water, so in order to differentiate among different types, the co-author's preference is to define the type of hammer relative to its inducer. Such focus identifies the cause and enables specific recommendations to help mitigate the hammer. There are three types of hammer that are considered: condensate-induced, steam-induced, and condensation-induced. Their differences and typical circumstances are explained as follows:

1. Condensate-induced water hammer (CWH) is what happens in a steam system line. It is when unremoved condensate travels at relatively high speed within a steam line and strikes an immovable object (such as a vertical line or valve):
    a. Usually this is caused by insufficient drainage to remove disentrained condensate from the steam line.
    b. Mitigation efforts require properly sized, installed, and maintained to proper function steam traps.

2. Steam-induced water hammer (SWH) occurs in a condensate return line. It is when too much steam in the condensate header collapses, creating a void that the condensate rushes in to back-fill, commonly with violent results.
    a. Typically, this is caused when there are leaking steam traps, opened bypass valves, or leaking control valves on the outlet of level pots that are leaking large amounts of steam into a condensate header. It also occurs when a flashing condensate header has an unwarranted vertical lift (all should be gravity drain).
    b. Mitigation efforts require reducing the amount of steam in the header and maintaining gravity drainage.
    c. Such hammer may be strongest when the condensate temperature is just a little lower than that of steam for the same pressure
3. Condensation-induced water hammer (CIWH) occurs in a steam system line when steam suddenly encounters a huge heat loss/transfer causing it to collapse virtually immediately. It can be similar in effect to steam-induced hammer, but the main difference is that steam hammer is what occurs in a condensate line – there is an expectation that condensate will be in the line. Condensation-induced hammer occurs in a steam line when the steam collapses rapidly with violent results. Commonly this can occur when rapidly rising, relatively cold water (such as a rising water table in a manhole or trench) surrounds a steam line, causing so much rapid condensation that the steam collapses and condensate fills the void with extremely violent results. However, it can also occur when steam is distributed in a steam line that is filled with pooled/undrained condensate. Steam lines are not expected to be filled with pooled condensate.
    a. Typically, this needs some circumstances as a rising water table such in a manhole with a heavy rainfall/flash flood event and/or blocked drain, a trench, an underground steam line such as a tunnel or under a roadway.
    b. Sometimes the same effect is experienced when steam is directed into an undrained loop that has filled with condensate in a steam header. Rapid condensation occurs, collapsing the steam and generating violent shock as water rushes to fill the void.
    c. Mitigation efforts involve design to keep low levels in steam lines drained, or to handle a rising water table so that the steam line is not submerged.

Conclusion

This paper reviewed the root cause of a 600 psig high pressure steam release that occurred in a Gulf Coast Ethylene Production Facility, and discussed the differences between condensation-induced, condensate-induced, and steam-induced water hammer. Key learnings can be summarized below:

- Assume that steam, unless superheated, begins wet and gains entrained water as it travels to the process units. This is an important consideration in steam trap maintenance.

- Remove the water from the supply line in a timely and effective manner to help mitigate water hammer events.
- Understand the role of steam traps- to discharge noncondensable gases and condensate without leaking steam.
- Identify CDL's that should have a higher priority for frequent checking and servicing on critical equipment.

References:

1. Risko, J. R., "Stop Knocking Your Condensate Return," *Chemical Engineering Progress,* 112 (11), pp. 27-34.
2. Risko, J. R., "Beware of the Dangers of Cold Traps," *Chemical Engineering Progress,* 109 (2), pp. 50-53.
3. Risko, J. R., "Steam Trap Management: Do Something; Anything. Please!," *Chemical Engineering Progress,* 113 (10), pp. 64–72.
4. Risko, J. R., "Allocate New Plant Focus to Steam System Design-Part 1," Hydrocarbon Processing, 105 (5), pp. 39 – 43.
5. Cane, B., "Risk Based Methodology for Industrial Steam Systems," *Inspectioneering Journal,* 23 (3), pp. 38-42.
6. Hou, G., and Mita, T., "Advanced Steam System Optimization Program," Hydrocarbon Processing, 97 (5), pp. 45 – 49.
7. Wikipedia, "Gramercy Park Asbestos Steam Explosion", San Francisco, CA.
8. Wikipedia, "2007 New York City Steam Explosion", San Francisco, CA.
9. New York Times, Haag M., and Gomez, M., "Asbestos Confirmed in Steam Pipe That Exploded in Manhattan", New York Times, New York, NY. https://www.nytimes.com/2018/07/19/nyregion/steam-explosion-pipe-flatiron-nyc.html (Jul. 19, 2018).

AIChE PAPER NUMBER: 120b

CASE STUDY OF HIGH PRESSURE STEAM HEADER FAILURE RCFA

David Golding
Plant Manager - RLOC
Qatar Chemical Company, Ltd.

Syed Naeem Shah
Technical Superintendent - RLOC
Qatar Chemical Company, Ltd.

Ali Mahmoud
Operations Specialist, Ethylene Unit – RLOC
Qatar Chemical Company, Ltd.

Tadigadapa V Phanikiran
Process Engineer – RLOC
Qatar Chemical Company, Ltd.

Linda Robinson, PE
Assistant Chief Engineer
KnightHawk Engineering, Inc.

Prepared for Presentation at the 2019 Spring Meeting
New Orleans, LA, March 31-April 4, 2019

CASE STUDY OF HIGH PRESSURE STEAM HEADER FAILURE RCFA

David Golding
Plant Manager - RLOC
Qatar Chemical Company, Ltd.

Syed Naeem Shah
Technical Superintendent - RLOC
Qatar Chemical Company, Ltd.

Ali Mahmoud
Operations Specialist, Ethylene Unit – RLOC
Qatar Chemical Company, Ltd.

Tadigadapa V Phanikiran
Process Engineer – RLOC
Qatar Chemical Company, Ltd.

Linda Robinson, PE
Assistant Chief Engineer
KnightHawk Engineering, Inc.

Abstract:

This paper presents a case study on two separate failure incidents at an ethylene cracking facility in which steam leaks were discovered at the same piping "tee" connection. The failed tee is located where Very High-Pressure Steam (XS) is let down to a High-Pressure Steam (HS) header. Upon investigation it was determined that the leaks were due to cracks in the circumferential welds of the tee to the HS header and letdown piping. To avoid further failures, or a significant process safety event, each steam leak incident resulted in a plant wide shut down, and repairs of the steam line weld cracks were carried out.

Qatar Chemical Company Ltd. (Q-Chem) and KnightHawk Engineering, Inc. (KHE) performed a root cause failure analysis (RCFA) on the HS tee, which included metallurgical analysis, process assessment, field measurements, and numerical analyses (finite element analysis and computational fluid dynamics). This RCFA concluded that the tee failures were caused by off-design cool temperatures in the XS-HS letdown piping, which allowed for condensate to develop in the letdown line. High thermal stresses developed at the tee when this condensate met with a flow of superheated steam that is present in the HS header during normal operations. Several short-term remediation measures were taken to effectively address the root cause of the failures. A long-term solution plan was also developed. Finally, a plant wide survey was conducted on other letdown stations to check if similar conditions existed.

Introduction

Ras Laffan Olefins Company, Ltd. (RLOC) has a 1.3 million MTA ethylene cracker unit located in the Ras Laffan Industrial City of Qatar that is operated by Qatar Chemical Company Ltd. (Q-Chem). At this facility, two separate failure events occurred over several months at the same location on a steam header. During routine rounds an operator observed water dripping from an elevated pipe rack and reported it as an incident through the RLOC Event Processing System (EPS). Further investigation identified a steam leak under insulation in a High-Pressure Steam (HS) header. Both leaks occurred at weld cracks in a piping "tee" where Very High-Pressure steam (XS) is let down to the HS header. A quick assessment of the failure was done, which ruled out a temporary repair (including clamping) owing to a significant risk to personnel and process safety. Plant management decided to shut down the plant, thus preventing the possibility for a failure that could lead to a high consequence process safety event, and the repair of the failed welds was attempted.

KnightHawk Engineering, Inc. (KHE) was contracted to work with Q-Chem to perform a root cause failure analysis (RCFA) on the HS header tee failures, and to develop short and long-term solutions to address the identified cause.

RCFA Investigation

First Failure

At the time the first leak was detected the plant was reported to be operating normally with no upset conditions. The XS-HS letdown is used periodically and the most recent XS-HS letdown was six months earlier. There were no reports of unusual noises or process behavior around the time the failure was detected. Upon removal of insulation, a crack was found in the circumferential weld that attaches the tee to the downstream (i.e. south) portion of the HS header. Figure 1 provides a diagram indicating the location of the cracks for both the first and second failures.

When the insulation was removed, steam could be observed exiting from the crack. Two stills from the video footage of the leak (Figure 2) demonstrate the intermittent bursts of visible steam that were observed. Figure 3 show photographs of the 1st failure crack after application of dye penetrant.

Upon discovery of the first HS header leak, the plant was shut down so that the cracked tee could be repaired. The failed weld was cut out, and a repair weld was made. Upon NDT inspection of the weld repair, an additional crack was detected at the weld seam of the HS header pipe on the south side of the tee (also shown in Figure 1.) This was a partial crack that had not fully propagated through the thickness of the metal. The portion of the header pipe that was cracked (approximately 3 m) was removed and replaced with new pipe.

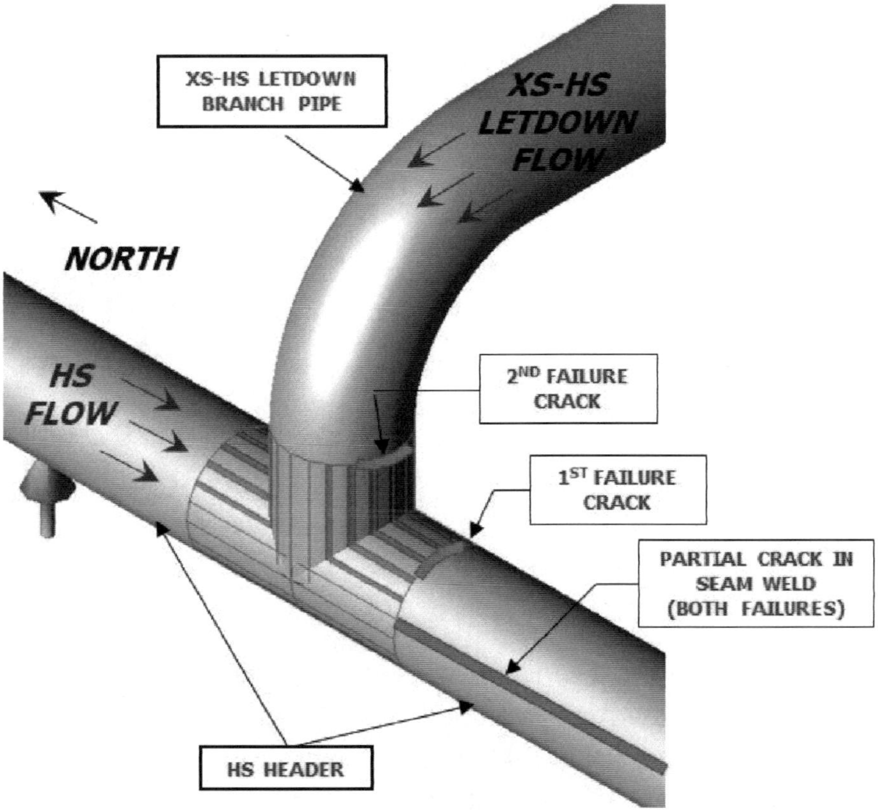

Figure 1: Diagram of Failed Tee with Crack Locations Indicated

Figure 2: Stills from HS Header Leak Video – Time = 3s (Top), Time = 4s (Bottom)

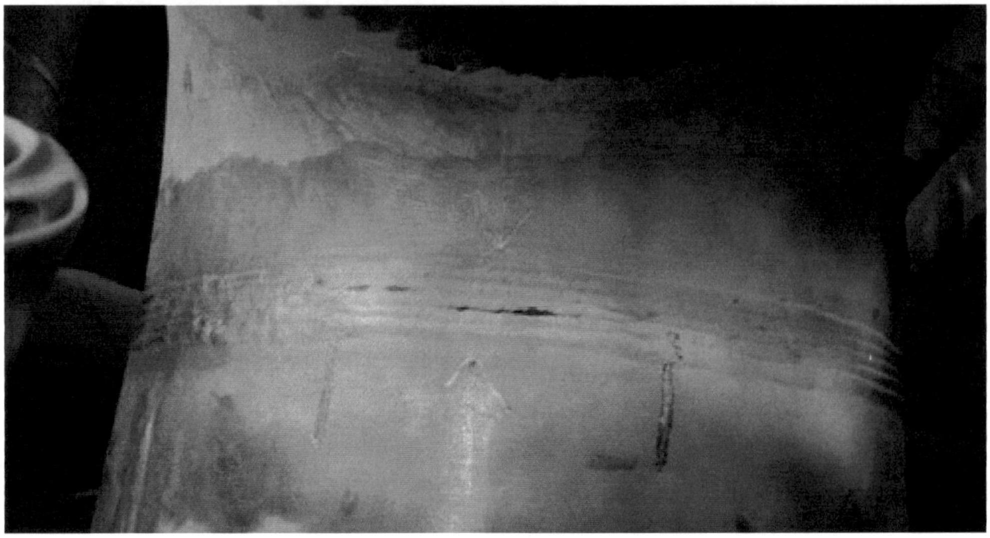

Figure 3: Photo of 1st Failure Crack – Top View

Second Failure

Following the repair from the 1st failure, the plant was started up and returned to normal operation. Seven months later steam was discovered leaking once again from the same general area. Upon shutdown and removal of the insulation it was determined that the same tee had failed; however, this time the crack occurred in the circumferential weld that connected the XS-HS letdown piping to the top of the tee (refer to Figure 1.) A photo of the 2nd failure crack is provided in Figure 4.

While the crack occurred in a different weld than the 1st failure, both cracks were located on the south side of the tee and were generally centered on the vertical plane that bisects the tee. It was later discovered that a new crack had initiated in the same location as the 1st failure but had not yet propagated through the thickness of the pipe wall. A new longitudinal crack near the weld seam had also started to form in the recently replaced downstream HS header pipe.

As was the case with the first failure, there were no reports of unusual noises, and the plant was running normally, at the time the failure was detected. There had also be no XS-HS letdown events since the start-up following the repair of the 1st failure. Following the 2nd failure the failed tee along with an attached section of the cracked downstream HS header piping was cut out and replaced.

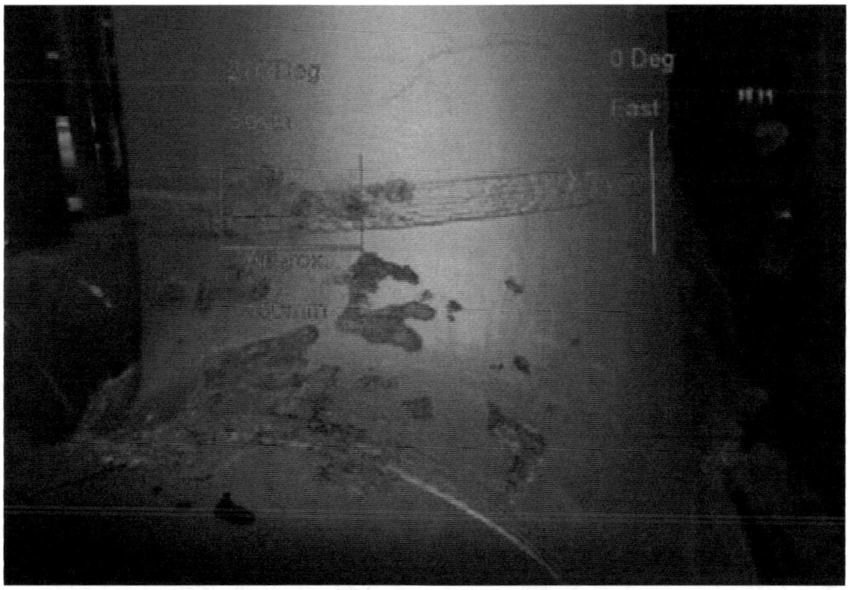

Figure 4: Photo of 2nd Failure Crack

Metallurgical Analyses

The parts removed after the 2nd Failure were sent to the KHE Materials Laboratory for metallurgical failure analysis. The crack causing the 2nd Failure leak was found to have initiated at the inside surface of the tee, in the root of the weld (see Figure 5). The partial crack at the 1st Failure location also initiated on the inside surface at the weld root. This partial crack had propagated approximately 75-80% through the thickness. Optical Emission testing confirmed the materials met specifications. Hardness measurements were below 200 HV10, indicating the welds were not susceptible to stress corrosion cracking. Fine crack branching was observed along the crack propagation path in both cracked weld regions.

The conclusion of the metallurgical analysis was that the main failure mechanism identified for both weld cracks was thermal induced fatigue. Crack initiation occurred at stress concentration locations at the fusion line between the base metal and weld root. Fine crack branching observed along the crack path was filled with an iron-oxide scale, most likely due to exposure to steam after the cracks propagated. Since the fracture surface also had an iron-oxide scale, KHE did not believe that the fine cracks branching was the result of chemical attack. KHE suspected the fine crack branching was the result of local thermal shock. Crack branching in steels can occur at very high loading rates. KHE also performed complete visual and dye penetrant inspections of the two tee welds and found no indications for additional cracks other than those already identified.

Following the 1st Failure, the section of the HS Header that was removed (i.e. the section with a partial crack in the weld seam) was sent to a third party for metallurgical analysis. That analysis also concluded that those cracks had characteristics of thermal fatigue.

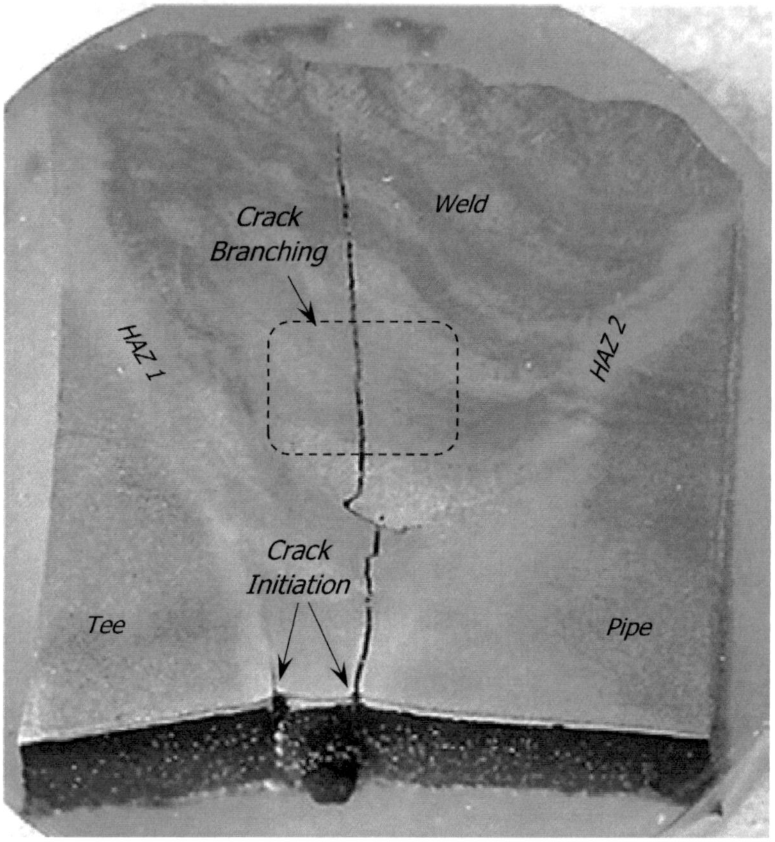

Figure 5: Cracking Details in Tee-to-Vertical Pipe Weld

XS-HS Letdown System Description

Typical operating conditions for the XS and HS systems are summarized in Table 1. During normal operations all the XS is consumed by the plant processes; however, during certain upset conditions it is necessary to let down the XS generated in the cracking furnace to the HS system. The XS-HS letdown system design includes various instrumentation to provide a controlled letdown. Figure 6 shows a simplified schematic of the XS-HS letdown controls that identifies key instrumentation for the discussion that follows.

Table 1: XS and HS Approximate Typical Operating Conditions

Piping	Pressure (kg/cm^2)	Temperature (°C)
XS	105	500
HS	40	375

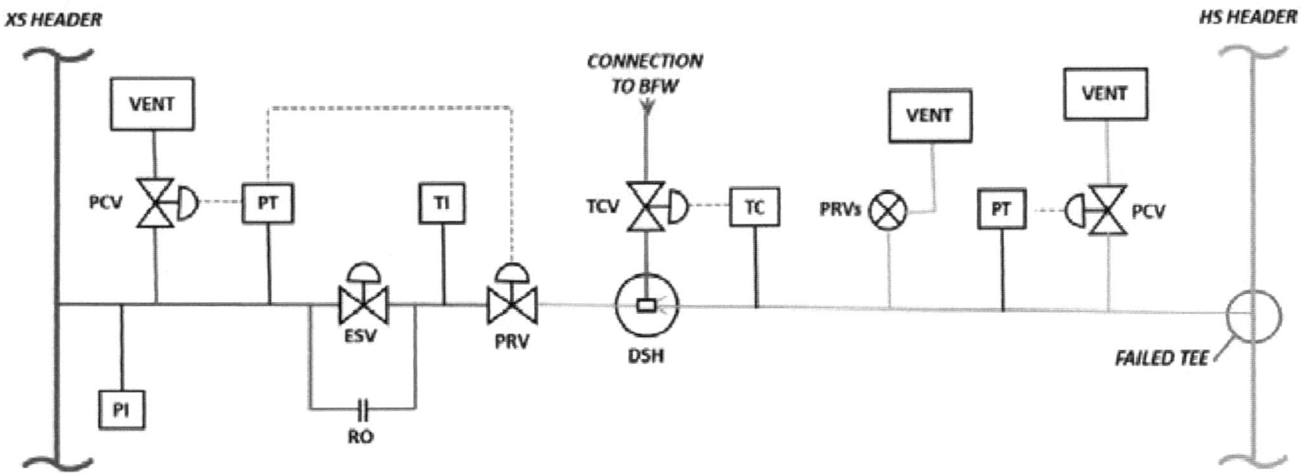

Figure 6: Simplified Instrumentation Schematic for XS-HS Letdown System

Downstream of the XS header connection there is a branch pipe with a Pressure Controlled Valve (PCV) used to relieve excess pressure from the XS header to atmosphere. Further downstream is an Emergency Shut-off Valve (ESV). During normal operation the ESV is closed, and the only steam flowing through the XS-HS letdown is what flows through the restricted orifice (RO) in the bypass around the ESV. The multi-hole plate RO was designed to provide a low flow of steam (approximately 0.350 t/hr.) to keep the XS-HS letdown line warm during normal operation. When a letdown from XS to HS is required, the ESV will open and the flow of steam through the letdown can reach up to 580 t/hr.

Downstream of the ESV is the Pressure Reducing Valve (PRV). The opening of the PRV is controlled by the pressure measured at the upstream pressure transducer. There is also a temperature indicator located downstream of the ESV and upstream of the PRV. When a letdown event occurs, the reduction from XS to HS pressure occurs at the PRV. However, during normal operation, the pressure reduction occurs at the RO. Therefore, the TI located between the PRV and RO is at HS pressure during normal operation, and XS pressure during letdowns.

Downstream of the PRV is the desuperheater (DSH). The DSH injects a cooling spray of boiler feedwater (BFW) into the steam flow coming from the PRV. The flow rate of BFW is controlled by a Temperature Control Valve (TCV), the opening of which is modulated based on measurement of the downstream temperature. Typically, the TCV opening is set to maintain a downstream temperature of 375 °C (the HS design operating temperature).

Downstream of the DSH a branch pipe connects to pressure relief valves (PRVs), and another branch pipe connects to a PCV used to relieve excess HS header pressure to atmosphere. Downstream of the pressure relief vent, the letdown line joins the main HS header at the failed tee location.

Process Assessment with Process Data Review

Review of historical data from normal operating conditions indicated that most parameters were effectively at design operating values. However, the temperatures in the XS-

HS letdown line were well below expected values (see Table 2). This was the case even though the XS header was at its design operating temperature. There was also notable variability in the temperature at the TI upstream of the DSH, as evidenced by occasional spikes from the typical value of approximately 290 °C up to 350 °C (see Figure 7). These spikes were eventually found to coincide with openings of the PCV that vents excess XS header pressure to atmosphere.

The 255 °C temperature measured downstream of the DSH was at the saturation temperature for the HS pressure. In comparing the trend in downstream temperature with the coincident pressure, as well as the upstream temperature (see Figure 7), one can see that the variation in the upstream temperature has effectively no impact on the downstream temperature, while the coincident pressure correlates significantly with the downstream temperature. This indicated that the steam in the XS-HS letdown line was at saturation with some condensate very likely present.

KHE also reviewed operating data for XS-HS letdown events. That review indicated that the control of the DSH valve at the initiation of XS-HS letdown events may not be optimized; however, KHE did not suspect that this issue was related to the failure events.

Table 2: Expected and Actual XS-HS Letdown Station Temperatures during Normal Operation

Parameter	Operating Avg. (°C)	Design/Expected (°C)
XS Header Temp.	498	500
XS-HS Letdown Temp Upstream of DSH	294	468
XS-HS Letdown Temp Downstream of DSH	255	370

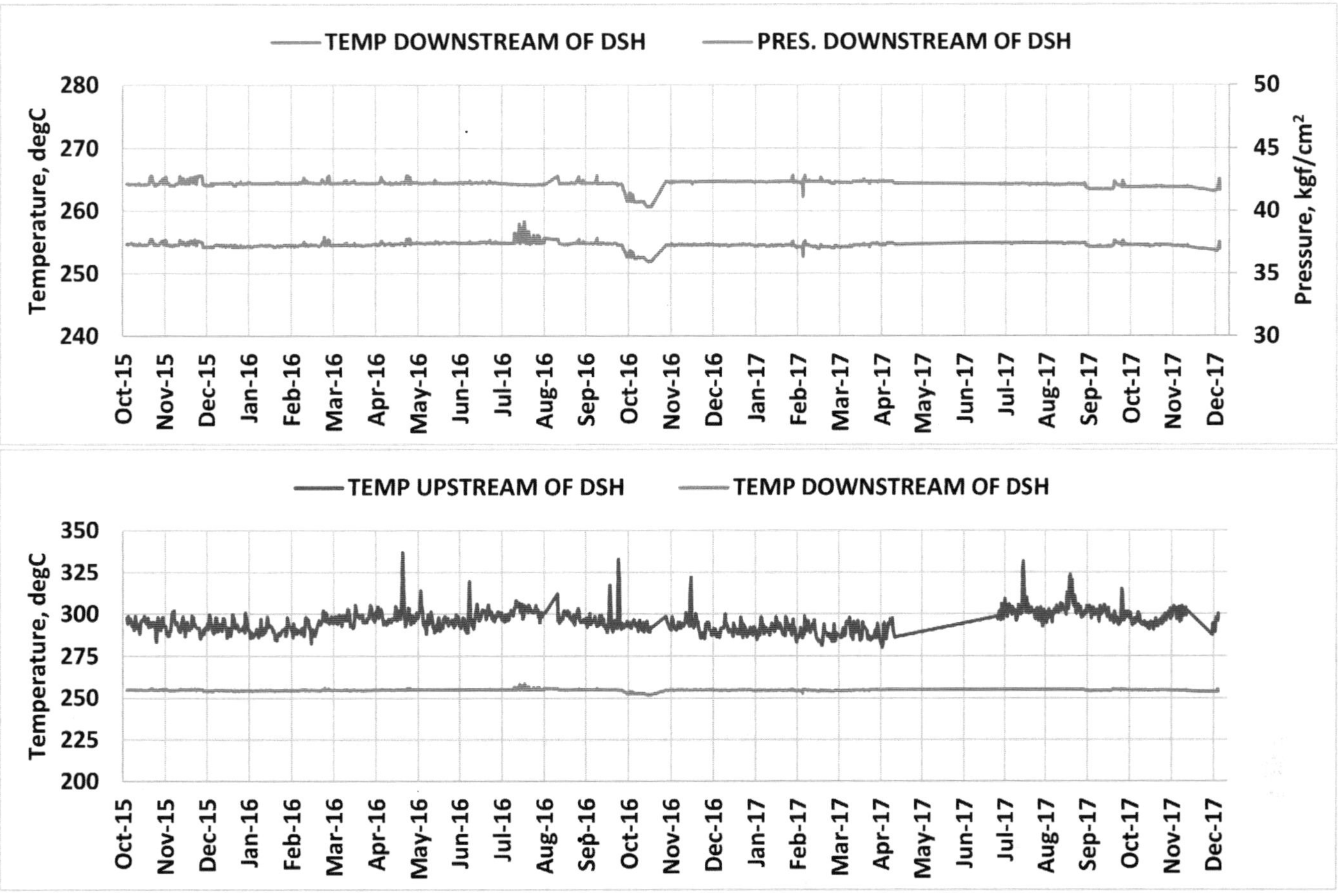

Figure 7: Temp. Downstream of DSH w/r to Pressure (above) and Upstream Temp. (below)

Root Cause Assessment

A successful RCFA is dependent on proper characterization of the failure by careful examination of the available evidence. A summary of significant evidence collected during the RCFA investigation is provided below and is tabulated in an Is/Is Not format in Table 3.

- Failure Mechanism: Per the metallurgical analysis the primary failure mechanism for the 2nd failure was thermal induced fatigue. A 3rd party metallurgical lab that evaluated the longitudinal crack in the HS header pipe from the 1st failure, concluded it had typical features of thermal fatigue. Neither of the metallurgical analyses (KHE's or the 3rd party) supported corrosion fatigue or chemical attack.

- Failure Locations: All the butt welds cracks associated with the tee failures occurred on the south side of the tee, roughly centered on the vertical plane that bisects the tee. Dye penetrant testing did not indicate cracks initiating in other locations. All the cracks examined, including the longitudinal crack in the downstream piping, originated on the inner surface of the tee or pipe.

- Failure Timing: The 1st failure occurred in April 2017, six months after the most recent letdown event. The failed parts were fully repaired or replaced. The plant was restarted in May and operated until the 2nd failure was discovered in November 2017. In addition to the new crack causing the 2nd Failure leak, cracks redeveloped in both the 1st failure location and in the downstream pipe. These redeveloped cracks initiated and propagated during the six-month operating between the two failures, during which time no XS-HS letdowns occurred.

- Presence of 2-Phase Flow: The Process Assessment found temperatures in the XS-HS letdown line were lower than expected under normal operating conditions. In fact, downstream of the DSH the process appears to be at saturation temperature, with the likely presence of condensate. Video footage of the 1st failure leak showed "puffs" of steam escaping from the crack in the pipe. While this footage makes it appear that the flow of steam through the leak was intermittent, in all actuality it is likely that the steam flow was constant, but that flow alternated between being invisible superheated steam and visible wet steam.

Considering the evidence reviewed above and in Table 3, KHE considered the following when developing a hypothesis for the root cause of the failure:

- The failure was caused by thermal loading. This is supported by:
  - The conclusions of the metallurgical analysis.
  - The presence of cracks exclusively on the south (and top) side of the tee rule out pressure/acoustical pulsations, which would not favor a given side.
  - The fact that the tee weld cracks initiate from the inside surface of the pipes, and the longitudinal crack in the HS header, ruled out all dynamic loading scenarios (e.g. steam hammer) considered.
- KHE concluded it was unlikely that the root cause of the failures was associated with conditions during a letdown event due to the following:
  - There were no letdown events after the startup following the 1st failure and the 2nd failure.
  - The data indicates that during letdown events the flow through the tee is superheated steam coming down from the letdown system with the flow typically splitting (i.e. some steam going to the north, and some to the south.) These conditions do not present an opportunity for thermal stresses which favor one side of the tee.
- KHE did not identify any conditions during start-up, or other operating conditions, that would result in local thermal fatigue stresses at the tee.

- The presence of two phase flow in the letdown piping during normal operation provides an opportunity for local thermal stresses to develop at the failed tee. Video footage of the first failure leak indicated an intermittent presence of wet steam. Intermittent flow of condensate from the letdown station into the tee was a likely explanation for the bursts of wet steam.

After development of the considerations above, KHE hypothesized that the relatively cool, off-design, two-phase flow coming down into the tee from the letdown line would be pushed to the south side of the tee by the predominant north-south flow of steam in the HS header present at the tee during normal operating conditions. This could cause local cool spots on the inner surface of the south side of the tee, and thereby local thermal stresses. The latent heat associated with the two-phase flow could result in the persistence of local cool metal temperatures into the downstream HS header pipe.

KHE further hypothesized that due to the low flow rate in the letdown during normal operation the two-phase flow could be intermittent with condensate pooling in the horizontal stretches of letdown piping, and only flowing into the tee when a certain amount accumulated. The intermittent introduction of condensate into the tee would create cyclical, local thermal stress in the tee and pipe walls as condensate is pushed against the downstream side and evaporates.

Table 3: Is/Is Not Table for HS Header Tee Failure RCFA

	PROBLEM PROFILE	IS	IS NOT	DISTINCTION/DEDUCTION
WHAT?	What Failed?	Two separate leak events at a tee in the HS system.	Other process piping	Conditions (process, mechanical, etc.) specific to HS steam system are related to this failure.
WHERE?	Where did failures occur	At a tee connecting XS-HS PRDS to HS header.	In other locations in in the HS system.	Failure caused by HS conditions only present at the tee (or most magnified at the tee) such as conditions related to the letdown.
	Where are cracks located?	Southside/Top of tee Downstream HS piping	Northside of tee Bottom side of tee Upstream HS piping	Failure is not driven by internal pressure pulsations, as these would not favor one side. Predominant north-south flow in HS header during normal operation may be a factor.
WHEN?	When first seen?	First leak detected in April 2017. Exact time it occurred is unknown.	No failures prior to first failure.	The crack either took ~ years to develop into a leak, or a change in conditions at tee occurred which caused, or amplified, the stresses that drove the failure.
	What pattern since?	Leak in same tee (different weld) in November 2017. No letdown events occurred between startup and leak.	Leak at 2nd location was not coincident with 1st failure.	Failure is not driven by letdown event conditions. At least not primarily. Stresses at 2nd leak location likely lower than in first leak location since it took longer to fail.
	What pattern since?	Redevelopment of first failure crack from (75-80% thru-wall) as well as longitudinal cracks in downstream HS piping in ~6 months.	Repaired tee and replaced pipe did not remain undamaged.	The speed at which the cracks are forming and propagating appears to have increased.
HOW?	How did failure occur? (Failure Mechanism)	Thermal fatigue, with branching	Ductile overload Low cycle fatigue Weld Defect Inadequate PWHT No signs of chemical attack	Local thermal stresses responsible for failure. Branching due to high loading rates, possibly thermal shock.

Numerical Analysis Validation

KHE identified the potential for thermal stress to develop at the failed tee location due to the presence of condensate in the XS-HS letdown line when the letdown system was not in use. However, given the high flow rate of HS that flows through the tee during normal operation, it was uncertain if the introduction of a relatively small amount of condensate could actually generate thermal stresses of sufficient magnitude to cause a failure. To validate the feasibility of the proposed failure scenario, KHE utilized qualitative Computational Fluid Dynamics (CFD) and Finite Element Analysis (FEA) to simulate the fluid dynamics and predict thermal stresses.

KHE used a CFD model to simulate the flow conditions at the failed tee under normal operating conditions. Given the limitations of the CFD model this analysis was considered qualitative, with the purpose of studying general flow and temperature trends. While the CFD solution provided explicit quantitative results (i.e. wall temperatures) these were understood to be rough estimates.

KHE ran a transient analysis simulating the normal operating conditions at the failed tee. A flow of 0.55 TPH of saturated steam/water with a quality of 0.5 was simulated flowing through the letdown line into the tee. The dominant HS flow from north to south through horizontal part of the tee was assumed to be 230 TPH at 370°C and 42.2 kgf/cm^2. The predicted temperature distribution at the tee is plotted in Figure 8 below. These results qualitatively show that the surface temperatures on the downstream (i.e. south) side of the tee are significantly cooler. The cooler temperatures were predicted to persist downstream into the HS header. For comparison purposes a steady-state CFD analysis was also performed without the introduction of wet steam from the letdown line (Figure 9). It is evident from Figure 9 that the temperature at the tee is significantly more uniform if no condensate enters the tee from the letdown line.

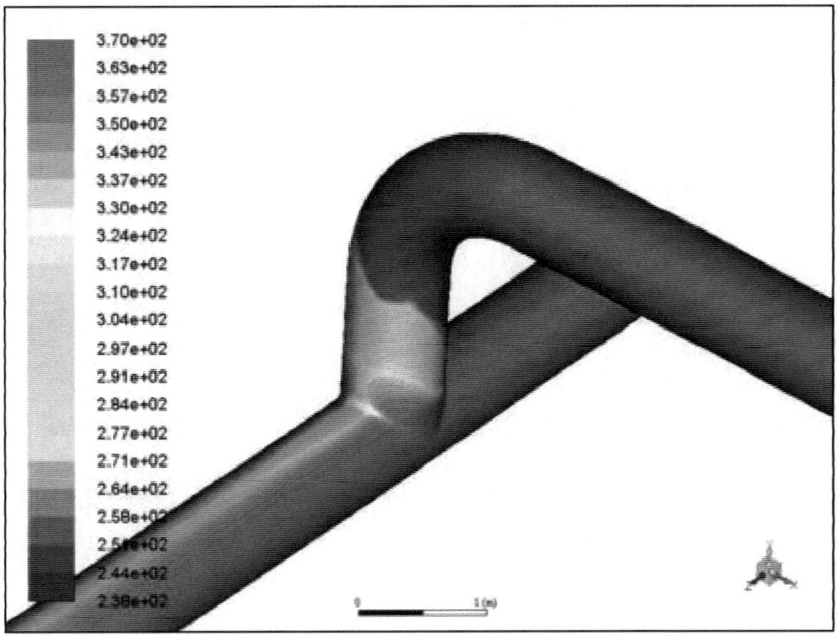

Figure 8: Temperature (°C) Distribution at Tee with Condensate Flash Off – Transient (t = 120s)

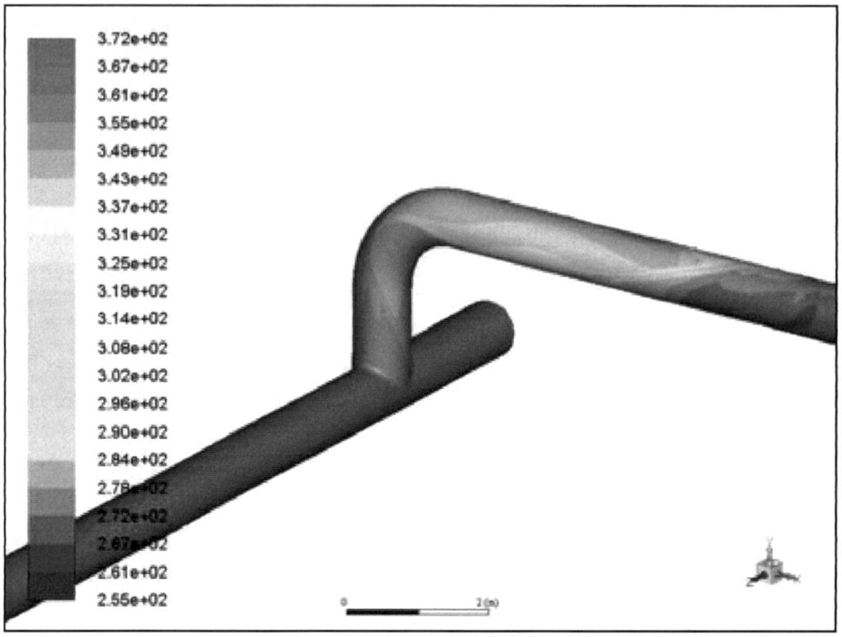

Figure 9: Temperature (°C) Distribution at Tee without Condensate Flash Off – Steady State

FEA was conducted to estimate the stresses resulting from a local cold spot on the tee. The analysis assumed that the tee was operated at 42 kg/cm^{2}g and 370 °C, initially. A local cold spot at 254 °C (saturation temperature) was applied to the model as shown in Figure 10, simulating the introduction of cold condensate from the letdown piping. The resulting stress distribution is shown in Figure 11. The maximum predicted stress was 227 MPa. Per the ASME Boiler and Pressure Vessel code, the yield stress for the tee material at 250 °C is 198 MPa.

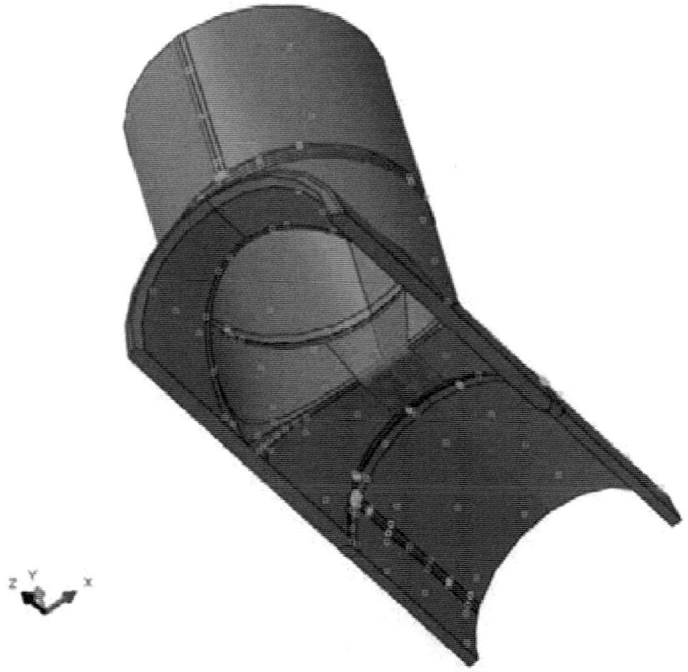

Note: Only a portion of the model is displayed in this figure for visibility purposes.

Figure 10: An Assumed Cold Spot Location and Size

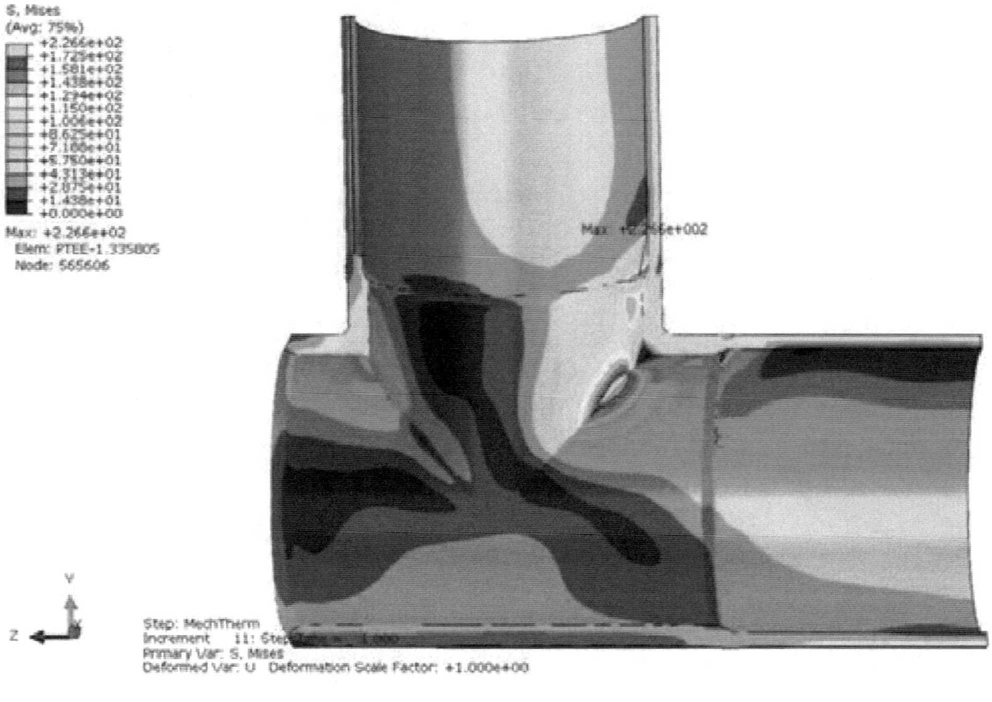

Note: Only a portion of the model is displayed in this figure for visibility purposes.

Figure 11: Von Mises Equivalent Stresses (MPa)

The results of the CFD and FEA analyses validated that it is feasible cracks were formed due to local thermal stresses resulting from the presence of condensate in the letdown line during normal operating conditions. This is particularly true at discontinuities present at the weld region, which act as stress intensification regions. While KHE did not explicitly model the downstream HS piping, it is reasonable to conclude that the relative cool spots that were predicted in the CFD analysis for the downstream HS piping could also result in thermal stresses that could cause a crack to form, particularly at the inside surface of the pipe at the seam weld discontinuity.

Field Verification

Following the 2nd failure KHE recommended that RLOC temporarily install thermocouples on each of the three branches of the failed tee. These thermocouples allowed for monitoring of tee skin temperatures during operation. Following start-up after the repairs were completed, RLOC reported that skin temperature readings at the tee were significantly lower than the expected temperature at this tee. These readings further support the RCFA conclusion that the cause of the tee failures was the off-design cool temperatures in the XS-HS letdown piping.

Short Term Remediation Measures

Several short-term remediation measures were taken to address the off-design, cool temperature in the XS-HS letdown piping. First, Q-Chem identified that the BFW control valve to the XS-HS letdown DSH was passing water when it was closed. This valve was repaired. However, following the repair, the process temperatures downstream of the DSH XS-HS letdown line were not sufficiently increased to eliminate the opportunity for condensate to develop. Furthermore, the low metal skin temperatures at the tee noted above were taken after the repair to the DSH BFW valve. Therefore, the concern regarding thermal stresses at the tee remained.

Q-Chem next established a reverse flow of superheated HS steam up into the XS-HS letdown line by opening a bypass vent valve near the PRVs (reference Figure 6). Once this reverse warming flow was established skin temperatures at the tee increased to match the HS operating temperature.

Q-Chem conducted a field survey of insulation of letdown piping, and found some improvement opportunities, but no major findings. All steam traps in the letdown piping were also checked and found to be functioning properly.

Long Term Remediation Measures

Q-Chem implemented permanent temperature and vibration sensors at the failed tee to record any abnormality during a process upset to validate the RCFA findings that conditions during letdown events did not likely contribute to the tee failures. A subsequent design review revealed that the flow through the RO in the XS-HS letdown system was insufficient to keep the XS-HS letdown adequately warm during normal operations. Measures were identified to ensure sufficient flow is constantly maintained including installation of a new orifice in the XS-HS letdown line, which is currently planned for the next turnaround (2020).

A plant wide survey was conducted on other letdown stations in the plant to check if similar low temperature conditions existed. None were identified. Detailed Inspection of the XS-HS let down piping is also planned for the next turnaround to verify the effectiveness of the short-term remediation measures.

AIChE Paper Number 120c

FITNESS-FOR-SERVICE STEAM SUPER HEATER COIL

Dan Evans
Sr. Fixed Equipment Engineer
Chevron Phillips Chemical Company, LP

Tom Garza
Manager – Fixed Equipment, Quality and Inspection
Chevron Phillips Chemical Company, LP

Cliff Knight
President and CEO
KnightHawk Engineering, Inc.

Carlos Corleto
Professor of Practice
J. Mike Walker '66 Department of Mechanical Engineering
Texas A&M University, College Station, TX

Prepared for Presentation at the 2019 Spring Meeting
New Orleans, Louisiana, March 31-April 4, 2019

FITNESS-FOR-SERVICE STEAM SUPER HEATER COIL

Dan Evans
Sr. Fixed Equipment Engineer
Chevron Phillips Chemical Company, LP

Tom Garza
Manager – Fixed Equipment, Quality and Inspection
Chevron Phillips Chemical Company, LP

Cliff Knight
President and CEO
KnightHawk Engineering, Inc.

Carlos Corleto
Professor of Practice
J. Mike Walker '66 Department of Mechanical Engineering
Texas A&M University, College Station, TX

Abstract:

Chevron Phillips Chemical Company LP (CPChem) operates a steam super heater at the ethylene plant in Port Arthur, Texas. The super heater receives saturated steam from the ethylene cracking furnaces and superheats it so that it can be used to drive high horsepower turbines. The super heater experienced a temperature excursion in the 1 ¼ Chrome radiant coils and was evaluated to determine if the tubes were acceptable for further safe operation. The coil outlet temperature increased to 1350° F, which would correspond to a tube metal temperature inside the super heater more than 1400° F during the excursion. This temperature is above what is allowed by ASME Section I.

This paper presents the results of a fitness-for-service assessment per API 579/ASME FFS-1 2007, that was conducted to ensure the integrity of the pressure boundary is maintained while the super heater continues to operate until the next scheduled opportunity for inspection or replacement. The fitness-for-service included a damage assessment of the coil using several non-destructive evaluation techniques including a visual assessment, metallographic replicas, thickness measurements, and hardness measurements of the coil. The remaining life of the coil was also assessed.

Background

The radiant coil layout of the steam super heater can be seen in **Figure *1***. The steam flows from the convection section of the heater into the radiant coil. After nine passes through the coil, the superheated steam flows out of the heater and into a collection header. The material of construction is 6" schedule 80 SA-335 P11 pipe (1-1/4 Cr). Normal operating conditions of the outlet steam is 830 psig @ 800°F with design conditions of 930 psig @ 950°F.

The steam super heater experienced a temperature excursion due to the unit upset. Based upon the coil outlet temperature, the 1-1/4 Cr tubes were exposed to elevated temperatures above the design temperature for approximately 90 minutes. The maximum coil outlet temperatures reached 1376°F in coil 7. This would correspond to a tube metal temperature inside the radiant section in excess of 1400°F. The ASME Boiler and Pressure Vessel Code allows a maximum service temperature of 1200°F.

With the coil having been exposed to temperatures above what is allowed by the ASME code, it was determined that a fitness-for-service assessment per API 579/ASME FFS-1, 2007 (FFS) needed to be performed. The following sections summarize the result of this FFS assessment.

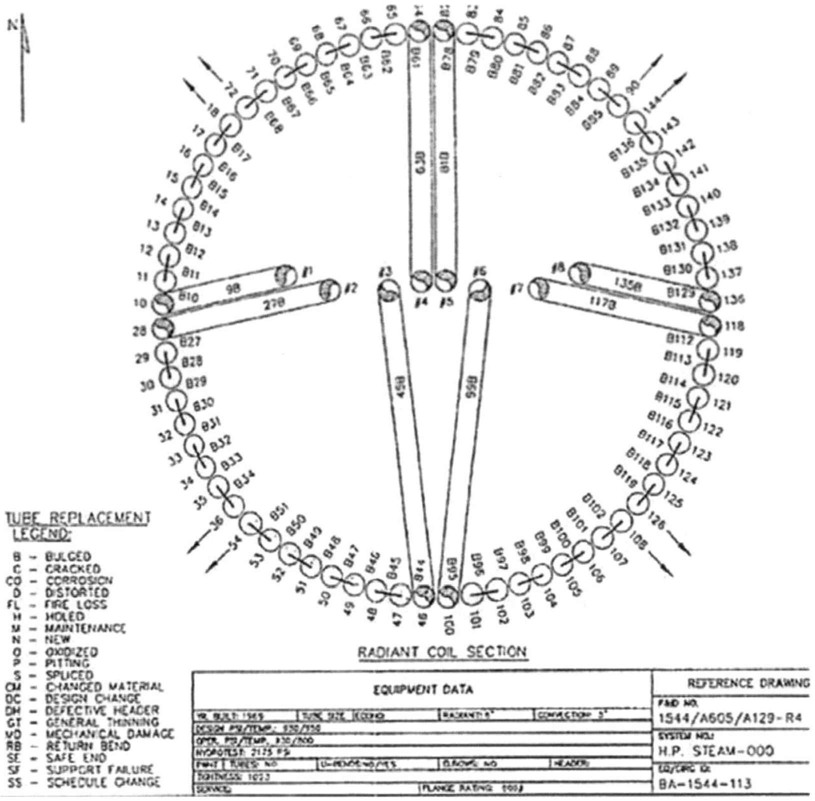

Figure 1. Radiant coil layout of the steam super heater.

Coil Damage Assessment

Visual Assessment

A complete visual assessment was made of the radiant coil. Any bulging or other distortion could indicate areas that were compromised. No evidence of bulging or distortion was observed. In addition, the outside diameter of the tubes was gaged. All of the outside diameters of the coil were within 2% of the nominal diameter of the 6" tubes, indicating minimal creep damage.

Thickness Measurements

A thickness measurement was taken on every straight section and on every return bend. Thicknesses on the straight sections ranged from 0.39 to 0.44 inch and the return bends ranged from 0.40 to 0.54 inch. These measurements are all within the fabrication tolerance of 6" SCH 80 seamless pipe. These thickness measurements do not indicate any problems with deterioration or thinning.

Hardness Measurements

Field hardness testing was performed on a representative sample of the radiant tubes. Hardness was measured at the top, middle, and bottom at multiple locations in the heater. The average hardness is 160 HB with a range of 152 HB to 176 HB. The minimum tensile strength of SA 335 P11 is 60 ksi, which corresponds to a hardness of 123 HB. Based on the hardness measurements the radiant coil did not experience any strength reduction due to the temperature excursion.

Metallurgical Replicas

Metallurgical replicas were collected on selected locations of the radiant coil. The areas were prepared using standard in-situ metallographic techniques. Replicas were extracted from the bottom and top regions of the coils near the return bends.

The replicas were observed using optical microscopy and Scanning Electron Microscopy (SEM). As illustrated in **Error! Reference source not found.**, some grain coarsening and carbide precipitation on the grain boundaries were observed. The extent of grain coarsening was higher at the bottom of the coil as compared to the top. Void formation due to the early stages of creep was not observed on the replicas indicating that there is no evidence of stage III creep.

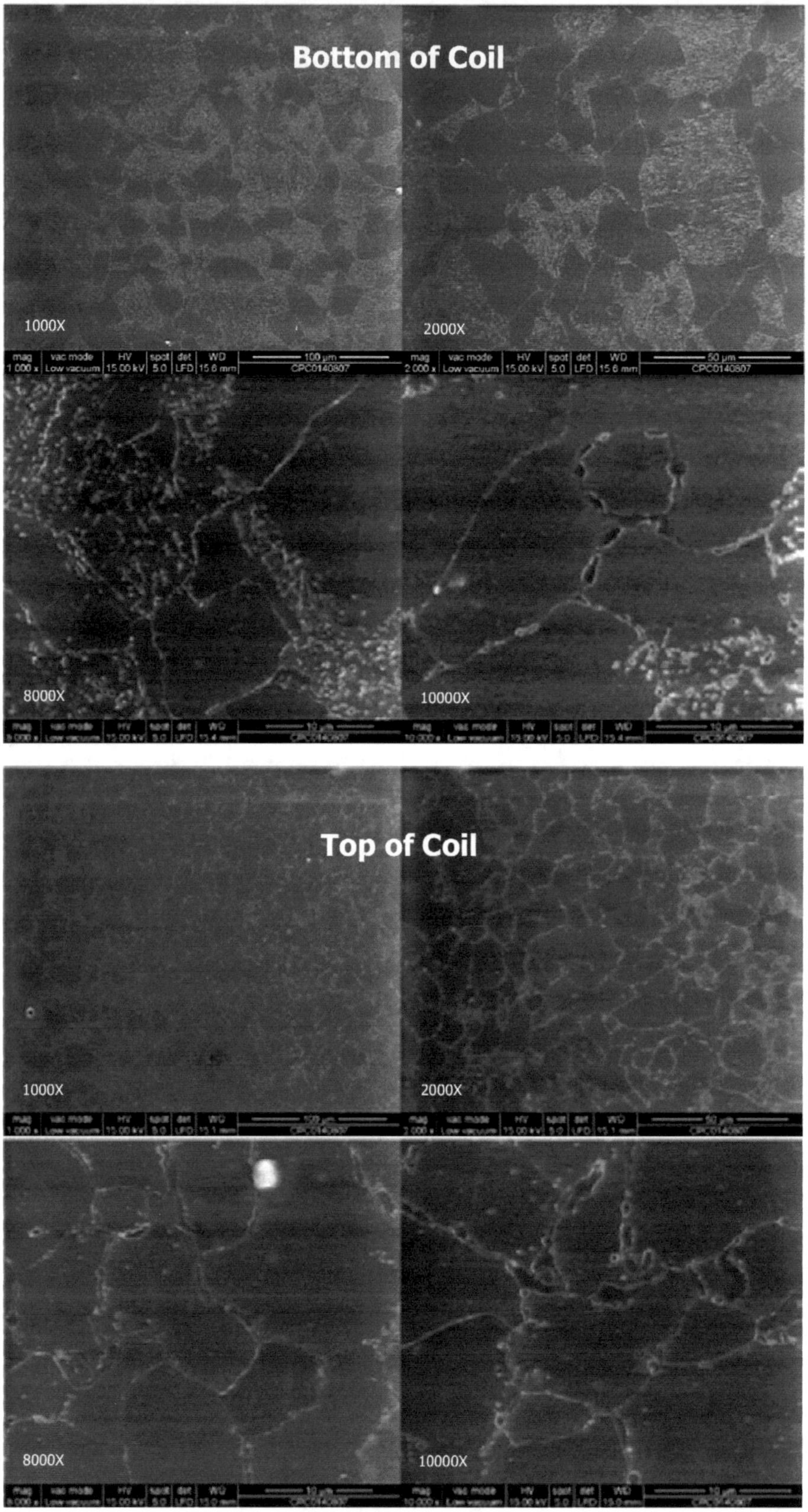

Figure 2. SEM micrograph on replica at the top and bottom of Coil 119.

Operational History

Representative historical pressure and temperature data of the radiant coils are shown in **Figure *3*** and **Figure *4***. According to these data, the average operating pressure experienced by the coils is 876 psig. The highest average operating temperature recorded is 811 °F for Coil 8. Note that only one shutdown took place during this time. Assuming this is representative of operating life conditions to date, only 4 shutdowns have taken place since the coils were installed in 2000.

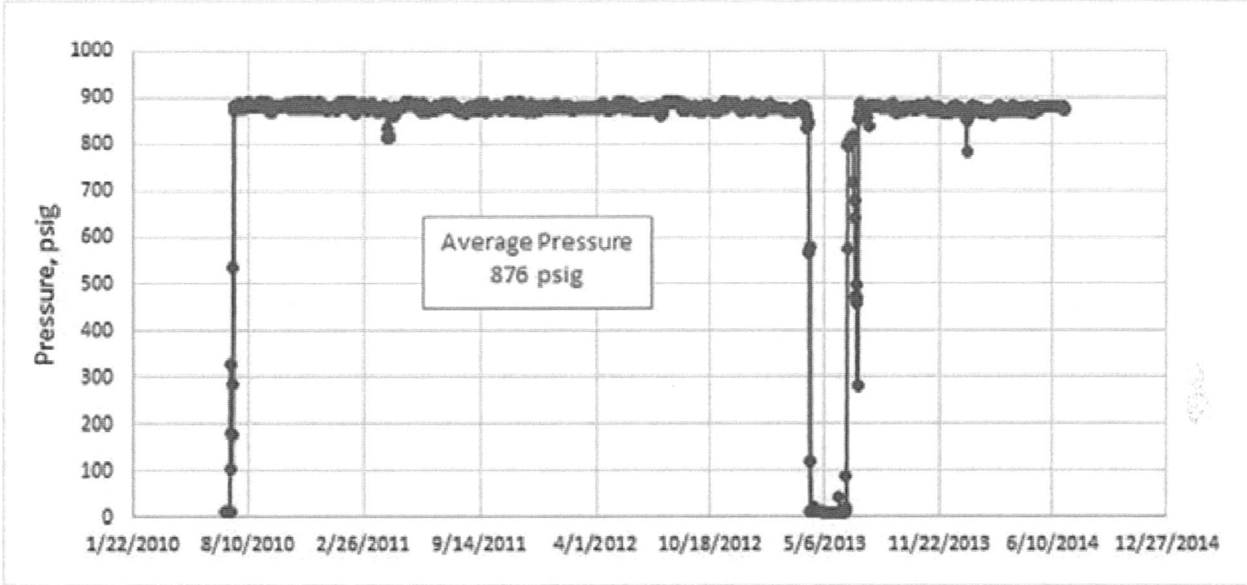

Figure 3. Representative historical steam super heater BA-113 pressure.

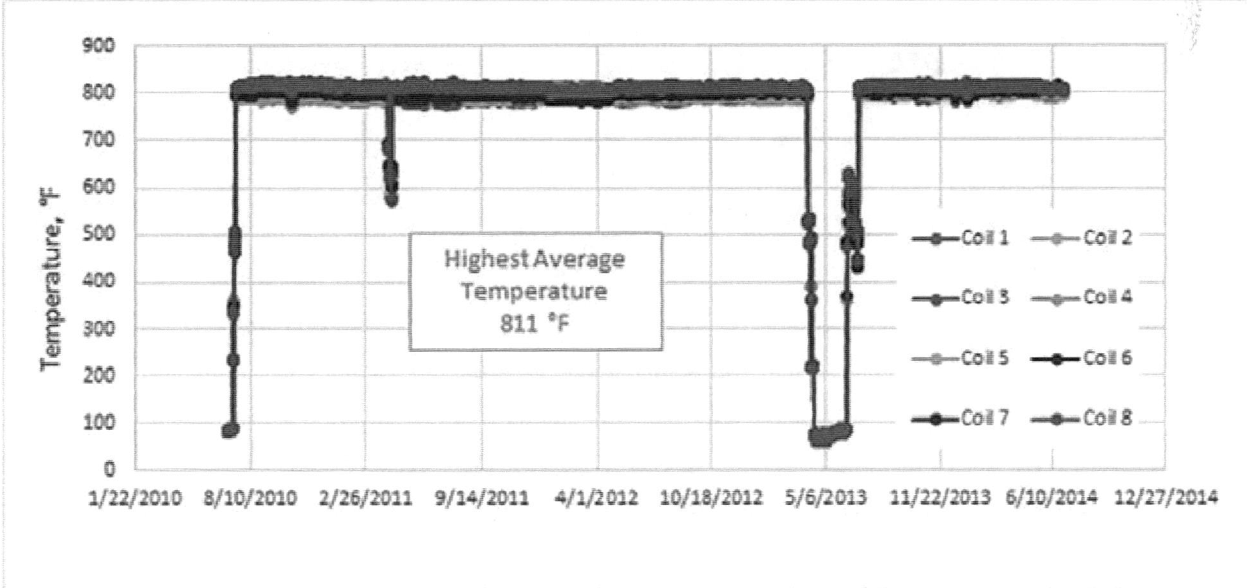

Figure 4. Representative historical steam super heater BA-113 coil temperature.

Operational data during the temperature excursion in BA-113 can be seen in **Figure *5*** and **Figure *6***. During the event, the pressure dropped to 491 psig while the temperature increased to a maximum of 1376 °F in coil 7 over a period of approximately 1 hour and then decreased to about 800 °F over approximately 2.5 hours.

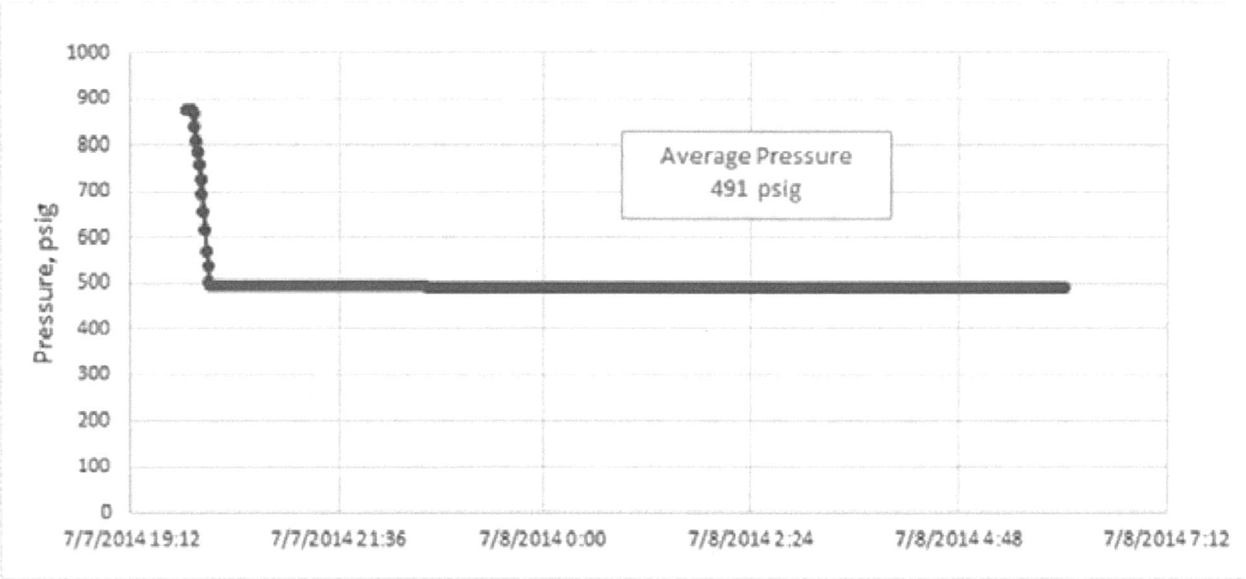

Figure 5: Pressure during temperature excursion in July 2014.

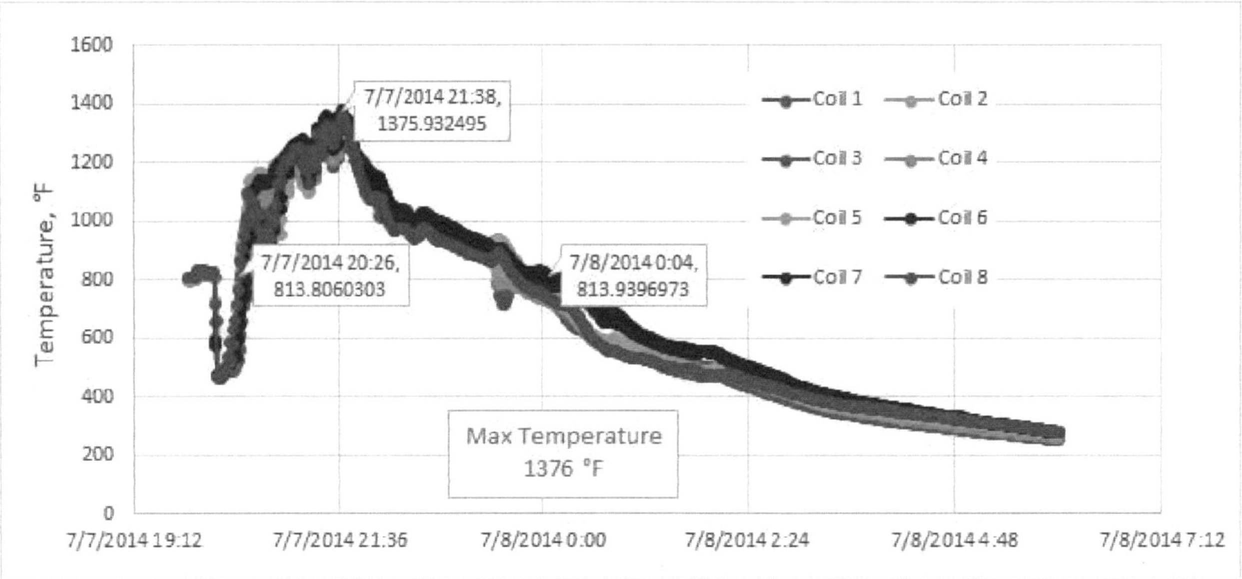

Figure 6: Temperature during temperature excursion in July 2014.

Determination of Remaining Life

Part 10, "Assessment of Components Operating in the Creep Range", of the FFS code was used to evaluate the remaining life of the coil. A level two assessment per Section 10.5.2.5 was used since it is specifically for fired heater and boiler tubes. Under these guidelines, tube life is assumed to be consumed when the cumulative life fraction, D_c^{total}, for all operating periods exceeds the allowable creep damage value, D_c^{allow}. If D_c^{total} is less than D_c^{allow}, the remaining life of operation is determined by analyzing future operating cycles until $D_c^{total} = D_c^{allow}$. D_c^{allow} is assumed to be 0.8 according to paragraph 10.5.2.3 for this application due to operation cycles at temperatures higher than the maximum allowable temperature of the tube material (SA335 P11).

Relevant Operating Cycles

Based on the data shown in **Figure *4*** through **Figure *5***, four operating cycles were conservatively identified to calculate the remaining life of BA-113 radiant coils, namely, one cycle representing operation from 2000 to 2014 and 3 operating cycles associated with the temperature excursion as shown in **Figure *7***. Pressure and temperatures for these cycles are shown in

Table 1.

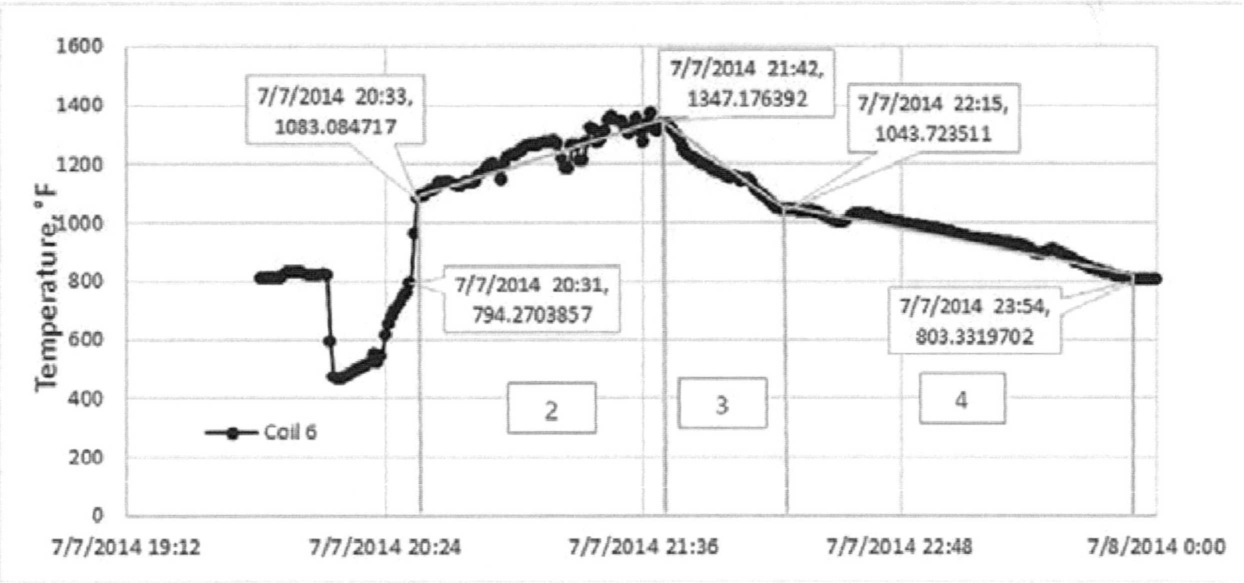

Figure 7: Operating cycles during temperature excursion – Coil 6.

Creep Cycle, (*n*)	Cycle Time (hr)	Pressure (psig)	Temperature (°F)
1	127104	876	811
2	1.15	491	1215
3	0.55	491	1195
4	1.65	491	924

Table 1: Creep operating cycles for determination of remaining creep life.

Assuming only four shutdowns since the startup of service and nearly constant temperature and pressure during operation, a fatigue analysis is not necessary for the radiant coils using Method A for fatigue analysis screening outlined in paragraph B1.5.2.3 of the FFS code. Therefore, creep-fatigue interactions that affect the remaining life of the coils tubes are not considered.

Creep Damage Accumulation

According to 10.5.2.5, D_c^{total} is given by,

$$D_c^{total} = \sum_{n=1}^{N} {}^{n}D_c = \frac{{}^{n}t}{{}^{n}L} \tag{1}$$

where ${}^{n}D_c$, ${}^{n}t$ and ${}^{n}L$ are the life fraction, cycle time and rupture life for the nth period of operation, respectively, and *N* the total number of cycles. To calculate D_c^{total}, ${}^{n}L$ was determined using the Larson-Miller method provided in Annex F Paragraph F.7.1.2 of the FFS code with the following equations:

$$\log_{10} {}^{n}L = \frac{1000 \cdot LMP({}^{n}\sigma)}{({}^{n}T + 460)} - C_{LMP} \tag{2}$$

$$LMP(^n\sigma) = A_o + A_1 \cdot {}^n\sigma + A_2 \cdot {}^n\sigma^2 + A_3 \ln(^n\sigma) \qquad (3)$$

$$A_o := 4.1444290 \times 10^1 \quad A_1 := -1.6608091 \cdot 10^{-3} \quad A_2 := 0 \quad A_3 := -2.5842632 \quad C_{LMP} := 20$$

^nT and $^n\sigma$ are the tube temperature and stress for the nth period, respectively. C_{LMP} and A_o, A_1, A_2 and A_3 are creep material parameters for calculating minimum lifetimes for A 335 P11 found in Table F.31 of the FFS Code. *LMP* is the Larson-Miller Parameter.

The stress on the tube was calculated based on the internal pressure loading only. This assumes the tubes are free to expand longitudinally and properly supported to minimize bending stresses due to weight. As indicated in Table 10 and paragraph 10.5.2.5 b) of the FFS Code, the stress for the heater tubes is given by the effective tube stress S_{eff},

$$^n\sigma = {}^nS_{eff} = 0.94 \frac{^nP \cdot D_{mean} \cdot L_f}{2 \cdot t_{comp}} \qquad (4)$$

^nP is tube pressure for the nth cycle; D_{mean} is the mean tube diameter and t_{comp} tube thickness adjusted for metal loss and corrosion allowance. L_f is the Lorenz factor given by,

$$L_f = \frac{\frac{R_b}{R_m} - 0.5}{\frac{R_b}{R_m} - 1} \qquad (5)$$

L_f accounts for the increase in stress at the tube bends. R_b is the tube bend radius and R_m the mean tube diameter.

lists the data and calculated D_c^{total} for the four creep cycles considered. Stresses were calculated using the minimum measured thickness shown in **Error! Reference source not found.** and the temperatures and pressures given in

Table **1**. Note that no corrosion allowance was included in the calculations. D_c^{total} is 0.017, which indicates the damage accumulation on the radiant coils due to the temperature excursion event is negligible.

Creep Cycle (*n*)	***^{n}t* *(hr)***	***^{n}P* (psig)**	***T* (°F)**	**$^n\sigma$ (ksi)**	***^{n}L***	***^{n} D$_c$***
1	127104	876	811	10.7	5.931(10^7)	0.00214
2	1.15	491	1215	6.0	93.8	0.012
3	0.55	491	1195	6.0	172.9	0.00318
4	1.65	491	924	6.0	3.91(10^6)	4.22(10^{-7})
					D$_{total}$	0.017

Table 2: Cumulative life fraction data for Type I tubes

R_b = 6-in ; R_m = 3.215-in ; t_{comp} = 0.325-in ; D_{mean} = 6.43-in

Remaining Life

There is no creep life limitation for radiant coil tubes due to negligible creep damage accumulation (D_c^{total}). This is primarily because the exposure to the high temperature event was short and there is essentially no creep damage during normal operating conditions.

Conclusions

This Part 10 Level 2 FFS assessment indicates that the temperature excursion did not cause any significant deterioration of the coil to the extent that it was unacceptable for further operation. In addition, there is not any creep life limitation due to the negligible creep damage accumulation calculated for the steam super heater coil.

AIChE Paper Number 120d

Transfer Line Exchanger – Internal Diameter Tube Inspection

Gary Kroner
Vice President – Technical Services
Versa Integrity Group, Inc.

Daniel Panek
Operations Supervisor/Engineer
Versa Integrity Group, Inc.

Prepared for Presentation at the 2019 Spring Meeting
New Orleans, Louisiana, March 31-April 4, 2019

Transfer Line Exchanger – Internal Diameter Tube Inspection

Gary Kroner
Vice President – Technical Services
Versa Integrity Group, Inc.
Daniel Panek
Operations Supervisor/Engineer
Versa Integrity Group, Inc.

Abstract:

Transfer Line Exchangers (TLE's) are a critical part of an ethylene producing unit. Similar to a high temperature, high pressure boiler, TLE's convert heat from the effluent gas coming from the radiant coils of the ethylene cracking heater to steam by exchanging the hot temperature of the gas to a feedwater source that produces steam. The effluent gas flows through the inside of the tube while the feedwater is on the outside of the tube. As with any high-pressure steam system, a leak in the tube causes a breach between the gas and feedwater side ultimately resulting in a shutdown of the unit. Helping to assist in reliable operation through inspection is always the goal especially when the industry can benefit from sharing experiences.

Routine maintenance cycles are common and set on typical frequencies most often for cleaning of the process tube ID's. Build-up of coke products from the process effluent on the tube ID's is a common occurrence at the inlet tubesheet in gaseous feedstocks. Unbolting and removing the heads allows for tube ID access for cleaning and NDE inspection. In general, Non-Destructing Examination (NDE) inspection needs cleaning conducted prior to the inspection to improve accuracy and is especially the case with electromagnetic and ultrasonic techniques.

Introduction

This paper presents various inspection techniques for some of the most common tube degradation mechanisms found in TLE's. Two of the most common degradation mechanisms we have seen in our inspection history are ID erosion and OD corrosion. To detect these mechanisms, we focus the presentation on two techniques and compare them with each other to allow the operators to decide based on their actual situation which inspection technique to be used. The first is an Ultrasonic technique known as IRIS (Internal Rotating Inspection System) and the second Remote Field Testing known as RFT is an electromagnetic technique. Both techniques are volumetric inspections with defect sensitivity originating either from the ID or OD. However, capabilities and limitations exist for both techniques which will be presented as well as sharing results of over 7 years of testing in-situ TLE components.

This paper presents our experience with the two techniques that we have used and is not intended to be used as a grading scheme to apply in sorting degradation mechanism versus NDE method

Ultrasonic Technique known as IRIS (Internal Rotating Inspection System)

Internal Rotating Inspection System, commonly known as IRIS, is an ultrasonic pulse-echo technique with a B-scan representation (cross section display) to determine wall thickness, degradation type, and identification of defect orientation. Typically, testing ranges from .049" to .375" wall thickness. The IRIS system was originally designed to measure the wall thickness of an internally corroded fin fan air cooled heat exchanger tubes. However, IRIS is now being performed in many different types of piping and tubing used in process pipelines, boilers and heat exchangers throughout the power generation industry, off shore installations, petrochemical plants, ships and paper mills. The IRIS technique is able to test ferrous and non-ferrous materials that transmit sound energy.

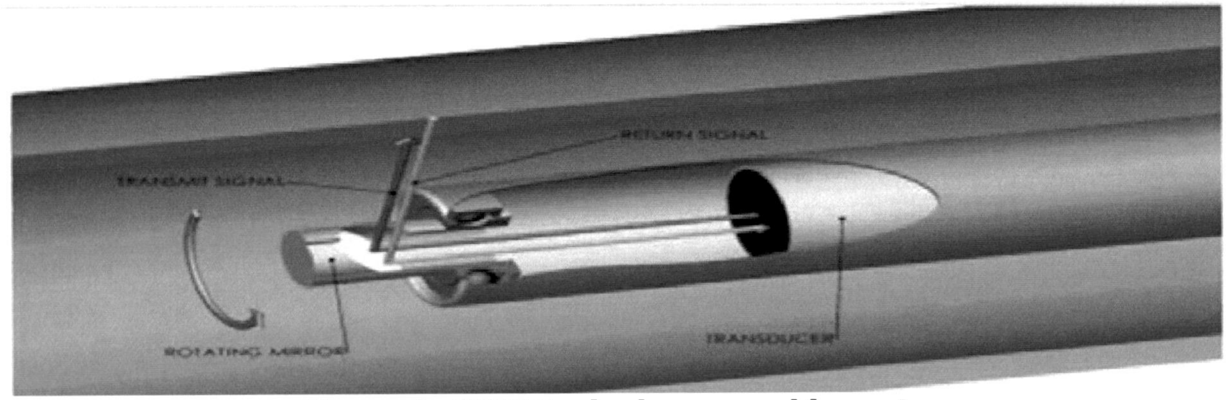

Figure 1: IRIS Probe immerged in water

The operating principle of the system is based upon the ultrasonic pulse-echo technique for measuring wall thickness. A transducer converts a pulse of electrical energy into ultrasonic sound vibration. Since the transfer of vibrational energy from a solid to a gas is extremely inefficient, it is necessary to us a couplant, such as water, to increase the energy transfer efficiency to a usable level. In our experience, the TLE tubes are filled with water so we have the proper couplant to perform the inspection. The pulse is a longitudinal wave otherwise known as a straight beam that travels perpendicular through the water to the tube wall where most of the energy of the pulse is reflected to the transducer. Two straight beam pulses which are reflected back to the transducer are electronically processed through a series of advanced digital gates. These received pulses are produced from the inner and outer wall and are called the front wall and back wall reflections. In addition to the wall thickness, the tube diameter (ID or OD) can be calculated by measuring the time between the mirror and the front wall reflection. As shown in Figure 1, the ultrasonic transducer is housed inside a probe which is placed and centered within the tube or pipe. The longitudinal pulses are emitted along a path parallel with the probe axis. These pulses are reflected by a revolving 45-degree mirror, so they are directed normal to the tube wall. Reflections from the inner and outer walls follow this same path back to the transducer. The mirror is powered by a water turbine that spins about its axis. While the probe is stationary, the mirror rotation coverage is such that the longitudinal pulses will be contiguous, and no part of the tube wall circumference will be missed.

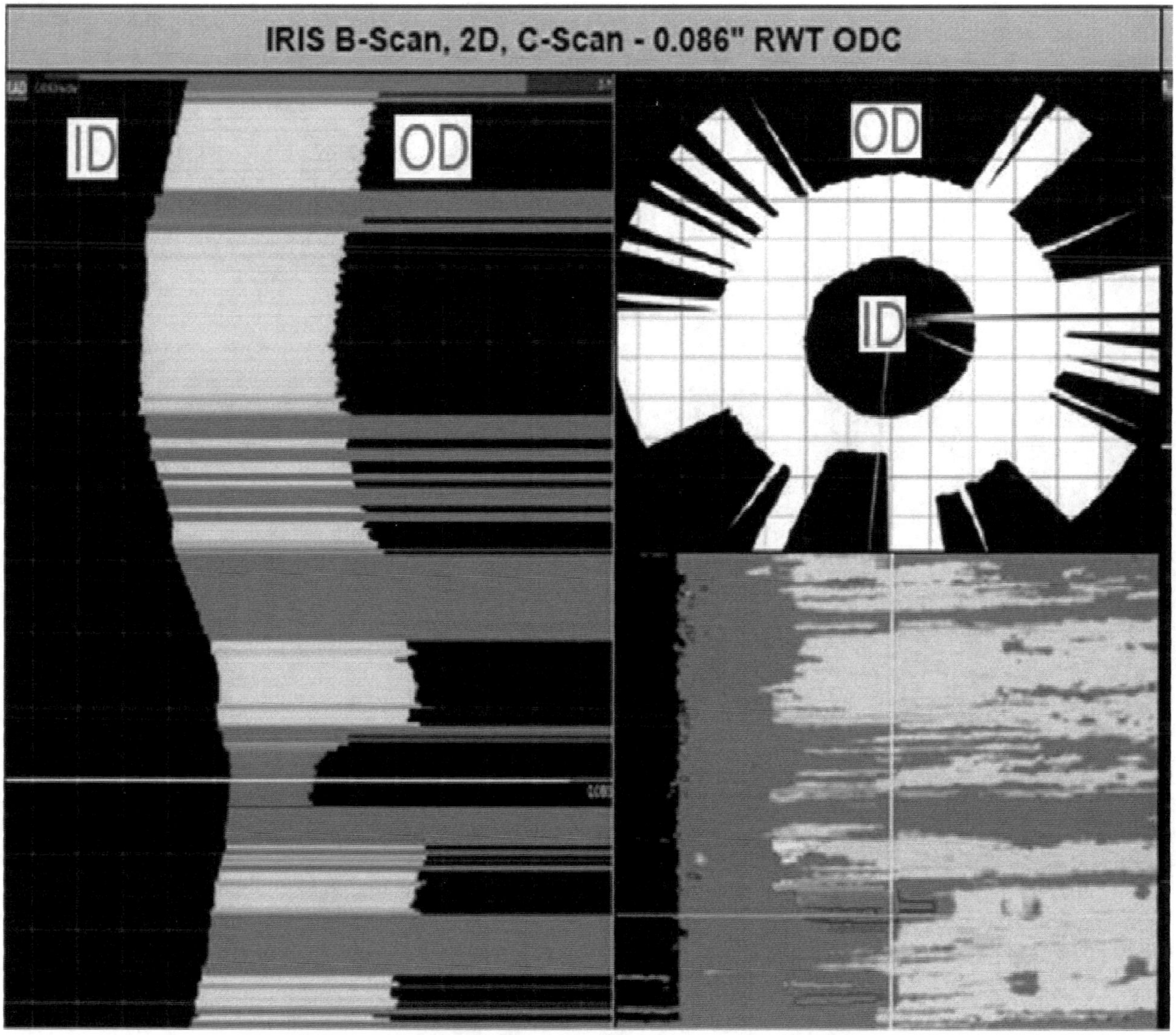

Figure 2: IRIS OD Defect Screen Capture

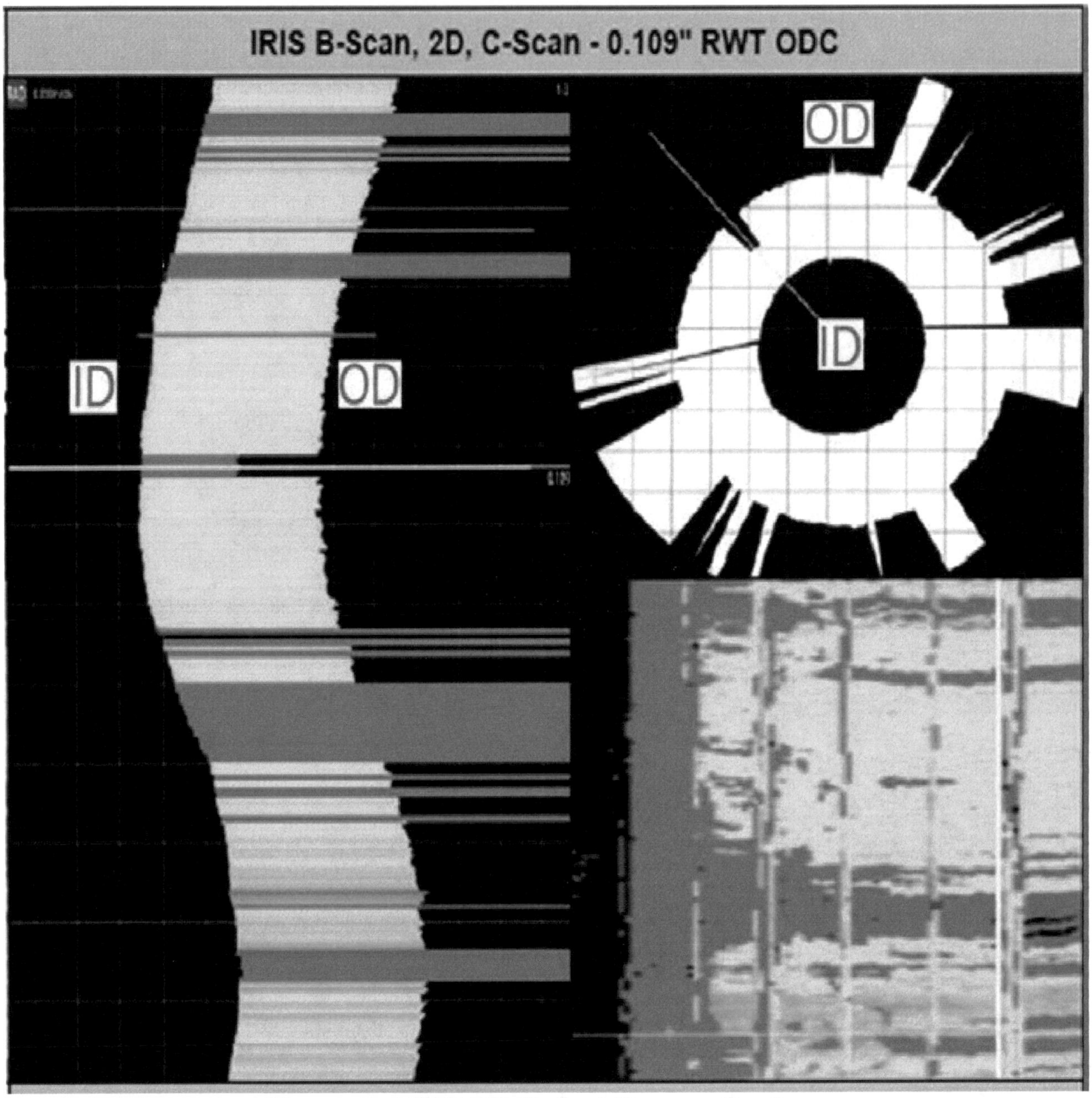

Figure 3: IRIS OD Defect Screen Capture

IRIS data can be displayed in three forms and presentations by measuring the time between the inner and outer wall reflections and the time from the mirror axis to the tube wall. These measurements are processed electronically, then all the measurements made during one full revolution of the mirror are displayed on a screen where they produce a stationary picture of the circumferential cross section of the tube wall. The picture must conform to the tube geometry thus, false and missed measurements are easily recognized in the circular view. The circular view is a cross section view displayed as if you are looking at the end of the tube from any individual slice. Lastly is the C-scan presentation which is made by aligning a series of continuous A-Scans to map the thickness in a two-dimensional area. The various displays help to quantify the profile of the degradation by allowing the depth, width and length to be measured. The accuracy of the technique is dependent on the transducer, tube diameter, wall thickness, and test conditions. Figures 2 and 3 show a screenshot of the IRIS data showing OD defect, while Figure 4 shows an ID defect. Figure 5 shows a combination of an OD and ID defect at the same location.

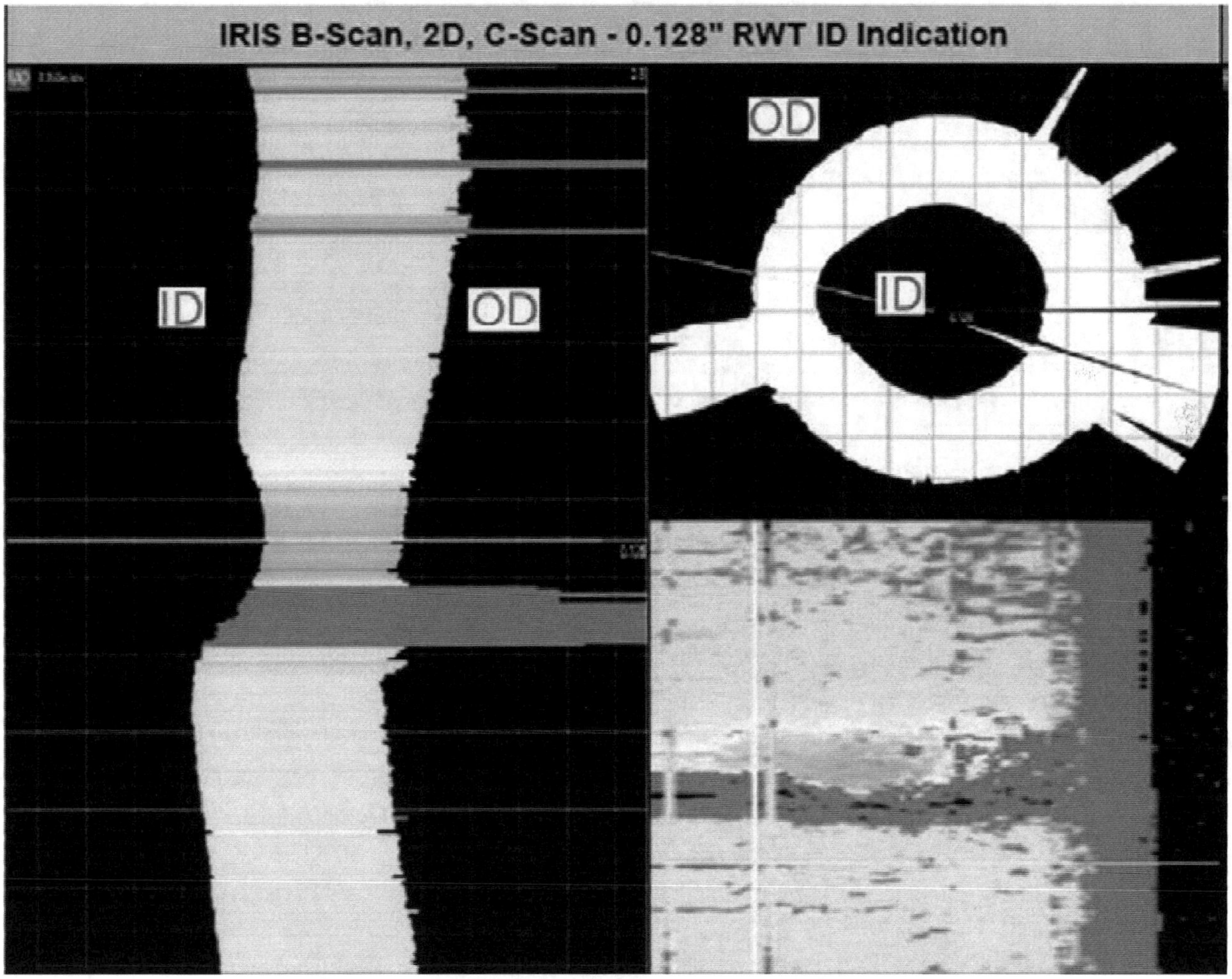

Figure 4: IRIS ID Defect Screen Capture

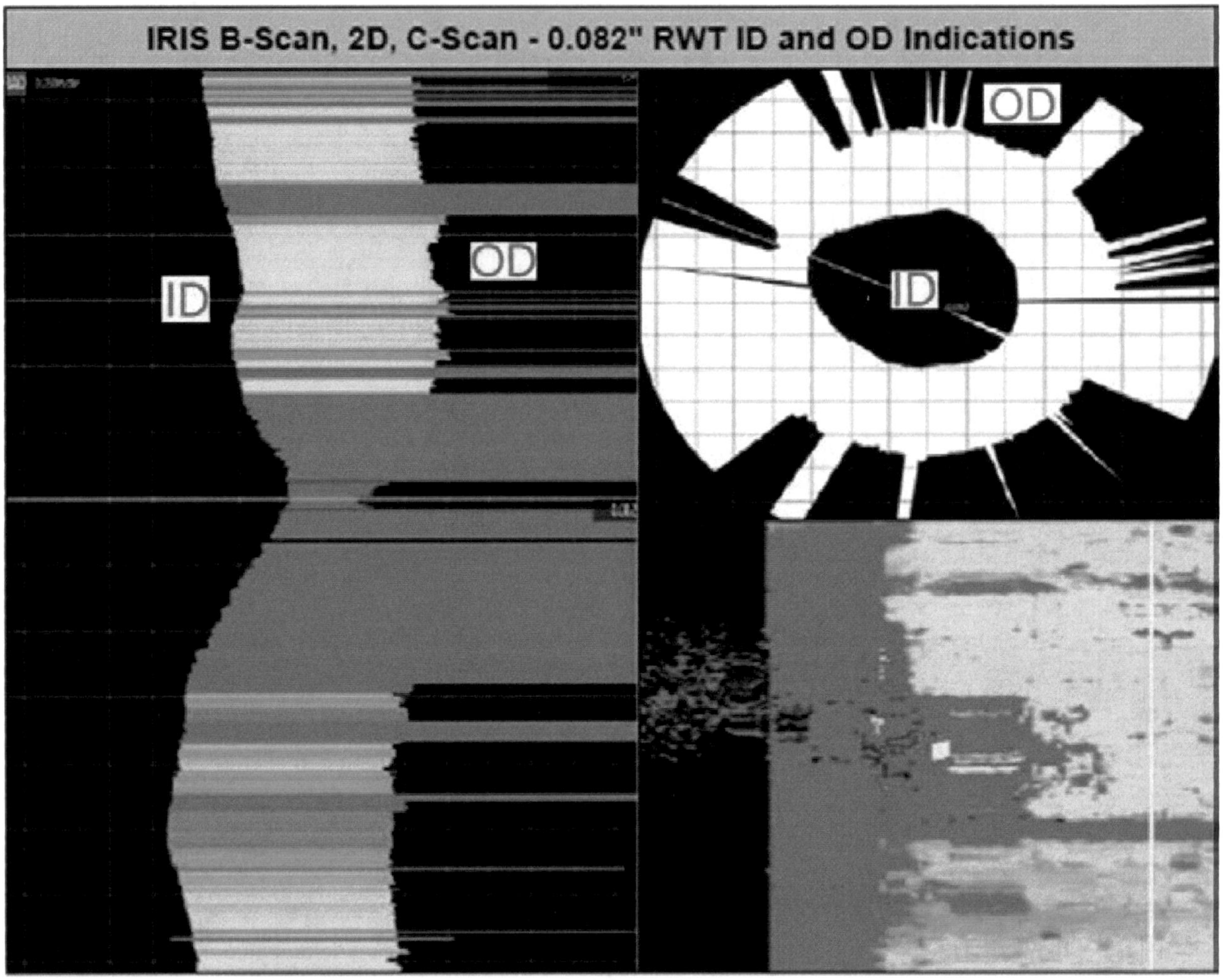

Figure 5: IRIS ID and OD Defect Screen Capture

Typically, the estimated depth of any detectable degradation can be accurately sized to within ±0.005". However, the further away the transducer is from the tube wall, the wider the transmitted signal fans out and the area of inspection becomes larger reducing accuracy. Sizing thinning of the tube wall down to a few thousandths of an inch can be resolved depending on operating settings and technician field adjustments. External corrosion that produces a rough surface will cause back wall scatter and lower resolution. Overall this technique, under ideal conditions, offers good resolution for the majority of degradation mechanisms and does very little to no averaging. However, not all types of degradation are detectable. As with all inspection techniques, there are some limitations to IRIS inspection. Centering of the probe is critical because the technique relies on the sound travel path to be normal to the tube wall. Depending on the size of the tubes, there are multiple centering devices used in order to ensure the probe is properly centered in the tube. Otherwise, the sound scatters and is not reflected back to the transducer which also occurs when the tube geometry changes. Tube environment such as the cleanliness of the tube ID surface is a critical factor in the field performance of the exam. If there are loose, rough hard deposits or scale then the sound energy may not be transmitted through the tube/pipe wall without significant loss. Sometimes the reduction in the sensitivity and coverage due to the degree of cleanliness can only be determined once the test is underway. Certain degradation such as tight axial and circumferential cracking are not discernable nor are through wall holes in many field conditions. Cleanliness of the water is also a factor if the sound energy bounces off debris floating in the water prior to it reaching the tube wall then the energy is lost. Air bubbles suspended in the water cause a similar effect.

Electromagnetic Technique known as RFT (Remote Field Testing)

The Remote Field Technique is an electromagnetic examination which utilizes the through-transmission effect to produce a resultant field that is affected by anomalies and is measured a few or more tube diameters away from the AC excitation source without any tube magnetization or saturation. The inspection technique is used primarily for ferromagnetic tubing/piping. The inspection consists of a full-length examination performed from the inside of the tube with equal inner and outer surface discontinuity detection and sizing. An alternating current is applied to an excitation coil which generates a strong electromagnetic field that propagates out of and extends away from the exciter along the tube length, see Figure 6. The field then re-enters the tube wall axially from the exciter. This is referred to as through-transmission, a term and effect commonly used in the Ultrasound Testing Method. The field has three distinct areas that are known as the Near Field Zone, Transition Zone, and the Remote Field Zone. During the field's propagation, the field intensity decays exponentially in the near field zone and linearly in the remote field zone.

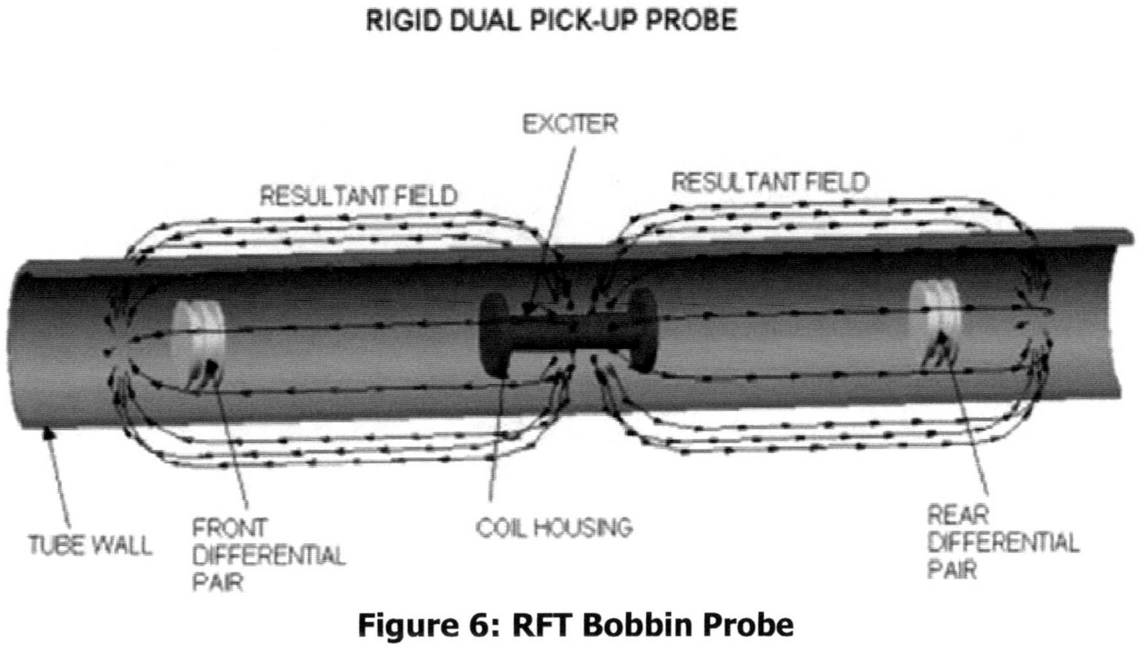

Figure 6: RFT Bobbin Probe

The resultant Remote Field Zone's strength decreases to $1/1{,}000{,}000^{th}$ the strength of the near zone, or as microvolts of the induced voltage. The induced voltage is measured through the use of a detector coil, bobbin or pancake, that is placed in the remote field zone providing for a volumetric inspection with equal sensitivity to ID and OD anomalies of the tube. The sensor coils pick up the disturbance of the field due to the presence of an anomaly caused by corrosion, stress, foreign object or mechanical vibration. Any unwanted noise such as welding in the area will affect the data which would produce inaccurate results. Inherent permeability variations are easily discerned from tube degradation signals. The technology has the ability to display the resultant signal in a vector (X-Y) format. This form of two-dimensional signal display allows the data analyst to use pattern recognition for characterizing signals. Typically, the analysis parameters are established where the phase or angular relationship of the signal is measured relative to the X-axis, in degrees, which equates to flaw depth. The amplitude in volts or size of the signal relates proportionally to the volume loss in a material. However, the phase of the signal is volumetrically dependent. The ID and OD defect phase planes are equivalent. Figures 7A/B show the data screenshot using the RFT array technique which shows 6 different array coils and the effect of the defect detection on each coil.

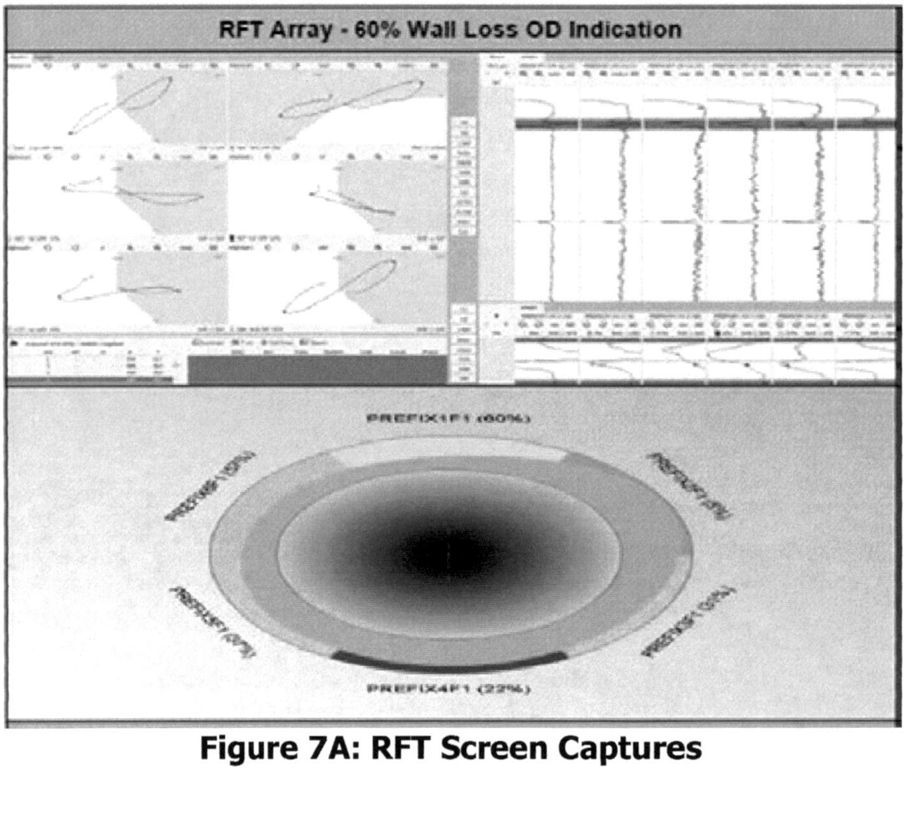

Figure 7A: RFT Screen Captures

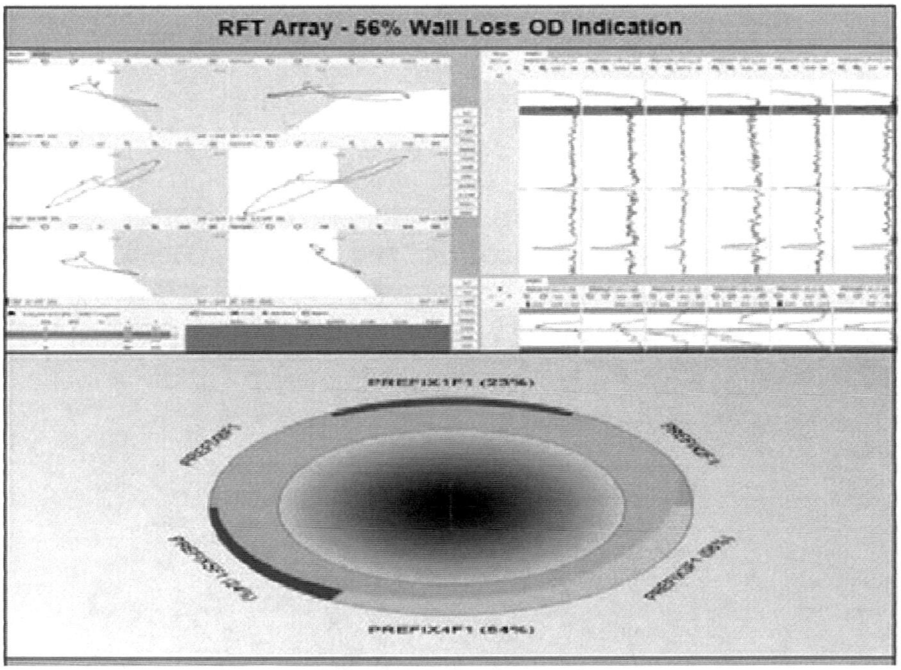

Figure 7B: RFT Screen Captures

Resolution is the phase separation between flaws of varying depths and will vary with excitation frequency. If lower operating frequencies are used, an increase in the depth of penetration results in an increase in magnitude of the resultant field which causes a larger signal response with reduced resolution. Conversely, higher frequencies have less depth of penetration reducing the magnitude of the resultant field, but an increased flux density provides for greater resolution. Therefore, a multifrequency tester provides greater capability for a broad range of tubing applications.

Sensitivity is dependent on the type of material, grade, dimensions, wall thickness, residual stresses, operating conditions such as aging, ambient temperature, tube cleanliness and location. Location affects the sensitivity due to the fact that as the field propagates outside of the tube wall, external material can short or shunt the field that prevents the field from reaching the pickup coils. Examples of shunting material would include tube support plates, tubesheets, and welded attachments. Some limitations of the RFT bobbin technique is that it cannot discern ID from OD indications. Tube cleanliness is a critical factor to prevent poor data signal quality. Brackets or another external structure could limit the probe from fully traversing through the tube length. Another probe design, RFT Array, is more suitable for TLE's since most defects occur near the tube ends.

Baseline/Preventative Maintenance Inspection

Performing an IRIS and/or RFT inspection is recommended after installation of a tube before the process is started. This will accurately determine a baseline inspection in order to provide a precise trending inspection and life expectancy. Furthermore, a routine inspection should take place during routine maintenance cycles in order to replicate a trending inspection compared to the original baseline inspection. This inspection will give us a timeframe of degradation decay and thus will be able to extrapolate to present the predicted life cycle of the tube.

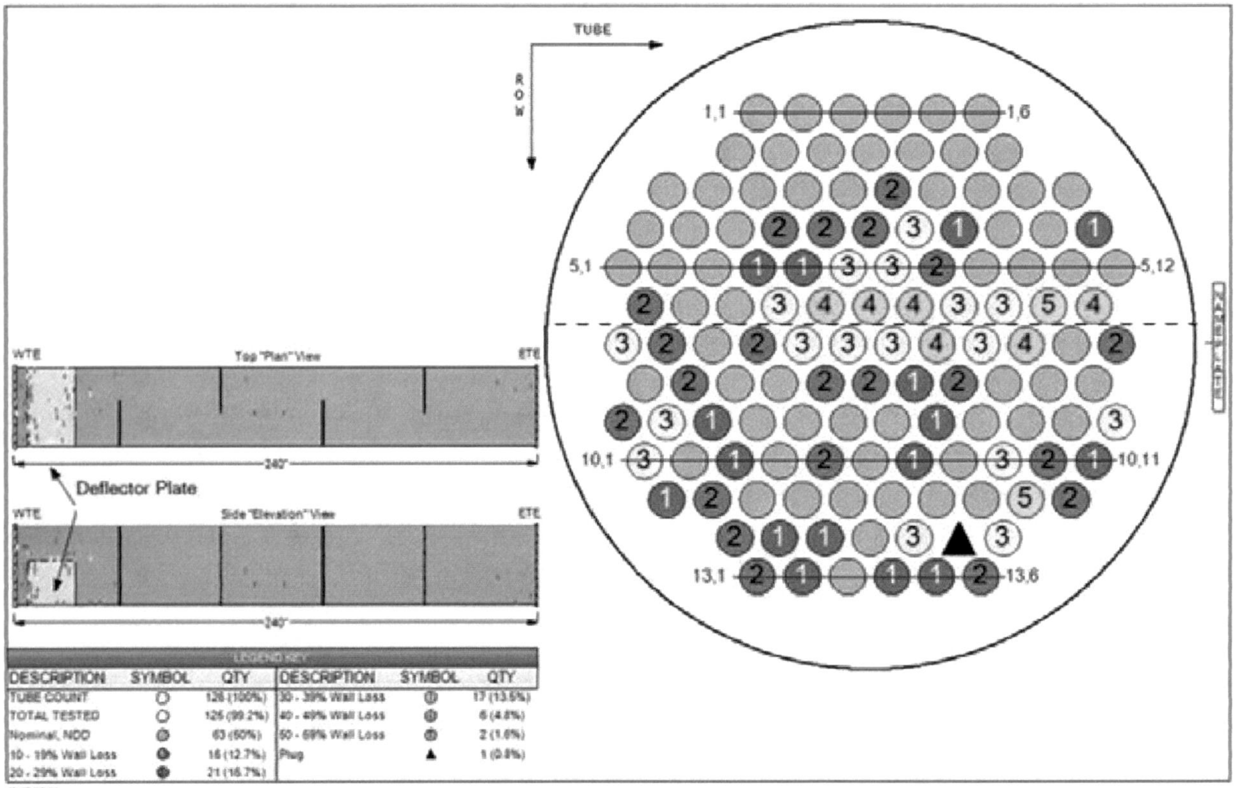

Figure 8: Exchanger Report Map Results

Figure 9: ID Erosion Found by IRIS

Figure 10: OD Corrosion Found by IRIS

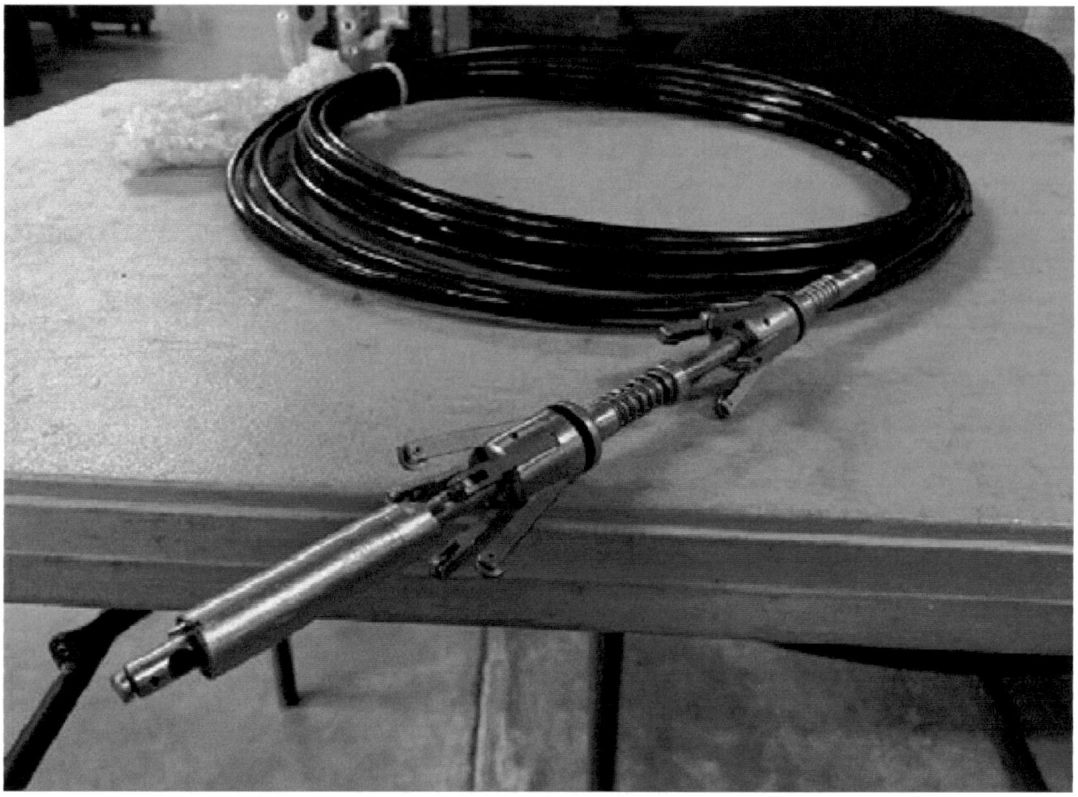

Figure 11: IRIS Probe with Centering

Figure 12: RFT Array Probe

Conclusion

This paper illustrates two successful inspection techniques to find some of the most common tube degradation mechanisms we have found in TLE's. Both the IRIS and RFT inspection techniques assist to help with the reliability of TLE's through inspection to identify and locate tube degradation. The main difference between RFT and IRIS inspection is that IRIS does require good clean water and pressure to perform the inspection while RFT does not require a couplant such as water. However, RFT will give a percent wall loss of the wall thickness based on the dimensions of the tubes given by the end user and IRIS will give the actual remaining wall thickness measured during the inspection. In addition, IRIS will differentiate whether ID or OD damage has occurred, whereas RFT cannot detect if the wall loss is ID or OD. Typically, depending on the area and length of inspection RFT tends to be a faster inspection because of the ability to acquire data at a quicker rate. Results are easily attainable for quick turnaround reports to the end user. Trending inspections provide the results attained are extremely repeatable. Figure 8 is a typical map results of an exchanger inspection. Figures 9 and 10 are ID and OD damage found by IRIS inspection. Figures 11 and 12 are pictures of each of the probes used for the inspection.

AIChE Paper Number: 120e

OVERVIEW OF ASME PCC-2 PIPING REPAIR METHODS AND ASSOCIATED NOVEL INSPECTION TECHNIQUES

Dr. Charles Becht IV, PhD, PE
President
Becht Engineering Co., Inc.

Esteban Cesan
General Manager Americas
Sonomatic, Ltd.

Prepared for Presentation at the 2019 Spring Meeting
New Orleans, Louisiana, March 31-April 4, 2019

OVERVIEW OF ASME PCC-2 PIPING REPAIR METHODS AND ASSOCIATED NOVEL INSPECTION TECHNIQUES

Dr. Charles Becht IV, PhD, PE
President
Becht Engineering Co., Inc.
Esteban Cesan
General Manager Americas
Sonomatic, Ltd.

Abstract:

Advances in technologies for in situ repair of piping, has created the possibility that the anticipated service life of pipe repairs may equal or exceed the expected life of the repaired components. ASME PCC-2 guideline does not differentiate between permanent and temporary repairs, leaving the Owner to assess the associated risk of continuing service.

Certain repair methods allowable under ASME PCC-2 guidelines cannot be monitored with conventional Non-Destructive Examination (NDE) methods and present a need for novel and proven field inspection tools and methods to monitor the integrity of the underlying component or repair over time.

Novel inspection techniques, including Dynamic Response Spectroscopy (DRS), Chime and Multiskip have been developed and found useful in evaluation of composite overwrap repairs and corrosion under pipe supports (CUPS). These techniques and methods have been evaluated and correlated in several blind trials.

Overview of ASME PCC-2, Repair of Pressure Equipment and Piping

ASME PCC-2 *Repair of Pressure Equipment and Piping* was developed to provide Recognized and Generally Accepted Good Engineering Practice (RAGAGEP) for repair methods. The first edition was published in 2006 and it is now on the fifth edition. It now provides over 30 articles describing repair methods and testing procedures, organized into sections covering welded repairs, mechanical repairs, nonmetallic and bonded repairs, and examination and testing.

The drivers for establishing standards for equipment that have been in service are two-fold. First, plants are ageing, and it is not feasible to replace them. Standards to assess and maintain mechanical integrity for these aging plants are important to the industry. Second, OSHA 1910.119j was made into law in the early 1990's, and it requires that RAGAGEP be followed to maintain the mechanical integrity of facilities handling certain highly hazardous chemicals. As such, there became an industry need to define RAGAGEP. With respect to repair other than very limited repair methods covered in post construction codes such as NBIC and API standards, most of the repair methods in industry were not covered by a consensus standard establishing RAGAGEP.

Table 1 Guide for the Selection of Repair Technique

Article Number and Title	General Wall Thinning	Local Wall Thinning	Pitting	Gouges	Blisters	Laminations	Circumferential Cracks	Longitudinal Cracks	Other
2.1 Butt-welded Insert Plates in Pressure Components	Y	Y	Y	Y	Y	Y	Y	Y	Note (1)
2.2 External Weld Overlay to Repair Methods for Internal Thinning	N	Y	Y	Y	N	N	N	N	Note (2)
2.3 Seal-Welded Threaded Connections and Seal Weld Repairs	NA	NA	NA	NA	NA	NA	NA	NA	Note (3)
2.4 Welded Leak Box Repair	N	Y	Y	N	N	N	R	R	Note (4)
2.6 Full Encirclement Steel Reinforcing Sleeves for Piping									
Type A sleeve	Y [Note (5)]	Y [Note (5)]	Y [Note (5)]	R	N	N	N	N	...
Type B sleeve	Y	Y	Y	Y	Y	Y	Y	R	...
2.8 Alternatives to Traditional Welding Preheat	Y	Y	R	Y	R	Y	R	R	Note (6)
2.9 Alternatives to Postweld Heat Treatment	Y	Y	Y	Y	S	Y	Y	Y	Note (7)
2.10 In-Service Welding Onto Carbon Steel Pressure Components or Pipelines	N	N	N	N	N	N	N	N	Note (8)
2.11 Weld Buildup, Weld Overlay, and Clad Restoration	N	Y	S	S	S	N	N	N	...
2.12 Fillet Weld Patches	N	Y	Y	Y	S	N	R	R	...
2.13 Fillet Welded Patches With Reinforcing Plug Welds	N	Y	Y	Y	S	N	R	R	...
2.14 Threaded or Welded Plug Repairs	N	Y	Y	Y	N	N	Y	Y	...
3.1 Replacement of Pressure Components	Y	Y	Y	Y	Y	Y	Y	Y	...
3.2 Freeze Plugs	NA	NA	NA	NA	NA	NA	NA	NA	Note (9)
3.3 Damaged Threads in Tapped Holes	N	N	N	N	N	N	N	N	Note (10)
3.4 Flaw Excavation and Weld Repair	NA	NA	Y	Y	Y	Y	Y	Y	Note (11)
3.5 Flange Refinishing	N	N	Y	Y	S	N	Y	N	Note (12)
3.6 Mechanical Clamp Repair	N	Y	Y	R	N	N	R	R	Notes (2), (13)
3.7 Pipe Straightening or Alignment Bending	N	N	N	N	N	N	N	N	Note (14)
3.8 Damaged Anchors in Concrete	NA	NA	NA	NA	NA	NA	NA	NA	Note (15)
3.11 Hot and Half Bolting Removal Procedures	N	N	N	N	N	N	N	N	...
3.12 Inspection and Repair of Shell and Tube Heat Exchangers	Y	Y	Y	R	Y	N	Y	Y	...
4.1 Nonmetallic Composite Repair Systems: High-Risk Applications	Y	Y	Y	R	Y	Y	R	R	...
4.2 Nonmetallic Composite Repair Systems: Low-Risk Applications	Y	Y	Y	R	Y	Y	Y	R	...
4.3 Nonmetallic Internal Lining for Pipe: Sprayed Form for Buried Pipe	Y	Y	Y	Y	Y	Y	Y	Y	...
5.1 Pressure and Tightness Testing of Piping and Equipment	N	N	N	N	N	N	N	N	Note (16)
5.3 Nondestructive Examination in Lieu of Pressure Testing for Repairs and Alterations	NA	NA	NA	NA	NA	NA	NA	NA	Note (17)

Y = generally appropriate
S = may be acceptable, but is not generally used for this condition
R = may be used, but requires special cautions
N = not generally appropriate
NA = not applicable

NOTES:
(1) This method may apply to replacement of nozzles, flat spots, and dents.
(2) See Part 2, Article 2.1, Limitations.

Table 1 Guide for the Selection of Repair Technique (Cont'd)

NOTES (Cont'd):
(3) This repair technique applies to seal welded threaded joints for leak tightness.
(4) Welded leak box repair applies to leaking flanges, valves, pipe components, and welded and mechanical joints.
(5) For internal defects, the cause of defect shall be understood or arrested, or a Type B sleeve is recommended.
(6) Alternative provisions for welded repair.
(7) Also may apply to new construction or to all methods requiring welding.
(8) Generally appropriate for surface welding on carbon steel substrates.
(9) This technique is used for isolating sections of pipes.
(10) Repair of damaged threads in stud holes.
(11) Also applies to the repair of hardened surfaces.
(12) Radial cracks.
(13) Gasket and packing leaks = Y.
(14) Pipe straightening.
(15) Repair of damaged anchor in concrete.
(16) This is not a repair technique; however, it may apply to most methods of repair.
(17) This is an alternative to pressure testing.

Table 1 of PPC-2 formerly provided a summary listing of potential repair procedures and guidelines on the applicability of each. This table has been dropped from the latest revision.

While some post construction codes characterize repairs as temporary or permanent, PCC-2 does not do so. This is because whether a repair is temporary or permanent should consider the service, damage mechanisms, and risk. For example, a patch plate on a water line could be considered permanent, but the same repair on a high temperature line may be considered temporary because of the likelihood of developing fatigue cracks around the perimeter of the fillet weld over time and possibly considering the relative hazard of a leak. One means to mitigate risk is continued inspection to monitor if there is further deterioration.

Nonmetallic wraps have been included in PCC-2 from the start, and found wide applicability. Repair of defects resulting from a variety of causes, including internal corrosion, were covered under both Article 401 Nonmetallic Composite Repair Systems: High-Risk Applications and Article 402 Nonmetallic Composite Repair Systems: Low-Risk Applications.

Article 401 requires inspection of the repair and provides limits to defects that may be found, as described in Table 401-5.2-1 of that article. However, the inspection methods that were available were fairly limited, visual and tap testing, tapping the repair with a blunt instrument and listening to try to identify delaminations and voids. Article 402 similarly requires inspection of the repair for defects.

In addition, para. 401-5.5.2 of Article 401 requires periodic inspection if the corrosion was internal to make sure that "the internal defect does not grow to a size greater than assumed in the design of the repair or that the repair laminate does not delaminate from the component substrate." However, no specific examination method is described in the article. A new examination technology has been developed, described herein, which provides accurate characterization of defects in the wraps, and also permits accurate monitoring of the thickness of the pipe under the composite repair wrap.

Novel Inspection Technique - Dynamic Response Spectroscopy (DRS)

Dynamic Response Spectroscopy (DRS) is an automated ultrasonic inspection technique which uses low frequency ultrasound (<1 MHz) to penetrate composite laminates and excite the steel substrate; the steel then vibrates at natural frequencies related to its thickness profile and this response is received by the transducer, as shown in Figure 1. Flaws in the laminate system inhibit transmission of the ultrasound, rendering them detectable by loss of vibration response from the steel at that location, as shown in Figure 2. DRS is capable of measuring the extent of flaws such as delaminations, voids or inclusions in the laminate and disbondment from the steel substrate. Where no flaws are present in the laminate, DRS provides measurements of the thickness profile of the steel substrate.

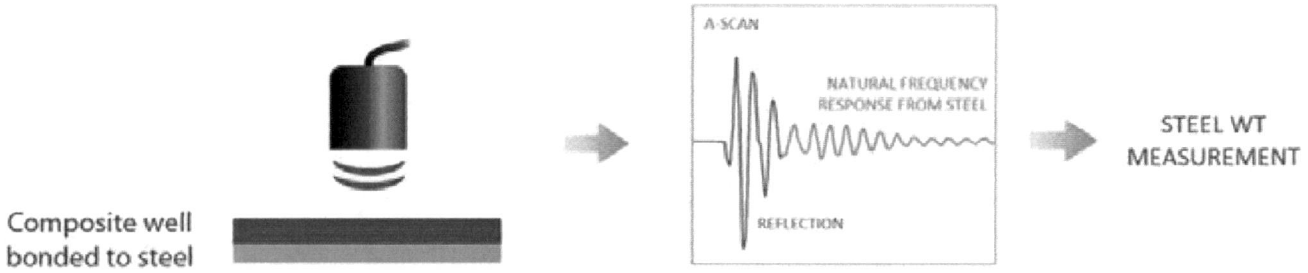

Figure 1. DRS inspection diagram. The orange part of the signal contains the natural vibration frequencies of the steel, which are used to determine its thickness profile.

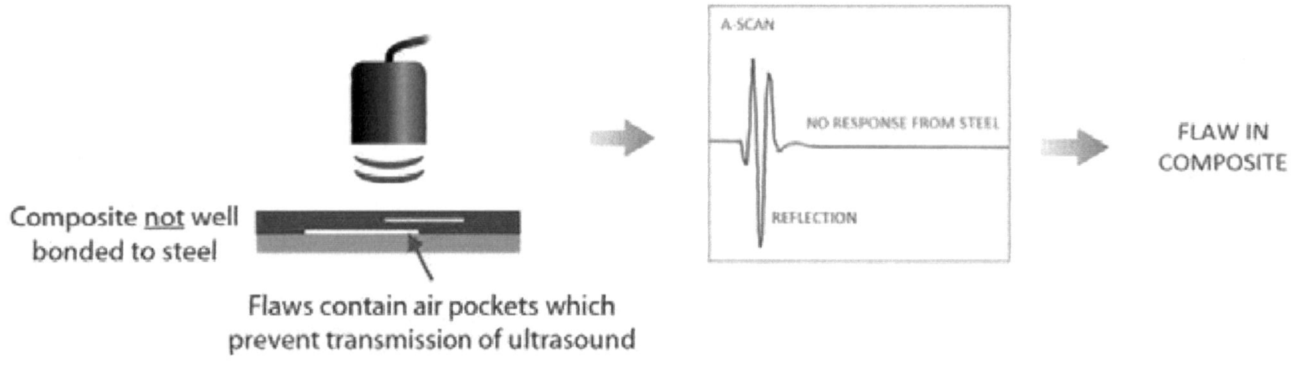

Figure 2. DRS inspection diagram. Flaws in the composite laminate prevent transmission of the ultrasound and are identified by the lack of response from the steel.

Figure 3. DRS inspection using a Nautilus system with a DRS transducer (red) in the scanning head. A low profile version of the system is also available; it can inspect areas where the clearance is only 5 inches.

Figure 3 shows the automated X-Y scanner which deploys a DRS transducer perpendicular to the steel substrate in a scanning head which conforms to any irregularities in the surface of the laminate. The transducer rasters across the surface of the laminate, emitting low frequency ultrasound which causes the steel substrate to vibrate; the vibrations are received by the transducer and collected at intervals determined by the inspection requirements. Advanced signal processing algorithms evaluate the data, identifying vibrations from the steel substrate and extracting their frequencies which are then used to create a thickness map.

DRS maps display steel wall thickness measurements in a range of colors, typically green for the nominal wall thickness and other colors showing wall loss. White pixels in the DRS map show where the ultrasound was not transmitted into the steel substrate, indicating flaws in the laminate.

Detection and Evaluation of Repair Flaws – Laboratory Tests

Prior to deploying the DRS technique in the field, extensive laboratory trials were carried out to determine which types of laminate flaws are detectable. Figure 4 shows DRS maps of four test plates manufactured with typical flaws. The air bubble, wrinkle, poor saturation and sand were all detected as they trapped air in the laminate; air prevents transmission of the ultrasonic DRS signals. Figure 5 demonstrates that DRS is equally capable of detecting natural flaws. The delamination in this sample was unintended and unknown prior to the DRS inspection; sectioning confirmed its location and dimensions.

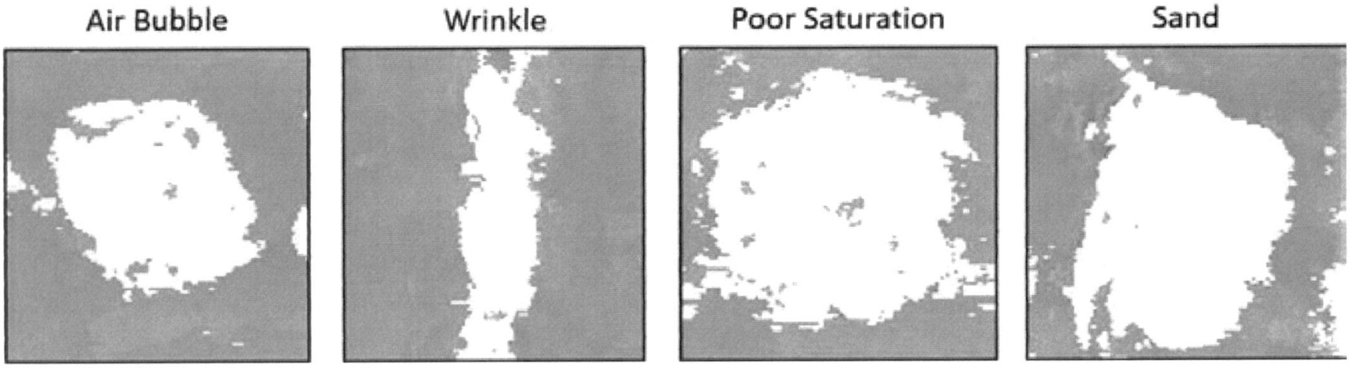

Figure 4. Manufactured flaws in composite repairs on steel plates of a uniform thickness. White areas show where flaws prevented transmission of the DRS signals into the steel. Green areas show the DRS measurements of the steel thickness where the repair is well bonded to the steel.

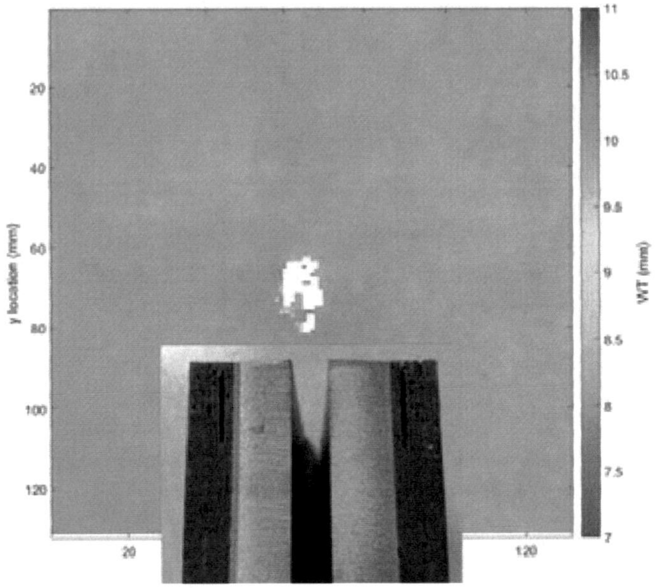

Figure 5. A natural flaw found by DRS inspection and confirmed by sectioning the sample.

Detection and Evaluation of Repair Flaws – Case Study

Carbon fiber composite repairs were to be installed on several miles of ageing pipelines. However, the existing coatings included asbestos fibers. Therefore, the repair was designed to encapsulate large regions of the original coating and only adhere to short zones of bare steel at regular intervals along the pipeline. This reduced the risks associated with removing all the old asbestos.

The integrity of these repairs relies on the bond quality in the 'adhesion zones', so the pipeline owner required an inspection technique to identify flaws in the laminate at these locations. They trialed several techniques and DRS was found to be the most suitable for this application.

It is difficult to eliminate flaws entirely, so it is important to understand the circumstances in which a 'flaw' (essentially an anomaly in the laminate) becomes a 'defect' which affects the integrity of the composite repair. Finite element modelling was carried out by Sonomatic to determine the effect of specific flaws, and their interactions, on the integrity of these repairs. This allowed flaw acceptance criteria to be developed and built into QA test software by Sonomatic. The software evaluates the flaws found in the DRS map and determines whether they interact across the entire length of the adhesion zone; if they do, it fails the QA test.

Figure 6 shows the design of the samples used in the inspection trials. DRS has been used to inspect the adhesion zones of many such samples. One of these was pipe sample A, shown in Figure 7. The DRS map of this sample is presented in the upper part of Figure 8. Where the composite laminate is well bonded, it shows the steel thickness measurements in shades of green. The slight thickness variation is due to the pipe manufacturing process. Flaws in the laminate are shown in white. The DRS map was then run through the QA test software which determined that few of the flaws interact with the encapsulated zone; interacting flaws are shown in black in the lower part of Figure 8. Therefore, pipe sample A passed the QA test. It subsequently passed a hydrostatic test.

A second example is pipe sample B, shown in Figure 9. The DRS map, in the upper part of Figure 10, shows large areas of flaws in the composite laminate. The QA test software confirmed that the flaws interact across the full length of both adhesion zones. The pixels resulting in the failure of the QA test are shown in red in the lower part of Figure 10. Pipe sample B then failed its hydrostatic test. Photographs of the leak locations are shown in Figure 11; they correlate well with the locations indicated by the DRS map.

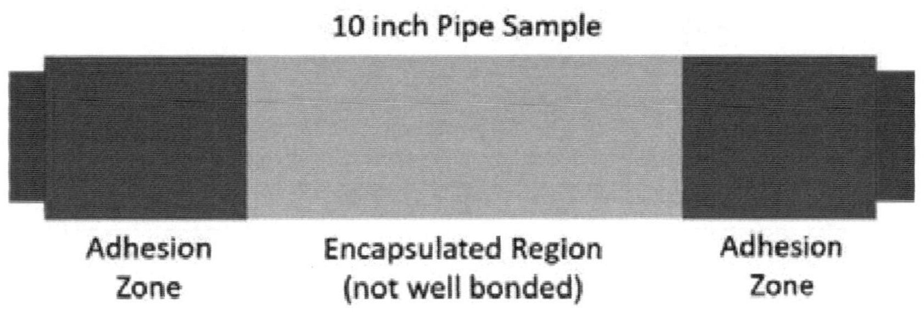

Figure 6. The layout of the composite repairs on pipe samples A and B. Holes were drilled in the steel substrate to simulate through-wall corrosion in the encapsulated region.

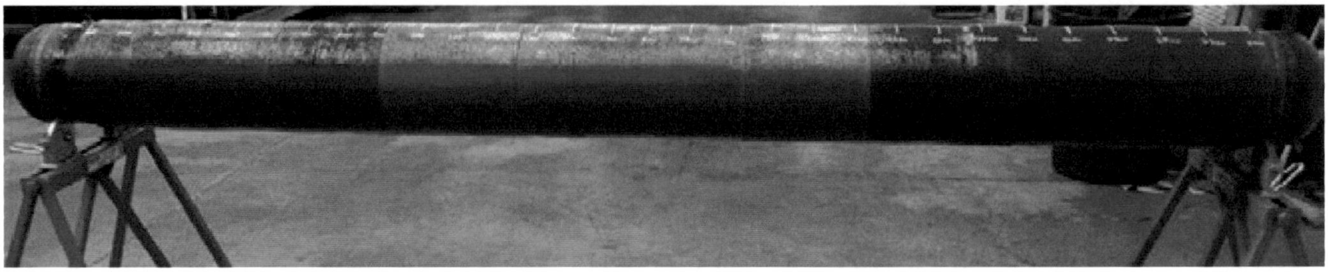

Figure 7. Pipe sample A with a carbon fiber composite repair; the red and grey areas denote the adhesion zones and encapsulated region respectively, as shown in

Figure 6.

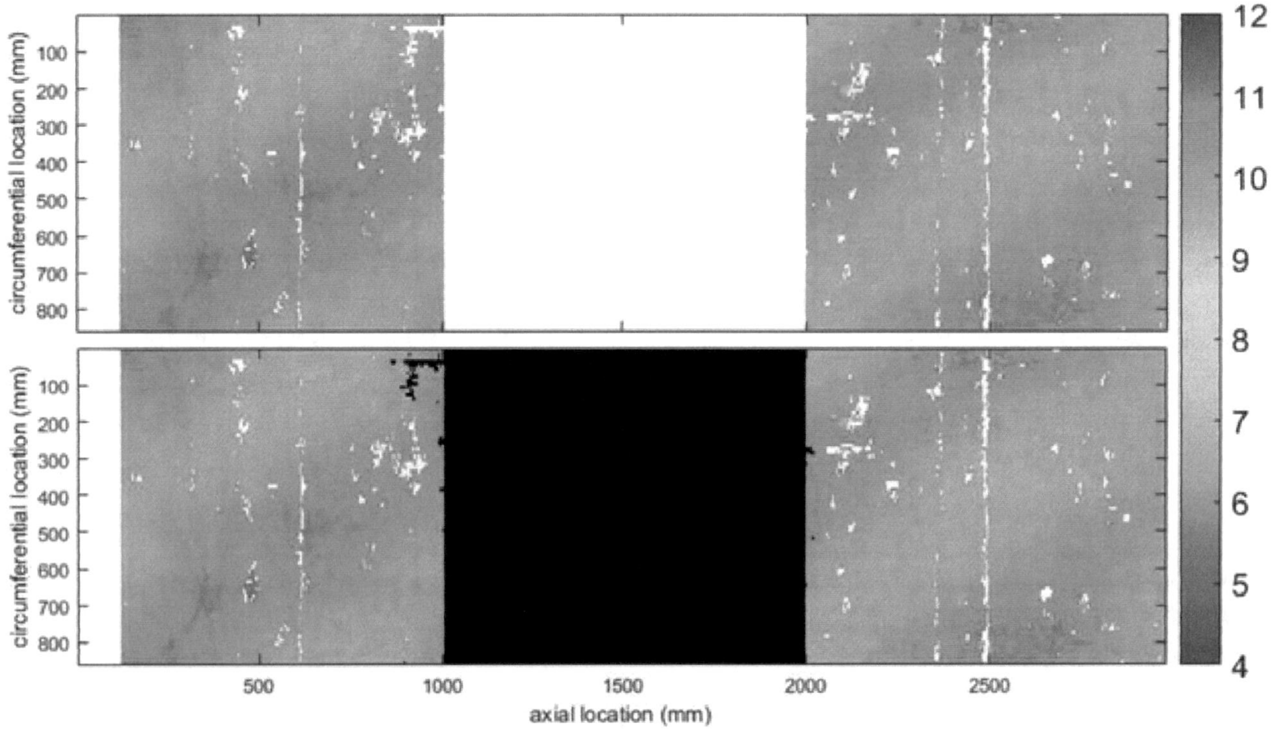

Figure 8. Maps of pipe sample A prior to a hydrostatic test. Upper: DRS map showing flaws in white. Lower: QA test map showing that few of the flaws interact (black pixels) and those that do interact do not extend across the full length of the adhesion zone. Therefore, both adhesion zones on pipe sample A passed the QA test; they also passed the hydrostatic test.

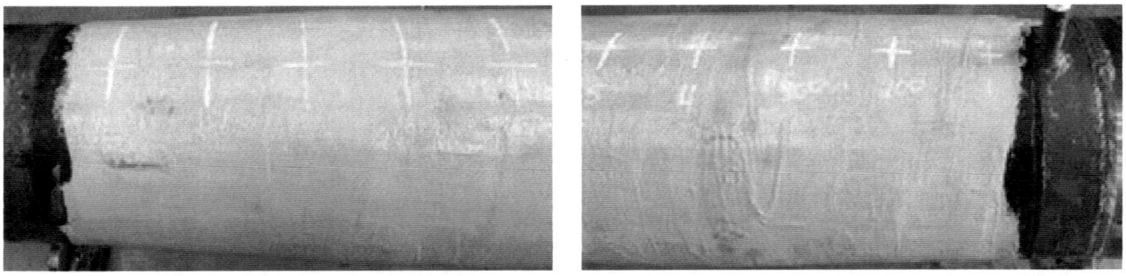

Figure 9. Pipe sample B with a carbon fiber repair. The adhesion zones and encapsulated region are arranged as shown in

Figure 6, although the top coat is blue in this case.

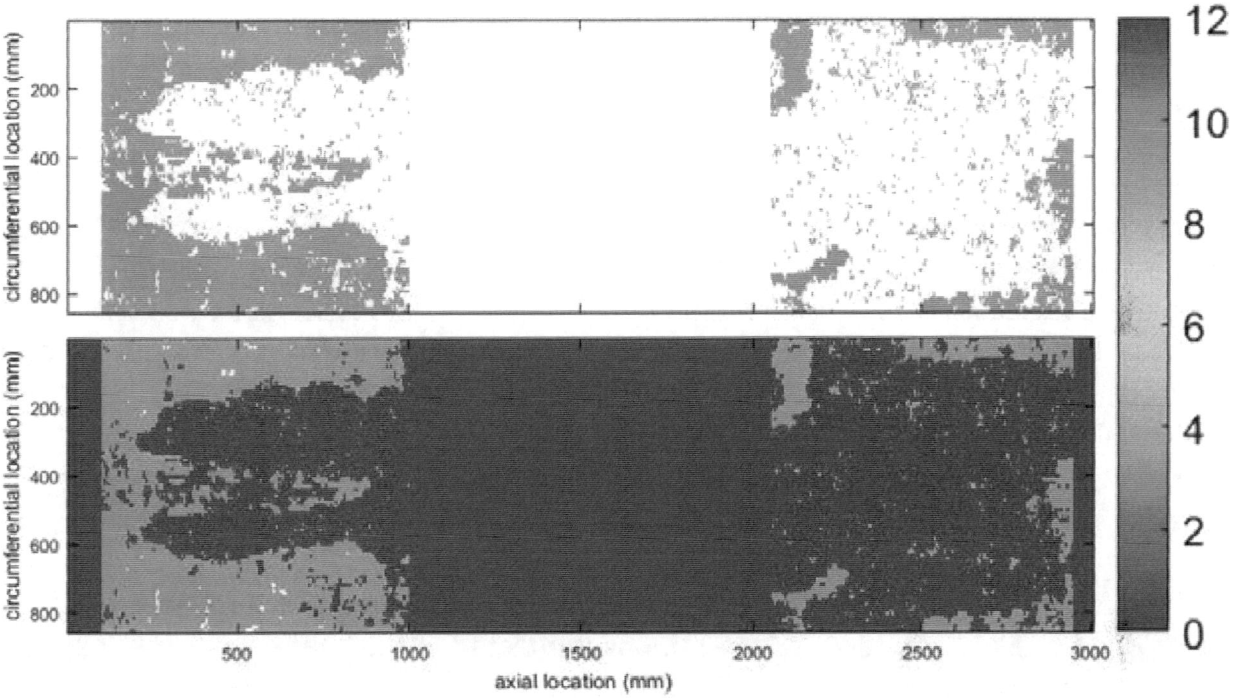

Figure 10. Maps of pipe sample B prior to a hydrostatic test. Upper: DRS map showing flaws (in white) covering a large area. Lower: QA test map showing that many of the flaws interact (red

pixels) and extend across the full length of both adhesion zones. Therefore, both adhesion zones on pipe sample B failed the QA test; they also failed the hydrostatic test.

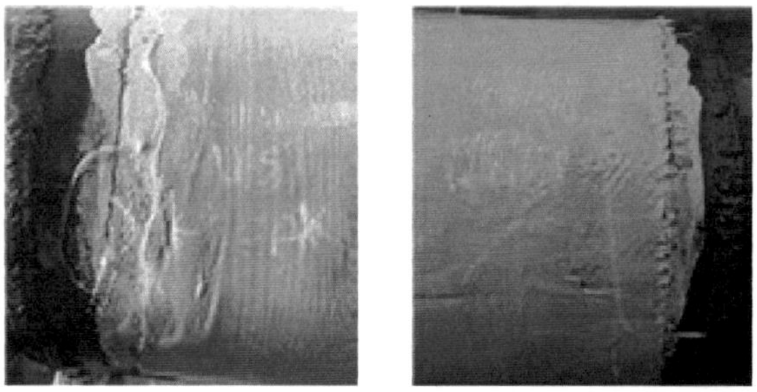

Figure 11. Photographs of pipe sample B, following the failed hydrostatic test, showing leak locations at the ends of both adhesion zones, labelled upstream (U/S) and downstream (D/S).

Steel Substrate Thickness Measurements – Trial

The steel thickness measurement capabilities of the DRS technique have been evaluated on a 30 inch ex-service pipe sample with extensive internal corrosion. The upper photographs in Figure 12 show the sample. Conventional 0° ultrasonic thickness measurements (UT) were made to create a baseline map. A combination of glass fiber and carbon fiber was then applied externally. The lower photograph in Figure 12 shows the DRS inspection in progress through the composite repair.

The results of both inspections are presented in Figure 13; the upper image shows the UT corrosion map made prior to the application of the composite repair and the lower image shows the DRS corrosion map made through the composite repair. No major flaws were identified in the repair and the two maps visually correlate well. As expected, DRS does not resolve the very small features found in the UT map due to the relatively large footprint of the DRS transducer. However, it does provide very accurate WT measurements for the majority of the inspected area. The DRS results lie within 0.5 mm of the UT results for most of the data.

Figure 12. Upper: 30 inch ex-service pipe with internal corrosion. Lower: DRS inspection of the pipe following application of a composite repair.

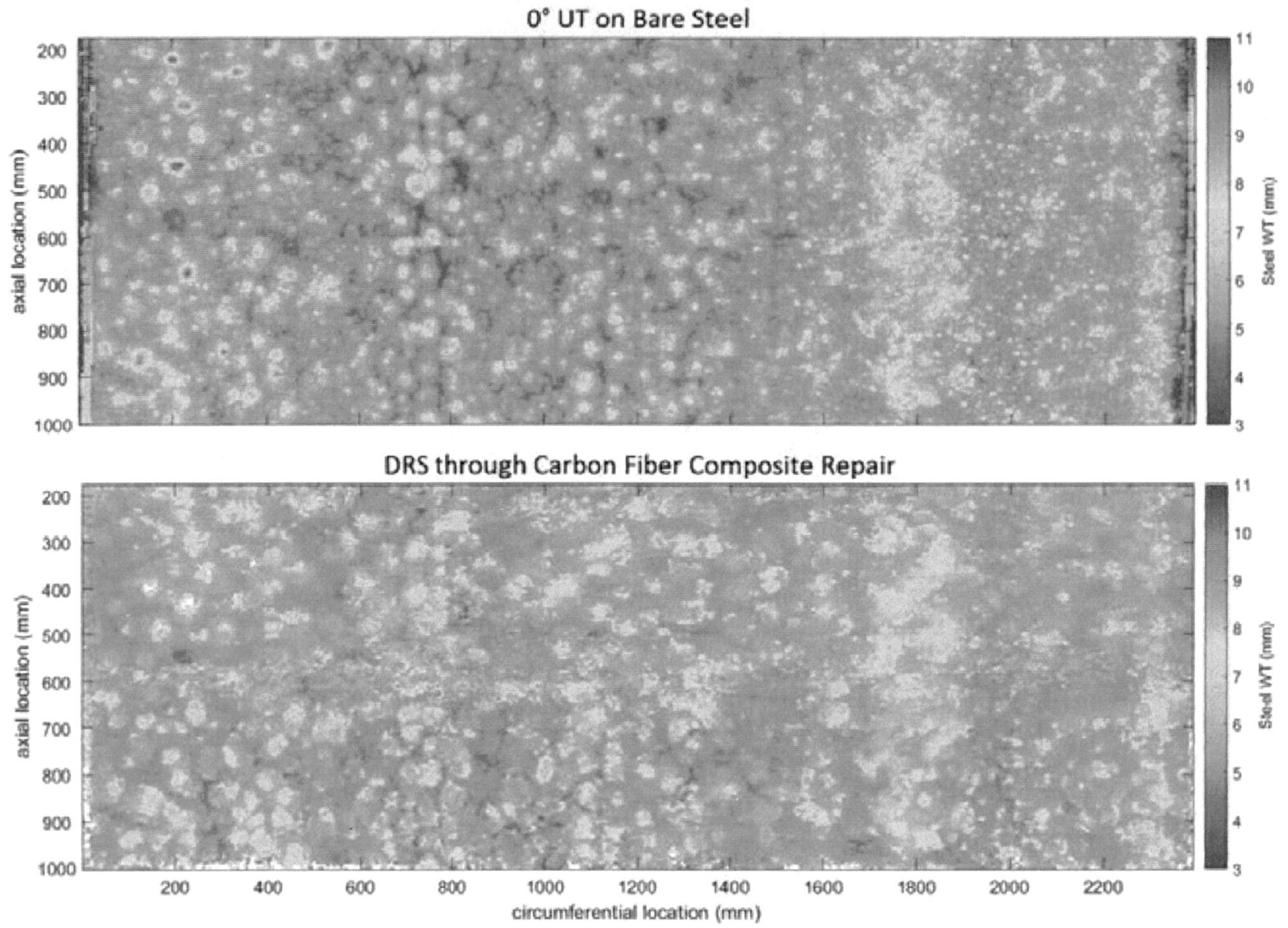

Figure 13. Upper: UT map prior to application of the composite repair. Lower: DRS map following application of the composite repair.

Multiskip and CHIME

Multiskip and CHIME (Creeping Head Wave Inspection Method) are both medium range twin probe pitch-catch techniques. The arrangement consists of a transmitter to inject the ultrasonic signal which then runs in the plate over a distance to the receiver. The transmitter and receiver both have to be placed on the steel substrate, hence access is required to either side of the composite repair.

The diagram below shows, in relatively simple terms, the theory behind the Multiskip technique. The transmitting ultrasonic probe (Tx) injects ultrasound, in the form of a shear wave, into the plate material. This covers a wide range of beam angle and then 'skips' through the material to be received by the second ultrasonic probe (Rx). The number of skips produced is dependent on the distance between the ultrasonic probes and the thickness of the material. For most composite repair applications on piping the probes would be separated axially and moved circumferentially, hence the distance axis would refer to circumferential position.

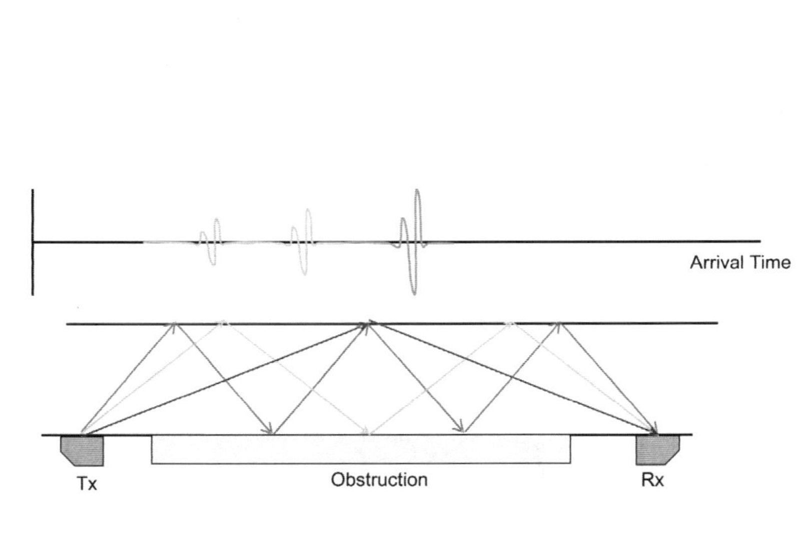	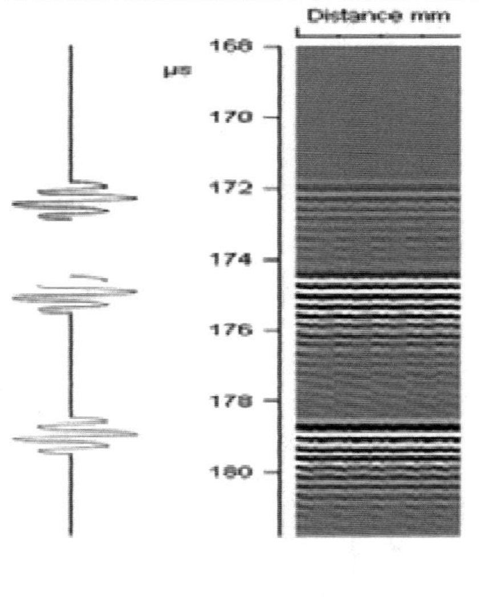
Multi-Skip Technique Probe Deployment	A-Scan and B-Scans

Changes in thickness are indicated by changes in the skip arrival times and analysis methods have been developed to allow quantitative thickness measurement. When damage is severe it is no longer possible to quantify the remaining thickness but the circumferential location and extent of the damage is identifiable.

CHIME uses a similar arrangement to that shown above but the nature of the ultrasonic signal is different. A combination of shear and compression waves is used and the signal pattern at the receiver changes according to the amount and nature of wall loss between the transmitter and receiver. A CHIME B-Scan is shown below, with the probe position (circumferential around the pipe) on the x-axis and arrival time (increasing downwards) on the y-axis. The pattern starts on the left for undamaged material and then covers corroded material for which the change in signal pattern is clearly evident. This is used to categorise the severity of the damage. Three categories are used, i.e. A <10% wall loss, B between 10 and 40% wall loss and C>40% wall loss.

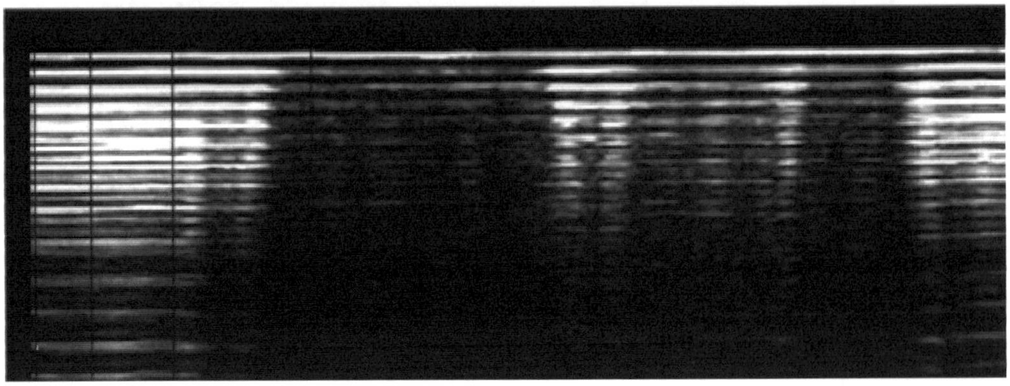

The signal quality for both Multiskip and CHIME can be affected by the presence of the repair as some of the ultrasonic signal will typically leak into the repair. As such the techniques would normally be applicable where the axial extent of the repair is less than 15″. Sonomatic's experience with these techniques over composite repairs is the primary value is with checking the extent of circumferential growth of degradation rather than in being able to quantify the depth of degradation and measure changes in this.

Conclusions

The DRS technique is an effective way of performing the Non-Destructive Examination of the composite repair system outlined in the ASME PCC-2 standard:

1. the inspection of the repair (composite) laminate
2. the inspection of the bond quality between the repair laminate and the substrate
3. the inspection of the substrate
4. periodic post repair inspection to monitor internal corrosion

This novel NDE method has been proven through several blind trials and deployed extensively in the field to assess the laminate and monitor the integrity of the underlying component and repair over time.

Multiskip and CHIME are both medium range ultrasonic techniques that allow inspection of the steel condition under composite repairs, primarily for identification of the circumferential extent of corrosion.

References

1. ASME PCC-2–2018 Repair of Pressure Equipment and Piping

AIChE PAPER NUMBER:

INDUCED DRAFT FAN RELIABILITY CASE STUDY

Brad Smith
Sales Manager
Twin City Clarage, LLP

Charles Smith
Rotating Equipment Engineering and Maintenance
BASF

Prepared for Presentation at the 2019 Spring Meeting
New Orleans, Louisiana, March 31-April 4, 2019

INDUCED DRAFT FAN RELIABILITY CASE STUDY

Brad Smith
Sales Manager
Twin City Clarage, LLP

Charles Smith
Rotating Equipment Engineering and Maintenance
BASF

Abstract:

Induced draft (ID) fans are normally installed on top of the furnace and are critical equipment in the operation of Ethylene furnaces as they are susceptible to vibration related and other reliability issues.

This paper discusses a case study involving an ethylene furnace ID fan design that suffered reliability issues resulting in unplanned furnace shutdowns leading to millions of dollars in lost opportunities over a 10-year period. We will explain some of the symptoms that the ID fan may have experienced that could have resulted in the unplanned furnace shutdown. Finally, we will also describe the design changes that were made to the new ID fan using the design features of API-673 fan specification and improved the ethylene furnace ID fan as well as overall reliability.

Introduction:

The induced draft (ID) fan is an important part of the operation of the ethylene cracking furnace as it draws the air through the furnace. If the ID fan shuts down for any reason, the entire furnace is shut down. With remote location of the fan, it is very important that the ID fan is properly designed for long-term operation and reliability. A typical ethylene cracking furnace has the ID fan installed on top of the furnace, which is approximately 30 m above grade. Figure 1 is an illustration of a typical ID fan installation on an ethylene cracking furnace.

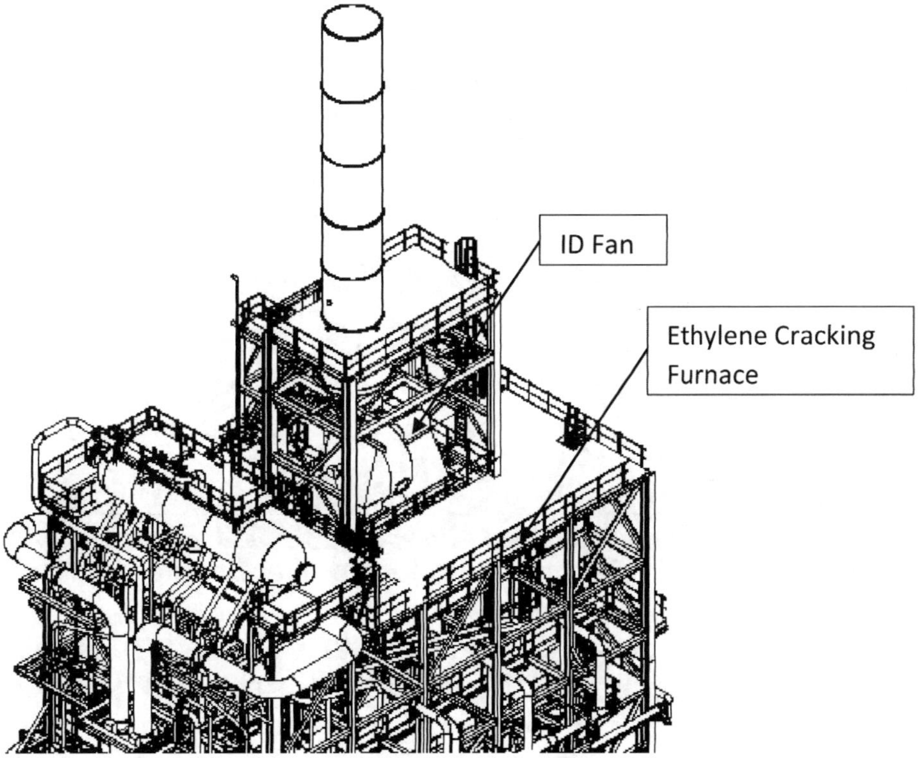

Figure 1: Typical drawing of Ethylene Cracking Furnace

This paper will discuss the reliability issues with the existing fan design that were causing furnace shutdowns, and then explain the changes that were made in the new ID fan design to improve the reliability and operation of an ethylene cracking furnace at a large olefins facility on the upper Gulf coast. Repetitive ID fan issues caused multiple furnace outages with year-on-year of production losses totaling millions of dollars. Finally, the general API-673 ID fan design features and important parameters to monitor to ensure long-term operation and reliability are discussed.

Description of Problem:

The original (ID) fan was experiencing reliability and fan bearing issues that led to the fan being shut down. This, in turn, would shut down the associated cracking furnace. The original ID fan design is located on top of the furnace and is directly mounted to the furnace structure. Because the furnace structure uses the actual furnace as part of the supporting structure, it is susceptible to thermal expansion. Thus, as the furnace reaches its normal operating conditions, the heat causes thermal expansion, twisting and distorting the fan base and causing misalignment in the fan assembly. As a result, the fan bearing temperature and vibration levels would increase to the point of failure.

Figure 2: Original fan mounted directly onto the furnace structure

There were other fan design issues that affected the reliability of the ethylene cracking furnace. These issues include: the fan's single width single inlet (SWSI) design; base mounted fan housing; and use of antifriction bearings. The SWSI arrangement increases the axial thrust loading on the fan bearings. Base mounting the fan housing onto the fan base only allows the fan housing to thermally expand in one direction. This makes it difficult to set the inlet cone in relation to the fan rotor. As a result, the fan rotor may rub the inlet cone transmitting vibration from the fan housing into the fan bearings. Antifriction bearings with a static oil bath and air cooling have an increased risk of higher temperatures and it is difficult to provide additional cooling for the fan bearings. All these symptoms compounded to increase the temperature and transfer high vibration levels to the fan bearings, leading to premature bearing failure and causing the unplanned ID fan/furnace shutdowns.

Fan Design Changes:

When a new furnace was approved for the site, the design of its ID fan contained many changes to increase the overall reliability and productivity of the ethylene cracking furnace.

The most important design change was to use an 18" concrete inertia base with spring isolators to support the fan assembly and expansion joints on the inlet and discharge connections to isolate the fan from the ductwork. The concrete inertia base makes the base considerably stiffer, and the weight of the concrete in the base lowers the ID fan assembly's center of gravity, which makes the fan assembly more stable. The other benefit of the concrete inertia base is that the stiffness helps to maintain the coupling alignment. Figure 3 shows the fan's general arrangement on the concrete inertia base. The spring isolators and the expansion joints isolate the fan from the furnace structure and the ductwork. Figure 4 shows a spring isolator installed under an inertia base. The isolation allows the furnace structure to expand and contract without distorting the fan base and creating any misalignment in the fan assembly.

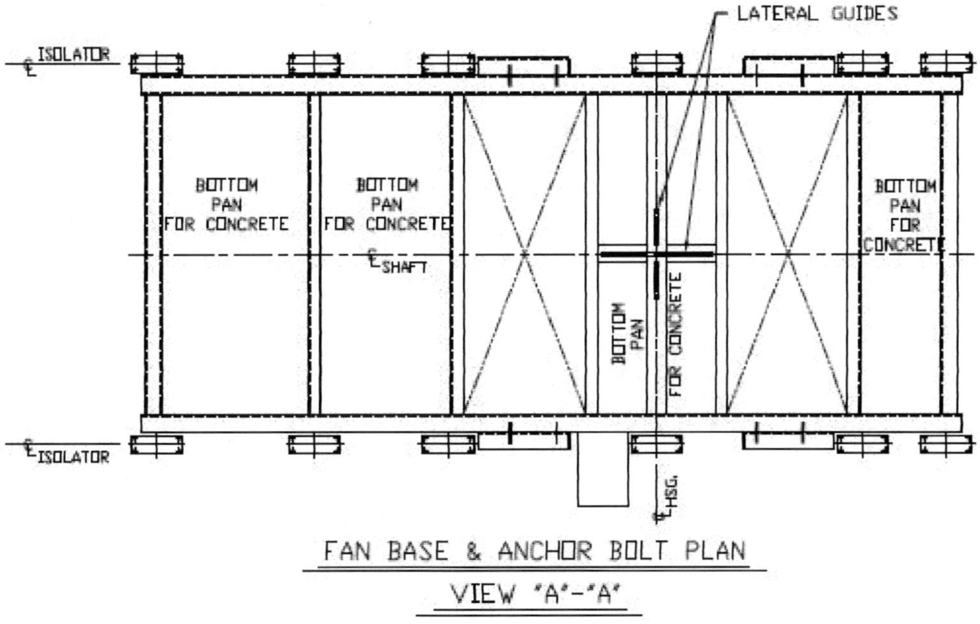

Figure 3: General arrangement drawing of ID fan base

Figure 4: The spring isolator mounted under fan inertia base

The fan arrangement and support were changed on the new ID fan design. The new ID fan was designed with a double width double inlet (DWDI) versus a SWSI and the fan housing support was changed to a centerline support from a base support. The DWDI arrangement greatly reduced the axial thrust loading on the fan bearings by balancing the suction-induced forces. Figure 5 shows how thrust loading affects the bearings. The new fan selection as a DWDI is also more efficient.

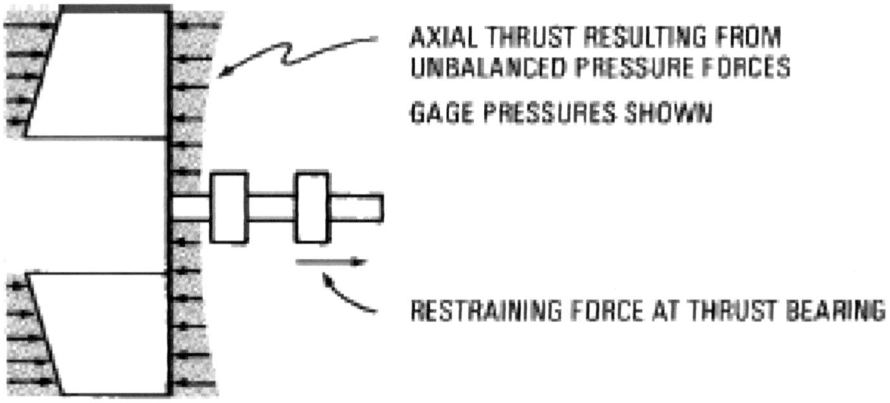

Figure 5: Axial thrust loading being transferred on the fan bearings from SWSI arrangement (1)

The centerline-supported fan housing allows thermal expansion in both directions instead of only growing vertically, as is the case with a base-mounted fan housing. Allowing the fan housing to grow in both directions makes it easier install and set up the fan rotor in relation to the inlet cone. The centerline-supported fan housing is also independent of the bearing pedestals, therefore ensuring that any vibration from the fan housing will not be transferred to the bearing pedestal. Figure 6 shows the fan assembly with the centerline-supported housing.

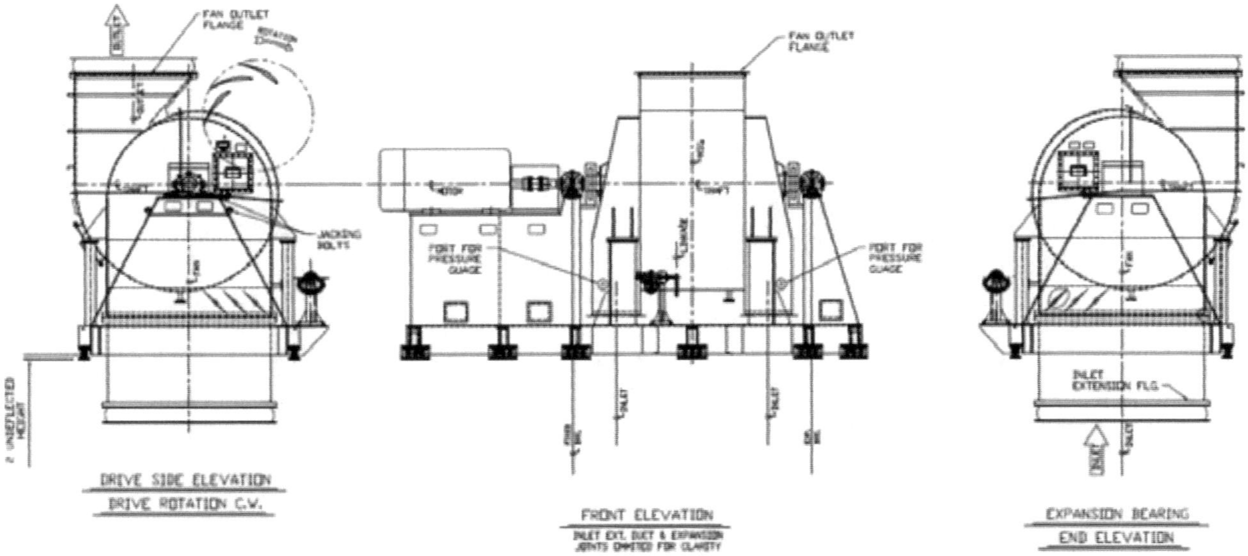

Figure 6: Front assembly illustrating the centerline-supported fan housing

The new ID fan design also implements fan specifications from API-673 2nd Edition. API-673 2nd Edition is a standard that covers the requirements for centrifugal fans that are intended for continuous duty in petroleum, chemical, or gas industry services. The design standards from API-673 that were applied to the ID fan, and that contributed to improving the reliability, are as follows:

- Changing the fan bearings from anti-friction bearings to hydrodynamic bearings. Hydrodynamic bearings operate using a film of oil between the fan shaft and the bearing sleeve so that there is very low friction. The low friction means that bearings will run cooler. Additionally, there is low wear when running on a film of oil, which leads to a longer life expectancy compared with standard anti-friction bearings.
- Hydrodynamic bearings are cooled in two ways: by water or by oil. Because the fan is located approximately 30 m above grade, it is difficult get cooling water with sufficient pressure on top of the furnace to the ID fan. It is much easier to cool the ID fan hydrodynamic bearings using oil. Thus, an oil lubrication system was added to the new ID fan design. The system, designed to API-614 specifications, consisted of oil pumps, a forced-air cooler, filtration system, and a reservoir. It improved the bearing reliability by providing sufficient oil flow at all times, continuous oil filtration, easy monitoring of the oil quality and temperature, proper cooling of the oil, and redundant lubrication (i.e. backup pumps). Finally, the oil lubrication system extends the ID fan operation time between the service intervals and reduces the risk of unplanned shutdowns. Figure 7 shows the oil lubrication system installed next to the ID fan.

Figure 7: Illustration of the oil lubrication system

- The rotor was dynamically balanced to a tolerance API 4W/N standard.

$$U = 4W/N$$

Where: U = Balance tolerance, in oz-in (for each correction plane)
W = Bearing journal static weight at each end of the rotor
N = Maximum service speed of the rotor

The API-673 standard requires that the fan rotor to be dynamically balanced to this standard. The rotor dynamic balance is essential for minimizing vibration and increasing the life of the fan bearings. Industry standard for a general industrial fan is ISO 21940-1 G2.5 and the API 4W/N standard is equivalent to ISO 21940-1 G0.7. Therefore, a smaller balance tolerance means a more finely balanced rotor, which in turn reduces the amount of vibration that can be transferred to the fan bearings.

- The ID fan motor and fan bearing pedestals were machined to the API-673 standard, which requires that the maximum surface roughness shall be 3 µm (125 µin). This will ensure that the motor and fan bearings mounting areas are flat and will prevent any 'soft foot' between the motor and motor pedestal. If the motor and bearing pedestals are uneven it will lead to misalignment between the motor and fan rotor.

With the above design changes implemented, the new ethylene cracking furnace did not experience any unplanned shutdowns resulting from the ID fan. The new design has a higher upfront capital cost, but with no unplanned shutdowns or repair costs for the ID fan, the payback time is very short. Figure 8 shows the new ID fan design installed.

Figure 8: New ID fan design installation

General ID Fan Reliability Recommendations:

Generally, the recommended method to improve the reliability of ethylene cracking furnace ID fans is to spend the time upfront to implement API-673 design features and to design the right fan for each specific installation. Additionally, investing in instrumentation to monitor the temperature and vibration of the fan bearings is recommended.

The most common method of monitoring the temperature in the fan bearings is using a resistance temperature detector (RTD). These are small and compact, and they are often used to detect temperature increases in bearings, thrust bearing plates, shafts, and motor windings. Monitoring bearing temperatures is important for reliability because as bearings fail, they get hotter. Trending their temperatures over time will help identify the early stages of a problem with the bearing before a total failure occurs which will cause damage to the bearing and fan shaft. Also, early detection allows the operating company to address the issue in a controlled, planned manner and avoid tripping a furnace prior to decoking.

Vibrations can appear in fan bearings in numerous forms: misalignment, soft foot, rotor imbalance, instabilities resulting from flow or pressure differentials, and pedestal or foundation distortion. See Figure 9 for an illustration of the sources of vibration. It is important to monitor the vibration in fans or motor bearings for reliability. The common methods of monitoring vibration are using a vibration switch, sensor, or probe. The vibration switch is mounted on the fan bearing pedestal and measures the vibration that is transmitted from rotating components through the bearing to the pedestal. The vibration switch is a simple device, but it does not have the sensitivity of a sensor probe. It is commonly used in smaller HVAC or general industrial fans that are arrangement 1 or 8 and in fans with anti-friction bearings. Vibration sensors (seismic sensors) are installed in the fan bearing housing, and they measure the level of vibration transmitted from rotating components. The sensors can provide a signal that can be sent to the control room for monitoring and recording. Typically, one vibration sensor is mounted on each bearing. It is more sensitive than the vibration switch, but not as sensitive as a vibration probe. Seismic sensors are commonly used in general industrial and petrochemical fans with anti-friction bearings.

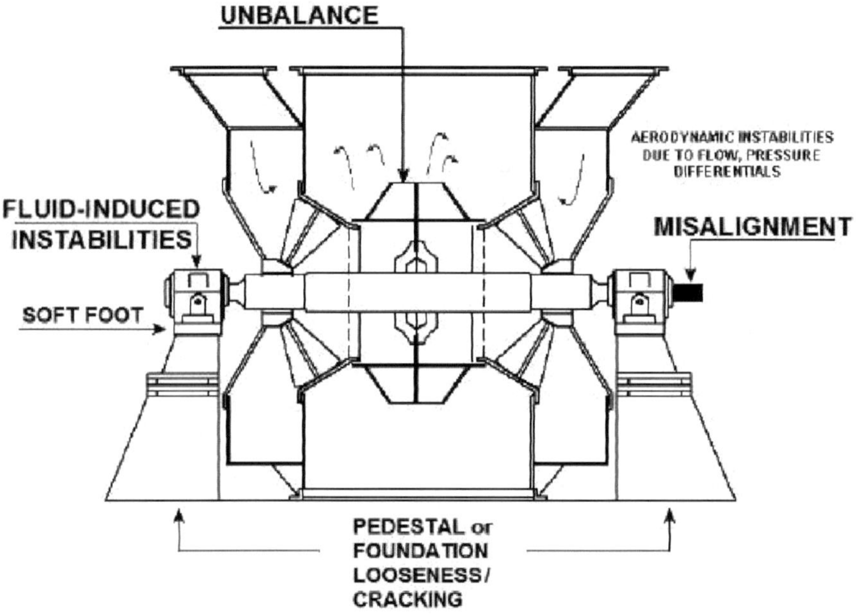

Figure 9: Common sources of vibration [2]

Vibration proximity probes (probes) used with hydrodynamic bearings are mounted on the bearing housings that directly contact the outside of the hydrodynamic sleeve. Generally, there are two probes per bearing: one in the X plane and the other in the Y plane. Figure 10 shows a common installation of the probes. The probes are the most sensitive vibration monitoring device; they produce a signal that can be sent to the control room, and they are commonly used in large general industrial or petrochemical fans that are a critical component of the system.

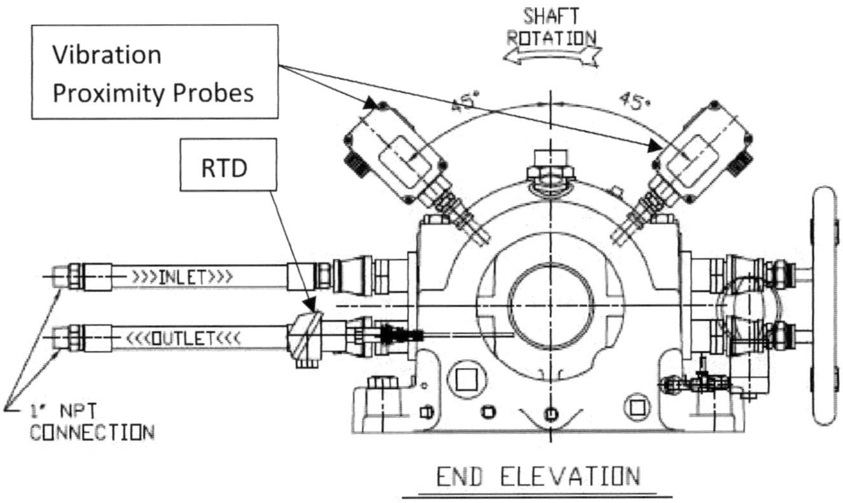

Figure 10: Typical hydrodynamic bearing drawing

The API-673 standard is now in its third edition, and the standard covers best industry practices for centrifugal fan equipment and fan accessories in two applications: general purpose and special purpose. General purpose application fans are usually spare or non-critical service equipment, whereas special purpose fans are designed for uninterrupted, continuous operation in critical service, and there is usually no installed spare equipment. For most ethylene cracker furnaces, the API-673 3rd edition special purpose application is the highest standard to apply for the fan design because most ID fans are installed in a remote location on top of the furnace, there are no installed spares, and the ID fan is a critical piece of equipment in the system (i.e. if the ID fan shuts down, the entire ethylene furnace system is shut down and if the cracking furnace is tripped unexpectedly, significant equipment damage may occur). The API-673 standard comes with a detailed fan data sheet that works in conjunction with the standard. It is strongly recommended that the fan data sheet be filled out to ensure the correct standard options and accessories are applied to provide the most reliably designed fan equipment for the given system.

References:

1. Jorgensen, Robert and The Buffalo Forge Co. (1983), "Fan Engineering; An Engineer's Handbook on Fans and Their Applications - Ninth Edition", page 17-3.
2. Shannon PE, Bob (2008), "Vibration Measurement Systems and Guidelines for Centrifugal Fans – A Field Perspective", page 17.

121a
Golf meets a steam cracking coil: dimpled reactor technology for the production of light olefins

Jens N. Dedeyne
ir.
Laboratory for Chemical Technology, Ghent University, Ghent, Belgium

Marco Virgilio
ir.
The von Karman Institute for Fluid Dynamics, Brussels, Belgium

Tony Arts
Prof. Dr. ir.
The von Karman Institute for Fluid Dynamics, Brussels, Belgium

Jens Floré
ir.
BASF Antwerp, Antwerp, Belgium

Heinrich Laib
Dr. ir.
BASF SE, Ludwigshafen am Rhein, Germany

Guy B. Marin
Prof. Dr. ir.
Laboratory for Chemical Technology, Ghent University, Ghent, Belgium

David Van Cauwenberge
Dr. ir.
BASF Antwerp, Antwerp, Belgium

Kevin M. Van Geem
Prof. Dr. wir.
Laboratory for Chemical Technology, Ghent University, Ghent, Belgium

Prepared for Presentation at the 2019 Spring National Meeting, Ethylene Plant Technology Fundamentals session,
New Orleans, Louisiana, March 31 – April 4, 2019

Golf meets a steam cracking coil: dimpled reactor technology for the production of light olefins

Jens N. Dedeyne
ir.
Laboratory for Chemical Technology, Ghent University, Ghent, Belgium

Marco Virgilio
ir.
The von Karman Institute for Fluid Dynamics, Brussels, Belgium

Tony Arts
Prof. Dr. ir.
The von Karman Institute for Fluid Dynamics, Brussels, Belgium

Jens Floré
ir.
BASF Antwerp, Antwerp, Belgium

Heinrich Laib
Dr. ir.
BASF SE, Ludwigshafen am Rhein, Germany

Guy B. Marin
Prof. Dr. ir.
Laboratory for Chemical Technology, Ghent University, Ghent, Belgium

David Van Cauwenberge
Dr. ir.
BASF Antwerp, Antwerp, Belgium

Kevin M. Van Geem
Prof. Dr. wir.
Laboratory for Chemical Technology, Ghent University, Ghent, Belgium

Abstract: In petrochemical industry, heat transfer limitations are known to have a detrimental effect on the process economics. In steam cracking, heat transfer limitations from a cracking coil to the process gas imply an increase in tube metal temperatures and reduces the time in between two decoking procedures. To enhance heat transfer, wall modifications such as fins or ribs were previously proposed and implemented. These rudimentary wall modifications however introduce an excessive pressure drop which is detrimental to the light olefin selectivity. In this work, a novel reactor design, i.e. a spherically dimpled coil, is proposed where the wall modifications are indented into the reactor inner wall. As the current concave modifications exhibit excessive drag, negatively impacting light olefin yield, the dimpled tube with convex modifications is introduced to achieve a comparable heat transfer at a lower pressure drop penalty. In this work, Computational Fluid Dynamics (CFD) is used to study the flow behavior inside the dimpled tube and heat transfer on the reactor inner wall. The transient simulations show the presence of asymmetric vortex structures within the spherical dimples. It is expected that this leads to a "self-cleaning" effect, as the vortex structures induce fluctuating shear forces on the coke depositions, aiding the removal of coke from the reactor wall. It was found that the geometric parameters of the design can have a significant impact on this effect as well as on pressure drop and heat transfer. Furthermore, implementing such a dimpled cracking coil leads to a low pressure drop penalty whilst the improved radial mixing leads to a more uniform temperature distribution, favoring the production of propene. Finally, the reduction of the wall temperature can increase run length by up to 40 %.

Introduction

Over the last few years, chemical industry has increased its efforts to reduce emissions by increasing process efficiency. In this respect, the steam cracking process is very relevant as it is one of the most energy-intensive processes in chemical energy and responsible for the emissions of over 200 million ton of CO_2 annually [1, 2].

In this process, a hydrocarbon feedstock is cracked into light olefins, aromatic compounds and various other product at temperatures between 900 and 1200 K. At the prevailing temperatures, undesired secondary reactions lead to the formation and deposition of coke, i.e. a solid residue composed of carbon which decreases heat transfer and induces a higher pressure drop due to the lower cross-sectional area of the reactor. This leads inadvertently to a reduced selectivity towards the desire light olefins. Next to this, coke has a low thermal conductivity, requiring the fuel flow rate to be increased to maintain the same level of production. Consequently, the rate of coke deposition increases and the tube metal temperatures (TMT) are elevated, which may reach the maximum allowable operating temperature. When heat transfer resistance or pressure drop becomes excessive, operations are halted and coke is burned off during a decoking procedure. It is clear that a decrease in the rate of coke formation can therefore increase the time between consecutive decoking operations, or decrease the amount of fuel to be burned. This decrease in rate of coke formation can be achieved by proper selection of reactor material[3] or by enhancing the heat transfer from the reactor coil to the process gas.

In heat exchanger applications [4], the enhancement of heat transfer has been accomplished by implementing a variety of three dimensional (3D) elements at the wall or in the center of the flow, e.g. pin fin arrays [5], rib turbulators, swirl chambers, etc. Such elements typically increase turbulence and radial mixing of the fluid. For reactors however, the implementation of these wall elements is not straightforward as they are associated with a pressure drop penalty which, in the case of steam cracking, can also negatively impact the selectivity towards the desired products [6]. Nonetheless, a few enhanced coil designs have been proposed an implemented, e.g. the mixing-element radiant tube (MERT) by Kubota [7, 8], SCOPE by Schmidt & Clemens

[9], Swirl Flow Tube (SFT) by TechnipFMC [10, 11] or Intensified Heat Transfer Technology (IHT) by Lummus and Sinopec [12].

In his review on heat transfer augmentation technologies, Ligrani [4] mentions that dimpled surfaces are notable for their low pressure drop and that dimpled cavities typically produce multiple vortex pairs, which can greatly enhance heat transfer relative to a bare tube. These vortices are associated with secondary flows, ejecting hot fluid from the wall region towards the colder center of the flow. It was found that a positive correlation exists between the magnitude of heat transfer enhancement and dimple depth, as the magnitude of turbulence production and transport increases for deeper dimples and the strength and intensity of the vortices and associated secondary flows increases as well. An increase in dimple depth can also negatively impact the pressure drop penalty [13].

As literature shows that heat transfer enhancement is often achieved at the cost of a higher pressure drop penalty, a metric is introduced to compare different designs and weigh the benefit of heat transfer enhancement versus the detrimental effect of the pressure drop penalty. In this respect, the Thermal Enhancement Factor (TEF) can be used and is defined as the ratio of Nusselt number ratio to the friction factor ratio:

$$TEF = \frac{\left(\frac{Nu}{Nu_0}\right)}{\left(\frac{f}{f_0}\right)^{1/3}} \qquad \text{Eq. 1}$$

In which the Nusselt number ratio is the ratio of the Nusselt number averaged over the entire length for the enhanced design to the averaged Nusselt number for the corresponding non-enhanced design. The friction factor ratio is similarly defined, with the friction factor being the Fanning friction factor, f, which can be related to the pressure drop, Δp, in a tube with radius R over a distance L by:

$$\Delta p = f \frac{L}{R} \rho {u_b}^2 \qquad \text{Eq. 2}$$

In this work, the effect of various dimple dimensions as well as dimple positioning is studied with a 3D large eddy simulation (LES) framework as proposed by Van Cauwenberge et al. [14] as the LES methodology is not only able to accurately predict flow and heat transfer in enhanced geometries [15], but also allows to investigate the transient behavior of oscillating vortex pairs inside dimples. Finally, the impact of this reactor design on product yields is investigated by reactive simulations.

Materials and Methods

Geometry description

The basis for the dimpled tube is a cylinder with radius R along a central axis perpendicular to the cross section and intersecting the cross section at the point labeled with C. The spherical dimples are formed by spheres with a radius r intersecting the cylindrical tube. The dimensions of the spherical dimple are uniquely defined by two parameters e and w as the radius r of the sphere is calculated as

$$r = \frac{e^2 + \frac{w^2}{4}}{2e} \qquad \text{Eq. 3}$$

where e is the dimple depth and w is a measure for the dimple width.

To uniquely define dimpled tubes with multiple spherical dimples, a number of additional parameters are required describing the relative positioning of these indentations. First, the number of dimples per cross section is denoted as nspan. Second, the distance between an upstream and a downstream row is denoted by I. For I = 0, the dimples are placed together as closely as possible in the streamwise direction, i.e. there is no distance in between dimples of row n and the dimples of row (n+1).

Third, the phase difference, α, is introduced to allow inline configurations shown in Figure 1 (left), i.e. dimples placed along lines parallel to the tube centerline, as well as staggered configurations shown in Figure 1 (right), i.e. dimples in row (n+1) are placed in between two dimples in row n when looking along the centerline axis.

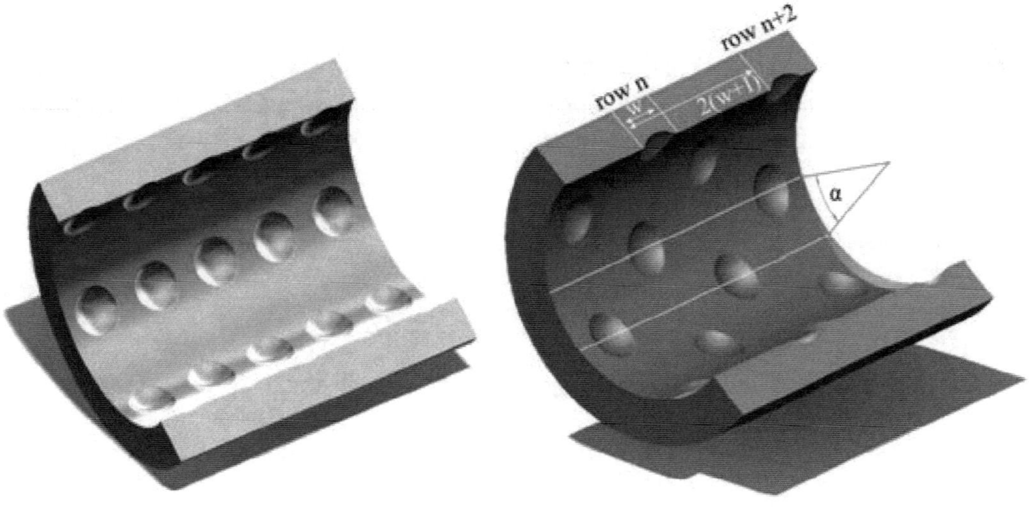

Figure 1. Dimpled tube with an inline configuration (left) and a staggered configuration (right) with n_{span} showing geometrical parameters w, I and α are shown.

CFD model set-up

Governing equations

The flow was modeled incompressible and temperature was treated as a passive scaler, i.e. viscous dissipation and buoyancy effects were not taken into account. The filtered governing equations for conservation of mass, momentum and energy are given below.

$$\frac{\partial \tilde{u}_i}{\partial x_i} = 0 \qquad \text{Eq. 4}$$

$$\frac{\partial \tilde{u}_i}{\partial t} + \frac{\partial \tilde{u}_i \tilde{u}_j}{\partial x} = -\frac{\partial \tilde{p}}{\partial x_i} + \frac{1}{Re_\tau}\frac{\partial^2 \tilde{u}_i}{\partial x_j^2} - \frac{\partial \tau_{ij}^R}{\partial x_j} \qquad \text{Eq. 5}$$

$$\frac{\partial \tilde{\theta}}{\partial t} + \frac{\partial \tilde{\theta}\tilde{u}_j}{\partial x_j} = \frac{1}{Re_\tau Pr}\frac{\partial^2 \tilde{\theta}}{\partial x_j^2} + \frac{\partial q_j}{\partial x_j} \qquad \text{Eq. 6}$$

Where $\tilde{}$ denotes a variable filtered with the OpenFOAM 'simple' LES filter, which is a simple top-hat filter for dynamic LES models. The subgrid-scale (SGS) turbulence was modelled with the dynamic k-equation eddy-viscosity model [16].

Boundary conditions

As LES is computationally demanding, the computational domain is limited by assuming fully developed flow and applying streamwise periodic boundary conditions, implying for any variable θ that:

$$\theta(x, y, z + nP, t) = \theta(x, y, z, t) \qquad \text{Eq. 7}$$

Where P is the periodic module length. For variables that are not intrinsically periodic such as temperature and pressure, the method of Patankar et al [17] is used in which both fields are separated into a fluctuating periodic part and a non-periodic part with a constant gradient. For more information on this method, the reader is referred to the work of Van Cauwenberge et al [14] who translated this concept to the field of steam cracking.

For all studied geometries, a flow rate corresponding to a Reynolds number of 11000 was imposed. This is well below typical values for steam cracking applications, but it was found that the effect of Reynolds number on local Nusselt numbers is minimal and overall values such as Nusselt number ratio and pressure drop ratio is negligible if the flow is completely turbulent.

On the fluid-solid interface, a no-slip boundary condition is imposed for the momentum equation, a zero-flux boundary condition for the mass conservation equation and an isoflux heating boundary condition for the energy conservation equation such that a bulk temperature of 1000 K was imposed.

Computational grid

The commercial meshing software Pointwise V18.0 [18] is used to create body-fitted fully structured grids. A butterfly topology is applied, as this permits the use of high quality hexahedral cells, whilst maintaining good control over the mesh spacing. For all grids, the y^+ values were between 0.6 and 0.8 allowing wall-resolved simulations. A grid dependency study was performed to determine an acceptable grid refinement level. This led to meshes with approximately 7 million cells for a tube with a diameter of 5 cm and a streamwise length of 0.24 m.

Data reduction

The maximum Courant number is kept below 0.8 according to the CFL condition, leading to time steps of the order of magnitude of 10-5 s. Ten flow-through times are simulated to reach fully developed turbulent flow, after which data is averaged over 50 flow-through times to reach statistical steady-state. These time-averaged values are then used for further data reduction.

It was previously mentioned that the overall efficiency of an enhanced geometry can be assessed by means of the Thermal Enhancement Factor (TEF), which is a combination of the Nusselt number ratio and the friction factor ratio. These ratios are calculated based on time-averaged simulation results.

For all simulations, the local Nusselt number is calculated as

$$Nu_x = \frac{\left.\frac{\partial T_x}{\partial n}\right|_w}{\frac{T_{w,x} - T_b}{D}} \qquad \text{Eq. 8}$$

To allow comparison between different designs, a spatially averaged Nusselt number is calculated as

$$Nu = \frac{\int_A Nu_x dA}{\int_A dA} \qquad \text{Eq. 9}$$

The spatially averaged Nusselt number for the bare tube, Nu_0, was calculated with this methodology as well rather than calculated by the use of correlations.

The Fanning friction factor was calculated by the use of Eq. 2, where the pressure drop is the pressure drop over the entire simulation domain.

Experimental set-up

To support the numerical findings in this work, a Stereo-Particle Image Velocimetry (S-PIV) campaign was performed in which the flow field was measured inside a dimpled tube at a Reynolds number of 20000.

An illustration of the testing facility is given in Figure 2. For a detailed description of the settling chamber, in- and outlet channel and discharge chamber, the reader is referred to Mayo et al. [19] who discusses S-PIV measurements for a ribbed geometry. For this study, the ribbed geometry was replaced with a dimpled test section with a length of 980 mm and a diameter of

150 mm which is made through vacuum casting and is fully transparent to provide optical access to the flow field. The test section contained 10 streamwise rows of 5 dimples which were placed on one half of the cylinder to ensure optical access without distortion. Water was selected as the working fluid and a Nd:YAG laser was used to illuminate the seeding particles.

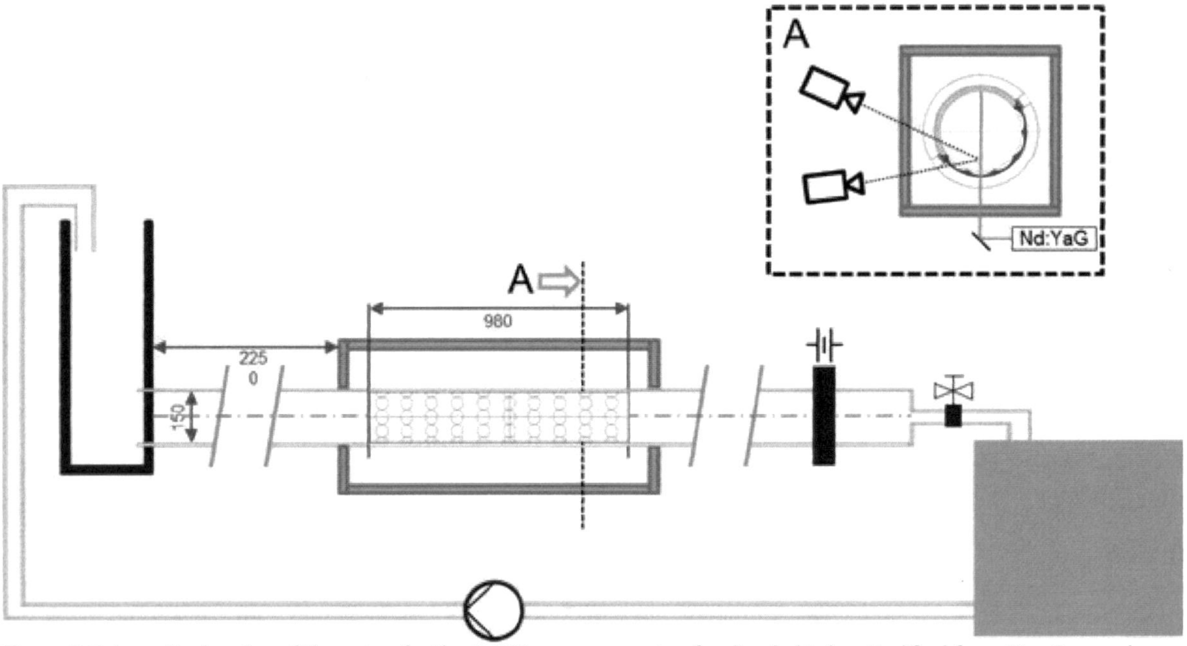

Figure 2 Schematic drawing of the setup for the S-PIV measurements of a dimpled tube. Modified from Van Cauwenberge et al. [15]

Measurements are performed in the second to last row of dimples and focused on the second dimple in the row, i.e. a dimple with one other indentations on one side and three on the other side. This to take into account the effect of neighboring dimples on both sides perpendicular to the mean flow direction.

Results and Discussion

Velocity profiles

The measuring plane for the S-PIV measurements was aligned along the mean streamwise direction and through the center of the dimple and its results are summarized in Figure 3. It shows a large swirling recirculation zone in the upstream segment of the dimple. In the downstream segment of the dimple, streamlines are focused near the wall indicating high near-wall velocities. This will have a positive impact on the heat transfer capabilities of this design as it decreases the thermal boundary layer. Finally, streamlines further away from the wall, i.e. r/R < 0.95, are almost perpendicular indicating that the effect of the dimple is focused in the near-wall region. This implies that little to no shear is introduced into the bulk of the flow due to the dimples, indicating a low overall pressure drop for this kind of wall modification.

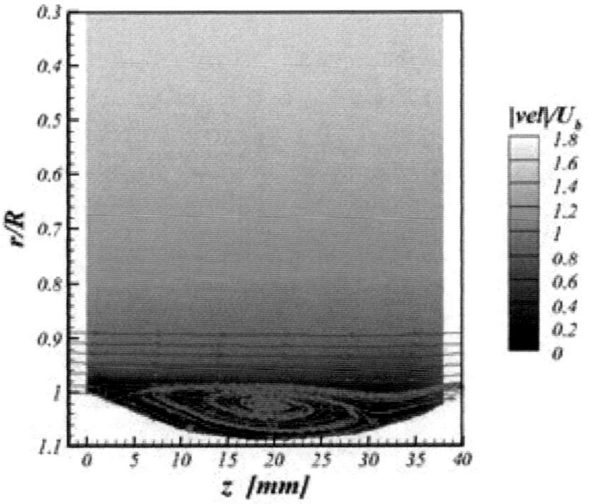

Figure 3 Normalized streamwise velocity and inplane streamlines in the streamwise direction through the center of the dimple. Mean flow direction from left to right.

Validation against experimental data

The experiments reported by Turnow et al. [20] on flow in a square channel with a dimpled wall at ReH = 13042 is used to validate the outcome of the simulation technique. In these experiments, the channel height is 15 mm with rows of dimples with a width of 23 mm and a depth of 6 mm, placed in a staggered arrangement. A sufficient number of dimple rows are used in the streamwise and spanwise direction to ensure fully developed flow, allowing to simulate the flow on a small domain with periodic boundary conditions in the streamwise and spanwise direction.

The Laser Doppler Velocimetry (LDV) data provides information on both mean and rms values for the velocity in the averaged flow direction and is compared to the simulation results obtained by Turnow and obtained through the methodology described above in Figure 4.

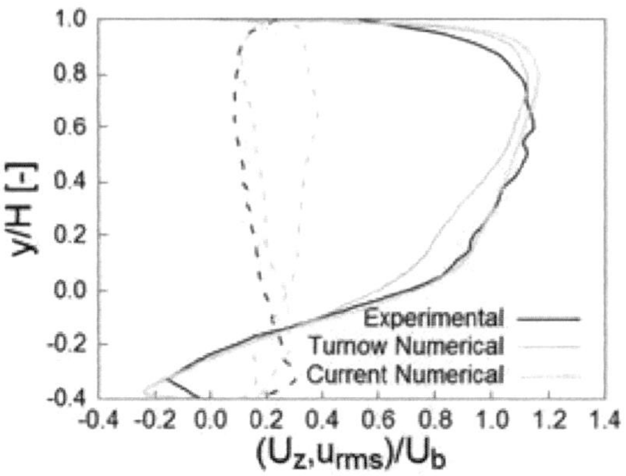

Figure 4 Mean (full lines) and rms (dashed lines) velocity in the averaged flow direction across the channel at ReH = 13042. y/H 0.0 corresponds to the reactor wall.

Similar to Turnow, the simulated mean velocity magnitude near the dimpled wall is slightly higher than the experimental results. Overall the agreement between experimental and numerical results is satisfactory, especially in the relevant bottom half of the channel where dimples are implemented. Next to this, the size of the recirculation zone, defined by negative velocities, is captured accurately as well as the second order statistics. This is important as both the recirculation zone as the second order statistics have a big impact on the heat transfer capabilities of the enhanced design.

Asymmetric vortices and the self-cleaning effect

It was previously reported by Turnow [20] that vortex structures are formed inside spherical dimples in a narrow channel. Furthermore, the vortex was reported to be asymmetric in nature, i.e. the ejecting direction oscillated between +45 ° and -45 ° relative to the mean flow direction.

To focus on this behavior, LES is performed in a channel with an imposed flow rate corresponding to a Reynolds number of 13000 and instantaneous streamwise velocity is measured downstream of the dimple at two points symmetrically opposite of the flow axis through the center of the dimple.

The result, shown in Figure 5 (left), demonstrates the oscillatory behavior of the jet ejecting from the dimple. The Fast Fourier Transform in Matlab R2015a was used to translate the velocity oscillations to the frequency spectrum, which is shown in Figure 5 (right). Both signals have a dominant frequency of around 2.4 Hz, corresponding to an oscillatory cycle of around 0.4 seconds, i.e. the ejecting jet switches between an angle of +45° and -45° relative to the mean flow direction in 0.2 seconds and returns to this initial position after 0.2 s.

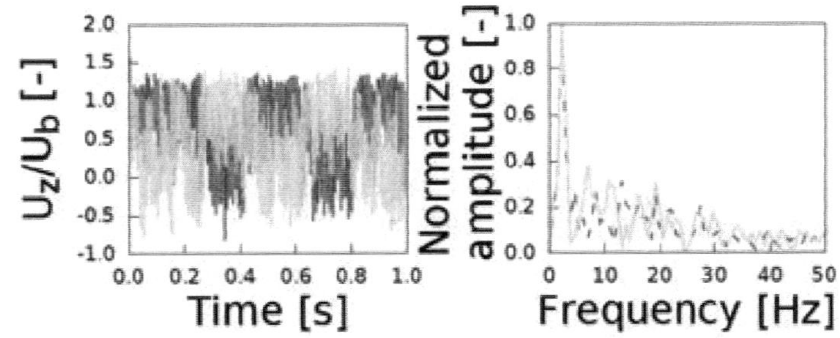

Figure 5 Instantaneous streamwise velocity at two probe positions lying symmetrically opposite of the flow axis as a function of time (left) and corresponding spectrum for both points (right).

As the time scale of these oscillations is much smaller than the time scale for coke formation, it is expected that small particles of cokes can be spalled off continuously, achieving a sort of "self-cleaning" effect of the surface. This behavior can be very desirable during decoking, as a faster coke removal will allow faster return to operations and hence less down time. During operation this can be desirable as well, as this further reduces the rate of coke formation. However, it should be noted that the downstream equipment should be able to handle the fragments of spalled coke.

To assess the effect of the dimple size on this oscillatory behavior. Simulations were performed for a range of dimple depths and the resulting spectra are given in Figure 6.

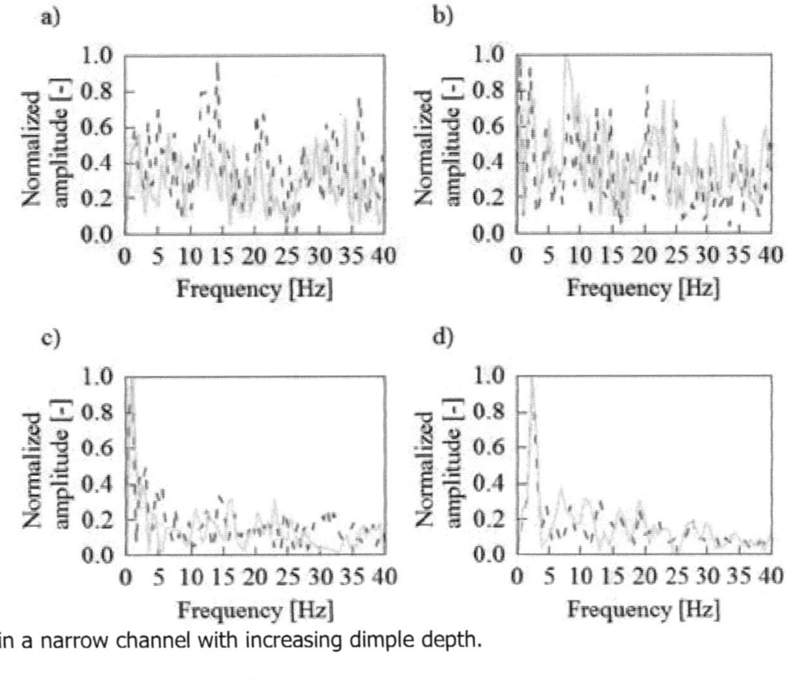

Figure 6 Spectra for flow in a narrow channel with increasing dimple depth.

For cases with shallow dimples, i.e. cases a and b, no peak frequency is found, corresponding to the findings of Turnow who reported that shallow dimples do not allow formation of vortices and their associated oscillating ejecting jets. Additionally, it is found that the dominant frequency is higher for deeper dimples, i.e. in deep dimples oscillations occur faster.

Impact of dimple dimensions

As the oscillating vortices and their ejecting jets have an important effect on heat transfer, the impact of dimple depth on overall heat transfer and pressure drop in pipe flow is now investigated. The dimple depth e is varied over the transition region in which the oscillatory behavior will start to occur.

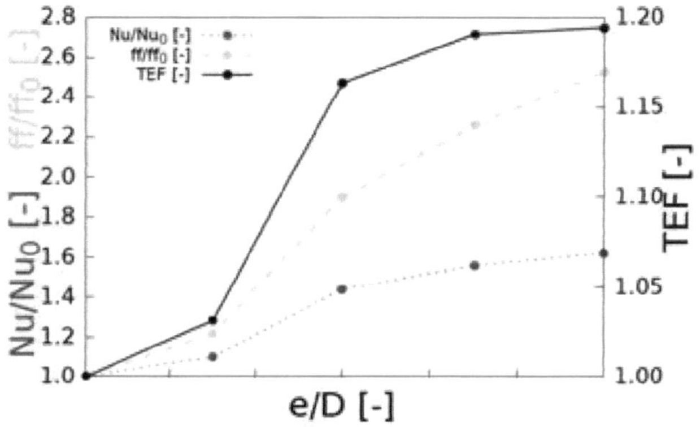

Figure 7 Influence of dimple depth (e) on Nusselt number ratio (blue dotted line), friction factor ratio (yellow dashed line) and thermal enhancement factor (black full line) for Re = 11000.

It can be seen in Figure 7 that the effect of dimple depth is most pronounced in the region in which the vortex structures first originate, indicating the impact of the swirling motion inside the dimple on the overall performance of the coil. As an increase in dimple depth is associated with both an increase in heat transfer as in pressure drop, the thermal enhancement factor is little affected by the e/D ratio once the dimple is deep enough to sustain a swirling vortex.

The effect of dimple width is less pronounced as demonstrated in Figure 8.

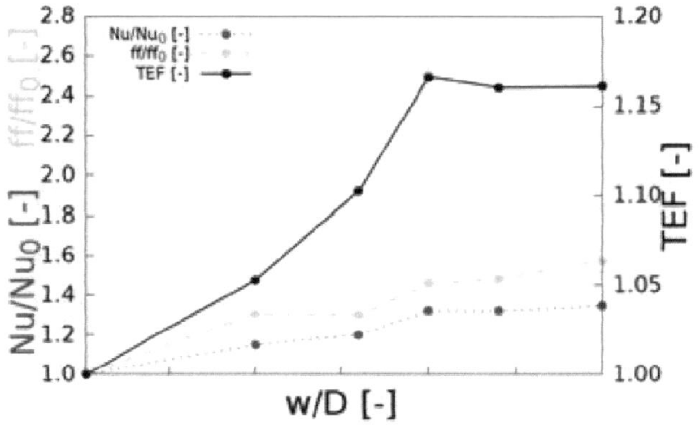

Figure 8 Influence of dimple width on Nusselt number, Fanning friction factor and thermal enhancement factor for Re = 11000.

Effect of wall modifications on olefin yields

As a final step, the impact of the spherically dimpled design on product distribution is assessed by simulation of a Millisecond reactor with a propane feedstock through a well-established methodology [6, 21, 22].

A pure propane feedstock at a flow rate of 118.54 kg/h is fed to a reactor of 11 m and an internal diameter of 30.2 mm with a steam dilution of 0.326 $kg/kg_{hydrocarbons}$ at a temperature of 903 K. The coil outlet pressure is fixed to 170 kPa and the heat flux profile is scaled to obtain a conversion of 80.6 %.

A deep dimple design was selected as it believed that the deeper dimples will achieve a more uniform process gas temperature which will positively affect light olefin yield. Next to this, the deeper dimples will also lead to a high heat transfer enhancement. As a side effect, a high initial pressure drop is to be expected however, it was previously shown by Vandewalle et al. [22] that a high initial pressure drop penalty can be negated over time due to a reduced rate of coke formation. The pressure drop over the reactor was 63 kPa for the enhanced tube which is double of the pressure drop over a bare tube and is lower than the pressure drop of a continuously ribbed geometry calculated as 88 kPa [22].

The effect of the enhanced heat transfer is demonstrated in Figure 9, which shows the temperature at the metal/gas interface. Even though a radial difference in temperature exists for enhanced geometries, the hottest regions are still well below the temperature of the bare tube. As this interface temperature is relevant for the rate of coke formation it indicates the increase in run length, as the maximum temperature is approximately 50 degrees lower than for the bare tube. It is important to note that the enhanced heat transfer also implies a decrease in required heat supplied to the reactor for equal conversion. For this design, a significant reduction in fuel consumption of 1.06% can be achieved as the heat flux was decreased uniformly throughout the reactor by this factor.

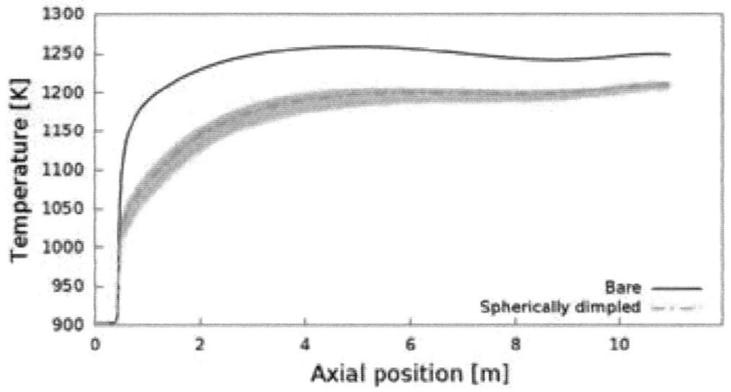

Figure 9 Inner reactor wall temperature along the reactor for a bare tube (black) and a spherically dimpled tube (green).

This lower wall temperature reduces over-cracking in the near wall region, which is stated to have a positive impact on propylene yield. It should be noted that the over-cracking is not limited to this region as the process gas temperature distribution is much more narrow throughout the reactor, as can be seen in Figure 10.

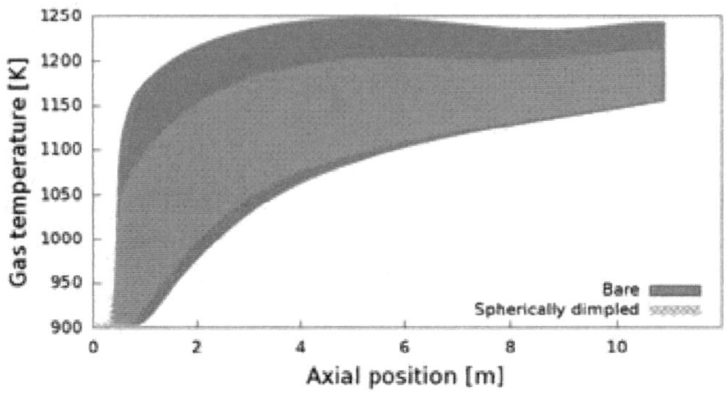

Figure 10 Process gas temperature distribution along the reactor for a bare tube (gray) and a spherically dimpled tube (green).

The implementation of dimpled enhancements leads to a decrease in ethylene production. However, as stated previously, the reduced maximum temperature reduces the tendency for over-cracking which has a positive impact on propylene yield. This leads to an overall increase in light olefin production, i.e. the sum of ethylene and propylene, of 0.5 wt%.

Table 1 Yields for some relevant products for bare and spherically dimpled tube.

Yields [wt%]	Bare	Spherically Dimpled
C3H6	17.4	18.3
C2H4	34.0	33.6
C4H6	1.7	1.6
CH4	19.3	19.3
C3H4	0.7	0.6
Benzene	1.5	1.4

Conclusions

The implementation of spherical dimples as heat transfer enhancing wall modifications was investigated using CFD. It was found that swirling vortices are formed inside the dimple when it is sufficiently deep and that this is an important method of heat transfer enhancement. The ejecting jets associated with the vortices are oscillating leading to an increased heat transfer downstream of the dimples. Furthermore it was found that deeper dimples have faster oscillating ejecting jets.

Not only does the dimple depth impact the oscillating flow behavior, a deeper dimple is also associated with a higher heat transfer and pressure drop. Overall, the dimple needs to be deep enough to sustain a swirling vortex after which limited effect was found on the thermal enhancement factor. It was found that dimple width has limited effect.

Finally, reactive simulations are performed showing that spherically dimpled tubes can decrease the temperature at the metal/gas interface and consequently decrease coking rate and increase run length. Next to this, a more uniform process gas temperature also leads to an increase in total light olefin yield. It can therefore be concluded that the implementation of spherically dimpled tubes in steam cracking furnaces can greatly improve process economics.

Acknowledgment

JND gratefully acknowledges financial support from the Fund for Scientific Research Flanders (FWO), in association with the Pi-Flow project and from Ghent University (Belgium) through GOA project BOF16/GOA/004.
The computational resources and services used in this work were provided by the VSC (Flemish Supercomputer Center), funded by the Research Foundation-Flanders (FWO) and the Flemish Government – department EWI. The authors would also like to acknowledge the resources provided by STEVIN Supercomputer Infrastructure at Ghent University.

References

[1] T. Ren, M. Patel, K. Blok, Olefins from conventional and heavy feedstocks: Energy use in steam cracking and alternative processes, Energy 31 (2006) 425-451.
[2] I. Amghizar, L.A. Vandewalle, K.M. Van Geem, G.B. Marin, New Trends in Olefin Production, Engineering 3 (2017) 171-178.
[3] A.E. Munoz Gandarillas, K.M. Van Geem, M.F. Reyniers, G.B. Marin, Coking Resistance of Specialized Coil Materials during Steam Cracking of Sulfur-Free Naphtha, Industrial & Engineering Chemistry Research 53 (2014) 13644-13655.
[4] P. Ligrani, Heat transfer augmentation technologies for internal cooling of turbine components of gas turbine engines, International Journal of Rotating Machinery 2013 (2013).
[5] P. Promonge, Thermal augmentation in circular tube with twisted tape and wire coil turbulators, Energy Conversion and Management 49 (2008) 2949-2955.
[6] C.M. Schietekat, D.J. Van Cauwenberge, K.M. Van Geem, G.B. Marin, Computational Fluid Dynamics-Based Design of Finned Steam Cracking Reactors, AIChE Journal 60 (2014) 794-808.
[7] K.M.C. Corporation, MERT - Technology to improve Cracking Furnace Performance, pp. Temperature distribution in gas for bare tubes and MERT technology.
[8] M. Györffy, M. Hineno, K. Hashimoto, S.-H. Park, M.-S. You, Mert performance and technology update, AIChE Spring Meeting: Ethylene Producers Conference, Tampa Bay, USA, 2009.
[9] Schmidt+Clemens, Strengthen the heart of your plant! SCOPE Fusion HTE - the steam cracker tubing of the future, 2018.
[10] M.W.M. van Goethem, E. Jelsma, Numerical and experimental study of enhanced heat transfer and pressure drop for high temperature applications, Chemical Engineering Research & Design 92 (2014) 663-671.
[11] C.M. Schietekat, M.W.M. van Goethem, K.M. Van Geem, G.B. Marin, Swirl flow tube reactor technology: An experimental and computational fluid dynamics study, Chemical Engineering Journal 238 (2014) 56-65.
[12] A. Carrillo, F. Bertola, G. Wang, L. Zhang, Intensified heat transfer technology–CFD analysis to explain how and why IHT increases runlength in commercial furnaces, AIChE and EPC 2010 Spring National Meeting, San Antonio, TX, 2010.
[13] C.M.J. Tay, B.C. Khoo, Y.T. Chew, Mechanics of drag reduction by shallow dimples in channel flow, Physics of Fluids 27 (2015) 22.
[14] D.J. Van Cauwenberge, L.A. Vandewalle, P.A. Reyniers, K.M. Van Geem, G.B. Marin, J. Flore, Periodic Reactive Flow Simulation: Proof of Concept for Steam Cracking Coils, AIChE Journal 63 (2017) 1715-1726.
[15] D.J. Van Cauwenberge, J.N. Dedeyne, K.M. Van Geem, G.B. Marin, J. Floré, Numerical and experimental evaluation of heat transfer in helically corrugated tubes, AIChE Journal (2017).
[16] W.-W. Kim, S. Menon, A new dynamic one-equation subgrid-scale model for large eddy simulations, 33rd Aerospace Sciences Meeting and Exhibit, American Institute of Aeronautics and Astronautics1995.
[17] S.V. Patankar, C.H. Liu, E.M. Sparrow, Fully Developed Flow and Heat Transfer in Ducts having Streamwise-periodic Variations of Cross-sectional Area, Journal of Heat Transfer-Transactions of the Asme 99 (1977) 180-186.

[18] Pointwise Mesh & Grid Generation Software, Pointwise Inc., Texas, 2015.
[19] I. Mayo, B.C. Cernat, M. Virgilio, A. Pappa, T. Arts, Asme, Aerothermal Investigation of the Flow and Heat Transfer in a Helically Corrugated Cooling Channel, Proceedings of the Asme Turbo Expo: Turbine Technical Conference and Exposition, 2016, Vol 5b (2016) 11.
[20] J. Turnow, N. Kornev, V. Zhdanov, E. Hassel, Flow structures and heat transfer on dimples in a staggered arrangement, International Journal of Heat and Fluid Flow 35 (2012) 168-175.
[21] P.A. Reyniers, C.M. Schietekat, D.J. Van Cauwenberge, L.A. Vandewalle, K.M. Van Geem, G.B. Marin, Necessity and Feasibility of 3D Simulations of Steam Cracking Reactors, Industrial & Engineering Chemistry Research 54 (2015) 12270-12282.
[22] L.A. Vandewalle, J.N. Dedeyne, D.J. Van Cauwenberge, K.M. Van Geem, G.B. Marin, Computational Fluid Dynamic design of steam cracking reactors: extrusion method for simulation of dynamic coke layer growth, AIChE Spring MeetingHouston, TX, USA, 2016.

Paper 121b

Conversion of Low-cost Feedstock to Petrochemical Products

Radaelli, Guido

Executive Vice President, Engineering & Process Technology

Siluria Technologies

Rafique, Humera

Product Manager

Siluria Technologies

Prepared for Presentation at the 2019 Spring National Meeting

New Orleans, LA, April 2, 2019

Conversion of Low-cost Feedstock to Petrochemical Products

Radaelli, Guido

Executive Vice President, Engineering & Process Technology

Siluria Technologies

Rafique, Humera

Product Manager

Siluria Technologies

Abstract: Siluria Technologies has been pioneering in olefin production via natural gas. It has successfully built and run the largest Oxidative Coupling of Methane (OCM) reactor producing ethylene and propylene via natural gas. Siluria's pre-commercial unit has been running for over four years, gathering sufficient data to build robust models to design commercial scale OCM units.

The Gemini Process (Ethylene production via OCM) and Orion Process (Propylene production via OCM) provides the flexibility of producing olefins at small scale (50-300kTa olefin production) as well as large scale (>1000 kTa) and offers the ability of consuming low-cost feedstock such as natural gas, flared gas, cracker off-gas, Propane De-hydrogenation unit (PDH) offgas, refinery fuel gas etc. This enables Siluria to provide olefin expansion solutions to a wide variety of existing facilities. One such product application that has gained much traction is the integration of OCM with new and existing PDH facilities. This paper will provide case studies of the product applications, specifically detailing the integration of OCM with PDH offgas for propylene production.

Siluria is also working towards its first commercial project, a small-scale olefins unit primarily producing propylene in the US Gulf Coast Region. It is integrated with an existing facility delivering olefins with low capital intensity and high returns.

Introduction

Siluria technologies has pioneered the development of a commercial process for converting natural gas to olefins using its proprietary Oxidative Coupling of Methane (OCM) catalyst. To enable a commercial process, Siluria established a world-class scale-up platform that screened and validated the catalyst for commercially critical parameters from bench to commercial scale.

The first step of the scale-up process included synchronized rapid screening of over 250 catalyst formulae for methane conversion and C2 selectivity over a wide range of operating temperatures and compositions. Catalyst compositions down-selected from this process were then placed through secondary screening in 1-2mm diameter and 4-8mm diameter fixed bed reactors. Data obtained from this stage of screening help build complex kinetic models. High performance catalyst from the secondary screening are then tested at pilot scale. Pilot reactors are generally 50-100mm in diameter. Detailed description of transport phenomenon were obtained and added to the kinetic models developed during the secondary screening. Also, lifetime and aging of the catalyst is also evaluated prior to testing the catalyst at the demonstration scale.

The pre-commercial demonstration unit built by Siluria includes the largest OCM reactor ever built. This unit operates in campaign mode and is used to validate data from the complex models previously produced at a range of commercially operating temperatures and pressure, gas velocities, residence times, and inlet gas compositions. Complex reactor geometry and fluid dynamics are added to the kinetic and transport models to produce a fully validated model for a commercial scale OCM reactor (1).

Siluria's OCM demonstration unit has been running for over 4+ years and validation of commercial conditions has allowed Siluria to develop optimized and integrated process designs that convert not only natural gas but any methane-containing fuel to high value olefins, primarily ethylene and propylene. Siluria Technologies has currently branded two of its main olefin producing processes; Gemini Process (Ethylene production via OCM process) and the Orion Process (Propylene production via OCM). The input stream for both processes is low value methane-containing streams such as, but not limited to, natural gas, offgas from various petrochemical facilities (e.g. steam crackers), propane de-hydrogenation units, and refineries. All processes can produce either ethylene or propylene or a mix of both (and additional co-monomers) depending on the end use of product.

OCM Reactor

At the core of both the Gemini and Orion process designs is the OCM adiabatic reactor that consists of two main sections: the OCM Reaction Section, which contains the axial-fixed OCM catalyst bed, and the small chained alkane (primarily ethane/propane) conversion section, an adiabatic empty chamber designed to perform Post Bed Cracking (PBC).

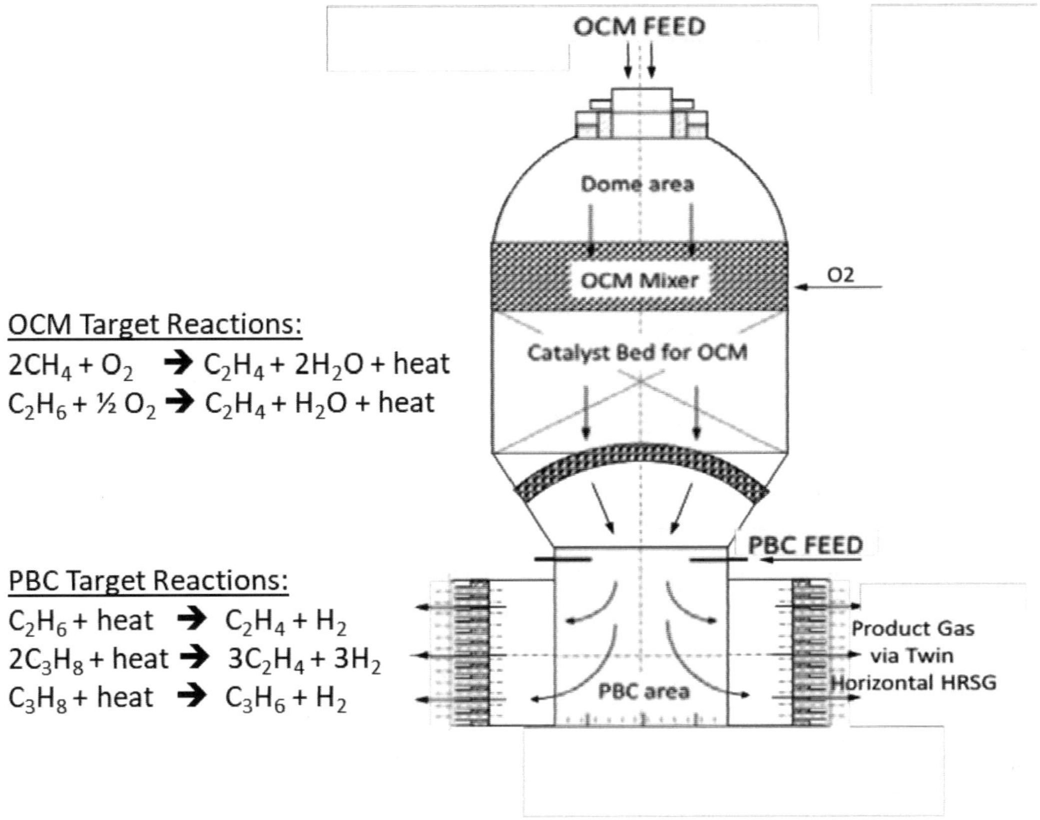

Figure 1 Core of Siluria Process: OCM Reactor

The OCM feed to the reactor, primarily consisting of methane with minor amounts of ethane, is mixed with oxygen in the OCM mixer and sent to the catalyst bed. OCM reactions alongside other side reactions begin to occur. The main reaction that takes place in the catalyst bed is shown in reaction 1.

$2CH_4 + O_2 \rightarrow C_2H_4 + 2H_2O$ + heat Reaction 1

In addition to methane oxidation, Siluria's OCM catalyst simultaneously promotes Oxy-De-Hydrogenation (ODH) of the ethane present in the feed according to the following reaction:

$C_2H_6 + 1/2O_2 \rightarrow C_2H_4 + H_2O$ + heat Reaction 2

OCM's Product Applications

Both the Gemini and Orion process include unit operations that can process and purify a wide variety of feed streams, including, but not limited to fuel value, alkane and alkene containing off-gases from refineries and petrochemical facilities. Due to the nature of the process and the ability to produce incremental olefins (between 40-300kTa olefins), Siluria is able to pursue a wide variety of product applications as shown in Figure 2. These application can be applied to new and existing facilities allowing plants to increase plant production with low overall capital investment time to time.

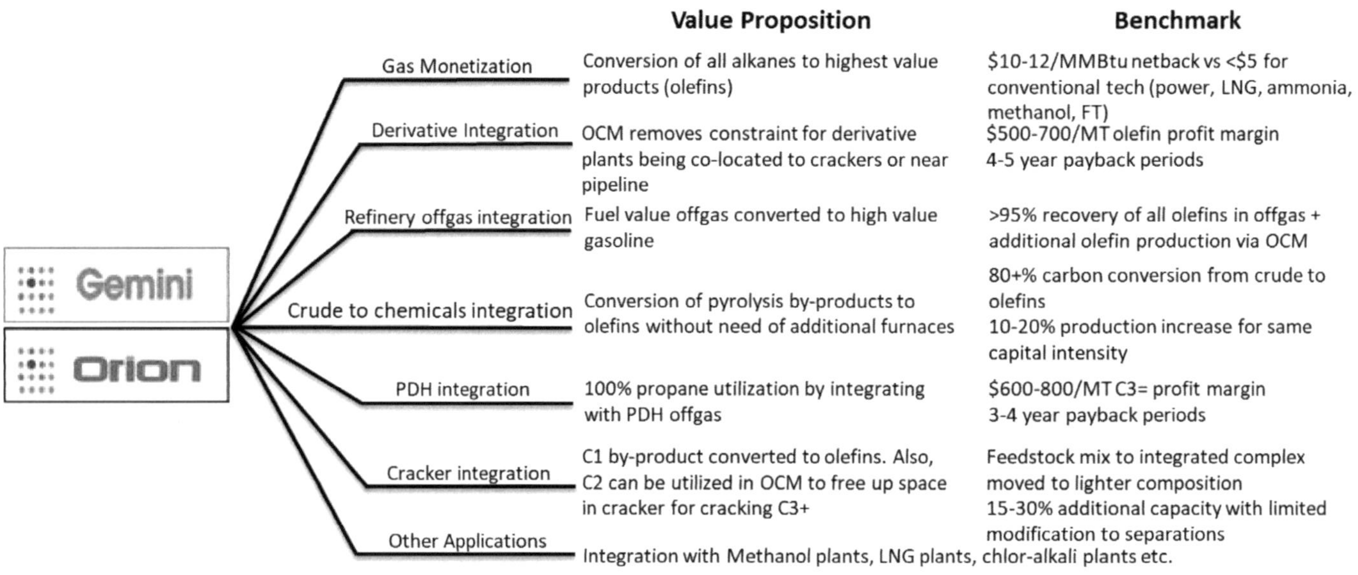

Figure 2 OCM Product Applications and their advantages

Gemini and Orion Process Overview

The Gemini and Orion process consist of two and three main sections respectively:

1. The Hot Section which consist of the methanation unit, the OCM reactor and the Heat Recovery unit;
2. The Cold section which consists of the compression for separations and fractionation to polymer grade ethylene and specifically for the Orion process;
3. The third section is the dimerization and metathesis unit which converts ethylene to polymer grade propylene (only present in Orion).

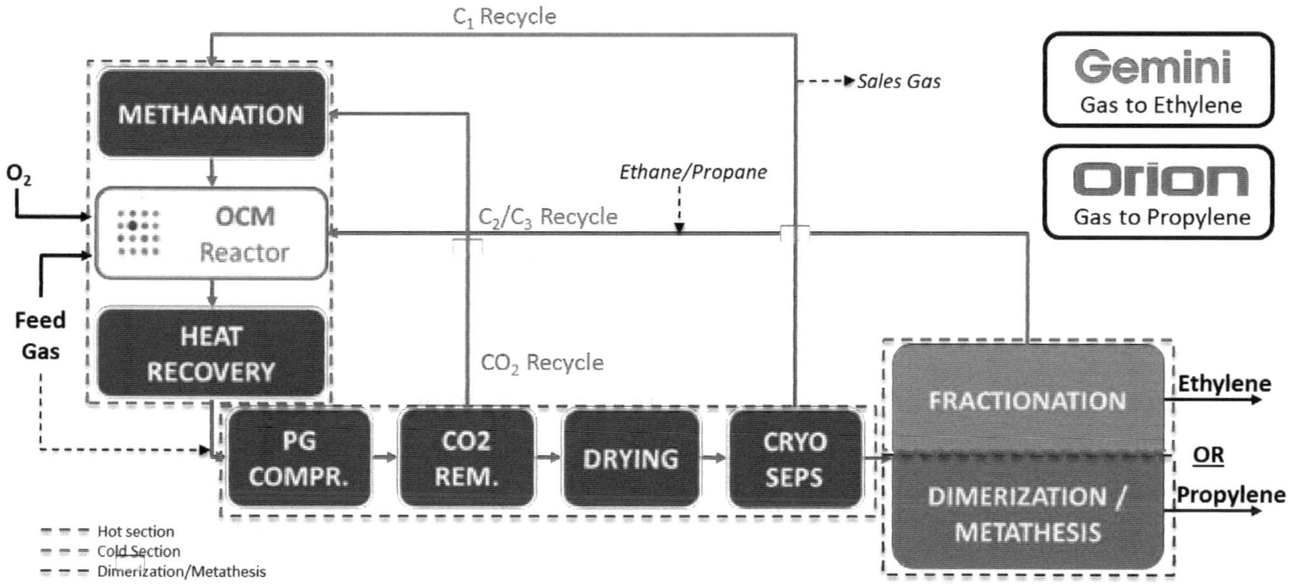

Figure 3 Block-Flow Diagram of the Gemini and Orion Process

Hot section: The C1 recycle stream – consisting of unreacted CH_4, and by-product H_2 and CO from the OCM reactions and coming from the cryo unit – is mixed with a specified amount of CO2 from the CO2 removal unit and methanated to methane in the methanation reactor system. The recycle stream is then mixed with incoming feed gas. Assuming the feed gas does not have to be separated and can be fed straight to the OCM reactor, it is mixed with the C1 recycle stream before being sent to the OCM reactor. Oxygen and the C1 recycle stream are then mixed in a proprietary mixer within the OCM reactor before being sent to the catalyst bed. Post the OCM reactions on the catalyst bed, the heavier alkanes, specifically C2 and C3, are cracked to additional olefins in the post bed cracking section as described previously. The heat from the OCM reactor is then recovered by producing high pressure steam, which is used in the downstream rotating equipment and reboilers, and by heating up cold OCM inlet streams in a series of heat exchangers.

Cold Section: Once the process effluent is cooled, it is compressed to 30-50 barg and sent to an amine and caustic unit to remove CO2 from the effluent. The effluent is dried in a typical mol sieve drier and sent to a de-methanizer (or cryo separations unit). For small scale projects, the de-methanizer can be designed to be a typical modular off-the-shelf gas processing plant de-methanizer unit for cost and project time saving purposes. The C2+ stream from the bottom of the de-methanizer is further fractionated in a de-ethanizer, acetylene hydrogenation unit, and a C2 splitter to obtain 99.95mol% (polymer grade) ethylene.

Dimerization and Metathesis Unit: For the Orion process specifically, the product ethylene is converted to propylene using a conventional dimerization/metathesis technology. The technology essentially converts a portion of the polymer grade ethylene to butene which reacts with the rest of the ethylene to form polymer grade propylene.

OCM process integrated with Propane De-hydrogenation Unit

PDH offgas (consisting of methane, ethane, and hydrogen) can be sent to the OCM plant, which converts the offgas, otherwise used as fuel, to olefins via the OCM process. Moreover, the hydrogen in the offgas can be utilized in the methanation section, thus utilizing all the CO2 by-product produced by OCM and the all the offgas from the PDH unit, leading to a 100% utilization of the propane fed to PDH. A high level block flow diagram showing the integration of a PDH facility with OCM is shown in Figure 4 below.

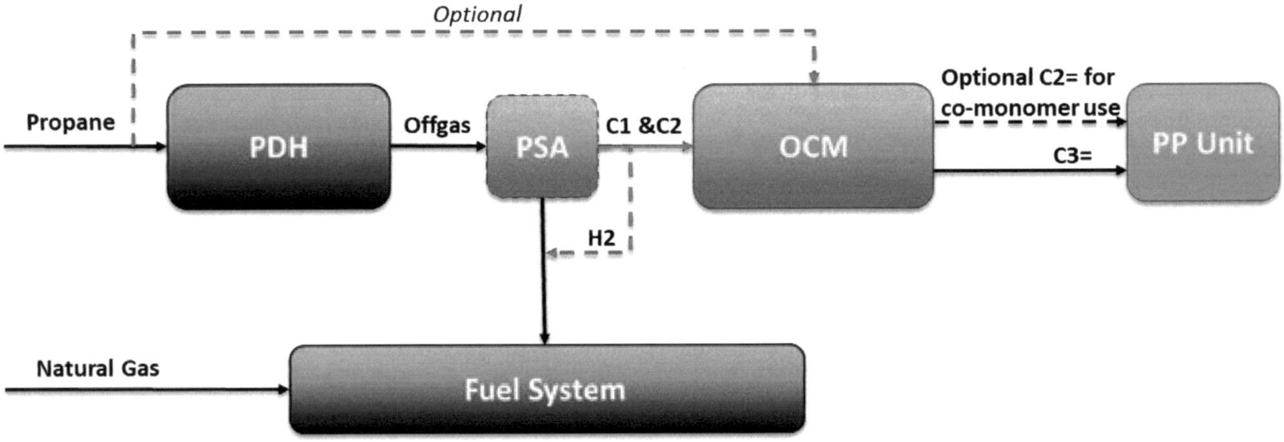

Figure 4 Integration of a PDH unit with OCM

Case Study: OCM process integrated with 750kTa PDH unit

OCM integrated with existing PDH facilities is most beneficial when the petrochemical facility is looking to expand its propylene production incrementally (by 40-150kTa C3=), at which scale addition of small PDH units become uneconomical. Integration of a small propylene unit within a larger complex also provides the advantage of integrating with the OSBL facilities of the existing petrochemical facilities (steam, power etc.). The Orion process uses a modular based approach for small scale plants using off-the-shelf units for most of its unit operations as previously discussed.

A 750kTa PDH facility (or 89 t/h C3= facility) typically produces 20-25% of offgas on a mass basis. For this case study, we assume the offgas from a 750kTa PDH facility has the following mass flow, conditions, and composition:

PDH offgas	
Mass Flow (kg/h)	23,100
Temperature (C)	45
Pressure (barg)	3
Mole Fraction	
Hydrogen	0.7183
Methane	0.0996
Ethylene	0.0122
Ethane	0.1488
Propene	0.0000
Propane	0.0030
C4+	0.0000
H2O	0.0211

Table 1 Typical PDH off-gas composition

The offgas from the PDH facility is fed through a hydrogen PSA removing 80-85% of the H_2 from the offgas. The C1/C2 stream is mixed with the OCM effluent stream and sent to the Process Gas Compressor (PGC) of the OCM plant.

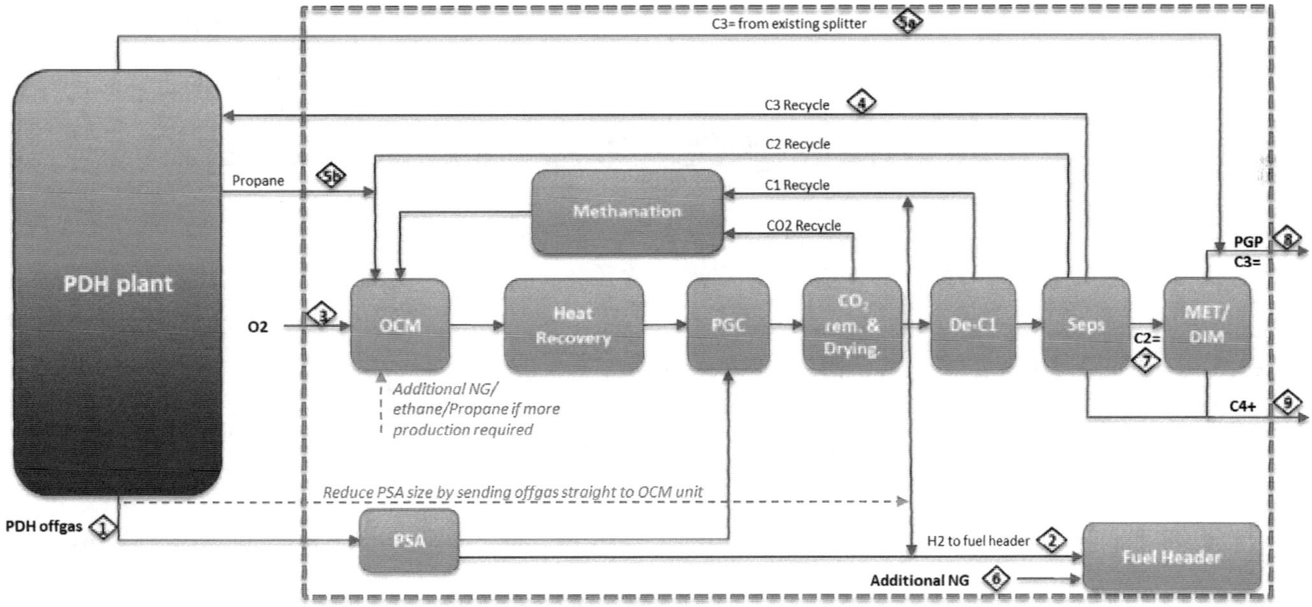

Figure 5 Detailed Block Flow Diagram for integration of a PDH unit with OCM for incremental propylene

The gas is compressed and purified through the CO_2 removal and drying process before being sent to the cryogenic separations unit. The overhead of the cryogenic unit is mixed with compressed CO_2 by-product and methanated before being fed to the OCM unit. Because of the high H_2 content, all the CO2 can easily be consumed in the methanation unit making the OCM system essentially 100% carbon efficient. For the case study presented, no additional NG is added to the OCM process for a production of 106kTa additional C3= from the process. However, if there is a requirement for additional C3= production, additional NG can be added to the C1 recycle feed to the OCM. The rest of the process for C3= production is the same as described in the process overview section above. All OSBL facilities are assumed to be integrated with the existing PDH facility. Operating costs of the additional utilities are added to the cash costs of the project. The material balance from the process is as follows:

Stream		
1	PDH offgas (kTa)	194
2	Excess H2 back to fuel gas (kTa)	22.5
3	Oxygen (kTa)	174
4	C3 Product to PDH C3 splitter (kTa)	10.6
5a	C3= wt% *(kTa)*	89 *(9.4)*
5b	C3 wt% *(kTa)*	11 *(1.2)*
6	Additional NG to FH (MMSCFD)	23
7	C2= product (kTa)	107
8	C3= Product (kTa)	106
9	C4+ product (kTa)	19.5

Table 2 OCM-PDH Integration Material Balance

As stated earlier, the process can also be used to generate an optional stream of ethylene and also butene-1 as co-products which enables the polypropylene plant to fulfill its co-monomer requirements for all grades of homo polymers from a single plant site.

Techno-economics for the case study shows over $630/MT C3= cash costs with a 3-4 year payback period making the product application very attractive for expanding polypropylene sites.

Feedstock		Pricing Assumption	$/MT C3=
Off-gas stream	$/MMBTU	2.5	195
Oxygen	$/MT O2	40	66
Other Variable Costs			
Cats and Chems + Utilities and Power			121
Fixed Costs			
Operations, Maintenance, and OH			74
Product credits			
C4s stream	$/MT C4s	453	(43)
C5+ stream		520	(47)
C3= cash cost			366
Profit Margin	$/MT C3=	1000	634
ROM Capex	$/MT C3=	2000-3000	

Table 3 Overall Techno-economics for OCM-PDH Product Application

Siluria has already executed a license with the PDH offgas integration product application. The application uses both polymer grade ethylene as well as propylene from the OCM unit, expanding its polypropylene capacity as well as removing its dependence from the market on the requirement of for small amounts of co-monomer ethylene.

Conclusions

Siluria Technologies has built a world class OCM catalyst scale-up platform that enabled the generation of an extremely accurate model to effectively determine catalyst performance on a commercial scale. Siluria has then developed two OCM-based industrial processes that convert low value feedstock to high value olefins, namely ethylene and propylene. Siluria has also developed various product applications that integrate small scale OCM plants within larger facilities providing incremental olefin production for a low capital investment.

One particular product application that has gained considerable traction in the recent past is the integration of the OCM plant with PDH facilities. Since PDH facilities are economical only in large scale propylene production, incremental propylene required for expanding polypropylene units are easily achievable by integrating a small scale OCM process with an existing PDH facility. Not only does the process generate a high return on a low capital investment but also enables the petrochemical facility to internally produce all the co-monomers required for the production of high-grade polypropylene and a plant that is essentially 100% carbon efficient.

References:

1. Radaelli, Guido, Iyer, Rahul, Rafique, Humera. (2017) "Building with Natural Gas," presented at the 2017 AIChE Spring Meeting, San Antonio, TX

AIChE Paper Number: 121C

IMPROVED ETHYLENE HEATER OPERATION USING IN-FURNACE, SPATIALLY-RESOLVED COMBUSTION MEASUREMENTS

Philippe Van den Bussche
Process-Asset Manager Steamcracker Furnaces
BASF

Eric Huelson
Product Manager
John Zink Company LLC

Eric Gebhard
Sales Manager
John Zink Company LLC

Prepared for Presentation at the 2019 Spring National Meeting
New Orleans, LA April 2, 2019

Session 121 Ethylene Plant Technology Fundamentals

IMPROVED ETHYLENE HEATER OPERATION USING IN-FURNACE, SPATIALLY-RESOLVED COMBUSTION MEASUREMENTS

Philippe Van den Bussche
Process-Asset Manager Steamcracker Furnaces
BASF
Eric Huelson
Product Manager
John Zink Company LLC
Eric Gebhard
Sales Manager
John Zink Company LLC

Abstract:
Ethylene cracking furnaces are often a challenge for operators to maintain well-balanced, uniform combustion throughout the furnace due to their size, the quantity of burners requiring adjustment and changing process conditions. This paper reviews results of integrating a ZoloSCAN Combustion Monitoring and Diagnostics System to optimize products of combustion to significantly extend furnace run lengths.

The combustion monitor employed uses laser based technology to measure Temperature, O_2, CO and H_2O across multiple paths in the combustion zone of the furnace. This data was used to adjust combustion equipment which resulted in lower Tube Metal Temperatures (TMT), increased run lengths, higher feed rates and reduced NOx emissions.

Ethylene Cracking Furnaces Operational Challenges

Operators are often challenged to maintain well balanced, uniform combustion in Ethylene Cracking Furnaces in order to optimize run-time, efficiency, reliability, productivity, safety and emissions reduction. While most industrial furnaces have several individual point-of-use sensors available to monitor temperature, O_2, and CO at the furnace exit, such sensors are not typically available for use directly inside the firebox.

This lack of in-furnace combustion data allows for local combustion non-uniformities and may result in Tube Metal Temperature (TMT) hot spots, high NOx and CO emissions and in inefficient combustion even with an appropriate stack analyzer excess O_2 setpoint. *Figure 1* illustrates this issue. Alternatively, with zone based in-furnace data, actions can be taken to reduce TMT hot spots and minimize emissions and maintain efficiency.

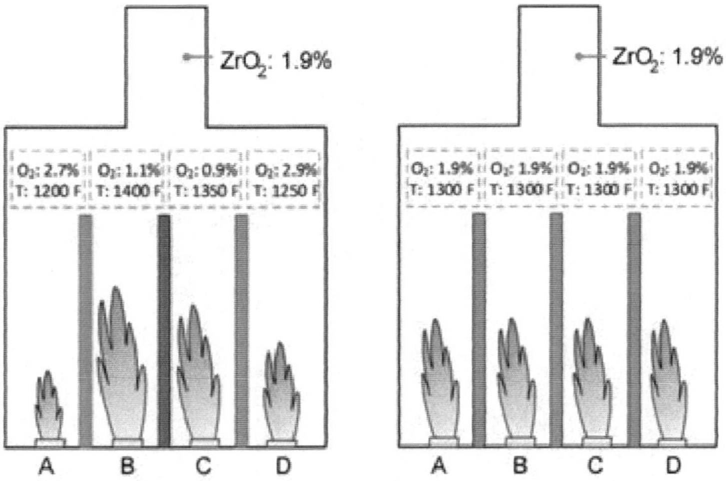

Figure 1: Two identical heaters are depicted both at an excess O_2 setpoint of 1.9%. Left: Imbalanced combustion resulting in TMT hot spots and excess NOx and CO Right: Balanced combustion for uniform TMTs and reduced emissions

Non-uniform combustion can adversely affect furnace operation in many ways but is most problematic for Process Efficiency, Combustion Efficiency, Reliability and Safety. Process efficiency is impacted by temperature non-uniformities causing localized hot spots and non-uniform coking rates that result in short production cycles. Combustion efficiency can be reduced by variations in fuel to air ratios across the furnace that may produce both higher NOx and CO emissions. Reliability may be limited by premature coil and tube failures due to flame impingement or higher than desired flux on the process tubes resulting in unplanned outages. Finally, safety may be a concern for plant personnel due to CO excursions and tube leaks. Optimized in-furnace combustion thus provides a significant benefit to efficiency, reliability and availability of the furnace.

Multiplexed TDLAS Laser Combustion Measurement System

This paper describes a case study installation on an ethylene cracker furnace deploying a wavelength multiplexed laser-based monitoring and diagnostic technology (multiplexed TDLAS system) that measures temperature, oxygen, carbon monoxide, and water directly within an industrial furnace. This technology allows operators to tune burner air controls to extend run lengths, improve fuel savings, increase production yields, lower excess O_2, and reduce NOx and CO emissions through well balanced temperature and oxygen profiles across the entirety of the furnace. API has designated Tunable Diode Laser Absorption Spectroscopy (TDLAS) systems as best practice for large fired heater applications.

Figure 2: Left: Laser Heads Mounted on Furnace Wall Right: In-furnace laser grid example

Multiplexed TDLAS systems have unique capabilities that make them a cost-effective solution for large industrial furnace control. These features include:

- *Multiple species from a single path*
    - Simultaneous Temperature, O_2, H_2O and CO measurements on each laser path. Typical TDLAS stack analyzers will measure only O_2 or CO.
- *Multiple paths from a single system*
    - Up to twenty-four paths can be measured on a single system providing for a cost-effect measurement of multiple zones and furnaces.
- *Reliable Reporting*
    - Automated path alignment allows for continuous reporting of all paths even with furnace wall warpage associated with startup and shutdown.
    - Minimal maintenance is achieved by positioning key electronics away from the furnace.

- Rapid Measurement
    - A complete furnace profile is often obtained every three to five minutes.

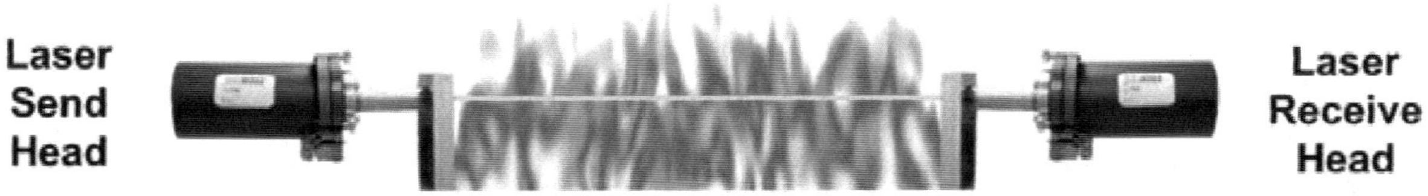

Figure 3: Typical Send and Receive laser path configuration. Temperature, O_2, CO and H_2O lasers are all on the same beam across the furnace.

BASF Ethylene Furnace Case Study

BASF Antwerp installed a ZoloSCAN system in 2015 with the primary objective of extended run lengths on one of their ethylene cracking furnaces. Data from the system would be used to balance combustion across the twin cell cracker and thus mitigate local hot spots and associated coke formation to extend the run length of the unit.

Installation and integration of the ZoloSCAN was jointly performed by Zolo Technologies (a group within John Zink Company) and BASF. Zolo Technologies specified a laser path design, provided the ZoloSCAN equipment, commissioned the laser system and worked with the plant to integrate the data within the plant DCS. BASF specified the coil, burner, and radiant section design, installed the sight tubes, and conduit and finalized DCS connections and visualization of the data. A laser grid layout was selected to best provide combustion measurement feedback for the burners and controls available on the cracker. At BASF this layout included an 8x2 laser grid in each cell with paths aligned to burner rows and burner columns as depicted in *Figure 4*.

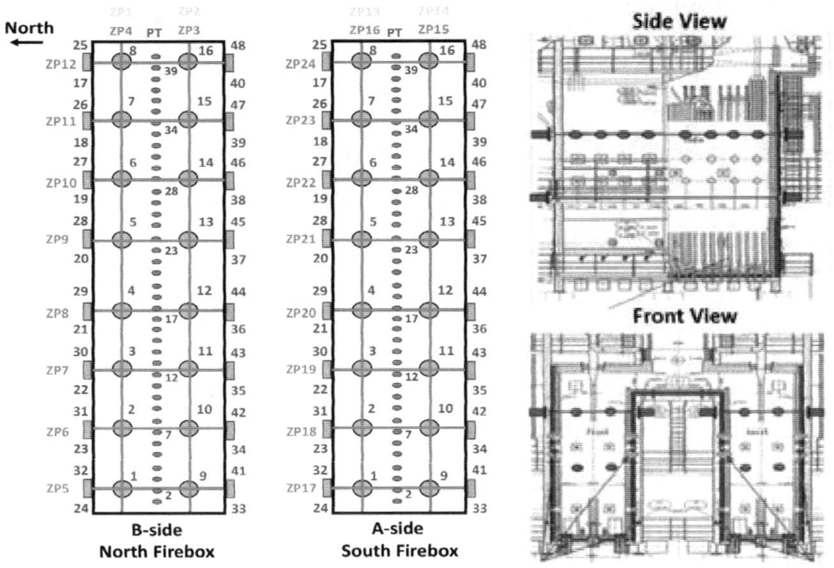

Figure 4: BASF Antwerp ZoloSCAN layout - Top, Side and Front view

Integration of the measurement system at BASF was done in a four-step process to best generate value from the data provided:

1. Determine Objectives and Constraints of the Furnace
2. Conduct Baseline Testing
3. Initiate Furnace Optimization Procedure
4. Integrate Optimized Processes into Furnace Operations

1. Determine Objectives and Constraints of the Furnace

The first task was a collaborative effort with BASF and Zolo Technologies to identify the objectives and constraints of achieving target furnace operation. While the top-level objective was extended run length, auxiliary objectives included NOx reduction and increased combustion efficiency. Constraining variables included emission limits (instant and continuous), coil venturi pressures, TMT's (both calculated and measured) and in-furnace and stack pressure readings.

To meet the objectives ZoloSCAN species measurements were correlated with in-furnace performance as described below:

Temperature
- Provide indication of flue gas temperature uniformity
- Balance temperature variations and extend decoking intervals

O_2
- Balance O2 distribution and optimize fuel usage
- Reduce furnace draft limitations
- Reduce NOx emissions

CO
- Alert for fuel rich zones in the furnace for improved safety

H_2O
- Monitor for leaks in the system (safe start up and shutdown)

2. Conduct Baseline Testing

In early 2016, BASF's Research Team, Operations Team and Zolo Technologies worked together to quantify typical furnace operation from historic furnace data (data from late 2015 and early 2016) and map out the impact of floor burner Primary and Secondary Air damper changes and sidewall burner flaps (secondary air adjustment) on in-furnace laser species measurements. The objective of the parametric testing was to correlate in-furnace changes of Temperature, O_2, CO, or H_2O with damper changes so that one could easily adjust damper positions to achieve uniformity of combustion products in the furnace. Results of these efforts produced the Operational Guidelines in *Table 1* specifying the degree combustion measurements would be impacted based on a change to any damper position. As an example, if the operator wanted to achieve temperature uniformity in a cell by increasing the temperature above a pair of burners by 10°C, the Guideline table would indicate that this might be achieved by a 25% increase in the Primary Air damper setpoint to both burners.

Table 1: Operational Guidelines for Multiplexed TDLAS ZoloSCAN Data at BASF

Description	Control (Range 0-100)	Movement	Unit	# Burners Manipulated	Average Δ Zolo
Temperature Burner Row Balance	Primary Air % Damper move	12,5	%	2	5,66
	Secondary Air % Damper move	-75	%	2	0,71
	% of Sidewall burner flaps open 1 flap = 25%	50	%	4	-2,38
O_2 Burner Row Balance	Primary Air % Damper move	12,5	%	2	0,09
	Secondary Air % Damper move	-75	%	2	-0,38
	% of Sidewall burner flaps open 1 flap = 25%	50	%	4	0,27
CO Burner Row Balance	Primary Air % Damper move	12,5	%	2	-0,01
	Secondary Air % Damper move	-75	%	2	0,23
	% of Sidewall burner flaps open 1 flap = 25%	50	%	4	-0,06
H_2O Burner Row Balance	Primary Air % Damper move	12,5	%	2	-0,12
	Secondary Air % Damper move	-75	%	2	0,14
	% of Sidewall burner flaps open 1 flap = 25%	50	%	4	-0,14

3. Initiate Furnace Optimization Procedure

Once parametric testing was completed, the actual furnace optimization procedure was a relatively quick process (one to two hours) to make damper position adjustments and balance the products of combustion based upon the Operational Guidelines. In general, the only additional consideration relative to the Operational Guidelines table is that air dampers should be equally opened and closed during optimization of the furnace so that the overall draft pressure is conserved, and one is not working against DCS adjustments of the furnace when opening or closing dampers. Typical in-furnace balancing benefits are shown below.

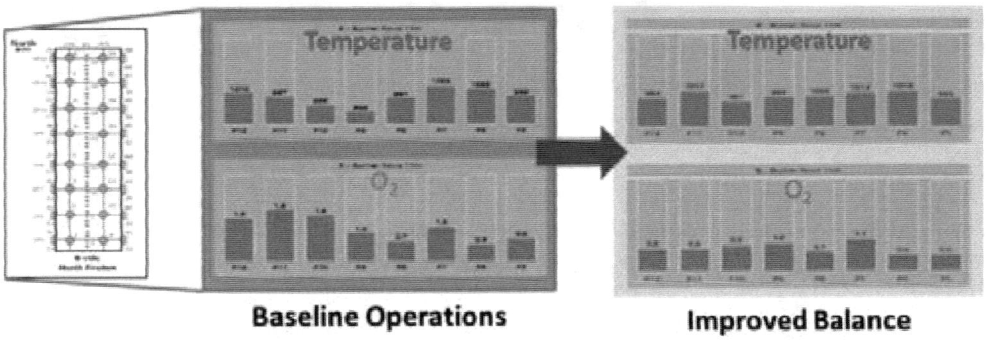

Figure 5: Typical combustion uniformity improvement after balancing combustion within the furnace.

4. Integrate Optimized Processes into Furnace Operations

Once the Optimization Procedure was realized and validated, the process was integrated with the Operations Team for sustained combustion control benefits. At BASF this was accomplish via the addition of visualization tools on the DCS controls screen and training of one of the four of the operations shifts on how to tune combustion when imbalances were observed.

BASF Ethylene Furnace Data Analysis

BASF conducted a comprehensive study to determine the benefit of the integration of the ZoloSCAN system. Data was evaluated from November 2015 to September 2018 with multiple feedstocks (LPG, Naphtha, Ethane) and baseline and optimized test cases to compare before and after results of the integration. The evaluation showed multiple benefits including the critical goal of the project, increased run lengths.

Run Length Extension via lower TMT

Both Recycle Feeds (Naphtha) and Gas-feed (LPG and Ethane) runs showed a significant extension in the run length of the furnace. *Figure 6* provides a perspective on the improvement of run length in the furnace relative to baseline operations. Prior to the integration of the ZoloSCAN the furnace was constrained by venturi pressures or TMT maximums that necessitated a de-coke prior to the expected run time of the cracker. Post the ZoloSCAN integration, TMT values were kept well below TMT de-coke trigger points and the furnace run lengths were correspondingly increased in duration.

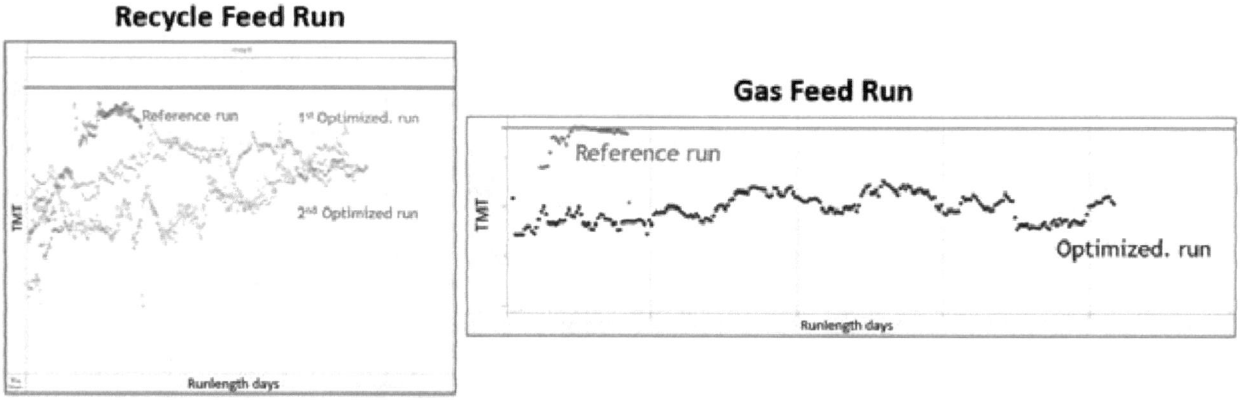

Figure 6: Left: Recycle Feed Run comparison of Reference run to two Optimized runs. Both Optimized runs showed over 2x the run length of the Reference run. Right: Gas Feed Run comparison of Reference run to Optimized run, Optimized runs showed over 3x the run length of the Reference run.

Higher Feed Rates (LPG)

Using in-furnace data, a furnace draft pressure limitation was identified that limited furnace firing and ultimately the total production of the furnace. Once the problem was identified, air could be redistributed across the furnace via damper adjustments to avoid the pressure limitation which resulted in an increase of total feed of 7% for LPG runs.

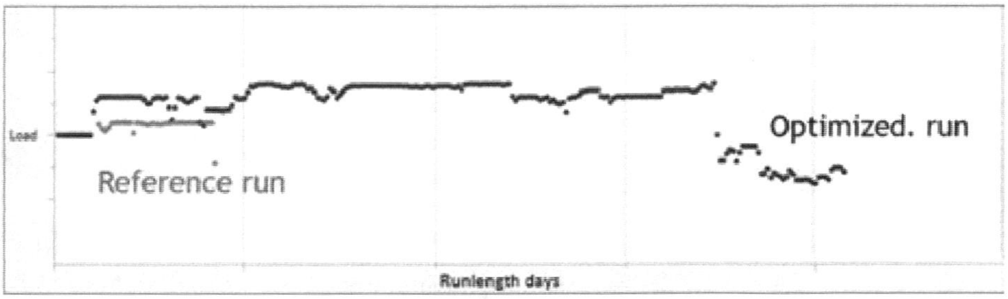

Figure 7: LPG run comparison of Reference run to Optimized run producing increased load. The Optimized run ran at consistently higher loads up until the last quarter of the run.

Reduced NOx Emissions

It is common to see NOx reduction via in-furnace combustion balancing activities by creating uniform O_2 profiles in the combustion zone and then reducing overall O_2 in the furnace. The work at BASF was no exception and NOx reduction was achieved and sustained for the duration of the Optimized runs.

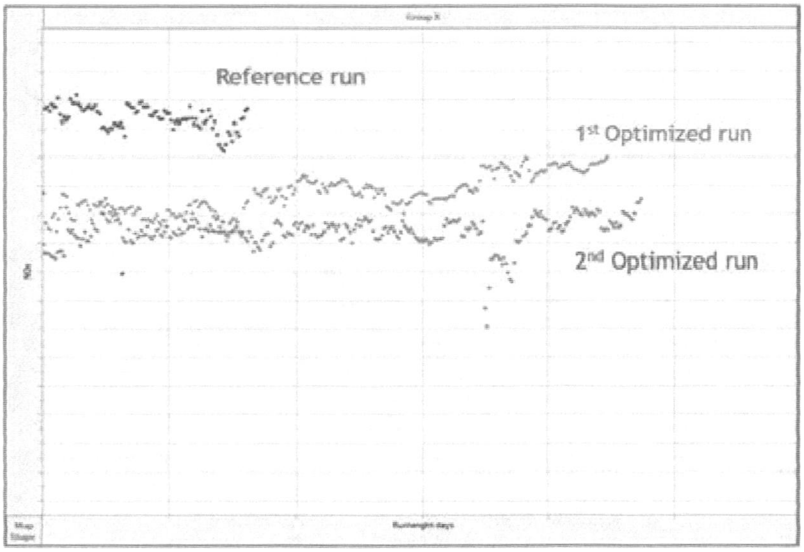

Figure 8: Recycle Feed-runs comparison of NOx Reference run to Optimized runs. Both Optimized runs ran at consistently lower NOx levels.

Conclusion

BASF successfully deployed a Multiplexed TDLAS Combustion Monitor to increase run lengths of an ethylene cracker. The ZoloSCAN system measured Temperature, O_2, CO and H_2O in the furnace and was used to achieve uniform products of combustion. Results of the integration included lower Tube Metal Temperatures, increased run lengths across multiple feedstocks, higher feed rates, and reduced NOx emissions.

Session 121: Ethylene Plant Technology Fundamentals
Paper 121d.

An update on the
Pilot and Field Trials to Separate Light Olefins and Paraffins

Dr. Hannah Murnen

Chief Technology Officer

Compact Membrane Systems

Dr. Sudip Majumdar

Membrane Expert

Compact Membrane Systems

William Charlton

VP Separations

Compact Membrane Systems

Prepared for Presentation at the 2019 Spring National Meeting

New Orleans, LA, April 2, 2019

Separation of Olefin-Paraffin Mixtures with Optiperm™ Customized Amorphous Fluoropolymer Membranes

Hannah Murnen

Chief Technology Officer

Compact Membrane Systems

Sudip Majumdar

Membrane Expert

Compact Membrane Systems

William Charlton

VP Separations

Compact Membrane Systems

Abstract: The separation of light olefins from paraffins is one of the most energy intensive in the world. The state of the art separation method is distillation, a technology that has remained unchanged over decades. Membranes hold the promise of a more modular, energy efficient and capital efficient method for separation. However, they have historically been hampered by low performance and stability.

Compact Membrane Systems has developed a novel membrane material using a silver embedded fluoropolymer material to transport olefins across the membrane. This material has demonstrated over 2 years of longevity in laboratory operation and has shown high selectivity and permeance under realistic operating conditions.

During the past year, CMS had operated a pilot utilizing this membrane material at the Delaware City Refining Company. We will share the results from the pilot plant operation as well as the update on the scale up of our membrane modules. The Delaware City pilot was operated on a low olefin stream (i.e., ~10-15% propylene concentration in the feed) with the goal of achieving HD5 quality propane and an enriched propylene stream. In addition, CMS has entered into a project with Dow Chemical through RAPID, a manufacturing institute funded by the Department of Energy. The goal of the work with Dow is to operate at the other end of the concentration spectrum where the feed is greater than 60% propylene and the target is an upgraded olefin-rich stream from the permeate. A pilot unit is installed in a Dow facility and we will share the latest results from our tests there.

CMS will also discuss initial results on laboratory testing of C_4 streams. This includes mixed gas separation of n-butane and 1-butene with permeances between 200-500 GPU and selectivities greater than 20. We will share results for a number of operating conditions and will highlight several exciting C_4 separation applications including those with isomer streams. In addition, longevity results for >20 days of C_4 operation will be shared demonstrating the stability of the membrane in realistic streams.

Background

Ethylene and propylene are major chemical industry raw materials and consume a great deal of energy in their production. The separations of these materials are some of the costliest, most energy intensive, and most technically challenging separations in the industry due to the very similar size and nature of the molecules being separated. The technology currently employed for the separation of ethylene and propylene from paraffins is distillation and is estimated to consume 250 trillion BTU/year of energy.

In addition, distillation columns to separate these molecules are very capital-intensive. A membrane based olefin-paraffin separation technology has the potential to provide substantial economic benefit to the petrochemical industry and to drastically reduce the energy required. In addition, membrane systems are scalable, allowing their implementation for smaller, stranded streams where the investment in distillation columns have not previously been installed due to limited economic value. Membrane processes utilizing facilitated transport membranes have been extensively studied for separating olefin/paraffin mixtures. While separations have been demonstrated in the laboratory, problems with membrane stability have prevented development of commercial systems.

Membrane Technology

Compact Membrane Systems (CMS) has developed Optiperm™ membrane, a customized amorphous fluoropolymer (CAF) containing silver ions that selectively transport olefin molecules from a mixture of olefin and paraffin. CMS Optiperm™ membranes are based on a silver-containing fluoropolymer material. The technology exploits a mass-transfer mechanism known as facilitated transport. Silver incorporated into the membrane polymer backbone as a binding site for the double bond in the olefins transporting them across the membrane as depicted in the cartoon below. Saturated paraffins do not have the same interaction. This difference in affinity allows for olefins to pass through the membrane at a significantly higher rate than do paraffins.

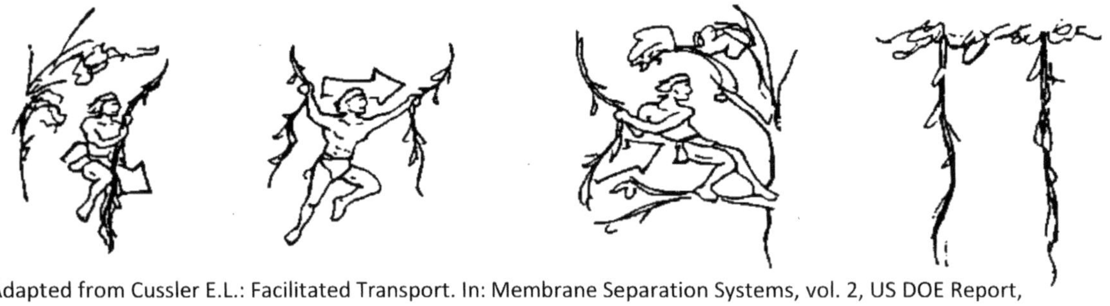

Adapted from Cussler E.L.: Facilitated Transport. In: Membrane Separation Systems, vol. 2, US DOE Report, DOE/ER/30133-H1 (1990)

Figure 1. *Illustration of Facilitated Transport mechanism – the vines represent the silver cations in the polymer backbone and Tarzan represents the olefin molecule*

The two key membrane properties that characterize the effects of facilitated transport are **permeance** and **selectivity**.

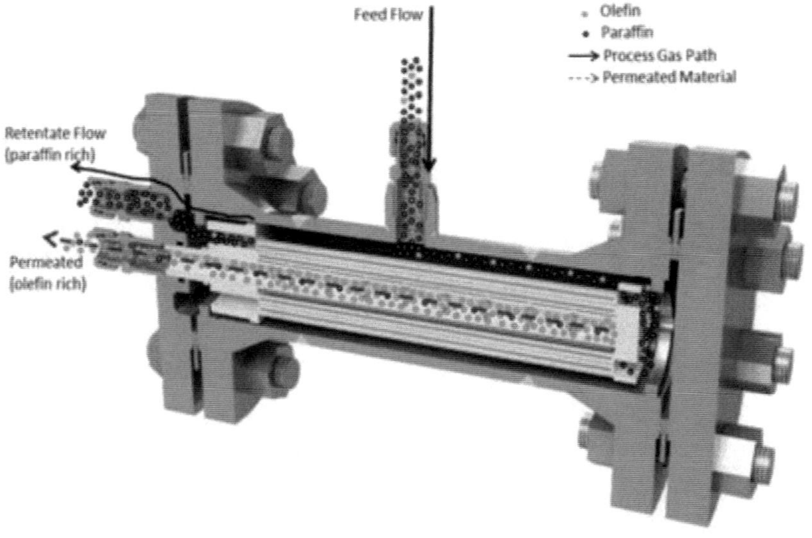

- **Permeance** is defined as the rate at which one component travels through a given surface area of membrane for a given driving force. This performance metric is typically expressed in GPU (gas permeation unit).
- The **selectivity** of one component to another is simply the ratio of the permeance of the two components. This tells you how much olefin is passing through the membrane as compared to paraffin.

Figure 2. *Illustration of Optiperm™ membrane module*

The Optiperm™ membrane has both high permeance and high selectivity, meaning high product purities can be reached with a reasonable membrane area. These characteristics yield very attractive process economics.

The Optiperm™ membrane can perform separations of olefins from paraffins over a wide range of operating temperatures, pressures, and compositions. CMS has demonstrated olefin removal at temperatures up to 70°C (158°F) and pressures up to 250 psig in the lab. For an efficient separation to occur, the olefin-paraffin mixture must be in the gaseous state. And optimal separation performance requires a highly humidified feed stream as water vapor improves the mobility of the olefin through the membrane. The membrane can remove olefin in both concentrated and dilute streams for compositions ranging 5-95% by volume of olefin. The membrane can process mixtures of C_2, C_3, and C_4 olefin and paraffin mixture.

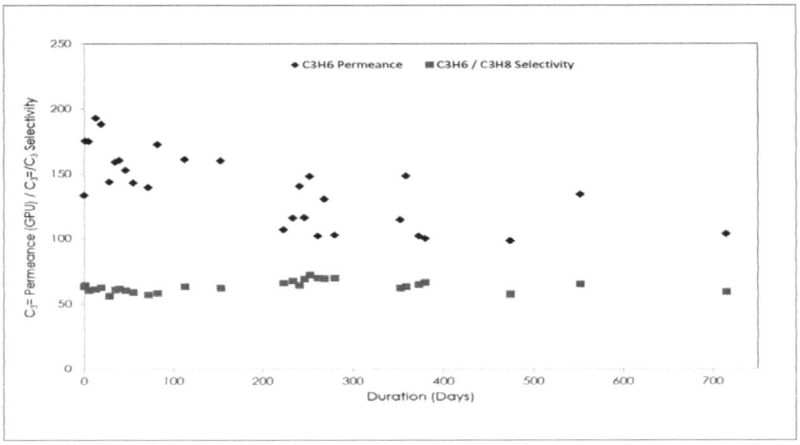

The chart displayed to the left shows a performance profile for a long term lab ageing study of an Optiperm membrane separating a 20% propylene /80% propane mixture at 60 psig and room temperature. As you can see, the membrane has shown stable performance over >650 days, a vast improvement over current state of the art materials.

Figure 3. *Optiperm™ membrane long term aging study*

Pilot Unit

The purpose of the Pilot Unit is to provide a platform for field demonstration of Compact Membrane Systems' (CMS) new Optiperm™ membrane technology for separation of olefins from paraffins in the gas state. The membranes utilize silver as a facilitating agent to improve the selective transport of olefins through the membrane relative to paraffins. These membranes provide a high olefin transport rate, which reduces the required membrane surface area, and high olefin/paraffin selectivity, which improves product split. Although various membranes for olefin/paraffin separation have been evaluated for many years, the CMS membranes are different in that they exhibit stable performance over a much longer time period than was ever achieved with earlier membranes. The improvement in membrane stability is made possible by incorporating the silver ion into backbone of the fluoropolymer which makes up the bulk of the membrane structure. As with most fluoropolymers, the fluoropolymer used in the CMS olefin/paraffin separation membrane is very resistant to chemical attack. It is this property that has been successfully utilized to create a membrane that provides stable separation performance in olefin/paraffin separations.

To date all of the development tests done with CMS Optiperm™ membranes have been carried out in the laboratory at small scales. The Pilot Test Unit enabled testing of membranes that are an order of magnitude larger in scale with actual refinery process streams. Moreover, membrane performance could be tested at operating conditions that are difficult to achieve in the laboratory. In particular, membrane performance in the separation of propylene/propane mixtures at feed pressures above 90 psig can be readily tested. Finally, and probably most importantly, through demonstration of the membrane technology in the field, the Test Unit will generate increased interest in the technology and promote its acceptance.

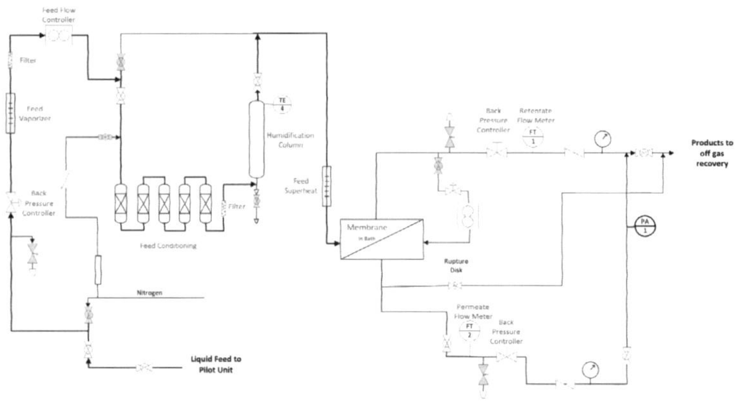

Figure 4. *Pilot Test Unit Process Flow Diagram*

Building upon the learnings of or membranes performance in the lab environment we began pilot field trial with our Optiperm™ membrane technology in late August 2017 utilizing a paraffin-rich mixed C_3 feed stream to gauge the technical performance and economic viability of a full scale system. The field pilot trials were conducted over 300 days of operations that spanned a period of approximately 15 calendar months at the Delaware City Refining Company facility.

Our primary test objectives:

- Demonstrate the ability of the Optiperm membrane to produce an HD5 quality propane (<0.05 propylene) from an 80% to 93% propane feed.
- Replicate and compare lab performance data developed on 47mm disks with field testing of spiral wound modules
- Demonstrate capability of operating at field operating conditions
- Demonstrate long term durability of facilitated transport membranes at process operating conditions

Our secondary test objectives:

- Map performance of prototype membranes over a range of operating conditions not achievable in our lab with propylene/propane mixtures
- Trial multiple membrane configurations to identify the optimal combinations of polymers, membrane thickness and materials of construction to enhance membrane performance

Prior to our field trials all of our membrane ageing and performance testing have been done in our lab facilities under controlled conditions with bottled gases and one of the things we did not know when we began the pilot trials was weather our membranes would be stable in a real process stream.

The major challenge in the operation of the Optiperm™ pilot unit at a refinery was to compare performance to continuous operation in a controlled laboratory environment. Testing of the membrane module after production is done in-house using a mixture of pure gases mixed using mass flow controllers to create the exact composition desired. Once mixed the gas then gets humidified to >90% relative humidity before entering to membrane housing. The feed, retentate, and permeate pressures are set using back pressure regulators. The flow rates are measured through calibrated bubble flow meters and stream compositions are measured through non-disruptive sampling by a gas chromatograph.

All the parameters are easily controlled and measured in a laboratory environment which is extremely difficult to replicate in the field due to numerous uncontrollable environmental conditions. With the unique behavior of facilitated transport membranes, small changes in feed composition and feed relative humidity can greatly swing calculated performance values. Process upsets from the unit and ambient temperature fluctuations can cause these parameters to vary without warning making it difficult to implement the appropriate responses.

Pilot Test Findings:

Optiperm™ membrane is stable in field operations. Presented on the next page in Figure 5, is field data from a membrane test conducted over a 50 day period. In this test the membrane was operating at a low stage cut with a 9% propylene feed to produce a permeate stream with a propylene concentration of 25%. Although there is some scatter in the data, the permeate stream composition is consistent over the 50 days of operation.

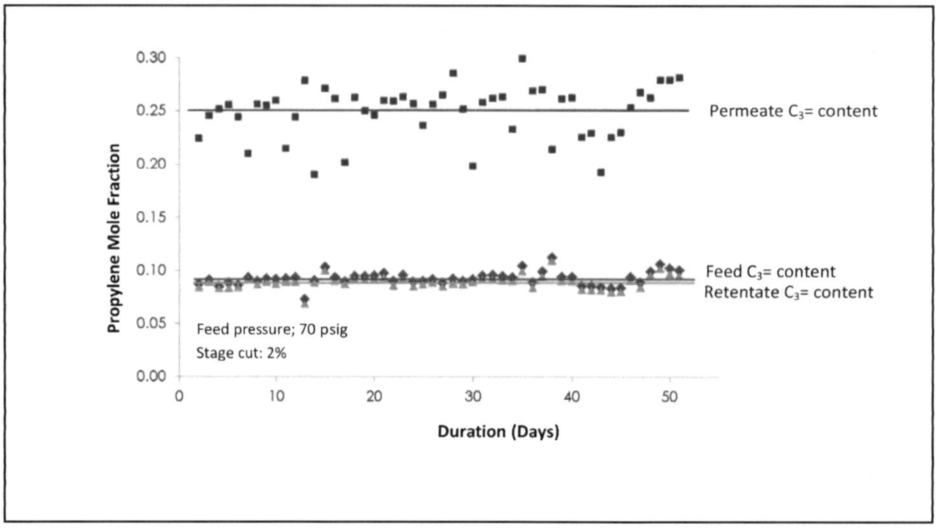

Figure 5. The results of one of our long term aging studies conducted over a 50 day period

As part of our standard procedures for deploying membranes for field testing we first conduct laboratory testing of the membranes to characterize the performance at the desired operating conditions to develop a baseline performance curve for the membrane. This performance curve is used both to optimize membrane field operating parameters and to also troubleshoot suboptimal field performance. The procedure also calls for returning the membrane to the lab at the midpoint of the run and then again at the end of run to complete performance testing under controlled conditions

The accompanying lab performance data for the 50 day membrane study shown previously in Figure 5 is presented here in Figure 6. While there is some variance in the permeance and selectivity measurements over the 50 day trial there is no significant change in the membrane performance

	Pre-field exposure	Mid-point exposure	50 day exposure
C_3= permeance (GPU)	128	174	154
C_3= / C_3 Selectivity	37	40	30

Figure 6. *Membrane lab performance validation during 50 day field trial*

Figure 7, presented on the following page, compares the observed field performance of membrane to its performance testing in the lab. Because the feed composition in the field is not constant, and our sampling was limited it is difficult to make a direct comparison to the lab composition data so we look at the membrane performance properties that are calculated for the data instead. The permeance and selectivity observed in the field are stable and agree well with the properties observed in the lab. In this set of tests, the feed composition averaged approximately 14% propylene, the retentate contained approximately 7% propylene and the permeate product was enriched to approximately 30% propylene.

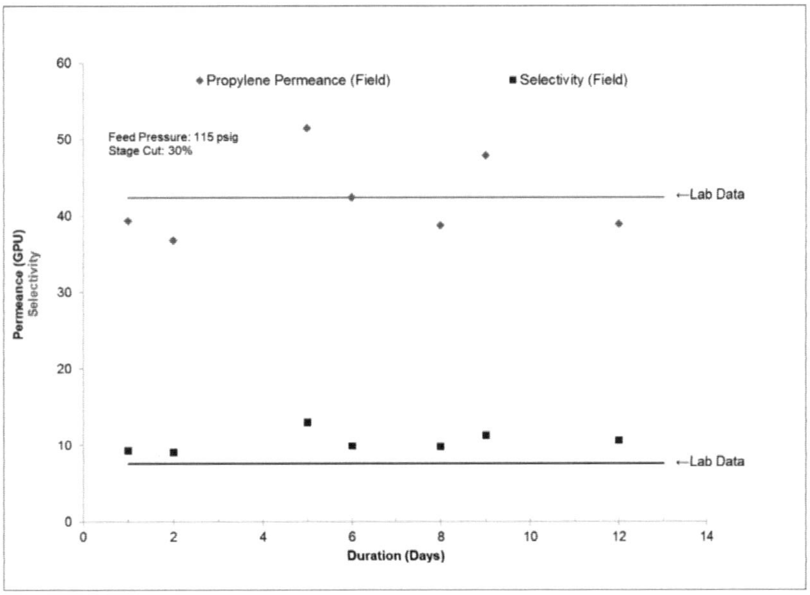

Figure 7. *Optiperm™ membrane is stable and compares well with laboratory data*

During the time interval plotted in Figure 8, the feed flow rate to the test membrane was adjusted to enable production of a retentate stream that meets HD5 propane specifications from a feed stream containing approximately 10% propylene. The membrane consistently produced HD5 over a period of 10 days, until the feed concentration increased above 15% and was out of range for the design of this pilot system. The propane recovery for this test was limited to 55% in this test which is lower than ideal and can be attributed to the inconsistent feed humidification levels we were able to achieve. The membrane performance is very sensitive to consistent humidification resulting in a lower than ideal recovery performance.

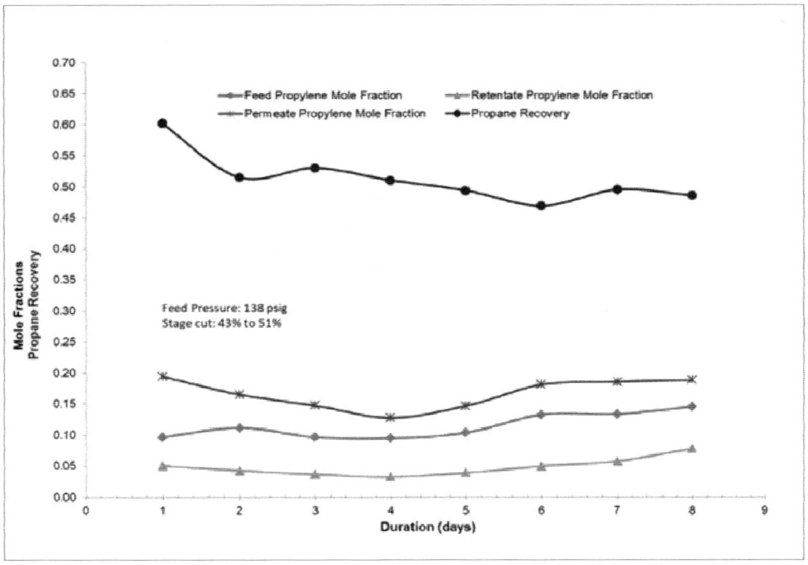

Figure 8. *HD5 propane production w/ Optiperm™ membrane*

As evidenced by the low recovery in the HD5 production test, one of the operational challenges for the pilot unit was maintaining constant feed humidity across a variety of ambient and operating conditions. During the course of the pilot testing, CMS developed a way to operate Optiperm™ membranes without humidifying the feed streams, potentially simplifying the operation of these systems. A novel method of humidification was developed that ensured uniform humidification throughout the entire module. CMS tested a simplified version of this new humidification method at DCRC and the membrane was able to produce HD5 propane from a feed containing 13% propylene with a much improved propane recovery; 82% vs. ~55% (Figure 9). We anticipate future installations incorporating this new method, simplifying operation and improving performance of the Optiperm™ membranes.

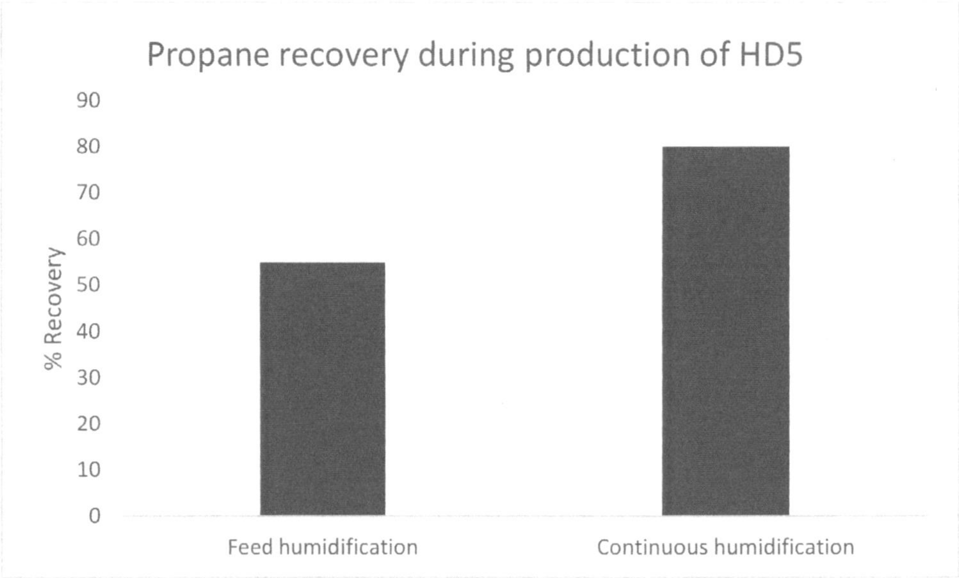

Figure 9. *Propane recovery during HD5 production*

Pilot tests key learnings:

- **Field membrane performance is stable**. Membrane pre-qualification and post-field performance tests confirm steady membrane performance parameters with permeance and selectivity matching the pre field testing values.
- **Membrane performance matches laboratory data** at similar conditions.
- We were able to achieve our primary goal of **producing an HD5 quality propane product**. A commercial scale system design will incorporate all of the design and control elements to continually produce on spec products including HD5 propane and propylene-rich streams.
- **Our spiral module membrane design and materials of construction performed as designed** with no mechanical failures noted in 300 days of field testing using multiple spiral wound membrane modules.

C_4 Separations

The separation and recovery of C_4 olefins is one of the most challenging separations in the petrochemicals industries as they require several integrated stepwise process to isolate the C4 isomers. These separation processes often include extractive distillation, fraction and absorption to effectively separate and concentrate the valuable components. With increasing customer inquiries around Butane separations, CMS has accelerated the study of Optiperm's™ effectiveness at separating C_4 olefins from paraffins and C_4 isomers. Initial olefin paraffin separation feasibility testing has demonstrated a stable membrane with 1-butene permeance of 430 GPU and a 1-butene/n-butane selectivity of 90 at feed conditions of 15 psig, 58 °C, and a 50/50 1-butene/n-butane mixture. Pressure and temperature dependencies were investigated to determine optimum operating conditions.

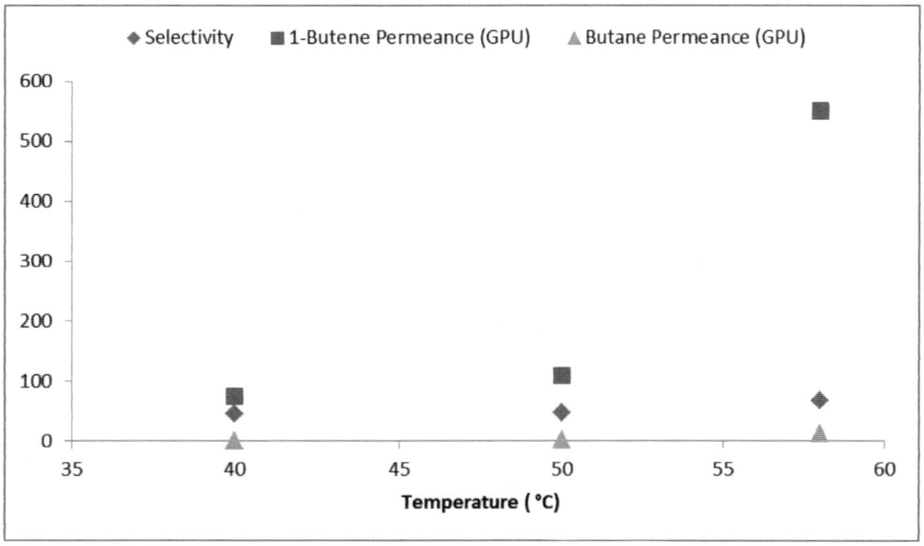

Figure 10. *Temperature influence on 1-Butene/Butane permeance and selectivity*

Figure 10. Illustrates the increased permeance that can be achieved with increased temperature. C_4 applications are high temperature, low pressure (<100psig) applications, which aligns with ideal operating conditions for Optiperm™. Testing continues to characterize performance in multicomponent mixtures containing n-butane, isobutane, 1-butene, 2-butene, butadiene, and/or isobutylene.

Economics:

Example Application – Polyisobutylene Reactor Purge

CMS's membrane technology gives polyisobutylene producers a direct method for recovering isobutylene conventionally lost as purged material while still preventing accumulation of inert isobutane. A simple process flow diagram illustrates the integration of an Optiperm™ membrane system into a polyisobutylene manufacturing process.

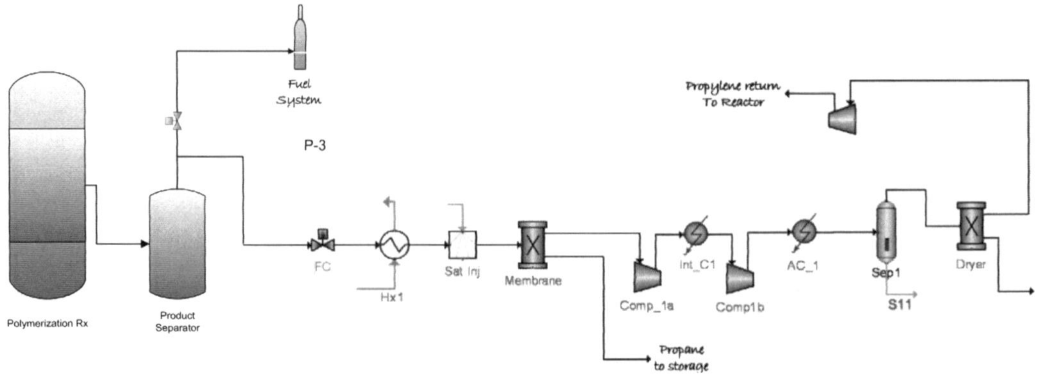

Figure 11. *Process Flow Diagram for polyisobutylene reactor purge recovery*

In this case study the Optiperm™ single-stage membrane process is retrofitted to the polymerization reactor purge stream of a polyisobutylene (PIB) plant. The purge stream is necessary to remove isobutane that builds up the reactor. The membrane process recovers isobutylene from the purge stream that otherwise would be lost from the process with the purged isobutane. The recovered isobutylene is then recompressed and returned to the reactor. Thus, the membrane process improves the efficiency of the polymerization process. The IRR for the retrofitted membrane process is very high, 346%, and the payback time is very short, less than 4 months. This is because a significant quantity of valuable feedstock is recovered from a stream that generally provides a low value to the PIB process.

Session 121e Ethylene Plant Technology Fundamentals

ON-LINE ANALYSIS OF NITROGEN CONTAINING COMPOUNDS IN STEAM CRACKING EFFLUENT

Marko R. Djokic
Dr. ir.
Laboratory for Chemical Technology, Ghent University, Ghent, Belgium

Nenad D. Ristic
ir.
Laboratory for Chemical Technology, Ghent University, Ghent, Belgium

Cato A.R. Pappijn
ir.
Laboratory for Chemical Technology, Ghent University, Ghent, Belgium

Guy B. Marin
Prof. Dr. ir.
Laboratory for Chemical Technology, Ghent University, Ghent, Belgium

Kevin M. Van Geem
Prof. Dr. ir.
Laboratory for Chemical Technology, Ghent University, Ghent, Belgium

Prepared for Presentation at the 2019 Spring National Meeting
New Orleans, Louisiana, March 31 – April 4, 2019

ON-LINE ANALYSIS OF NITROGEN CONTAINING COMPOUNDS IN STEAM CRACKING EFFLUENT

Marko R. Djokic
Dr. ir.
Laboratory for Chemical Technology, Ghent University, Ghent, Belgium

Nenad D. Ristic
ir.
Laboratory for Chemical Technology, Ghent University, Ghent, Belgium

Cato A.R. Pappijn
ir.
Laboratory for Chemical Technology, Ghent University, Ghent, Belgium

Guy B. Marin
Prof. Dr. ir.
Laboratory for Chemical Technology, Ghent University, Ghent, Belgium

Kevin M. Van Geem
Prof. Dr. ir.
Laboratory for Chemical Technology, Ghent University, Ghent, Belgium

Abstract: The petrochemical industry is strongly interested in finding new ways to economically apply heavy and alternative feedstocks from plastics in the steam cracking process. To accomplish this, it is of utmost importance to investigate more thoroughly how the steam cracking process reacts to the high amount of impurities present in these feedstocks. In this work the main focus is on the nitrogen containing compounds that are considerably present is feedstocks such as crude, shale oil or oil derived from plastic waste. These feeds contain an increased amount of nitrogen containing compounds compared to the conventionally used oil derived feeds. As nitrogen compounds have an influence on the operation of thermal processes occurring in coker units and steam crackers, and as some species are considered as environmentally hazardous or can lead to safety issues, a detailed analysis of the reactions involving nitrogen containing compounds under pyrolysis conditions provides valuable information. Therefore, a novel method for on-line analysis of nitrogen containing compounds has been developed and validated in a steam cracking pilot plant by feeding pyridine dissolved in heptane. The former being a representative compound for one of the most abundant classes of compounds present in crude and shale oil based feedstocks. The composition of the reactor effluent was determined via an in-house developed automated sampling system followed by immediate injection of the sample on a GC × GC coupled with a time-of-flight mass spectrometer (TOF-MS), flame ionization detector (FID) and nitrogen chemiluminescence detector (NCD). A novel method for quantitative analysis of nitrogen containing compounds using NCD and 2-chloropyridine as an internal standard has been developed and demonstrated. Consequently, the application of the NCD can be considered as a substantial improvement. The experimentally determined minimum detection level of 1.5 pgN/sec obtained using a reference mixture gives an indication that it is possible to trace even ppm and ppb concentrations of nitrogen containing compounds in the analyzed samples. The described work demonstrates the accuracy, repeatability and reproducibility of the methodology for on-line quantification of nitrogen containing products in an abundant hydrocarbon matrix. The high separation power of the GC × GC combined with a sensitive selective detector significantly goes beyond the present state of the art, resulting in a more detailed product detection and quantification. The described methodology is crucial for a proper identification of the formed nitrogen containing products and their subsequent tracing in the downstream ethylene unit equipment.

Introduction

The reserves of light sweet crude oils are gradually diminishing, and hence, alternative fossil resources are being considered to be used in the energy and petrochemical industry. Moreover, chemical recycling of plastic waste to liquid fuels and feedstocks for chemicals is being intensively considered by petrochemical industry. On the other hand, renewables such as bio-oils produced by fast pyrolysis of biomass are becoming a more attractive resources of bio-based fuels and chemicals. Heavy crude oil is a logical first choice because of the large proven reserves in Canada and Venezuela [1-3]. The latter are being recognized as the largest crude oil reserves in the world and their composition is similar to the composition of natural bitumen. Heavy crude oils differ from light crude oils by their high viscosity at reservoir temperatures, high density (low API gravity), and significant contents of nitrogen, oxygen, and sulfur containing compounds [4,5]. Another promising alternative is shale oil, derived from oil shale. Oil shale is a fine-grained sedimentary rock containing kerogen, a mixture of organic chemical compounds with a molar mass as high as 1000 Da [6]. Characterization methods have shown that the concentration of heteroatoms (S, O and N) in shale oil, plastic waste pyrolysis liquids and heavy crude oils is typically substantially higher than the specifications set for the products used in for example the petrochemical industry [6]. It is well documented that nitrogen containing compounds present in heavy conventional crude oil, plastic waste oils and shale oil have a negative effect on the catalyst activity in hydrocracking, catalytic cracking and reforming processes [7]. Similarly, it has been reported that the presence of nitrogen containing compounds are a safety concern because they promote gum formation in the cold-box of a steam cracker [8].

These processing and safety challenges are a strong driver to improve the current methods for offline and on-line characterization of nitrogen containing compounds in complex hydrocarbon matrices. Two-dimensional gas chromatography (GC × GC) coupled with a nitrogen chemiluminescence detector (NCD) is a superior characterization technique compared to one-dimensional gas chromatography (GC) for analyzing conventional diesels or liquefied coal samples [7]. Recently a method has been developed and applied to the offline characterization of nitrogen content in shale oil [6], the identification of extracted nitrogen compounds present in middle distillates [9], and the determination of the detailed composition of plastic waste pyrolysis oil [10]. It is thus clear that GC × GC analysis is a powerful offline processing technique for analyzing complex mixtures [11-17]. However, on-line application is more challenging due to the need for a reliable and non-discriminating sampling methodology. One of the first developed methodologies for comprehensive on-line characterization was demonstrated by analyzing steam cracking reactor effluents using a TOF-MS and a FID [18]. Information concerning kinetics of conversion processes of nitrogen containing compounds is scarce [19], partly due to the lack of an adequate method to identify and quantify nitrogen containing compounds in the reactor effluent. Establishing the methodology for offline and on-line analyses is thus a prerequisite before one can even attempt feedstock reconstruction [20-27] and kinetic modeling. One of the fields which would benefit from the accurate identification and quantification of nitrogen containing compounds is steam cracking or pyrolysis.

In hydrocarbon mixtures nitrogen is mainly present in aromatic structures, e.g. as pyridine or pyrrole; hence most experimental efforts have been dedicated to the decomposition of these structures. Hydrogen cyanide and ethyne were reported as major products for the thermal decomposition of pyridine studied in a temperature range of 1148-1323 K. Other products such as aromatics and nonvolatile tars were also detected in minor quantities [29]. The thermal decomposition of pyrrole was investigated in a broader temperature range of 1050-1450 K using shock wave experiments. The main products were 3-butenenitrile, cis and trans 2—butenenitrile, hydrogen cyanide, acetonitrile, 2-propenenitrile, propanenitrile, and propiolonitrile [30]. Additionally thermal decomposition shock tube experiments were performed for pyridine at elevated temperatures resulting in comparable product spectra [31,32]. Product yields in these studies have been determined by applying GC's equipped with a FID, a nitrogen–phosphorus detector (NPD) [31], a mass spectrometer (MS) [32] and a Fourier transform infrared (FTIR) spectrometer [32]. Simultaneous pyrolysis of toluene and pyridine showed a difference in soot formation tendency compared to pyridine pyrolysis, indicating the complex nature of the free-radical reactions [31,33].

In this work, first the offline comprehensive characterization results of the complex shale oil are reported. Next, the limitations of using an on-line GC × GC-TOF-MS/FID for the analysis of nitrogen compounds in a complex hydrocarbon matrix are discussed. Finally, the newly developed methodology for the on-line quantification of nitrogen containing compounds by GC × GC- NCD is demonstrated. The qualitative analysis of products was carried out using TOF-MS, while FID and NCD were used for quantification. The application of the NCD is a substantial improvement compared to using the FID because of its higher selectivity, lower detection limit and equimolar response.

Materials and Methods

Chemicals and Standards

All the analytical standards used in this work were procured with the minimum purity of 98%. Analytical gases were provided from Air Liquid with a minimum purity of 99.999%. For producing steam, deionized water was used.

Offline GC × GC-NCD analysis

GC × GC setup is equipped with a dual-stage cryogenic liquid CO_2 modulator [34]. The first dimension non-polar column (PONA, 50 m L × 0.25 mm I.D. × 0.5 µm df) is directly connected to the second dimension mid-polar column (BPX-50, 2 m L × 0.25 mm I.D. × 0.25 µm df). Applied GC × GC operating conditions optimized for detection of nitrogen containing compounds in shale oil are presented in Table 1. Response signal of NCD is equimolar [6,35,36,37], hence the procedure for quantification of nitrogen containing compounds in the sample is straightforward. Concentration of detected compounds can be calculated using the following equation:

$$W_i = \frac{V_i}{V_{st}} \cdot W_{st} \cdot \frac{M_i}{N_i \cdot M_{st}} \qquad \textit{Eq. 1}$$

where *Vi* and *Vst* are the peak areas of the nitrogen containing compound *i* and the internal standard, respectively. *Wst* is the concentration of the internal standard compound, *Ni* is the number of nitrogen atoms in compound *i*, *Mi* and *Mst* are the molar masses of the nitrogen containing compounds and of the nitrogen internal standard respectively [6].

Table 1. Overview of the GC × GC conditions applied to offline shale oil characterization.

Detector	**NCD**
Injector	573 K
Split flow	200 ml/min
Carrier gas	2.1 ml/min
Initial oven temperature	313 K
Heating rate	3 K/min
Final temperature	573 K
Modulation time	7 sec
Detector settings	
Temperature	1198 K
Range	1
Acquisition rate	100 Hz

On-line analysis in the pilot plant steam cracking unit

The pilot plant steam cracker consists of three sections: the feed section, the furnace/reactor section and the analysis section. The unit is explained in details elsewhere [18,38-40]. During the experiments, heptane was fed using a computer controlled rotary pump by setting the flowrate using the Programmable Logical Controller (PLC), while pyridine was dosed to the reactor convection section by manually setting the speed of the piston pump. The diluent used depends on the purpose of the test and can be varied. Depending on the requirements of the experiment, gas controllers are used for helium and nitrogen dilution, while water (steam) is feed using a computer controlled rotary pump. Prior to the reaction section of the furnace, the feedstock and the diluent are heated up and mixed in two separately heated zones kept at a temperature of 773 K. Reactor coil is made of Incoloy800HT, it is 12.4 m long and has an internal diameter of 9 mm. The process conditions applied in the experimental campaign are outlined in Table 2.

Table 2. Applied experimental conditions.

Detector	Pyrolysis	Steam Cracking	Method Test	Steam Cracking
Heptane flowrate, g/hr	2480	2880	98.89	4000
Water flowrate, g/hr	/	1440	2000	2000
Nitrogen flowrate, g/hr	/	/	4000	/
Helium flowrate, g/hr	507	/	/	/
Pyridine flowrate, g/hr	25.1	29.1	5.21	4.33
2-Chloropyridine flowrate, g/hr	/	/	4.21	0.85
Hexane flowrate, g/hr	/	/	85.91	83.63
Coil outlet temperature (COT), K	1073	1073	823	1123
Coil outlet pressure (COP), bar	1.7	1.7	1.7	1.7
Detector	TOF-MS	FID	NCD	NCD

Analytical methodology

The reader is referred to the following papers for the detailed explanation of the pilot plant analytical methodology [18, 38-40]. Here only the specifics related to this work are described. The refinery gas analyzer (RGA) equipped with three detectors (two TCD's and a FID) working on parallel channels is used to detect compounds such as CO, CO2, H2, and hydrocarbons up to four carbon atoms. For the GC × GC set-ups, a non-polar PONA (50 m L × 0.25 mm I.D. × 0.5 μm df) column is used as the first dimension and a mid-polar BPX-50 (2 m L × 0.15 mm I.D. × 0.15 μm df) column is utilized as the second dimension column. The GC × GC settings for the on-line analyses are given in Table 3. Electron impact ionization of TOF-MS detector is 70 eV, and detector voltage is 1700 V. The acquisition frequency of TOF-MS analysis is 30 spectra/s in a mass range of 15–400 am. For the NCD, flow rates for the H2 and O2 are set at 5 and 11 mL min-1 respectively, while keeping the burner temperature at 1198 K. The data acquisition rate is 100 Hz. For FID H2, air and N2 (make-up gas) flow rates are respectively 35, 350 and 35 mL min-1. The FID temperature is set at 573 K with a data acquisition rate of 100 Hz.

Table 3. Pilot plant on-line analysis section – GC × GC settings.

Detector	FID	TOF-MS	NCD
Injector	573 K	573 K	573 K
Split flow	30 ml/min	30 ml/min	10 ml/min
Carrier gas	2.1 ml/min	2.6 ml/min	2.1 ml/min

Initial oven temperature	233 K	233 K	233 K
Heating rate ramp I	5 K/min	5 K/min	5 K/min
Temperature hold, min	4	4	4
Temperature ramp	313 K	313 K	313 K
Heating rate ramp II	4 K/min	4 K/min	4 K/min
Final temperature	573 K	573 K	573 K
Modulation time	5 sec	5 sec	5 sec
Detector			
Temperature	573 K	473 K	1198 K
Range	10	n.a.	1
Acquisition rate	100 Hz	30 Hz	100 Hz

Internal standard addition

The pilot plant effluent is sampled on-line, i.e. during pilot plant operation, and at high temperature (673 – 773 K), using a valve-based sampling system, see Figure 1. 2-chloropyridine is used as an internal standard for the NCD analyses. The introduced amount of this compound depends on the purpose of the experiment (values illustrated in Table 2). To add 2-chloropyridine precisely and uniformly, it was dissolved in n-hexane (values illustrated in Table 2) and as such pumped using a peristaltic pump through a capillary transfer line, which is centrally positioned in the product stream before the sampling system.

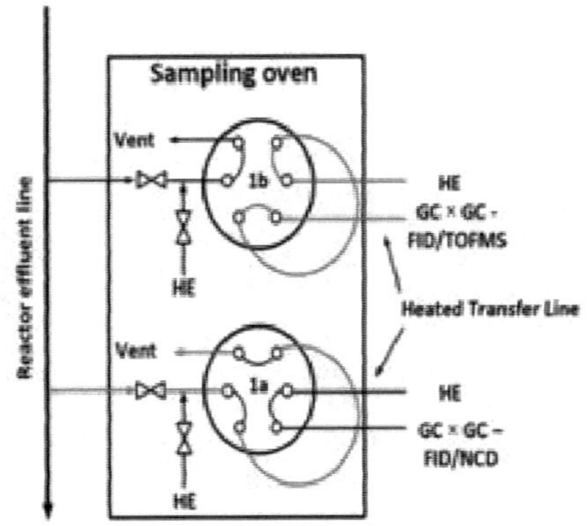

Figure 1. Detailed schematic representation of the GC × GC sampling oven and valves. Valve 1a is shown in the purging position, flushing the sample loop with effluent. Valve 1b is shown in injection position: carrier gas (helium) is rerouted to the sampling oven, where it is used to transport the effluent sample to the respective GC via a transfer line [18].

Data acquisition and quantification

The data obtained with TOF-MS are used for product identification and establishing the retention times of the reaction products. Acquisition and reprocessing of data obtained from FID is similar to the procedure applied to reprocessing of results gathered using the NCD, as explained in the section above. The product yields are calculated based on dosed internal standard [18]. A fixed amount of nitrogen (N2) is continuously introduced to the product stream and used as a reference compound as shown in Figure 2. The data obtained on the different instruments are combined according to the scheme. The calibration factors for each compound are applied depending on the detector where the methane calibration factor is set at 1, thus response factors relative to methane are used. Response factors relative for permanent gases detectable on RGA (C4- compounds) are determined by calibration. For C5+ compounds the FID calibration factors are taken from the work of Dietz [41]. If the response factors for a compound are not available, the group contribution method developed by Dierickx et al. [42] is used to calculated the same. Methane functions as a secondary internal standard, and its concentration is needed to determine the concentrations based on the peaks of the GC × GC-FID chromatogram. Methane mass flow rate is determined using equation 2:

$$\dot{m}_{CH_4} = \frac{f_{CH_4}}{f_{N_2}} \cdot \frac{A_{CH_4}}{A_{N_2}} \cdot \dot{m}_{N_2} \qquad Eq.\ 2$$

Where, *fCH4* and *fN2* are the calibration factors of methane and nitrogen on TCD respectively. *ACH4* and *AN2* are peak areas obtained during analysis, while $\dot{m}(N_2)$ is the nitrogen mass flow rate added to the effluent stream. Quantification of the nitrogen containing products is performed by applying the procedure similar to the one used of off-line GCxGC-NCD analysis (see equation 1).

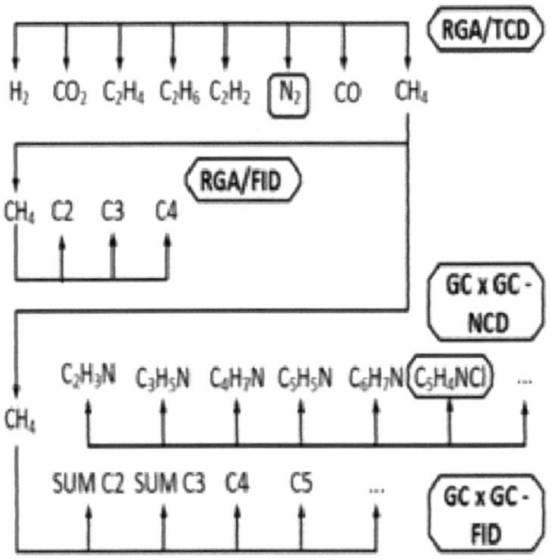

Figure 2. Schematic representation of the quantitative on-line effluent analysis method using a reference compound. The known amount of nitrogen added to the effluent stream is detected on the RGA/TCD and used to determine the concentration of methane which is a reference compound for reactor effluent analysis. Similarly a known amount of 2-chloropyridine is added to the effluent stream and used as internal standard for GC × GC – NCD analysis.

Results and Discussion

Offline GC × GC-NCD analysis

The chromatogram obtained using the offline GC × GC – NCD for characterization of nitrogen containing compounds in a shale oil sample is given in Figure 3. Since the NCD cannot be used for identification, the retention times of the observed species need to be established in advance by carrying out analyses on the GC × GC coupled to TOF-MS, based on the detailed procedure of carrier gas flowrate optimization for each detection method [18,43]. Flow rate adaptation leads to similar retention times of compounds on chromatograms obtained using different detectors [18]. In addition to the established TOF-MS retention times, literature data and Kovats retention indices were used for a tentative identification of compounds. To enable precise quantification of individual compounds, the internal standard should be chosen in such a way that peak overlap with peaks from other compounds in the sample is avoided. The bidimensional resolution should be higher than 1.515.

Repeat measurements show that the NCD has ±3% uncertainty for concentrations higher than 100 ppmw. The uncertainty increases to±10% as the concentration of the nitrogen containing compounds decreases to only a few ppmw [6]. The following classes were identified: pyridines, anilines, quinolines, indoles, acridines, and carbazoles. Moreover, detailed quantification of the individual compounds was possible. The gathered data was used to determine the individual compound concentrations, and the obtained values are presented in Table 5. The analyzed sample contains 4.21 wt.% of nitrogen containing compounds mainly belonging to pyridine class. From a processing point of view this high nitrogen content is a concern when shale oil is considered to replace traditional steam cracking feedstocks where nitrogen containing compounds are typically only present in ppm levels.

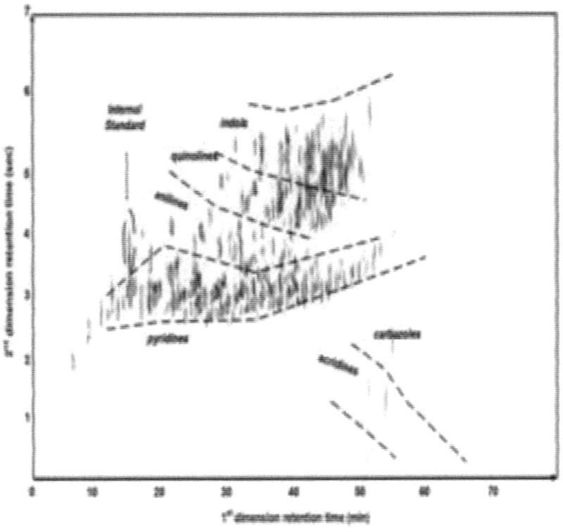

Figure 3. GC × GC – NCD chromatogram of the shale oil sample. The internal standard and separated nitrogen containing hydrocarbon group types, pyridines, anilines, quinolines, indoles, acridines, and carbazoles are illustrated.

Table 5. Concentration of nitrogen containing compounds in the shale oil.

Carbon number	pyridines, wt.%	anilines, wt.%	quinolines, wt.%	indoles, wt.%	acridines, wt.%	carbazoles, wt.%
5	0.01	0.00	0.00	0.00	0.00	0.00
6	0.04	0.00	0.00	0.00	0.00	0.00
7	0.11	0.02	0.00	0.00	0.00	0.00
8	0.26	0.05	0.00	0.01	0.00	0.00
9	0.47	0.07	0.01	0.06	0.00	0.00
10	0.15	0.11	0.08	0.17	0.00	0.00
11	0.18	0.11	0.17	0.28	0.00	0.00
12	0.12	0.08	0.19	0.30	0.00	0.00
13	0.18	0.03	0.12	0.16	0.00	0.02
14	0.17	0.00	0.03	0.14	0.01	0.02
15	0.13	0.00	0.00	0.00	0.02	0.00
16	0.10	0.00	0.00	0.00	0.00	0.00
17	0.05	0.00	0.00	0.00	0.00	0.00
18	0.03	0.00	0.00	0.00	0.00	0.00
19	0.00	0.00	0.00	0.00	0.00	0.00
20	0.00	0.00	0.00	0.00	0.00	0.00
21	0.00	0.00	0.00	0.00	0.00	0.00
22	0.00	0.00	0.00	0.00	0.00	0.00
23	0.00	0.00	0.00	0.00	0.00	0.00
24	0.00	0.00	0.00	0.00	0.00	0.00
25	0.00	0.00	0.00	0.00	0.00	0.00
26	0.00	0.00	0.00	0.00	0.00	0.00
27	0.00	0.00	0.00	0.00	0.00	0.00
28	0.00	0.00	0.00	0.00	0.00	0.00
29	0.00	0.00	0.00	0.00	0.00	0.00
30	0.00	0.00	0.00	0.00	0.00	0.00
31	0.00	0.00	0.00	0.00	0.00	0.00
32	0.00	0.00	0.00	0.00	0.00	0.00
33	0.00	0.00	0.00	0.00	0.00	0.00
Total, wt.%	1.98	0.46	0.59	1.10	0.04	0.04

Online analysis of pyridine-heptane steam cracking products

In a second step the focus was switched to the on-line detection of nitrogen compounds in an abundant hydrocarbon matrix. Samples of the reactor effluent were taken at an elevated temperature preventing condensation of molecules with a molecular weight higher than 128 Da [18]. The combination of gas chromatography columns enables a separation of products based on differences in the compound volatility in the first dimension and based on differences in the specific polarity of compounds in the second dimension. Hence, a truly comprehensive system is established. Nevertheless, the most volatile compounds present in the reactor effluent cannot be trapped and refocused by the cryogenic CO2 modulation. First the pyrolysis products of heptane with 1.0 wt.% of pyridine were analyzed using GC × GC–TOF-MS, enabling identification of the obtained products. Calibration of the MS detectors is shown to be strongly dependent on the tuning parameters [44], hence, using this type of detector for quantification is laborious and can give substantial uncertainty. Helium was chosen as the diluent for the tests, minimizing the background noise on the detector during analysis. In the complex reactor effluent nitrogen atoms are incorporated in several aromatic structures. Separation of these nitrogen containing compounds was possible in the second dimension column since their polarity is higher in comparison to the other obtained products, as shown in Figure 4a. Furthermore, the developed experimental procedure was applied to product quantification using the FID, see Figure 4b. The presented experimental work illustrates that detailed product characterization is possible using this method. Moreover, the same procedure will be applied in follow-up studies of pyrolysis reactions of nitrogen containing compounds.

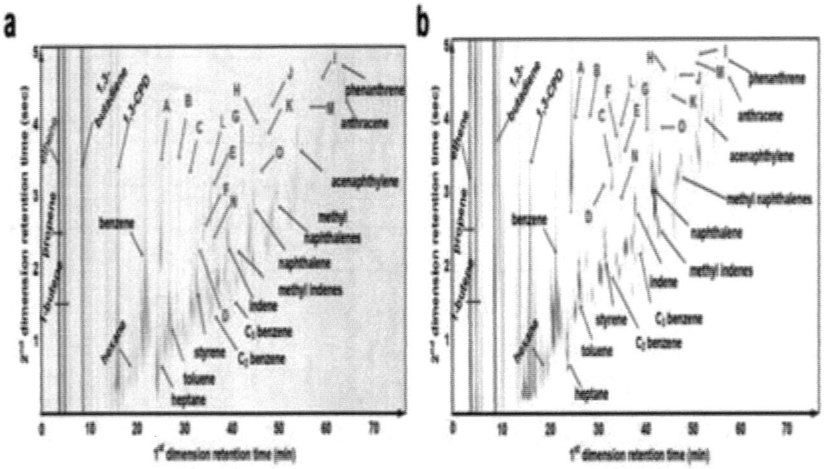

Figure 4. Analysis of the pyridine-heptane mixture pyrolysis products (a) GC × GC – TOF-MS chromatogram (b) GC × GC – FID chromatogram. The nitrogen containing products of the pyrolysis experiment performed at a COT of 1073 K and a COP of 170 kPa are presented with capital letters (A: pyridine, B: 2-methylpyridine, C: 3-methylpyridine, D: 4-ethylpyridine, E: 3-ethenylpyridine, F: 4-ethenylpyridine, G: 2-methylbenzonitrile, H: quinoline, K: isoquinoline, I: 1-H-indole-7-methyl, J: indole, L: benzonitrile, M: 4-methylquinoline, N: 5-ethenyl-2-methylpyridine, O: 7-methylindolizine).

On-line analysis of the reactor effluent during pyrolysis of a pyridine-heptane mixture at a coil outlet temperature (COT) of 1073 K and a coil outlet pressure (COP) of 170 kPa, performed with GC × GC–TOF-MS (see Figure 4a), was used for identifying the reaction products and establishing the compound retention times for a specific set of GC × GC operating conditions. GC × GC – FID (see Figure 4b) analysis was used for determining the reactor effluent composition while using steam as a diluent. The obtained product concentrations, normalized to 100%, are given in Table 5. The identified products in these chromatograms show that the addition reactions of pyridine are favorable compared to pyridine decomposition. Hauser and Lifshitz [29,30] reported the formation of light nitriles in pyrolysis experiments of pyridine and pyrrole. Since these molecules were not detected in the present set of experiments and the nitrogen molar balances in the experiments closed, it can be concluded that pyridine is not decomposing to a great extent at the selected process conditions.

Table 6. Concentrations of the compounds in the reactor effluent during cracking of heptane with 1.0 wt.% of pyridine (steam dilution 0.5 kg/kg, COT 1073 K, COP 170 kPa).

Compound name	Concentration, wt.%	St.Dev.
methane	14.21	0.75
ethene	40.46	2.06
ethane	5.50	0.10
ethyne	0.25	0.02
propane	0.50	0.04
propene	12.70	0.78
prpopyne	0.17	0.01
propadiene	0.06	0.01
1.3-butadiene	3.07	0.67
butane	0.08	0.03
trans-butane	0.23	0.07
1-butane	1.17	0.44
iso-butane	0.10	0.02
cis-butane	0.17	0.05
hydrogen	1.12	0.21
carbon monoxide	2.22	0.20
carbon dioxide	2.38	0.64
3-methylbutane	0.02	0.02
2-methyl-butane	0.02	0.01
1-pentene	0.09	0.01
2-methyl-1,3-butadiene	0.55	0.04
3-ethylcyclopropane	0.02	0.00
3-methyl-1-butene	0.08	0.01
2-pentene	0.05	0.01
1,3-pentadiene, (E)-	0.32	0.01
1,3-cyclopentadiene	1.54	0.01
1,3-pentadiene, (Z)-	0.18	0.01
pyridine	0.79	0.06
1,3-cyclohexadiene	0.45	0.03

iso-paraffins C6	0.07	0.01
hexane	0.04	0.01
cycloparaffins C6	0.02	0.01
fulvene	0.06	0.00
5-methyl-1,3-cyclopentadiene	0.24	0.02
1-methyl-1,3-cyclopentadiene	0.22	0.02
benzene	3.60	0.63
2-methylpyridine	0.01	0.01
methylcyclohexane	0.01	0.00
cycloalkene C7	0.02	0.01
cycloalkyne C7	0.02	0.03
heptane	2.46	0.63
cycloparaffin C7	0.30	0.03
toluene	0.88	0.16
2-ethylpyridine	0.01	0.00
2-ethenylpyridine	0.02	0.01
Benzonitrile	0.02	0.00
4-ethylpyridine	0.00	0.01
4-ethenylpyridine	0.01	0.01
3-ethenylpyridine	0.05	0.01
cycloparaffin C8	0.03	0.03
ethylbenzene	0.09	0.00
xylene (m, p)	0.06	0.01
monoaromatics C8	0.04	0.01
styrene	0.38	0.06
o-xylene	0.03	0.00
5-ethenyl-2-methylpyridine	0.03	0.02
2-methylbenzonitrile	0.02	0.02
3-methylbenzonitrile	0.02	0.03
7-methylindolizine	0.01	0.00
indole	0.01	0.00
monoaromatic C9	0.29	0.09
indene	0.16	0.02
quinoline	0.01	0.00
isoquinoline	0.01	0.01
methyl-1H-indole	0.02	0.00
3-methyl-1H-indene	0.52	0.60
monoaromatic C10	0.11	0.10
1-methyl-1H-Indene	0.04	0.00
naphthalene	0.41	0.39

4-methylquinoline	0.01	0.00
monoaromatics C11	0.19	0.05
1-methyl-naphthalene	0.10	0.11
2-methyl-naphthalene	0.01	0.02
monoaromatics C12	0.13	0.10
diaromaticss C12	0.17	0.04
acenaphthylene	0.03	0.01
1,2-dimethyl-naphthalene	0.01	0.00
diaromatics C13	0.27	0.15
naphthenodiaromatics C13	0.08	0.03
naphthenodiaromatics C14	0.08	0.12
phenanthrene	0.05	0.08
anthracene	0.01	0.02
triaromatics C15	0.02	0.03

Proof-of-test method for the on-line GC × GC–NCD method was performed in a separate experiment, at conditions that preclude pyridine decomposition, i.e. a temperature of 823 K and a COP of 170 kPa. A pyridine concentration of 841.4 ppmw was added to the nitrogen and water flow, and after addition of the internal standard, the reactor effluent sample was injected on the GC × GC. Using the obtained detector response and known concentration of the internal standard, a concentration of 819 ppmw pyridine in the effluent was measured. Hence, the relative error of the measurement was determined to be less than 3% (see Figure 5).

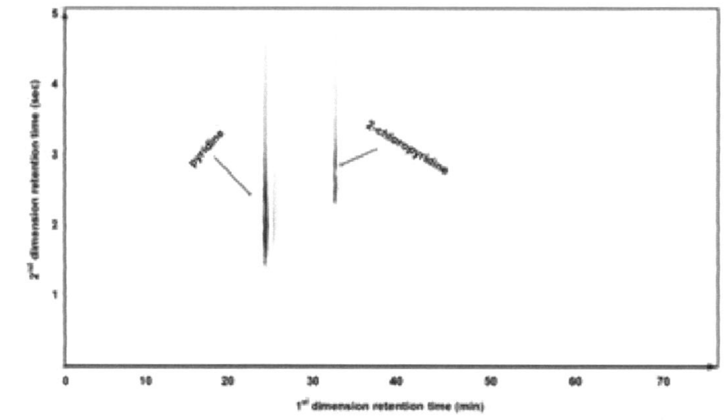

Figure 5. GC × GC – NCD on-line detection of pyridine in the reactor effluent stream under non-reactive conditions. The experiment is performed isothermally at 773 K and a COP of 170 kPa and used for the evaluation of the internal standard quantification method.

Finally a heptane steam cracking experiment under more severe conditions was conducted with a small amount of pyridine added to the n-heptane feed. The experiment was performed under typical steam cracking conditions, with a steam dilution of 0.5 kg/kg, a COT of 1123 K and a COP of 170 kPa. Figure 6 shows the resulting GC × GC–NCD and FID chromatograms. The compounds were identified based on retention times and data obtained from the TOF-MS. The following compounds were detected on the GC × GC–NCD chromatogram: acetonitrile, pyridine, 2-methylpyridine, 3-methylpyridine, 3-ethylpyridine, 3-ethenylpyridine, 3-methylbenzonitrile, and indole. Using their respective Kovats retention indices 2—butenenitrile and propanonitrile could be tentatively identified. The quantitative results are shown in Table 7. The mass flow rate of pyridine to the reactor was set to 1.2 mg/s, i.e. concentration of elemental nitrogen in the reaction mixture was 125.9 ppmw. After reprocessing of the acquired data, the nitrogen concentration in the reaction effluent was determined to be 124.5 ppmw, which corresponds to a nitrogen recovery of 98.5%. Considering the boiling point range and solubility in water of the nitrogen containing compounds present in the cracking effluent, one can expect that 3-methylbenzonitrile and indole are ending up in pyrolysis fuel oil fraction (BP >200 °C), while the remaining (acetonitrile, pyridine, 2-methylpyridine, 3-methylpyridine, 3-ethylpyridine, 3-ethenylpyridine) will be distributed between the pyrolysis gasoline and condensed water.

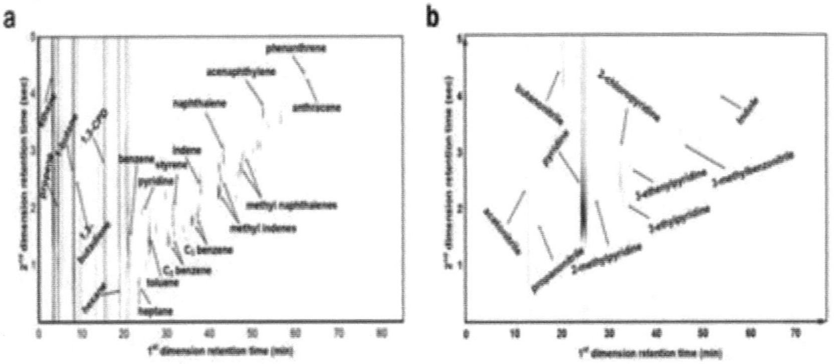

Figure 6. Product analysis during steam cracking of heptane with traces of pyridine (a) GC × GC – FID chromatogram (b) GC × GC–NCD chromatogram. Detection of major steam cracking products with FID and minor nitrogen containing steam cracking products with NCD. The experiment is performed with steam dilution of 0.5 kg/kg, a COT of 1123 K, and a COP of 170 kPa.

Table 7. Concentrations of the compound in the reactor effluent during steam cracking of heptane with traces of pyridine (steam dilution 0.5 kg/kg, COT 1123 K, COP 170 kPa).

Compound name	Concentration, wt. %	Std.Dev.
hydrogen	1.04	0.01
carbon dioxide	0.18	0.02
ethene	44.33	0.16

ethane	5.57	0.11
ethyne	0.27	0.03
methane	14.30	0.17
carbon monoxide	1.43	0.13
propane	0.53	0.01
propene	14.89	0.22
propadiene	0.18	0.02
iso-butane	0.01	0.00
normal-butane	0.10	0.00
trans-butene	0.36	0.02
1-butene	1.38	0.08
iso-butene	0.12	0.01
cis-2-butene	0.24	0.02
propyne	0.21	0.02
1,3-butadiene	5.16	0.06
1-pentene	0.21	0.02
cyclopentene	0.20	0.01
trans-1,3-pentadiene	0.26	0.01
2-methyl-1,3-butadiene	0.25	0.01
1,3-cyclopentadiene	1.81	0.03
1-methyl-1,3-cyclopentadiene	0.27	0.03
fulvene	0.04	0.01
benzene	3.52	0.13
5-methyl-1,3-cyclopentadiene	0.30	0.02
heptane	1.50	0.28
toluene	0.66	0.06
xylene (m, p)	0.03	0.01
ethylbenzene	0.05	0.01
o-xylene	0.01	0.01
styrene	0.21	0.06
monoaromatics C9	0.03	0.01
indene	0.06	0.01
1-methyl-1H-Indene	0.02	0.00
3-methyl-1H-Indene	0.02	0.00
1,2-dihydro-Naphthalene	0.01	0.00
1-methylene-1H-Indene	0.01	0.00
naphthalene	0.09	0.01
2-methyl-naphthalene	0.01	0.00
1-methyl-naphthalene	0.01	0.00

acenaphthylene	0.01	0.01
Compound name	**Concentration, ppmw**	**Std.Dev.**
acetonitrile	18.79	1.88
propanonitrile	13.42	1.34
butanonitrile	47.48	1.42
pyridine	517.82	15.53
2-methylpyridine	21.43	2.14
3-ethylpyridine	27.58	2.76
3-ethenylpyridine	5.51	0.55
3-methylbenzonitrile	6.56	0.66
indole	41.53	1.25

Conclusions

Quantification of the compounds masked by the hydrocarbon matrix is a challenge using only an FID. The application of the NCD can be considered as a substantial improvement. The experimentally determined minimum detection level of 1.5 pgN/sec obtained using a reference mixture [45] gives an indication that it is possible to trace even ppm and ppb concentrations of nitrogen containing compounds in the analyzed samples. Unlike the NPD detector, a NCD gives an equimolar response and it is not substantially affected by the matrix [46,47]. However, quantification of nitrogen containing compounds detected by the NCD requires introducing an internal standard to the reactor effluent. Therefore the method has been tested using 2-chloropyridine as the internal standard as this compound is not formed during pyrolysis and it does not overlap with nitrogen containing products on the NCD chromatogram (see Figure 6). The internal standard thus fulfills the criterion to have a peak bidimensional resolution that exceeds 1.5. The obtained relative experimental error - less than 3% obtained under nonreactive conditions – illustrates the accuracy and reliability of the method. The same procedure applied to quantification of nitrogen containing products in a steam cracking effluent resulted in a nitrogen recovery of 98.5% in the analysis. The application range of the method could be further broadened by enhancing the separation power using a differential flow modulator [48] for the GC × GC, enabling also a two dimensional separation of the most volatile products.

The described work demonstrates the accuracy, repeatability and reproducibility of the methodology for on-line quantification of nitrogen containing products in an abundant hydrocarbon matrix. The high separation power of the GC × GC combined with a sensitive selective detector significantly goes beyond the present state of the art, resulting in a more detailed product detection and quantification. The described experimental procedures enabled a successful comprehensive off-line and on-line identification and quantification of nitrogen containing compounds in the studied samples.

Acknowledgment

The SBO project "Bioleum" (IWT-SBO 130039) supported by the Institute for Promotion of Innovation through Science and Technology in Flanders (IWT) and the 'Long Term Structural Methusalem Funding by the Flemish Government' are acknowledged.

References

Meyer, R.F., Witt, W. J. Definition and World Resources of Natural Bitumens. U.S. Geological Survey. 1944 (1990).

2 Dusseault, M. B. Comparing Venezuelan and Canadian Heavy Oil and Tar Sand. Petroleum Society's Canadian International Petroleum Conference. 2001-061 (2001).

3 Hernández, R., Villarroel I. Technological Developments for Enhancing Extra Heavy Oil Productivity in Fields of the Faja Petrolifera del Orinoco (FPO), Venezuela. AAPG Annual Convention and Exhibition. Search and Discovery Article 20205 (2013).

4 Escobar, M. et al. The organic geochemistry of oil seeps from the Sierra de Perijá eastern foothills, Lake Maracaibo Basin, Venezuela. Org. Geochem. 42, 727-738, doi: 10.1016/j.orggeochem.2011.06.005 (2011).

5 Shafiei, A., Dusseault, M. B. Geomechanics of thermal viscous oil production in sandstones. J. Petrol.Sci.Eng. 103, 121-139, doi: 10.1016/j.petrol.2013.02.001 (2013).

6 Dijkmans, T., Djokic, M. R., Van Geem, K. M., Marin, G. B. Comprehensive compositional analysis of sulfur and nitrogen containing compounds in shale oil using GC × GC – FID/SCD/NCD/TOF-MS. Fuel. 140, 398-406, doi: 10.1016/j.fuel.2014.09.055 (2015).

7 Adam, F. et al. Comprehensive two-dimensional gas chromatography for basic and neutral nitrogen speciation in middle distillates. Fuel. 88, 938-946, doi: 10.1016/j.fuel.2008.11.032 (2009).

8 Charlesworth, J. M. Monitoring the products and kinetics of oil shale pyrolysis using simultaneous nitrogen specific and flame ionization detection. Fuel. 65, 979-986, doi: 10.1016/0016-2361(86)90208-5 (1986).

9 Lissitsyna, K., Huertas, S., Quintero, L. C., Polo, L. M. Novel simple method for quantitation of nitrogen compounds in middle distillates using solid phase extraction and comprehensive two-dimensional gas chromatography. Fuel. 104, 752-757, doi: 10.1016/j.fuel.2012.08.054 (2013).

10 Toraman, H. E., Dijkmans, T., Djokic, M. R., Van Geem, K. M., Marin, G. B. Detailed compositional characterization of plastic waste pyrolysis oil by comprehensive two-dimensional gas-chromatography coupled to multiple detectors. J. Chromatogr. A. 1359, 237-246, doi: 10.1016/j.chroma.2014.07.017 (2014).

11 Phillips, J. B., Beens, J. Comprehensive two-dimensional gas chromatography: a hyphenated method with strong coupling between the two dimensions J. Chromatogr. A. 856, 331-347, doi: 10.1016/S0021-9673(99)00815-8 (1999).

12 Dallüge, J., Beens, J., Brinkman, U. A. T. Comprehensive two-dimensional gas chromatography: a powerful and versatile analytical tool. J. Chromatogr. A. 1000, 69-108, doi: 10.1016/S0021-9673(03)00242-5 (2003).
13 Adahchour, M., Beens, J., Vreuls, R. J. J., Batenburg, A. M., Brinkman, U. A. T. Comprehensive two-dimensional gas chromatography of complex samples by using a 'reversed-type' column combination: application to food analysis. J. Chromatogr. A. 1054, 47-55, doi: 10.1016/j.chroma.2004.07.084 (2004).
14 Marriott, P., Shellie, R. Principles and applications of comprehensive two-dimensional gas chromatography. TrAC, Trends Anal. Chem. 21, 573-583, doi: 10.1016/S0165-9936(02)00814-2 (2002).
15 Dutriez, T. et al. High-temperature two-dimensional gas chromatography of hydrocarbons up to nC60 for analysis of vacuum gas oils. J. Chromatogr. A. 1216, 2905-2912, doi: 10.1016/j.chroma.2008.11.065 (2009).
16 Dutriez, T., Courtiade, M., Thiébaut, D., Dulot, H., Hennion, M.C. Improved hydrocarbons analysis of heavy petroleum fractions by high temperature comprehensive two-dimensional gas chromatography. Fuel. 89, 2338-2345, doi: 10.1016/j.fuel.2009.11.041 (2010).
17 Vendeuvre, C. et al. Characterisation of middle-distillates by comprehensive two-dimensional gas chromatography (GC × GC): A powerful alternative for performing various standard analysis of middle-distillates. J. Chromatogr. A. 1086, 21-28, doi: 10.1016/j.chroma.2005.05.106 (2005).
18 Van Geem, K. M. et al. On-line analysis of complex hydrocarbon mixtures using comprehensive two-dimensional gas chromatography. J. Chromatogr. A. 1217, 6623-6633, doi: 10.1016/j.chroma.2010.04.006 (2010).
19 Van de Vijver, R. et al. Automatic Mechanism and Kinetic Model Generation for Gas- and Solution-Phase Processes: A Perspective on Best Practices, Recent Advances, and Future Challenges. Int. J. Chem. Kinet. 47, 199-231, doi: 10.1002/kin.20902 (2015).
20 Van Geem, K. M., Reyniers, M.F., Marin, G.B. Reconstruction of the Molecular Composition of Complex Feedstocks for Petrochemical Production Processes. 7th Netherlands Process Technology Symposium (NPS7). doi: http://hdl.handle.net/1854/LU-681066 (2007).
21 Van Geem, K. M. et al. Molecular reconstruction of naphtha steam cracking feedstocks based on commercial indices. Comput. Chem. Eng. 31, 1020-1034, doi: http://hdl.handle.net/1854/LU-409468 (2007).
22 Van Geem, K. M., Reyniers, M. F., Marin, G. B. Challenges of modeling steam cracking of heavy feedstocks. Oil Gas Sci. Technol. – Revue d'IFP. 63, 79-94, doi: http://hdl.handle.net/1854/LU-409153 (2008).
23 Alvarez-Majmutov, A. et al. Deriving the Molecular Composition of Middle Distillates by Integrating Statistical Modeling with Advanced Hydrocarbon Characterization. Energy Fuels. 28, 7385-7393, doi: 10.1021/ef5018169 (2014).
24 Hudebine, D., Verstraete, J. J., Hudebine, D., Verstraete, J., Chapus, T. Reconstruction of Petroleum Feedstocks by Entropy Maximization. Application to FCC Gasolines Statistical Reconstruction of Gas Oil Cuts. Oil Gas Sci. Technol. – Rev.IFP Energies nouvelles. 66, 437-460, doi: 10.2516/ogst/2011110 (2011).

25 Verstraete, J. J., Schnongs, P., Dulot, H., Hudebine, D. Molecular reconstruction of heavy petroleum residue fractions. Chem. Eng. Sci. 65, 304-312, doi: 10.1016/j.ces.2009.08.033 (2010).
26 Neurock, M., Nigam, A., Trauth, D., Klein, M. T. Molecular representation of complex hydrocarbon feedstocks through efficient characterization and stochastic algorithms. Chem. Eng. Sci. 49, 4153-4177, doi: 10.1016/S0009-2509(05)80013-2 (1994).
27 Hudebine, D., Verstraete, J. J. Molecular reconstruction of LCO gasoils from overall petroleum analyses. Chem. Eng. Sci. 59, 4755-4763, doi: 10.1016/j.ces.2004.09.019 (2004).
28 Joo, E., S. Park, M. Lee. Pyrolysis reaction mechanism for industrial naphtha cracking furnaces. Ind. Eng. Chem. Res. 40, 2409-2415, doi: 10.1021/ie000774o (2001).
29 Houser T. J., Mccarville E. M., Biftu T. Kinetics of thermal decomposition of Pyridine in a Flow System. Int. J. Chem. Kinet. 12, 555–568, doi: 10.1002/kin.550120806 (1980).
30 Lifshitz, A., Tamburu, C., Suslensky, A. Isomerization and decomposition of pyrrole at elevated temperatures: studies with a single-pulse shock tube. J. Phys. Chem. 93, 5802-5808, doi: 10.1021/j100352a030 (1989).
31 Memon, H. U. R., Bartle, K. D., Taylor, J. M., Williams, A. The shock tube pyrolysis of pyridine. Int. J. Energy Res. 24, 1141-1159, doi: 10.1002/1099-114x(20001025)24:13<1141::aid-er653>3.0.co;2-s (2000).
32 Mackie, C.J., Colket, M. B., Nelson, P.F. Shock tube Pyrolysis of Pyridine, J. Phys. Chem. 94, 4099-4106, doi: 10.1021/j100373a040 (1990).
33 Alexiou, A., Williams, A. Soot formation in shock-tube pyrolysis of pyridine and toluene-pyridine mixtures. Fuel. 73, 1280-1283, doi: 10.1016/0016-2361(94)90302-6 (1994).
34 Tranchida, P. Q., Purcaro, G., Dugo, P., Mondello, L., Purcaro, G. Modulators for comprehensive two-dimensional gas chromatography. TrAC, Trends Anal. Chem. 30, 1437-1461, doi: 10.1016/j.trac.2011.06.010 (2011).
35 Yan, X. Sulfur and nitrogen chemiluminescence detection in gas chromatographic analysis. J. Chromatogr. A. 976, 3-10, doi: 10.1016/S0021-9673(02)01231-1 (2002).
36 Özel, M. Z., Hamilton, J. F., Lewis, A. C. New Sensitive and Quantitative Analysis Method for Organic Nitrogen Compounds in Urban Aerosol Samples. Environ. Sci. Technol. 45, 1497-1505, doi: 10.1021/es102528g (2011).
37 Kocak, D., Ozel, M. Z., Gogus, F., Hamilton, J. F., Lewis, A. C. Determination of volatile nitrosamines in grilled lamb and vegetables using comprehensive gas chromatography – Nitrogen chemiluminescence detection. Food Chem. 135, 2215-2220, doi: 10.1016/j.foodchem.2012.07.002 (2012).
38 Dijkmans, T., et al. Production of bio-ethene and propene: alternatives for bulk chemicals and polymers. Green Chem. 15, 3064-3076, doi: 10.1039/C3GC41097H(2013).
39 Pyl, P. S., et al. Biomass to olefins: Cracking of renewable naphtha. Chem. Eng. J. 176-177, 178– 187, doi: 10.1016/j.cej.2011.04.062 (2011).
40 Schietekat, M. C., et al. Catalytic Coating for Reduced Coke Formation in Steam Cracking Reactors. Ind. Eng. Chem. Res. 54, 9525–9535, doi: 10.1021/acs.iecr.5b02263 (2015).
41 Dietz, W. A. Response Factors for Gas Chromatographic Analyses. J. Chromatogr. Sci. 5, 68-71, doi: 10.1093/chromsci/5.2.68 (1967).

42 Dierickx, J. L., Plehiers, P. M., Froment, G. F. On-line gas chromatographic analysis of hydrocarbon effluents: Calibration factors and their correlation. J. Chromatogr. A. 362, 155-174, doi: 10.1016/S0021-9673(01)86965-X (1986).
43 Beens, J., Janssen, H. G., Adahchour, M., Brinkman, U. A. T. Flow regime at ambient outlet pressure and its influence in comprehensive two-dimensional gas chromatography. J. Chromatogr. A. 1086, 141-150, doi: 10.1016/j.chroma.2005.05.086 (2005).
44 Schoenmakers, P. J., Oomen, J. L. M. M., Blomberg, J., Genuit, W., van Velzen, G. Comparison of comprehensive two-dimensional gas chromatography and gas chromatography – mass spectrometry for the characterization of complex hydrocarbon mixtures. J. Chromatogr. A. 892, 29-46, doi: 10.1016/S0021-9673(00)00744-5 (2000).
45 Agilent Tech. Agilent Sulfur Chemiluminescence Detector and Nitrogen Chemiluminescence Detector. Specification Guide (2006).
46 Agilent Tech. Nitrosamine Analysis by Gas Chromatography and Agilent 255 Nitrogen Chemiluminescence Detector (NCD). Technical Overview (2007).
47 Agilent Tech. Agilent Model 255 Nitrogen Chemiluminescence Detector (NCD) Analysis of Adhesive Samples Using the NCD. Technical Overview (2007).
48 Griffith, F.J., Winniford, W.L., Sun, K., Edam, R., Luong, C.J. A reversed-flow differential flow modulator for comprehensive two-dimensional gas chromatography. J. Chromatogr. A. 1226, 116-123, doi: http://dx.doi.org/10.1016/j.chroma.2011.11.036 (2012).

AICHE PAPER NUMBER: 157a

ETHYLENE RISK AND TECHNOLOGY REVIEW WHAT'S THE OUTLOOK?

Steve Smith
Manager, Environmental Issues
Lyondell Chemical Company

Prepared for Presentation at the 2019 Spring National Meeting
New Orleans, April 3, 2019

ETHYLENE RISK AND TECHNOLOGY REVIEW WHAT'S THE OUTLOOK?

Steve Smith
Manager, Environmental Affairs
Lyondell Chemical Company

Abstract:

Per the court-mandated timeline, the EPA is currently working through the statutorily-required risk and technology review (RTR) for the ethylene NESHAP (40 CFR, Part 63, Subpart XX and Subpart YY). It is anticipated the proposed rule, due in mid-2019, will impact current ethylene operations. It is anticipated the proposed rule will mirror certain aspects of the refinery rule, such as the flare and PRD language and likely will address furnace stack emissions and decoke emissions (both decoke to the fire box and decoke to an atmospheric pot) and potentially waste water emissions.

This paper will discuss the potential rules and the potential impact on the ethylene facilities.

Ethylene production units are regulated, at the Federal level, by 40 CFR, Part 63, Subpart XX and Subpart YY, commonly referred to as EMACT (Ethylene Maximum Available Control Technology). Every eight years the MACT standards go through the risk and technology review (RTR). EPA failed to meet the eight-year review cycle and is now under a court-mandated timeline for completing the RTR. The proposed rule is expected in July or August 2019. The final rule, by Court order, should be issued in March 2020, however the date for the final rule may be delayed. The effective date is three years later, which could be as early as March 2023. It is anticipated the amended rule will impact ethylene operations. One potential outcome is the proposed rule will mirror certain aspects of the refinery sector rule (RSR), such as the flare and pressure relief device (PRD) language. Also, the furnace stack emissions and decoke emissions (both decoke to the fire box and decoke to an atmospheric pot) may be impacted. Industry trade groups have been engaged in meetings with EPA as this process has moved forward and the following discussion is based on those meetings.

Risk and Technology Review Process

The Clean Air Act requires a residual risk review be completed eight years after a Maximum Achievable Control Technology (MACT) standard is established. The risk review is done to determine if there is residual risk and if there is an unacceptable risk, then control technology is reviewed. The technology review looks for changes in the potential control technology and the costs associated with the control technology. The general process is as follows: The initial step in the risk review is hazardous air pollutant (HAP) emissions data collection. The emissions data may be gathered from historical emissions inventories or it may be collected through an information collection request (ICR). A basic air dispersion model is then used to model the community exposure. The community exposure data is then used in a risk model to determine if the risk is at or below an acceptable threshold. If the risk is unacceptable and a MACT floor has previously been established the technology is reviewed to determine if the ambient exposure levels can be reduced. If the risk is unacceptable and a MACT floor has not previously been established then EPA moves forward with setting a MACT floor which is based on the best performers. EPA uses the 99% confidence interval of the upper prediction limit (UPL) to analyze the emission data from the top performers. The UPL is calculated using historical data and is designed to equal or exceed a specified number of future comparisons. In this case the number of future comparisons is the number of future performance tests to demonstrate compliance. Generally, the top performers are represented by the top 12% of the data set. In the ethylene RTR nineteen furnaces were tested which resulted in review of the furnace data to find the three furnaces with the lowest HAP concentration for each of the HAPs. For example, furnaces A, D, and G had the lowest nickel concentration and furnaces B, C, and H had the lowest chromium concentration. The data review is looking for the lowest concentration for each individual HAP, not the furnace with the lowest aggregate HAPs. The UPL is then calculated using the three top performers for each HAP. The UPL is then compared to the representative detection limit (RDL) for the test method as calculated for each test run. If three times the RDL is greater than the UPL then the MACT floor is set at three times the RDL. If three times the RDL is less than the UPL, then the MACT floor is set at the UPL. If the data set is flawed, then EPA may move to a work practice in place of a numerical limit on the basis that it may not be feasible to prescribe or enforce an emissions standard.

The ethylene risk and technology review (RTR) started with the EPA Clean Air Act Section 114 information collection request (ICR) to a subset of the ethylene producers. The Component 1 ICR addressed all aspects of the ethylene production and was a paper exercise to complete an extensive database. EPA reviewed the Component 1 data and followed with a Component 2 ICR that required testing of the cooling towers; cracking furnaces when operating at design firing rate; cracking furnaces when operating at the reduced firing rate associated with decoke operation; and the decoke drum stack. EPA used the Component 2 data to develop model emission rates that are applied to the units that weren't part of the testing. Using the Component 2 data and the model emission rates, EPA completed the final model runs last year. A question on the table, the testing data set is from furnaces that were cracking ethane and ethane/propane blends, therefore can the data be considered representative of furnaces that crack heavier feeds? If the coke formation, both the speed of formation and type of coke, changes as the feed gets heavier, can this limited data set be considered representative of furnaces that crack heavier feeds? If the data set isn't representative, then can the data be used to set a MACT floor? These questions remain open, but may be addressed in the preamble to the rule or in comments on the proposed rule.

Potential Rule Outcomes Based on Information Available at this Time

Overview

EPA continues to review all potential options: work practices; surrogate based limits with monitoring; and potential numerical limits. The following is from CAA section 112(h): "...if it is not feasible in the judgment of the Administrator to prescribe or enforce an emission standard for control of a hazardous air pollutant or pollutants, the Administrator may, in lieu thereof, promulgate a design, equipment, work practice, or operational standard, or combination thereof...". Work practices were the outcome under Boiler and Process Heater MACT for devices that fire natural gas and other gas 1 fuels. In some MACT reviews it has been determined that the data shows an unacceptable risk, but ongoing monitoring of the HAP of concern isn't feasible and a surrogate has been selected as the parameter for continuous monitoring. When considering combustion devices, an example is using CO stack measurements as a surrogate for organic HAPs. If numeric limits are set, then it is possible some sources may need to be controlled. How sources control emissions to meet a potential numeric standard is generally determined by the source operator.

Cracking Furnaces

The work practice rule currently applicable to many combustion devices, other than cracking furnaces, is the Boiler and Process Heater MACT. To address an organic HAP standard the likely control would be tied to combustion controls to minimize CO. However, the work practice rule (Boiler and Process Heater MACT) currently applicable to combustion devices, other than cracking furnaces, is the preferred outcome.

If a work practice followed the Boiler and Process Heater MACT requirements, then the following would likely be required on an annual basis and would likely include:

- inspect the burners, and clean or replace any components of the burner as necessary;
- inspect the flame pattern and make any adjustments to the burner necessary to optimize the flame pattern consistent with the manufacturer's specifications;
- inspect the system controlling the air-to-fuel ratio, and ensure that it is correctly calibrated and functioning properly;
- minimize total emissions of CO consistent with the manufacturer's specifications;
- measure the concentration in the effluent stream of CO, before and after the adjustments are made; and
- maintain on-site an annual report containing the following information:
    - concentrations of CO and oxygen in the effluent stream, measured before and after the adjustments of the furnace;
    - a description of any corrective actions taken as a part of the combustion adjustment; and
    - the type and amount of fuel used over the 12 months prior to the annual adjustment.

Until a proposed rule is published for comment, all options remain on the table. Generally, if an option isn't presented in the proposed rule for review and comment, it isn't considered for the final rule. One potential outcome, if work practices, as outlined above, aren't accepted, is emission controls on the cracking furnace stacks. One approach is numerical limits established pollutant-by-pollutant, however, at this juncture, that appears unlikely. The alternative to metal-by-metal limits would be to control metal HAPs as a group, as in a total metals limit or find a surrogate. One approach that has been discussed is to limit the radiant coil useful life to seven years, basically using coil life as a predictor or surrogate for metals emissions. That suggestion has led to the question: how can an operator go beyond seven years? A possible approach would be to do a performance test to demonstrate metals emissions are below some threshold level. If the performance test demonstrates metals emissions are below the threshold then the run length could be extended. As stated above, at this point in the discussions it appears it is unlikely a numeric limit would be set for the organic HAPs. A potential surrogate could be a carbon monoxide (CO) limit on the furnace stack effluent. If there is a CO limit then it is possible a CO CEMS (Continuous Emissions Monitoring System) would be required to demonstrate continuous compliance.

Decoke Process

As discussed above, the risk review is built using the results from the testing. The decoke pot stack testing data set is flawed due to issues with maintaining isokinetic sampling rates. Therefore, it appears numeric limits cannot be established and work practices will be defined that address the potential HAP emissions. A potential work practice might address proper isolation of the cracked gas header from the decoke header for units that decoke to a pot which should minimize the potential for organic HAP emissions. A potential work practice to address the metal HAP emissions might be a designed decoke pot that includes a cyclone or an equivalent. The work practice might go further and include an operating procedure that includes steps to expedite air removal following coke removal. In this context coke removal includes both the removal of coke from the radiant tubes as well as the TLEs.

The decoke to the fire box stack data set does not have the isokinetic sampling issue, therefore it appears the data may be considered a valid representation of the emissions when the furnace is going through the decoke process. The cracking furnace work practice, outlined above, would address the organic HAP emissions from the furnace. The PM emissions data from the decoke to the firebox process have EPA's attention. In an early conversation EPA suggested a cyclone in the decoke line prior to the decoke stream entering the firebox. The cyclone would address the PM emissions and the firebox work practice would address the organic HAP emissions. There are several concerns with that suggested approach and those concerns were addressed in the conversation. For instance, the equipment arrangement and layout of the cyclone and associated piping, considering the existing furnace structure and maintenance access, how do you fit the new equipment into the existing footprint. What is the needed gas velocity for an effective cyclone and how would that velocity be achieved? The discussions are ongoing.

Flares

As with the cracking furnaces and decoke process the flare discussions are ongoing. It appears the current EPA position is the ethylene rule will point to the flare language in the Refinery Sector Rule (RSR). For perspective, the following is a summary of the RSR flare rule language.

When regulated material is routed to the flare:

- Operate each flare with a pilot flame present;
- Specify the smokeless design capacity and operate with no visible emissions, except for periods not to exceed a total of five minutes during any two consecutive hours, applicable when the flare vent gas flow rate is less than the smokeless design capacity of the flare
- Actual flare tip velocity must:
    - be less than 60 feet per second; or
    - be less than 400 feet per second and also less than the maximum allowed calculated flare tip velocity (equation based on the NHV of the vent gas)
- Net Heating Value in the combustion zone (NHV_{cz}) maintained at or above 270 Btu/scf, determined on a fifteen-minute block period when regulated material is routed to the flare for at least fifteen minutes
    - Current flare regulatory language – Net Heating Value, vent gas (NHV_{vg}) maintained at or above 300 Btu/scf
- Dilution operating limits for flares with perimeter assist air – maintain net heating value dilution parameter (NHV_{dil}) at or above 22 Btu/ft^2 determined on a fifteen-minute block period when regulated material is routed to the flare for at least fifteen minutes
    - If the only assist air provided is perimeter assist air intentionally entrained in lower and/or upper steam at the flare tip and the effective diameter is 9 inches or greater, then the only operating limit is NHV_{cz}
- Continuously monitor the presence of the pilot flame(s)
- Monitor visible emission
    - At least once per day and when visible emissions are observed, conduct visible emissions observations using an observation period of five minutes using Method 22, if during the observation there are visible emissions for a one minute the observation period will be extended to two hours; or
    - Use a video surveillance camera to continuously record (at least one frame every 15 seconds with time and date stamps) images of the flare flame and at a reasonable distance above the flare flame at an angle suitable for visual emissions observations
- Install, operate, calibrate, and maintain a:
    - monitoring system capable of continuously measuring, calculating, and recording:
        - the volumetric flow rate in the flare header as well as any supplemental natural gas used;
        - the volumetric flow rate of assist air and/or assist steam; and

    - if premix assist air and perimeter assist air are both used, monitor the volumetric flow rate of each separately
  - Flow rate monitoring systems must be able to correct for the temperature and pressure of the system and output parameters in standard conditions
  - Mass flow monitors may be used for determining volumetric flow rate of flare vent gas provided the molecular weight of the flare vent gas is determined using compositional analysis
  - Mass flow monitors may be used for determining volumetric flow rate of assist air or assist steam
  - Continuous pressure/temperature monitoring system(s) and appropriate engineering calculations may be used in lieu of a continuous volumetric flow monitoring system(s) provided the molecular weight of the gas is known
- Determine the concentration of individual components in the flare vent gas or alternatively, directly monitor the net heating value of the flare vent gas
- Determine flare tip velocity (V_{tip}) on a 15-minute block average
- Determine the net heating value of the flare vent gas (NHV_{vg}) based on the composition monitoring data on a 15-minute block average
- Determine the net heating value of the combustion zone gas (NHV_{cz}) on a 15-minute block average
- Determine the net heating value dilution parameter (NHV_{dil}) on a 15-minute block average
- If a flare that has the potential to operate above its smokeless capacity under any circumstance, then
  - Develop a flare management plan to minimize flaring during periods of startup, shutdown, or emergency releases and submit to the Administrator
  - Conduct a root cause analysis and a corrective action analysis for each flow event that contains regulated material and meets either of the following the criteria:
    - vent gas flow rate exceeds the smokeless capacity and visible emissions are present for more than 5 minutes during any 2 consecutive hours; or
    - vent gas flow rate exceeds the smokeless capacity and the 15-minute block average flare tip velocity exceeds the maximum flare tip velocity
  - A root cause analysis and corrective action analysis must be completed as soon as possible, but no later than 45 days after a flare flow event meeting the criteria
  - Implement the corrective action(s) identified in the corrective action analysis
  - Determine the total number of events for which a root cause and corrective action analyses was required during the calendar year

- Following events would be a violation of this emergency flaring work practice standard:
  - the root cause was determined to be operator error or poor maintenance;
  - two visible emissions exceedance events that were not caused by a force majeure event from a single flare in a three-calendar year period for the same root cause for the same equipment;
  - two flare tip velocity exceedance events that were not caused by a force majeure event from a single flare in a three-calendar year period for the same root cause for the same equipment;
  - three visible emissions exceedance events that were not caused by a force majeure event from a single flare in a three-calendar year period for any reason; or
  -
  - three flare tip velocity exceedance events that were not caused by a force majeure event from a single flare in a three-calendar year period for any reason

Atmospheric Pressure Relief Devices

The expectation is the atmospheric pressure relief device (PRD) language in the Refinery Sector Rule (RSR) may be moved into the ethylene rule or the ethylene rule may point to the RSR PRD language. In the final RSR atmospheric PRDs are allowed with certain provisions. The following is a summary of the key RSR PRD requirements.

- Except during a pressure release, operate each PRD in organic HAP service with an instrument reading of less than 500 ppm as detected by Method 21
- If the PRD doesn't include a rupture disk, then within five days of returning to organic HAP service verify, using Method 21, the PRD is operating with an instrument reading of less than 500 ppm
- If the PRD includes a rupture disk, then within five days of a pressure relief replace the rupture disk
- If the PRD consists only a rupture disk, within five days of a pressure relief replace the rupture disk and the equipment served by the rupture disk cannot be restarted until the rupture disk is replaced
- Pressure relief management plan – equip each affected PRD with a device or use a monitoring system capable of:
  - identifying a the pressure release;
  - recording the time and duration of the pressure release; and
  - notify operators of a release
  - the device or monitoring system may be specific to the PRD or may be associated with the process system or piping
- Pressure relief management plan – includes at least three documented redundant prevention measures for each affected PRD
  - Examples of prevention measures include:

- flow, temperature, level and pressure indicators with monitors, or automatic actuators;
  - documented routine inspection and programs and/or operator training;
  - deluge systems;
  - staged relief systems where initial PRD (with lower set pressure) discharges to a flare or other closed vent system and control device
- If an affected PRD releases to atmosphere, then a root cause analysis and corrective analysis must be completed
- PRD releases will be tracked on an annual basis and releases associated with *force majeure* events will be noted
- Certain PRD releases will be considered violations of the pressure release management work practice standard
  - a PRD release when the root cause is operator error or poor maintenance
  - a second PRD release from a single PRD in a three year period for the same root cause for the same equipment
  - a third PRD release from a single PRD in a three year period for any reason
  - the exception is PRD releases due to a *force majeure* event

Ethylene Process Vents

The current definition of ethylene process vents applies to vents from <u>continuous</u> unit operations. If the RSR model is followed then "continuous" may be dropped, and if that happens, then the definition would be much broader in applicability. The new definition of an ethylene process could be as follows:

> a gas stream with a flow rate greater than 0.005 standard cubic meters per minute containing greater than 20 parts per million by volume HAP that is discharged during operation of an ethylene production unit. Ethylene process vents are gas streams that are discharged to the atmosphere (or the point of entry into a control device, if any) either directly or after passing through one or more recovery devices. Ethylene process vents do not include: gaseous stream routed to a fuel system; PRD releases; fugitive emission leaks; and in situ sampling systems (online analyzers).

The potential impact is nonroutine or intermittent vents that previously were not covered would be covered as ethylene process vents. If the RSR change to the definition is followed, then the RSR work practice language should follow. There is a need for a de-minimis threshold below which emissions would not need to be controlled. Again, if the RSR model is followed, then a de-minimis threshold is part of the work practice. As an example, the de-minimis threshold allows for opening equipment to insert blinds for maintenance.

Start-up, Shutdown and Malfunction

A 2015 court decision set aside the start-up, shutdown and malfunction (SSM) language that allowed for the affirmative defense for excess emissions during SSM activities. The effect of the decision is there is no exemption from emission limits during SSM activities and SSM language is being removed from each affected rule. Therefore, because of the court decision the SSM language and exceptions embedded in individual definitions/standards for periods of startup, shutdown, episodic, maintenance or other transient periods will be removed. EPA has the latitude to establish alternative work practices to address periods of SSM.

Heat Exchange Systems

If the RSR heat exchange system language is adopted, then periodic sampling using the Texas Commission on Environmental Quality (TCEQ) modified El Paso stripper method will be required to monitor for heat exchanger leaks. The RSR requirement is to either monitor monthly using a leak action level defined as a total strippable hydrocarbon concentration (as methane) in the stripping gas of 6.2 ppm-v or monitor quarterly using a leak action level defined as a total strippable hydrocarbon concentration (as methane) in the stripping gas of 3.1 ppm-v. If a leak is detected, then it be must repaired to reduce the measured concentration to below the applicable action level as soon as practicable, but no later than 45 days after identifying the leak. The current ethylene heat exchange system monitoring monitors the concentration in the water and does not require the use of the modified El Paso stripper.

Fence Line Monitoring

Fence line monitoring is part of the refinery sector rule (RSR). Fence line monitoring was not required to address an unacceptable risk it is in the rule as a surrogate for additional wastewater and fugitive emission monitoring. The BWON rule addresses wastewater in the ethylene sector, therefore fence line monitoring should not be a surrogate for additional wastewater monitoring and should only be required if driven by the risk modeling. As background, RSR fence line monitoring is a passive method using sorbent tubes to measure the ambient benzene concentration. However, it doesn't appear fence line monitoring will be in the proposed rule.

AIChE PAPER NUMBER 157B

Proposed Stationary Source Testing and Requirements for Ethylene Production Facilities

David L. Elam, Jr.
Project Director and Vice President
TRC

Prepared for Presentation at the Spring National Meeting
New Orleans, LA March 31 – April 4, 2019

Proposed Stationary Source Testing and Requirements for Ethylene Production Facilities

David L. Elam, Jr
Project Director and Vice President
TRC

Abstract:

In April 2016, EPA issued an Information Collection Request (ICR) for ethylene production facilities under the statutory authority of Section 114 of the Clean Air Act (CAA). As required by Section 112 of the statute, EPA must examine Maximum Achievable Control Technology (MACT) standards with respect to residual risk and technological advancements every eight years and promulgate revised MACT standards if the Risk and Technology Review (RTR) indicates that changes are necessary.

MACT requirements for ethylene production plants were promulgated in 2002 (40 CFR Part 63, Subpart YY; Generic Maximum Achievable Control Technology Standards and 40 CFR Part 63, Subpart XX, Ethylene Manufacturing Process Units: Heat Exchange Systems and Waste Operations). It is anticipated that EPA's RTR process will result in a revision of these MACT standards, imposing new emission testing requirements on ethylene production facilities.

EPA is expected to propose new regulations in connection RTR activities in spring of 2019. At the time of publication, the proposed rules are under review at the Office of Management and Budget and not yet available to the public; however, it is anticipated that the proposed rule will not include many of the test methods required during the ICR. Instead, the proposed rule will likely include a requirement for carbon monoxide and metals measurements. The presentation will outline a strategy to conduct a successful emission testing program, focusing on the requirements for successful carbon monoxide and metals emission measurements.

Introduction

In April 2016, EPA issued an Information Collection Request (ICR) for ethylene production facilities under the statutory authority of Section 114 of the Clean Air Act (CAA). As required by Section 112 of the statute, EPA must examine Maximum Achievable Control Technology (MACT) standards with respect to residual risk and technological advancements every eight years and promulgate revised MACT standards if the Risk and Technology Review (RTR) indicates that changes are necessary. EPA uses the authority of Section 114 of the CAA to gather information that supports the RTR evaluation process.

MACT requirements for ethylene production plants were promulgated in 2002 (40 CFR Part 63, Subpart XY; Generic Maximum Achievable Control Technology Standards and 40 CFR Part 63, Subpart XX, Ethylene Manufacturing Process Units: Heat Exchange Systems and Waste Operations). The Ethylene MACT regulates emissions of the following organic hazardous air pollutants (HAPs):

- Benzene
- 1,3-Butadiene
- Cumene
- Ethylbenzene
- Hexane
- Naphthalene
- Styrene
- Toluene
- o-, m-, and p-Xylene

EPA's recent Ethylene Production ICR was issued to 20 separate facilities representing a universe of 191 ethylene cracking furnaces and 37 heat exchangers. Unless a facility conducted testing that met the requirements of the ICR in the preceding five years, each facility had to select **one** ethylene cracking furnace for emission testing during cracking and decoking operations and **all** identified heat exchange systems.

The ethylene cracking furnace test programs were particularly complex, requiring multiple sophisticated sampling trains to measure the aforementioned compounds and:

- Aldehydes (Acrolein, Acetaldehyde, Formaldehyde, Propionaldehyde)
- Ammonia
- Carbon Dioxide
- Carbon Monoxide
- Ethane
- Ethylene
- Filterable Particulate Matter
- Flow Rate
- Hydrogen Cyanide
- Metal HAPs (Sb, As, Be, Cd, Cr, Co, Pb, Ni, Mn, Hg, Se, and P)

- Methane
- Oxygen
- Polycyclic Aromatic Hydrocarbons
- Propylene
- Total Hydrocarbons

The heat exchanger test programs involved collection of water samples for laboratory analysis and the use of the El Paso method to determine the following strippable compounds:

- Benzene
- 1,3-Butadiene
- Chlorine
- Chloroform
- Cumene
- Dimethylformamide
- Ethylbenzene
- Ethylene
- Hexane
- Naphthalene
- Propylene
- Styrene
- Toluene
- Total Strippable Volatile Organic Compounds (VOC)
- *o*-, *m*-, and *p*-Xylene (and total)

Although EPA identified a range of compounds that had to be measured in connection with these test programs, it did not establish detection limits for the measurements. Instead, EPA identified sample methods and run times.

EPA required the use of FTIR for several of the target pollutants. In addition, continuous emission analyzers (CEMs), two isokinetic sample trains, and Reference Method 18 sampling procedures were required to quantify the numerous target compounds. Seven (7) test runs were required for the ethylene cracking furnaces. During decoking operations, which can span several weeks due to operational schedules, EPA required 9 test runs (3 runs for each of 3 cycles). All test data and results had to be submitted via the Electronic Reporting Tool (ERT). Test methods and run times are summarized in Tables 1 and 2.

Table 1: Furnace Stack - Test Matrix

Parameter	Test Method	# of Test Runs	Run Length	Units of Measure
Aldehydes[1]	ASTM D6348-03	7	60 minutes	lb/hr and ppm_{vd} @ 3% O_2
Ammonia	ASTM D6348-03	7	60 minutes	lb/hr and ppm_{vd} @ 3% O_2
Organic HAP[2]	EPA Method 18	7	60 minutes	lb/hr and

Table 1: Furnace Stack - Test Matrix				
Parameter	**Test Method**	**# of Test Runs**	**Run Length**	**Units of Measure**
1,3-Butadiene[3]	(ASTM D-6348-03)			ppm_{vd} @ 3% O_2
Polycyclic Aromatic Hydrocarbons (PAH)[4]	CARB Method 429	7	Minimum of 2 dry cubic meters (~2 hours)	lb/hr and µg/dscm @ 3% O_2
Total Hydrocarbons (THC)	EPA Method 25A	7	60 minutes	lb/hr and ppm_{vd} @ 3% O_2
Methane Ethane Ethylene Propylene	ASTM D6348-03	7	60 minutes	lb/hr and ppm_{vd} @ 3% O_2
Nitrogen Oxides (NOx)	EPA Method 7E	7	120 minutes, minimum	lb/hr and ppm_{vd} @ 3% O_2
Filterable PM and Metal HAP[5]	EPA Method 5 EPA Method 29	7	Minimum of 4 dry cubic meters (~4 hours)	lb/hr and µg/dscm @ 3% O_2
Hydrogen Cyanide	ASTM D6348	7	60 minutes	lb/hr and µg/dscm @ 3% O_2
Carbon Monoxide	EPA Method 10	Concurrent with all test runs	4 hours	lb/hr and ppm_{vd} @ 3% O_2
Flow Rate[6]	EPA Method 2 O_2-EPA Method 3A EPA Method 4 (CO_2-ASTM 6348-03)	Concurrent with all test runs (7)	4 hours (60 minutes)	

[1]Acrolein, Acetaldehyde, Formaldehyde, Propionaldehyde
[2]Benzene, Cumene, Ethylbenzene, Hexane, Styrene, Toluene, o-Xylene, m-Xylene, p-Xylene
[3]1,3-Butadiene measured by ASTM D6348-03
[4]The target analyte list is included as Table 1 of CARB Method 429
[5]Metal HAP includes compounds of Antimony, Arsenic, Beryllium, Cadmium, Chromium, Cobalt, Lead, Nickel, Manganese, Mercury, and Selenium and Phosphorus. Report concentrations and mass emissions rates of each pollutant separately.
[6]Measured in conjunction with Method 5/Method 29

Table 2: Decoking Pot Stack - Test Matrix				
Parameter	**Test Method**	**# of Test Runs**	**Run Length**	**Units of Measure**
Aldehydes[1]	ASTM D6348-03	3 runs per cycle	60 minutes	lb/hr and ppm_{vd} @ 3% O_2

Table 2: Decoking Pot Stack - Test Matrix				
Parameter	**Test Method**	**# of Test Runs**	**Run Length**	**Units of Measure**
Ammonia	ASTM D6348-03	7	60 minutes	lb/hr and ppm_{vd} @ 3% O_2
Organic HAP[2] (1,3-Butadiene[3])	EPA Method 18 (ASTM D6348-03)	3 runs per cycle	60 minutes	lb/hr and ppm_{vd} @ 3% O_2
Polycyclic Aromatic Hydrocarbons (PAH)[4]	CARB Method 429	3 runs per cycle	Minimum of 2 nominal actual cubic meters (~2 hours)	lb/hr and µg/dscm @ 3% O_2
Total Hydrocarbons (THC)	EPA Method 25A	3 runs per cycle	60 minutes	lb/hr and ppm_{vd} @ 3% O_2
Methane Ethane Ethylene Propylene	ASTM D6348-03	3 runs per cycle	60 minutes	lb/hr and ppmvd @ 3% O_2
Nitrogen Oxides (NOx)	EPA Method 7E	3 runs per cycle	120 Minutes, Minimum	lb/hr and ppm_{vd} @ 3% O_2
Filterable PM and Metal HAP[5]	EPA Method 5 EPA Method 29	3 runs per cycle	Minimum of 4 nominal actual cubic meters (~4 hours)	lb/hr and µg/dscm @ 3% O_2
Hydrogen Cyanide	ASTM D6348	3 runs per cycle	60 minutes	lb/hr and µg/dscm @ 3% O_2
Carbon Monoxide	EPA Method 10	Concurrent with all test runs	4 hours	lb/hr and ppm_{vd} @ 3% O_2
Flow Rate[6]	EPA Method 2 O2-EPA Method 3A EPA Method 4 (CO2-ASTM 6348-03)	Concurrent with all test runs (7)	4 hours (60 minutes)	

[1]Acrolein, Acetaldehyde, Formaldehyde, Propionaldehyde
[2]Benzene, Cumene, Ethylbenzene, Hexane, Styrene, Toluene, o-Xylene, m-Xylene, p-Xylene
[3]1,3-Butadiene measured by ASTM D6348-03
[4]The target analyte list is included as Table 1 of CARB Method 429
[5]Metal HAP includes compounds of Antimony, Arsenic, Beryllium, Cadmium, Chromium, Cobalt, Lead, Nickel, Manganese, Mercury, and Selenium and Phosphorus. Report concentrations and mass emissions rates of each pollutant separately.
[6]Measured in conjunction with Method 5/Method 29

EPA has examined the results of this exceedingly complex ICR test program in accordance with the requirements of the RTR process. Although the results of the RTR evaluation are not yet publicly available, it is anticipated that Ethylene Production facilities will be required to install continuous emission monitoring systems (CEMS) for CO and perhaps be required to perform periodic testing for metals. Other MACT standards have allowed the use of particulate matter as a surrogate for HAP metals. Test methods for CO and HAP metals are discussed in the following sections.

Carbon Monoxide Measurement

As noted, it is anticipated that EPA will require the installation of CO CEMS on ethylene furnace and/or decoking pot stacks. CO CEMS that are used for compliance demonstration purposes must meet the requirements of one of three EPA Performance Specifications (PS):

1. Performance Specification 4: *Specifications and Test Procedures for Carbon Monoxide Continuous Emission Monitoring Systems in Stationary Sources*
2. Performance Specification 4A: *Specifications and Test Procedures for Carbon Monoxide Continuous Emission Monitoring Systems in Stationary Sources*
3. Performance Specification 4B: *Specifications and Test Procedures for Carbon Monoxide and Oxygen Continuous Monitoring Systems in Stationary Sources*

PS4 is intended for use on CO CEMS with CO spans up to 1000 ppm_v. PS4A is intended for use on CO CEMS with CO spans up to 200 ppm_v. PS4B is intended for use on CEMS that incorporate an O2 analyzer and a CO analyzer that is spanned up to 200 ppm_v.

A performance specification test program is comprised of a calibration drift test and a relative accuracy test. Calibration Drift (CD) is defined as the difference in the CEMS output readings from the established reference value after a stated period of operation during which no unscheduled maintenance, repair, or adjustment took place. Relative Accuracy (RA) is defined as the absolute mean difference between the gas concentration or emission rate determined by the CEMS and the value determined by the reference method (RM), plus the 2.5 percent error confidence coefficient of a series of tests, divided by the mean of the RM tests or the applicable emission limit.

Once the CEMS is installed and determined to be functioning properly, the CD test is performed by allowing the CEMS to recalibrate daily. To pass the CD test, the CEMS calibration must not drift or deviate from the reference value of the calibration gas, gas cell, or optical filter by more than 5 percent of the established span value for 6 out of 7 test days. As an example, the established span value is 1000 ppm for Subpart J affected facilities.

Following completion of the CD test, the RA test is performed by conducting at least nine RM runs. In the case of CO, the RM is Method 10, *Determination of Carbon Monoxide Emissions from Stationary Sources (Instrumental Analyzer Procedure).* More than nine RM runs may be performed. If more than nine runs are performed, a maximum of three sets of the test results may be rejected so long as the total number of test results used to determine the RA is greater than or equal to nine. However, all data must be reported, including the rejected data.

RM 10 references Method 7E, *Determination of Nitrogen Oxides Emissions from Stationary Sources (Instrumental Analyzer Procedure),* for several operational requirements. Highlights of an instrumental RM test program are summarized below:

- A stratification test must be performed before sampling. The results of the stratification test will determine the number of sampling points for the RM traverse.
- A three-point calibration must be passed before beginning RM testing.
- RM runs must be at least 21 minutes long.
- A bias check must be performed before and after each run.

CO measurement technology is mature and reliable, and CO CEMS rarely have problems meeting PS4, PS4A, or PS4B requirements. Accordingly, CO CEMS problems are usually associated with the sampling locations for the CEMS or the RM or calibration gas issues. An efficient way to begin a CO CEMS RATA test is to introduce the same calibration gas to the CEMS and RM systems at the probe. If the RM system and CEMS register approximately the same value, RA testing can proceed. If the systems do not register the same value, troubleshooting typically begins with an investigation of calibration gases and proceeds to an evaluation of CEMS components including the probe, the sample conditioning and transport system, the analyzer, and the data acquisition and handling system (DAHS).

Because both RM and CEMS data are available onsite, relative accuracy should be calculated before the test team leaves the site.

The ICR required that CO data be reported in lb/hr and corrected to 3% O_2. If the proposed rule includes similar reporting requirements, the CEMS will also have to include O_2 and flow monitors, both of which will be subject Performance Specification testing. The O_2 CEMS will have to meet PS3, *Specifications and Test Procedures for O_2 And CO_2 Continuous Emission Monitoring Systems in Stationary Sources.* The flow CEMS will have to PS6, *Specifications and Test Procedures for Continuous Emission Rate Monitoring Systems in Stationary Sources.* CD testing for PS3 is identical to that for PS4. CD testing for PS6 is similar to PS4. RA testing for PS3 and PS6 are identical to that for PS4.

HAP Metals Measurement

HAP metals are measured by EPA Method 29, *Determination of Metals Emissions from Stationary Sources.* Although it is not known which metals will be included in the proposed rule, RM29 can measure Antimony (Sb), arsenic (As), barium (Ba), beryllium (Be), cadmium (Cd), chromium (Cr), cobalt (Co), copper (Cu), lead (Pb), manganese (Mn), mercury (Hg), nickel (Ni), phosphorus (P), selenium (Se), silver (Ag), thallium (Tl), and zinc (Zn).

Although RM 29 has been in use for years, it remains a difficult method to employ. It is an isokinetic sampling method based on the EPA Method 5 train for the measurement of particulate and is therefore subject to any of the problems associated with Method 5 including:

- Selection of a sample location that meets the requirements of EPA Method 1,
- Maintaining calibrations of the various temperature, flow, and pressure measurement sensors in the sampling train,
- Maintaining isokinetic sampling rate of 90-110%.
- Sampling train leaks.

In addition to the problems associated with Method 5, Method 29 is subject to additional problems including:

- Set-up and recovery of the sample train. The Method 29 sampling train includes a filter and various impingers designed to capture specific metals, yielding 8 fractions when recovered.
- Corrosive impinger contents. One impinger in the sampling train contains a mixture of 5% nitric acid and 10% hydrogen peroxide. Another impinger contains a mixture of 4% potassium permanganate and 10% sulfuric acid. The impinger containing potassium permanganate and sulfuric acid is particularly troublesome because the recovered solution evolves gases which can pressurize the sample container, making shipment of the recovered solution challenging.
- Selection of filter media. Some filter media has background metal levels that can obscure measurement results. It is important to establish acceptability of the filter media before use in the field.
- Source moisture levels and detection limits. If low levels of metals are expected, the sample run time will need to be extended to meet detection limit requirements. If extended run times are required on sources with high moisture levels, it will be necessary to add more impingers to the sample train to accommodate increased water volume.
- The detection limits vary depending on what analytical procedure is employed by the laboratory. Most labs will use ICAP (Inductively Coupled Argon Plasma emission spectroscopy) for most metals and CVAAS (Cold Vapor Atomic Adsorption Spectroscopy) for mercury. Lower detection limits for several metals can be achieved by employing GFAAS (Graphite Furnace Atomic Adsorption Spectroscopy) or ICPMS (Inductively Coupled Plasma Mass Spectroscopy).
- Data reduction. The various fractions and detection limits associated with each complicates data reduction and reporting.

The ICR required that metals data be reported in lb/hr and µg/dscm @ 3% O_2. Because the Method 29 incorporates O2 and volumetric flow measurements, no additional measurements will be required if the proposed rule requires reporting metals emissions in these units.

Electronic Data Reporting

For the past several years, new emission limits associated with 40 CFR Part 60 or 40 CFR Part 63 have required electronic reporting using EPA's Electronic Reporting Tool (ERT). ERT files are submitted through EPA's Central Data Exchange (CDX) using the Compliance and Emissions Data Reporting Interface (CEDRI). Ethylene producers should anticipate that similar reporting requirements will be associated with the forthcoming proposed rule.

It is recommended that plant personnel decide in advance which steps in this process will be carried out by the testing company and which by the facility staff. If a facility will be reporting for the first time with the ERT, the facility must to be registered with EPA's Central Data Exchange (CDX) for the Compliance and Emissions Inventory Data Reporting Interface (CEDRI) program and individuals must be assigned roles as Preparers, Certifiers, and Delegated Certifiers.

It is important to understand that the electronic data submittal is not simply an upload of a document to an EPA website. Although the ERT application is a Microsoft Access database, the ERT can be utilized using a downloaded free copy of Microsoft Access Runtime. Test program data are transferred into a series of Microsoft Access input screens. Completion of an ERT submittal is typically challenging for the first-time or infrequent user. For this reason, facilities are advised to assess capabilities and resources for electronic reporting and plan accordingly.

Conclusion

Although the ICR for ethylene production facilities was one of the most complex to date, the anticipated proposed rule is likely to include limited measurement requirements; however, emission limits associated with those measurement requirements may still prove challenging. As noted, CO CEMS represent a mature and proven technology that should pose minimum implementation problems to ethylene producers; however, there are several challenges associated with HAP metals measurement that may prove difficult for the industry. Although electronic reporting is not particularly difficult, it does take time to become both efficient and proficient with the process.

AIChE Paper Number: 157c

Building Water Resiliency by Balancing Salt Levels at Petrochemical Facilities

William Celenza

Technical Professional Leader – Process

KBR, Inc.

Sarah Shank

Process Engineer

Matrix Technologies, Inc.

Prepared for Presentation at the 2019 Spring National Meeting
New Orleans, Louisiana, April 1-4, 2019

Building Water Resiliency by Balancing Salt Levels at Petrochemical Facilities

William Celenza
Technical Professional Leader – Process
KBR, Inc.

Sarah Shank
Process Engineer
Matrix Technologies, Inc.

Abstract:

Building resiliency into a water supply enables industrial production to be maintained during droughts, floods, or other episodes that impact source water availability or quality. The quantity of water used in petrochemicals production is influenced by available source water quality; cooling system blowdown quantity and quality, and the resultant treated effluent quantity and quality from onsite wastewater treatment. Balancing elevated Total Dissolved Solid (TDS) or salt levels can provide for expanded reuse options within petrochemical facilities, and provide for increased resiliency.

This paper discusses the dependency of a petrochemical facility's largest water user, evaporative cooling towers, on source water quality and the ability to successfully manage elevated TDS levels within these streams. Developing reuse options helps ensure flexibility in water usage to better match source water variability, and approach the challenges of water resource management with integrated planning.

Introduction

Various water sources may be considered for use in petrochemical facilities. Sources of freshwater are categorized as surface water or groundwater. A majority of U.S. water is withdrawn from surface sources, while less than a quarter comes from ground sources (1). Local surface or groundwater, depending on quality, may be used directly in industrial applications with little or no preliminary or primary physical treatment. However, the driving forces encouraging water reuse such as regional water shortages, regulatory requirements, and/or corporate waste reduction goals have led to the consideration of lower quality sources. Water of lower quality potentially available to a petrochemical facility could include: seawater; brackish groundwater or surface water; reclaimed wastewater from onsite or another municipal or industrial facility; desalinated water such as from cooling tower or boiler blowdown; utility water such as condensate or softener backwash / rinse water, and / or remediated groundwater. Resiliency, or the ability to consistently meet performance requirements despite the occurrence of adverse events, can be achieved by providing critical infrastructure to manage these sources for reuse. Water resiliency can encompass developing new ways to deal with floods, maintaining industrial production when drought strikes, or finding new ways to guarantee a plentiful and affordable water supply.

The purpose of this paper is to help quantify and characterize makeup water supply needs for the main water user at a petrochemical facility, the evaporative cooling water system. The paper can be used to assist in the definition of potential water supply options, along with treatment requirements, to offset use of high quality freshwater sources. Options to supply cooling water makeup are characterized into the following two categories:

- Treatment and Reuse: Brackish Surface Water or Groundwater, on or offsite Wastewater Treatment Plant (WWTP) secondary treated effluent, or cooling tower blowdown.

- Direct reuse: Tertiary and filtered WWTP treated effluent, Reverse Osmosis (RO) Reject, Softener Backwash and Rinse (low conductivity), Boiler Blowdown, Condensate and Storm Water.

Less than 10% of the water supplied to an ethylene production process sees a direct process use. The supply water used as cooling tower makeup can account for 90% of water use in olefin production (2). Note also that cooling tower blowdown can contribute over 75% of the total process wastewater, or dry weather flow, in ethylene production. Even in petrochemical intermediates production, such as purified terephthalic acid (PTA), over 50% of the water supplied is reported to be used as cooling tower makeup and up to 40% used as deionized water system feed (3). With such a relatively large makeup requirement, the cooling towers offer the best possibility of reuse. Examples of existing water resiliency efforts at petrochemical and refining sites are presented in Table 1. Examples cover water reuse programs that limit freshwater feed by reusing water as cooling tower makeup or boiler feed water. The intent of Table 1 is to capture different examples of water reuse systems, such as reuse of municipal wastewater treated effluent, onsite wastewater treated effluent, storm water or combinations.

Table 1 Water Resiliency Programs Currently Practiced at Petrochemical / Refinery Sites

Site	Resiliency Program	Reuse Treatment Methods
	U.S.	
Toledo Refining Company, PBF Energy [4]	50% of treated onsite wastewater effluent recycled as cooling tower makeup.	API, DNF, biotreatment, clarification, sand filters.
Torrance CA Refinery, PBF Energy [5, 6]	Treated Title 22 water from municipal water district provides 70% of cooling tower water makeup.	3,000 gpm to nitrification plant for ammonia removal before reuse, although TDS double prior groundwater source.
Chevron Oil Refinery, El Segundo, CA [7]	Secondary treated municipal effluent used to supply BFW.	MF and RO produces 2,950 gpm, 80% of process demand
Flint Hills Resources Refinery, Pine Bend, MN [8, 9, 10]	Reuse of onsite treated wastewater and storm water to generate BFW.	API, DAF, Activated Sludge with PAC, clarifier, media filter, UF, RO.
	Outside U.S.	
CAPCO PTA Plant, Kaohsiung, Taiwan [10, 11] (rates are design capacities, operational rates have varied)	1,650 gpm combo of onsite secondary wastewater effluent / CT blowdown treated to become 1,200 gpm of feed to DI system.	ChemOx (NaOCl), dual media filtration, GAC adsorption, cartridge/UF, UV disinfection, RO and degasification.
Bahrain Petroleum Company (BAPCO) Refinery, Sitra [12, 13]	Reduce brackish water use and discharge to the sea, by increasing water reuse for cooling towers.	Four-stage 5,000 gpm MBR treats wastewater to industrial reuse standards.
Jazan Refinery and IGCC power plant, Saudi Arabia [14]	A 14,700 gpm RO desalination plant fed by combo of seawater and onsite treated wastewater.	Onsite wastewater 4,400 gpm biotreatment plant with MBBR, high rate clarifiers, discfilters.
Antonio Dovali Jaime Refinery, Oaxaca, Mexico [15]	Use of tertiary treated refinery wastewater for cooling water makeup.	Activated Sludge, chemical precipitation, single/dual media filtration, disinfection.
Shell Pearl GTL, Qatar [16,17] (Note: operates as a no liquid discharge facility)	1,400 gpm Cooling Tower and Boiler Blowdown treated for reuse as 1,200 gpm of makeup. Process wastewater treated effluent reused as 6,000 gpm of feed to DI system.	For Blowdown: UF and 3-stage RO with 200 gpm reject used for irrigation. For process: API/CPI, DAF, Equalization, Activated Sludge / MBBR, Clarifier, UF, 2-stage/2-pass RO.
Notes:		

API = API Oil/Water Separator	MBBR = Moving Bed Biofilm Reactor
BFW = Boiler Feed Water	MBR = Membrane Biofiltration Reactor
ChemOx = Chemical Oxidation	MF = Microfiltration
CPI = Corrugated Plate Interceptor	PAC = Powdered Activated Carbon
DAF = Dissolved Air Filtration	RO = Reverse Osmosis
DI = Deionized Water	UF = Ultrafiltration
DNF = Dissolved Nitrogen Filtration	UV = Ultraviolet Disinfection
GAC = Granular Activated Carbon	

Quantifying Cooling Makeup Water Demand

Cooling towers are used in the petrochemical and refining industry to dissipate heat load from production as shown in Figure 1. Even though the amount of heat that needs to be extracted from a process is independent of the weather, the design and operation of wet cooling towers is dependent on ambient weather conditions. The design of wet cooling towers needs to ensure adequate cooling for continuous production under expected atmospheric conditions. However 99% of the time, cooling towers operate below design conditions; creating an opportunity to conserve energy or water resources much of the year.

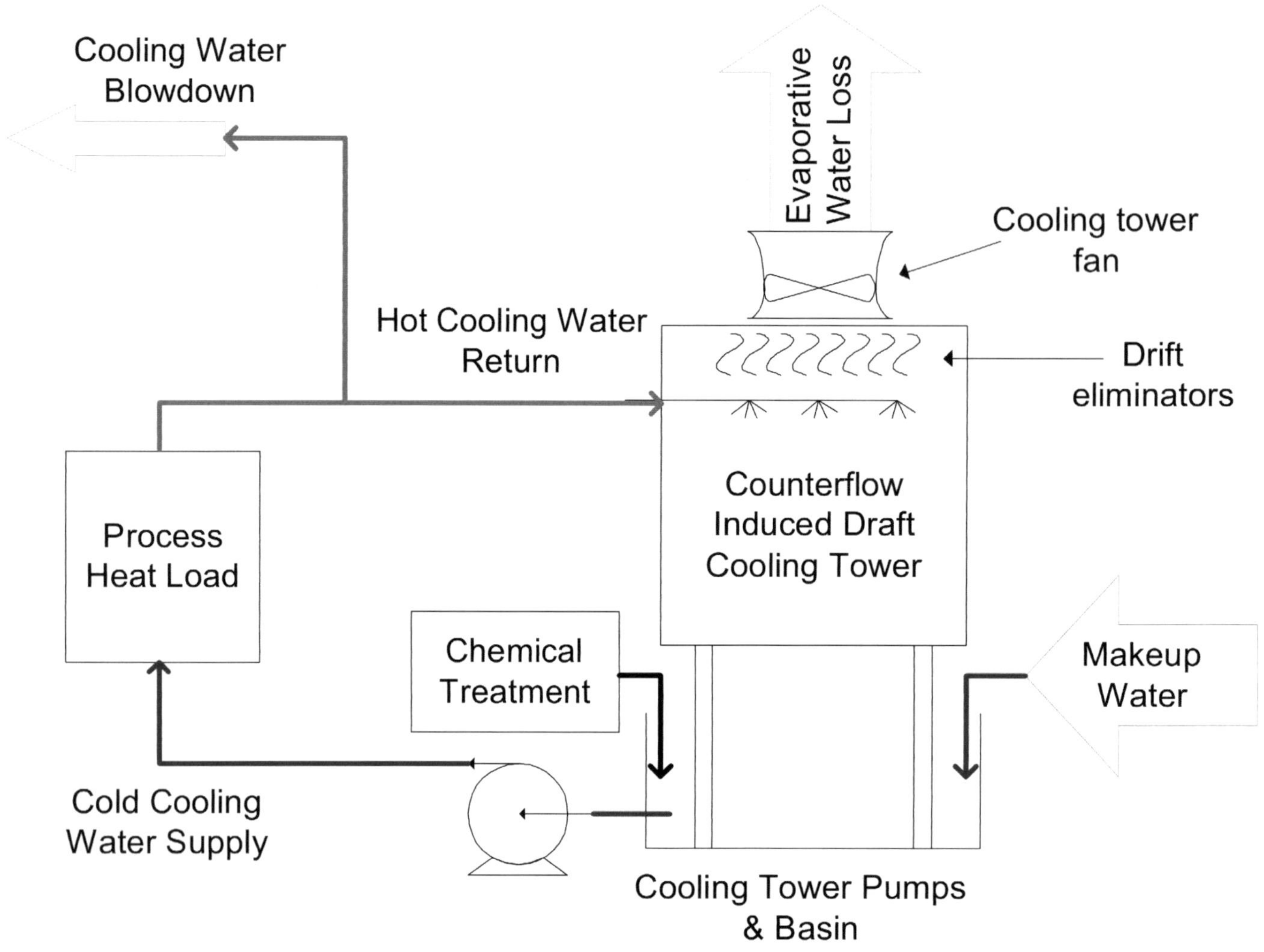

Figure 1 Diagram of an Evaporative Cooling Tower System at an Industrial Facility

Through convective, or sensible, heat transfer and evaporation, or latent, heat transfer the cooling water gives up heat and mass to the air. Even though the heat load imposed on the cooling tower will be constant, the ratio of latent to sensible heat will be determined by the entering air wet-bulb temperature. Seasonal weather changes and fan operating speed will affect the cooling tower operating conditions including the accompanying amount of water lost to evaporation. The net result is that less evaporation can be required for the same total heat load over the course of the year.

Table 2 lists the anticipated monthly average evaporation as a percentage of total cooling water recirculation rate. As indicated in Table 2, as weather conditions vary for a constant heat load, the minimum fan speed to achieve the required cooling will also change. The evaporation rates in Table 2 assume the fan energy use will be minimized by reducing air flow rates during lower temperatures, at the expense of maximizing the water use. The evaporation rates are lowest during winter months when sensible heat transfer is a larger part of the total cooling tower heat transfer and evaporation rates are highest during summer months when latent heat transfer is more dominant (sensible rates are lowest).

Table 2 Projected Seasonal Water Loss to Evaporation from Cooling Tower Operation with a Temperature Range of 20°F in Houston, TX

Month	Average Wet-Bulb Temp (F)*	% Total Heat Transfer Due To Sensible Heat Exchange	Evaporation as a % of Recirculating CT Water		
			% Design Air Flow Rate		
			90	95	100
Jan	48	31	1.03		
Feb	51	30		1.05	
Mar	56	25		1.12	
Apr	62	21		1.19	
May	69	16			1.25
Jun	73	12			1.31
Jul	75	10			1.35
Aug	75	10			1.35
Sep	72	13			1.30
Oct	65	18		1.23	
Nov	59	24		1.13	
Dec	51	30		1.04	
SUMMER DESIGN CASE	79 **	5.8			1.41
WINTER DESIGN CASE	29 **	44.2	0.84		

* "Engineering Weather Data Handbook" for Houston TX, Period of Record 1973-1996; compiled by the Air Force Combat Climatology Center (AFCCC). Correspondence via National Climatic Data Center (NCDC). Wet Bulb psychrometric data calculated from ASHRAE Fundamentals 2001 Chapter 6.

** For Summer Design Condition, the Wet Bulb Temp is 1.0% Occurrence Value. For Winter Design Conditions the Wet Bulb Temp is the Mean Coincident (Average) Value at the 99.0% Dry Bulb Temp Occurrence.

Values in Table 2 are based on a cooling temperature range of 20°F. The range is the temperature difference between the circulating water temperature returned to the tower (the red stream in figure 1) and the cooled water temperature leaving the tower basin to supply process users (the blue stream in figure 1).

Since evaporation of a portion of the circulating cooling tower water is the primary method of cooling, the concentration of dissolved solids and other impurities in the recirculated water will increase as the evaporation rate increases. To maintain the quality of the circulating water, a purge rate or blowdown that is directly proportional to the amount of evaporation is required. Table 3 lists anticipated blowdown rates of a cooling tower in Houston, Texas with monthly average wet bulb temperatures listed in Table 2.

Table 3 Blowdown rate as a percentage of Cooling Water Circulating Rate in Houston, TX

	Cycles of concentration (COC)				
Cooling water "Range," °F	**1.5**	**2.0**	**3.0**	**4.0**	**5.0**
10	0.7-1.6	0.4-0.8	0.2-0.4	0.1-0.3	0.1-0.2
20	2.4-3.3	1.2-1.7	0.6-0.8	0.4-0.6	0.3-0.4
30	4.5-5.4	2.3-2.7	1.1-1.3	0.8-0.9	0.6-0.7

The term "cycles of concentration" or COC refers to the ratio of minerals between the blowdown water and makeup water. COC is dependent on makeup water quality and indirectly proportional to the blowdown rate, and thus also a function of the evaporation rate. To operate at a COC of 4, the makeup water would be concentrated by a factor of 4 in the recirculating cooling water by control of the blowdown rate. To maintain a COC, the blowdown rate would need to be changed to compensate for changes in the evaporation rate.

Total makeup water is the sum of evaporation loss + blowdown + drift loss + process leaks; however, drift and leak losses are both typically one to two orders of magnitude less than evaporation or blowdown losses. Drift eliminators capable of achieving efficiencies from 0.002% of the recirculating water to high efficiencies of 0.0005% can be installed on a cooling tower. The use of high performance drift eliminators increases the operating horsepower of cooling tower fans as drift eliminators create pressure drop. However, the potential for particulate emissions from the cooling tower operation would also be reduced. The choice of drift eliminators is also subject to site air permit considerations, with a source water choice such as reclaimed treated municipal wastewater potentially subject to California Department of Public Health Services Administration Code Title 22 type requirements (18). Although process leaks are inevitable, dedicated maintenance programs can limit losses.

Table 4 presents total makeup water demand as a percentage of the cooling water recirculation rate for a cooling tower in Houston, Texas. Drift or windage loss and process leaks are not included.

Table 4 Makeup Water as a percentage of Cooling Water Circulating Rate in Houston, TX

	Cycles of concentration (COC)				
Cooling water "Range," °F	**1.5**	**2.0**	**3.0**	**4.0**	**5.0**
10	1.1-2.4	0.7-1.6	0.6-1.2	0.5-1.1	0.4-1.0
20	3.6-5.0	2.4-3.3	1.8-2.5	1.6-2.2	1.5-2.1
30	6.8-8.1	4.5-5.4	3.4-4.0	3.0-3.6	2.8-3.4

Cooling water recirculation rates for ethylene production facilities are typically measured in the 100,000's of gpm. For a 100,000gpm cooling water recirculation rate operating at a COC of 4 and a range of 20F° under summer wet bulb temperatures, the makeup water requirement is 2,200gpm. Under winter wet bulb temperatures the makeup water requirement would be 1,600gpm.

Qualifying Cooling Tower Makeup Water Demand

Makeup water quality objectives depend on the equipment to be cooled. Hardness, TDS, silica, COD and/or chloride concentrations are often the cycle-limiting factor for recirculating water quality. In the petrochemical and refining industry, maximum values of typical limiting constituents for cooling applications are listed in Table 5 along with treatment technologies. Dividing the values in Table 5 by the intended COC would indicate the required maximum concentration for source water to serve as cooling tower makeup. Note that the values in Table 5 are maximums and typical values would be less; however, extended operation at elevated levels of the constituents would serve to increase corrosion and deposition that would reduce equipment efficiency and increase the energy use of cooling tower systems.

Table 5 Recommended Limits on Recirculating Water Quality (19, 20, and 21)

Constituent	Maximum Value (ppm unless noted) (a,b,c)	Affected Area	Technology Application (4,7)
Ammonia	<50 if copper alloys present	I,III	F, I, K, L, M, N
Amine	<25	I,III	K,N
Biological Oxygen Demand (BOD) 5-Day	<200	III	A,B,D,F,H,I,K, L
Calcium hardness (as CaCO3)	<1,500	II,V	E,G,I

Chlorides	Galvanized Steel <750 300 Series Stainless Steel <1,500	I	H,I,J
Chemical Oxygen Demand (COD)	<200	III	A,F,H,I,O,P
Conductivity, Specific	<10,000 Micromhos / cm	I	G+H,I,J
Hydrocarbon - Total	<50 (or less with film type fill)	II,III	B,D,F,I,K
Hydrocarbon - light	<50	III	B,D,F,I,K
Hydrocarbon - heavy	<25 (or less with film type fill)	II,III	B,D,F,I,K
Hydrogen Sulfide	<10 (or less if safety considerations are of concern)	I,III	A,E,F,H
Iron	<5	II	A,B,E,G,I,J
Magnesium	<1,000	II	E,G,I,J
Manganese	<1 (or less due to copper and stainless steel concerns)	I,II	A,B,E,G,I,J
pH	6.5 to 9.0 standard units	I	Q,R,S
Phenols	<20	II,III	A,F,I,K,O
Phosphate, Total	<50	II	L
Silica (as SiCO2)	<200	II	E,H,I
Sulfates	<5,000. But, if Calcium >800, then <800	V	H,I,J
Temperature	< 150°F on Return Water to Tower	IV	T
Total Dissolved Solids (TDS)	<5,000	VI	S,I,J
Total Hardness	<2,000	II	E,G,I
Total Organic Carbon (TOC)	<200	III	A,F,H,I,N,O,P
Total Suspended Solids (TSS)	<200, <25 for film type fill	II	B,C,D

Notes:

I Corrosion II Fouling III Microbiological Fouling	IV PVC Components Degradation V Scaling VI Thermal Performance Degradation
A Chemical Oxidation B Filtration C Clarification D Physical Separation – API, DAF/IAF E Lime or Soda Ash Softening F Air / Steam Stripping G Cation Exchange H Anion Exchange I Reverse Osmosis J Electrodialysis reversal (EDR) K Biological / Secondary Treatment	L Nutrient Removal: aerobic, anoxic, anaerobic M Ion Exchange N Nitrification O Ozonation P Carbon Adsorption Q Equalization R Neutralization S Blending T Cooling

a) Limits are intended for recirculated water, not as measured in the tower.

b) Limits do not account for safety considerations, or environmental permit compliance.

Note that petrochemical and refinery cooling water quality criteria are generally less strict than the criteria for power plants as the quality can directly affect power plant performance.

Makeup Water Source Options

To address local or regional water resource limitations, petrochemical facilities can consider onsite sources for direct reuse without treatment, or indirect reuse with prior treatment. Raw wastewater at petrochemical facilities typically contain the constituents listed in Table 5 that require a combination of treatment methods to achieve an effluent suitable for reuse as cooling tower makeup. A typical petrochemical wastewater treatment system may include equalization, neutralization, coagulation/flocculation/precipitation, floatation/sedimentation/filtration, clarification and biodegradation (e.g., trickling filter, anaerobic treatment, and aerated lagoon, rotating biological contactor and activated sludge), with denitrification/nitrification and possibly post-denitrification. A final polishing step using filtration, ozonation, activated carbon, or chemical treatment may also be required based on environmental regulations. Advanced biological treatment methods, such as membrane bioreactors (MBR) or Moving Bed Biofilm Reactors (MBBR), now form a sizable portion of wastewater treatment equipment for petrochemical applications, due in part to a growing preference for non-chemical treatment processes (22) and improved micropollutant and solids management.

Figure 2 depicts average concentrations of the main constituents in petrochemical process wastewater streams (23) when considering reuse. Three values are listed for each constituent: typical concentration in "raw" or untreated petrochemical wastewater, a target concentration in order to reuse the water based on 4 COC and finally a maximum concentration limit of the component in recirculating cooling tower water. Although 4 COC are depicted in Figure 2, the treatment level required will depend on the COC to be used. Note the concentration axis on Figure 2 is logarithmic, with the average raw wastewater concentration of TDS shown as up to 10,000 ppm and the required treatment for 4 COC being around 1,500 ppm. Thus an average of 85% reduction would be required for TDS.

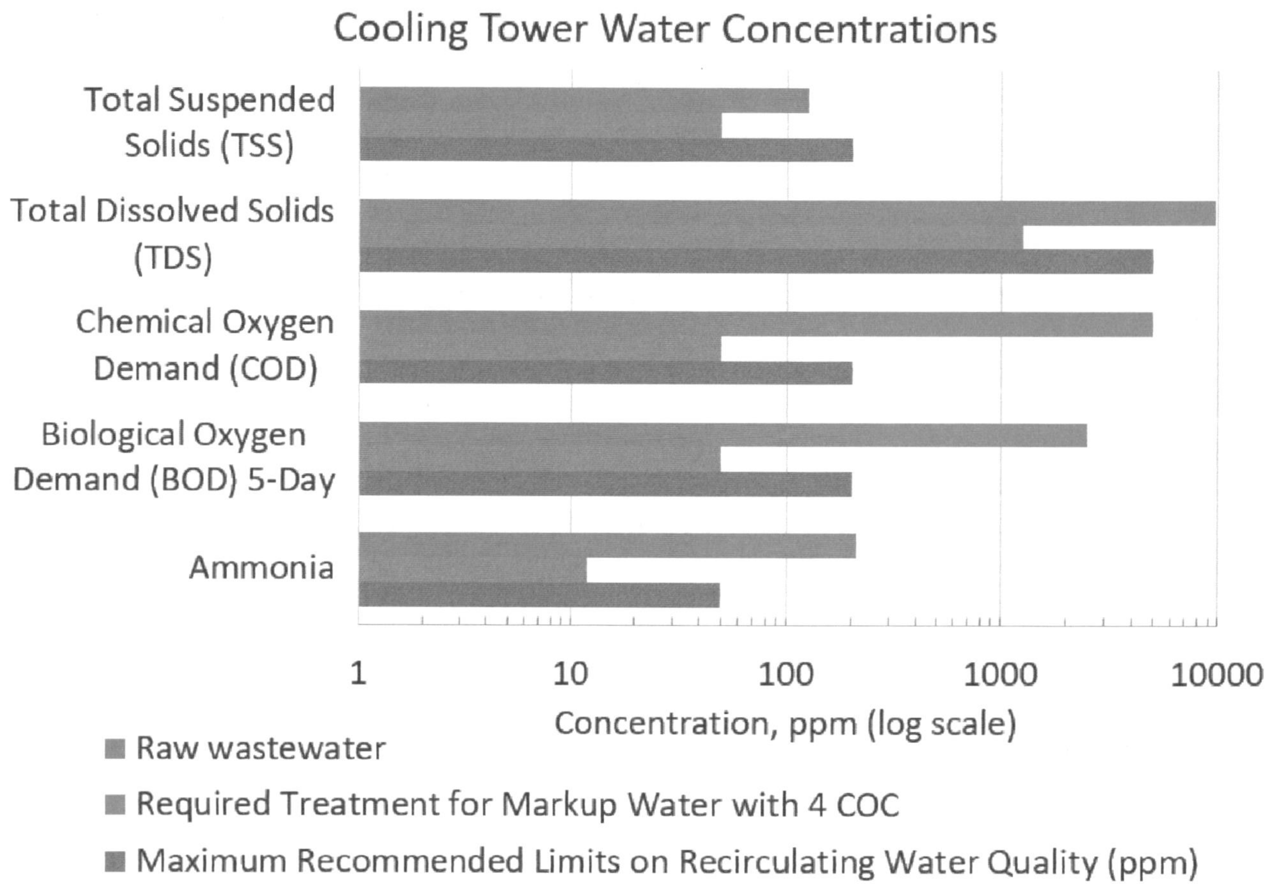

Figure 2 Petrochemical Plant – Raw Wastewater vs. Treatment Target Level for Reuse

With the exception of TSS, the COD, BOD and Ammonia would on average require over 90% reduction. Reducing constituent concentrations in onsite raw wastewater low enough to be suitable for cooling tower makeup would require the advanced biotreatment systems listed in the right column of Table 5. The TDS reduction itself would require post biotreatment to achieve reuse levels.

Other lower quality sources such as seawater, brackish groundwater (1,000 to 10,000 ppm TDS), surface water (>1,000 ppm TDS), or cooling tower blowdown would likely not require secondary biological treatment, but would need varying degrees of pretreatment and filtration for TDS reduction.

Petrochemical facilities can also consider onsite water resources for reuse directly, such as those listed in Table 6. Although direct reuse of a source such as RO Reject would be limited due to TDS and chloride, even a low percentage could be suitable for blending if the reject does not include chemical residues. Overall, storm water would seem to offer the most blending opportunities; assuming onsite storage capacity is available.

Table 6 Direct Reuse Water Sources for Petrochemical Plant vs. Target Level for Use as Cooling Tower Makeup Water (24, 25)

Characteristic (all values in ppm unless noted)	Storm Water	Reverse Osmosis Reject	Boiler Blowdown	Condensate Water	Tertiary Treated Municipal Wastewater	Target Level for Use as Makeup at 4 COC (*)
Ammonia	0	0	0	5	1	<12
Calcium hardness (as CaCO3)	2	0	0	0	150	<375
Chlorides	2	500	30	12	180	<185 to 375
Nitrates (as NO3)	0	0	0	0	55	<200
Phosphate, Total	0	3.5	30	0	4	<12
Silica (as SiCO2)	1.5	9	20	0	10	<50
Total Hardness	2.5	0	0	0	250	<500
Total Alkalinity (as CaCO3)	5	50	300	30	230	<500
Total Dissolved Solids (TDS)	35	1,500	1,000	50	700	<1,250
Total Organic Carbon (TOC)	8	5	20	0	12	<50
Notes:						
*Derived from Table 5 Maximum Values.						

As shown in Figure 3, relying on multiple sources and blending, subject to availability, to address the limitations of some treatment methods could also provide margin to manage upsets and variability.

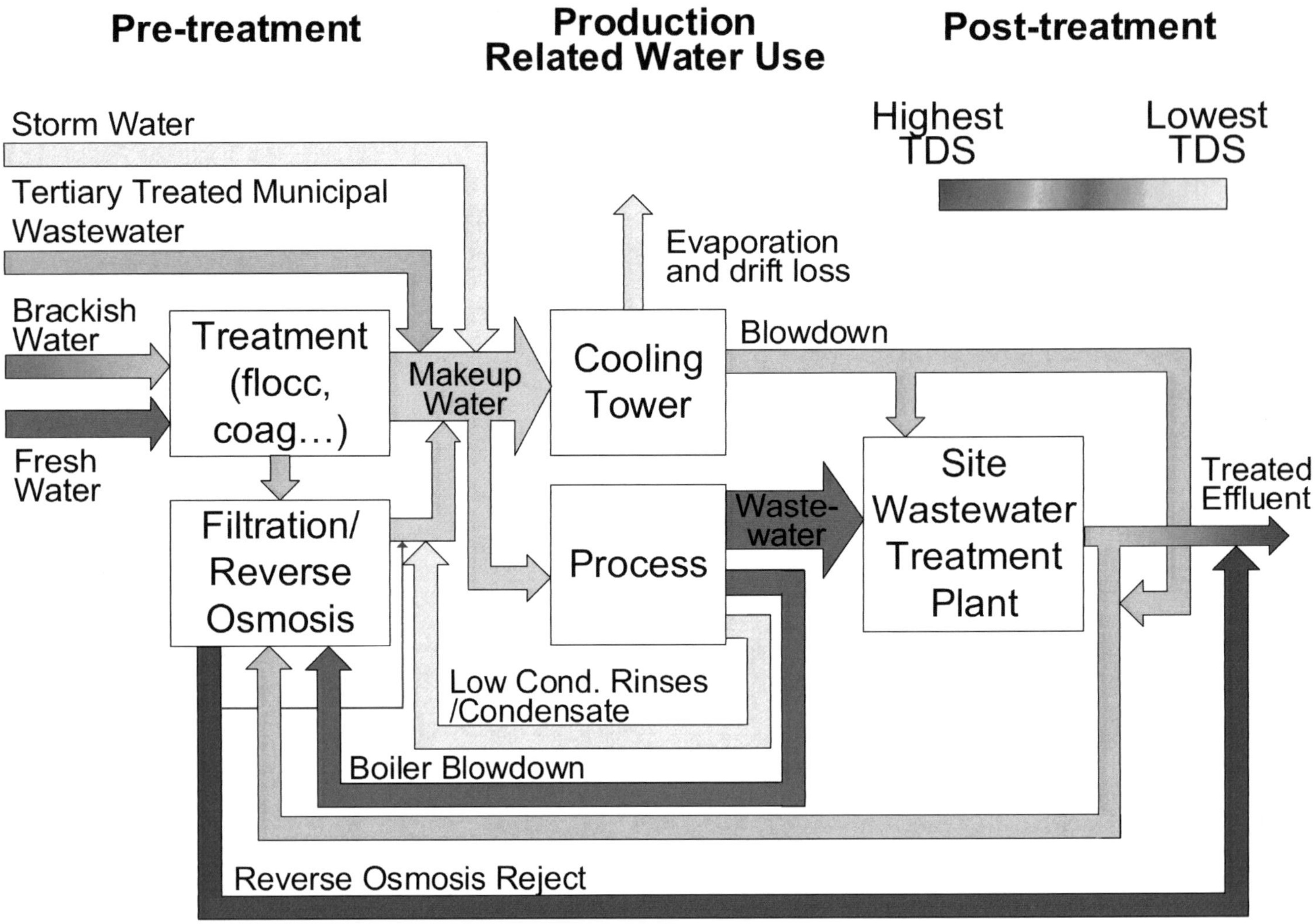

Figure 3 Salt Level Balancing for Various Reuse Streams

Figure 4 provides a schematic of how a filtration / RO system might connect into a site reuse network. RO is a pressure-driven desalination technique, requiring 1 psi for every 100 ppm of TDS removal. Most RO feed systems are designed for 75% recovery of the feed source water as product permeate. However, recovery can range from 40% for seawater and wastewater to as high as 85-90% for low TDS water (26).

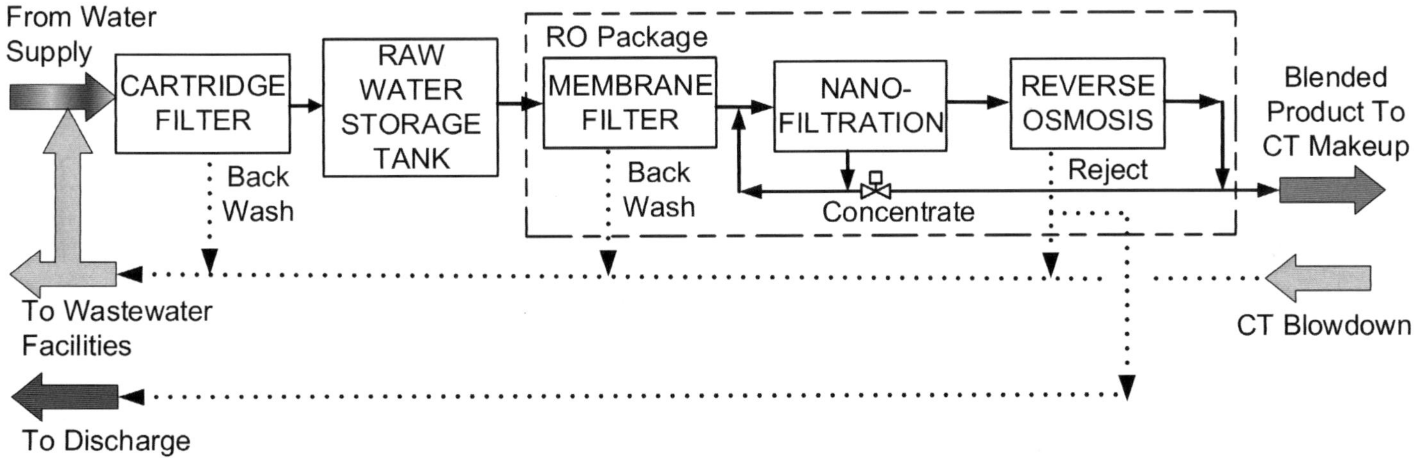

Figure 4 Example Water Filtration Treatment System

Potential Limiting Factors

The ability to reuse direct or indirect sources of water supply can be limited by the following conditions.

Recalcitrant COD / Nitrogen Compounds

The complexity of petrochemical wastewater can induce instability in a treatment system. Petrochemical wastewaters are characterized by the presence of:

- Recalcitrant, or non-biodegradable, organic substances
- Nitrogen compounds
- Heavy metals

These constituents can be present at toxic concentrations during biotreatment. A frequent result has been slower biokinetic rates due to biomass inhibition and suspended solids carryover in the effluent. However, advanced biotreatment systems have been shown to reliably improve effluent quality in some applications (12, 27).

Safety and Permitting

Safety considerations may prevent using sources that contain constituents which may produce toxic air emissions during handling, such as H2S, Mercaptans, and Ammonia. Reclaimed treated municipal wastewater subject to the California Department of Public Health Services Administration Code Title 22 includes disinfection and testing requirements in both the production and use of recycled water in cooling towers (18, 28). While risks related to microorganisms and particulate matter levels in cooling tower drift have received consideration in recent years, less attention has been given to contaminants of emerging concern (CECs), which are present in reclaimed water (29). If exposure is greater than health-based thresholds, such as minimum infective doses for pathogens, PM standards, or minimal risk levels (MRLs) for chemicals, then risks could be considered significant and require mitigation through additional treatment or greater setback distances from the towers.

RO Reject / Brine Management

Permit limits on discharge of high TDS reject streams may prevent reuse of even low to moderate TDS level streams.

Variable Quality of Source Water

Variable water quality is an inherent problem with any wastewater. Reliance on wastewater treated effluent sources would be subject to upset in both concentrations and flow rates. To compensate, a reuse system may need to operate at reduced capacity; or avoid a component stream and rebalance blending. Alternatively, the operational COC of the cooling tower water can be shifted to compensate.

Turndown for Makeup Water Demand

As shown in Tables 2, 3 and 4, the makeup water demand can average 40% less during the coldest month compared to the warmest month and over 60% between summer and winter design conditions.

Concluding Thoughts

This paper reviewed a methodology for approaching the design of water resilient systems at petrochemical facilities in order to provide solutions to fresh water availability constraints. As freshwater becomes increasingly scarce and expensive, designing to minimize fresh water consumption will involve use of alternative, lower quality source waters. More complex designs can offer the advantage of significantly less freshwater consumption; however, compared to traditional designs this advantage is at the expense of increased operating requirements and more equipment. Providing options for water reuse will become more of a concern for many facilities, if only to have a plan in place to remain in compliance with local water-source regulations. Approximately 60% of refining and petrochemical sites are subject to a water stress that is considered medium to high, and 25% are located in areas where precipitation is expected to decrease (30). The U.S. has not taken advantage of saline water as a source of freshwater, with only 14% of total U.S. water withdrawals coming from a saline source and most of that being from brackish groundwater. While the U.S. is starting to more confidently rely on the use of alternative, lower quality source waters such as through brackish water desalination, a majority of the world's desalination capacity is devoted to desalting seawater (1). Utilizing these largely untapped resources, reusing waste streams or building in source water variability can help keep a petrochemical facility operating during times of water stress.

References:

1. Bartels, Lauren. (2017) The Ocean Holds Solutions to Water Sustainability Challenges, Chemical Engineering Progress (CEP), AIChE, September.
2. Stone, John C. et al (1975), An Integrated Power Process Model for Ethylene Production, Water Resources Research Vol 11, No. 9 December.
3. Wong, Joseph, (2000) Testing and Implementation of an Advanced Wastewater Reclamation and Recycling System in a Major Petrochemical Plant, Water Science and Technology, September.
4. (2013) Sustainable Water Treatment Solution Reduces Costs, Conserves Water at Ohio Refinery, Industrial WaterWorld, September Volume 13 Issue 5.
5. Bresnahan, W.T. (1996) Water Reuse in Oil Refineries, NACE Annual Conference, Paper No. 567
6. McCarron, Suzanne (2016) Smarter Ways to Use Water, ENERGYFACTOR – ExxonMobil Perspectives Blog, https://energyfactor.exxonmobil.com/perspectives/smarter-ways-use-water/ , March 25.
7. Puckorius, Paul, et al (1995) Reclaimed Water as Cooling Tower Makeup for Refinery/Petrochemical Plants – Southern California's Activities and Time Table, Cooling Tower Institute Annual Meeting, Paper No. TP 59-15, February.
8. (2014) Environmental Assessment Worksheet, Flint Hills Resources – Tier 3 Clean Fuels Projects, Section 11. Water Resources, December.
9. (2018) National Pollutant Discharge Elimination System (NPDES) / State Disposal System (SDS) Permit Program Fact Sheet, Permit Reissuance MN0000418, Minnesota Pollution Control Agency, August.
10. Wong, Joseph, (2014) Advanced Water Reuse in Petrochemical and Refining Facilities: Global Perspectives and Case Studies, Texas Industrial Water Management Forum: Industrial Water Reuse and Desalination, Houston, TX November 6.

11. Wong, Joseph. (2012) Water Reuse for Petroleum Oil, Product Processing Industries, Industrial WaterWorld, March/April.

12. Arnsten, Brian, et al. (2016) The Use of Membrane Bioreactor Technology in Refinery Applications for Stringent Discharge Requirements and Reuse Applications, AFPM Annual Meeting Paper AM-16-79 March.

13. Zaffaroni, Carlo (2014) Treating Industrial Wastewater at Bahrain Petroleum Company's Refinery, CH2M Hill's Access Water blog http://www.ch2mhillblogs.com/water/2014/06/01/treating_industrial_wastewater/

14. (2016) Veolia Chosen to Engineer and Deliver a Large Scale Wastewater Treatment Plant for Aramco's Jazan Refinery in Saudi Arabia, Business Wire – Regulatory News, November 28.

15. Baron, C, et al. (2000) B.O.O. Case: water management project for the use of reclaimed wastewater and desalted seawater for the "Antonio Dovali Jaime" refinery, Salina Cruz, Oaxaca, Mexico, Water Science and Technology, Sept.

16. Migliavacca, Michele (2015) Shell Pearl GTL – Qatar Wastewater Reuse and Zero Liquid Discharge. Veolia Refinery-Petrochemical Seminar October 21st, Houston TX.

17. Onwusogh, U (2015) Feasibility of Produced Water Treatment and Reuse – Case Study of a GTL Plant, International Petroleum Technology Conference, Doha, Qatar 6-9 December, Paper 18354-MS.

18. California Department of Health Services. 2018. California Code of Regulations, Title 22, Division 4, Chapter 3, Article 3, §60306. Use of recycled water for cooling. California Department of Health Services, Sacramento, California. October.

19. Hall, Steven (2012), Rules of Thumb for Chemical Engineers, Elsevier, Chapter 10 Cooling Towers.

20. Eble, Karen; Jennifer Feathers (1992), Water Reuse Within A Refinery, 1992 NPRA Annual Meeting. March 22-24, New Orleans, Paper AM-92-08, Table 1.

21. Sampson, Dan. (2012), Think Water When Designing CSP Plants, Power Magazine, May.

22. Tong, Flora; Peter Aerts (2009), RO and Petrochemicals, Water & Wastes Digest, September.

23. Verlicchi, Paola, et al. (2011) Efficiency and Reliability of Upgraded Industrial Treatment Plant at Porto Marghera, near Venice, Italy, in Removing Nutrients and

Dangerous Micropollutants from Petrochemical Wastewaters, Water Environment Research, V83, N8, August.

24. Puckorius, Paul. (2016) Recycled Waters for Cooling Tower Systems – Good Idea or Bad – What You Need to Know, NACE International Corrosion Conference, Paper No. 7151.

25. Puckorius, Paul. (2013) Consider Recycled Water for Your Cooling Tower Makeup, Chemical Engineering Progress (CEP), AIChE, February.

26. Kucera, Jane (2019), The Essentials of Reverse Osmosis, Chemical Engineering Process, February.

27. Benedetti, Luigi, et al (1976) Prevention of Water Pollution in the Petrochemical Industry, Pure & Applied Chemistry, Vol. 45 pp. 147-151.

28. Robinette, Daniel (2008), The Impact of Reclaimed Water of Two Different Qualities on Two Similar Power Plants, NACE Corrosion Conference, Paper No. 08-373.

29. U.S.EPA (2012), Guidelines for Water Reuse, Appendix D – Assessing Contaminants of Emerging Concern (CECs) in Cooling Tower Drift, pp D-107 to D-109, Office of Wastewater Management, EPA/600/R-12/618, September.

30. Sentenac, Jean (2016), Improved Design Practices Can Aid Environment, Energy Transitions, Hydrocarbon Processing, September.

Further Reading:

31. Levine, Audrey; Takashi Asano (2002), Water Recycling and Resource Recovery in Industry, Chapter 2 Water reclamation, recycling and reuse in industry, IWA Publishing.

32. Hansen, Everton, et al. (2016) Wastewater Reuse in a Cascade Based system of a petrochemical industry for the replacement of losses in cooling towers, Journal of Environmental Management, June

AIChE Paper: 157d

TREATMENT OF SPENT CAUSTIC – CLEANING THE DIRTIEST WASTEWATER

Chad Felch

R&D Director

Siemens Energy Water Solutions

Prepared for Presentation at the 2019 Spring National Meeting City, New Orleans, LA, March 31 - April 4, 2019

TREATMENT OF SPENT CAUSTIC – CLEANING THE DIRTIEST WASTEWATER

Chad Felch

R&D Director

Siemens Energy Water Solutions

Abstract:

Spent caustic is arguably the world's most foul industrial wastewater. It is generated in the removal of acidic components—hydrogen sulfide (H_2S) and mercaptans, which are toxic, corrosive, and extremely malodorous—and carbon dioxide. These compounds make conventional wastewater treatment tough for several reasons:

- Toxicity to the biological wastewater treatment facility
- High concentrations of chemical oxygen demand and total dissolved solids
- Health, safety, and environmental concerns for people onsite and nearby

Additionally, the quantity of hydrogen sulfide and mercaptan can produce obnoxious odors that will extend far beyond the cracking facility's fence line.

Oxidation is the best approach to treating spent caustic because its effluent is non-reactive inorganic salts and low-molecular-weight carboxylic organics. That's why, after decades of proven, reliable use worldwide, cost-effective wet air oxidation (WAO) has become a preferred oxidation technology for treating spent caustic.

This presentation will outline a WAO process that Siemens has been using to produce high quality oxidized effluent from ethylene spent caustic over the past 25 years. The presentation will include a case study from SINOPEC Zhenhai Refining & Chemical Company (ZRCC). SINOPEC ZRCC is China's largest integrated refinery, with a crude oil processing capacity of 23 million metric tons a year.

Located in metropolitan Shanghai with residential areas just 6.6 kilometers (3.7 miles) away, the plant produces more than 50 refining and chemical products. Among them are gasoline of different grades, diesel, jet fuel, asphalt, polypropylene plastics, and ethylene. The plant uses WAO technology to process a spent caustic flow of 18.2 m^3/hr generated from its ethylene production.

This presentation will also briefly cover the most recent technology upgrades for treating spent caustic, including the recent coupling of WAO with a combination of powered activated carbon treatment (PACT®) and membrane bioreactor (MBR) technologies for a greatly simplified, end-to-end industrial water treatment solution.

MAIN BODY

Spent caustic is arguably the world's most foul industrial wastewater. It is generated in the removal of acidic components—hydrogen sulfide (H_2S) and mercaptans, which are toxic, corrosive, and carbon dioxide. These compounds make conventional wastewater treatment difficult for several reasons:

- Toxicity to the biological wastewater treatment facility
- High concentrations of chemical oxygen demand and total dissolved solids
- Health, safety, and environmental concerns for people onsite and nearby

Additionally, the quantity of hydrogen sulfide and mercaptan can produce obnoxious odors that will extend far beyond the cracking facility's fence line.

Spent caustic consists of biorefractory, hazardous, and inhibitory reduced sulfur compounds. It also contains high dissolved solids, and its salts mass fraction typically approaches 10 percent by weight. This composition makes treating spent caustic biologically quite difficult, but it can also create problems with human health and extreme odors in the biological treatment process area.

Another issue is reduced sulfur compounds that become volatile when the pH is reduced to a level necessary for biological treatment. This step releases hydrogen sulfide and mercaptans, both of which are malodorous and hazardous to human health even in small concentrations.

Ethylene spent caustic has an exceptionally high chemical oxygen demand (COD). It can range from 10,000 to 50,000 mg/L. With spent caustic flow rates between 5 and 20 m^3/h, the kilogram COD load can often surpass the combined amount found in a water treatment plant's other waste streams.

Oxidation is the best approach to treating spent caustic because its effluent is non-reactive inorganic salts and low-molecular-weight carboxylic organics. That's why, after decades of proven, reliable use worldwide, cost-effective wet air oxidation (WAO) has become a preferred oxidation technology for treating spent caustic.

The WAO process completely destroys sulfides while converting the remaining organics to biodegradable, short-chain compounds. The latter is done by heating the sulfide-laden caustic stream in a heat exchanger, putting it into a reactor vessel for a prescribed residence time, cooling it back down, and separating the off-gas and oxidized liquid effluent (Figure 1). It is a simple, reliable process with virtually no downtime.

In this process, oxygen reacts with sodium sulfide ($NaHS$ and Na_2S) to form sodium thiosulfate ($Na_2S_2O_3$), then further reacts to form sodium sulfate (Na_2SO_4), reducing odor problems. Mercaptans (C_2H_5SH or CH_3SH) are turned into sulfonates ($CH_3\text{-}SO_3Na$).

Both of these conversions eliminate odor issues and break down any larger organics into smaller ones that are easier to treat. By contrast, if the stream was not pretreated, neutralization and aeration in a biological wastewater system would release malodorous hydrogen sulfide and mercaptans into the air.

For further polishing, it's recommended that industrial producers consider combining WAO with biological powdered activated carbon treatment (PACT®) and membrane bioreactor

(MBR) technologies. This combination can treat the oxidized spent caustic in a dedicated fashion to discharge criteria thereby not requiring treatment in the facility wastewater treatment plant. In facilities that desire to reuse water, the dissolved solids (salts) in the spent caustic create challenges. By treating the spent caustic in a dedicated fashion, it allows more the facility wastewater to be amenable for reuse.

In practice, a PACT-MBR solution combines three distinct treatment processes — biological, physio/chemical, and ultrafiltration liquid-solids separation — in a single, compact, and robust treatment technology. It achieves some of the lowest chemical oxygen demand (COD) and total organic carbon (TOC) effluents in refining and petrochemical applications and can be used at high salt concentrations. Testing has shown that concentrations in excess of 10% solids can be effectively treated. It also yields best-in-class performance for treatment of water in recovery and reuse programs.

The PACT stage integrates the biological reaction steps of conventional activated sludge with the mass transfer kinetics of carbon adsorption. This reduces footprint and capital investment requirements, as compared to conventional discrete biological treatment coupled with activated carbon filtration. The PAC buffers the biomass from toxins including the high TDS environment of oxidized spent caustic, delivering high-quality effluent where conventional activated sludge treatment cannot satisfy.

The MBR stage is used to clarify the final water discharge. Biological treatment of highly saline solutions hinders gravity clarification for solids separation. Membrane filtration is key to complementing biological treatment of solutions with high salt concentrations in oxidized spent caustic. The membrane provides a physical barrier for complete solids removal without polymer.

WAO, A Case Study

WAO is the approach chosen by China's largest integrated refinery and chemical production facility, the SINOPEC Zhenhai & Chemical Company (ZRCC). The plant has a total annual crude oil processing capacity of about 23 million metric tons, producing more than 50 refining and chemical products that include more than 1 million metric tons of ethylene. Other products include gasoline of different grades, diesel, jet fuel, asphalt, and polypropylene plastics.

The ZRCC plant is located across Hangzhou Bay from Shanghai. From there, it ships ethylene and its other products to Hong Kong, Japan, the U.S., Korea, India, and many other places in the world. Despite the toxic nature of its industry, SINOPEC prides itself on its environmental, health, and safety record. In fact, the Chinese government awarded ZRCC one of the nation's earliest National Environment-Friendly Enterprises awards.

Good environmental stewardship is vital to the health of the region, including its citizens and the bay. Although the plant is located in a designated industrial zone that is also home to some 200 other refining and chemical plants, metropolitan Shanghai's 24 million people surround it. The city itself lies to the north across the bay, and residential areas to the south and west start just 6 kilometers away.

ZRCC has since won many more environmental awards. Helping it win these awards in the past decade has been the 2009 deployment of an extremely effective wet air oxidation (WAO) technology used to treat toxic and odorous sulfidic spent caustic wastewater from ethylene production.

The ZRCC plant uses solutions of water and sodium hydroxide to sweeten its ethylene product by removing sour components, such as malodorous and potentially dangerous sulfides and mercaptans. Because the spent caustic wastewater that is formed defies standard biological wastewater treatment, it must be pre-treated.

For these reasons, the ZRCC plant needed a way to pre-treat its spent caustic before it entered the facility's biological treatment stream. In searching for a solution, the ZRCC engineering team chose to evaluate oxidation treatment technologies, which are considered the best approach to addressing spent caustic treatment. Other treatment technologies either try to strip the sulfides out of the spent caustic, which results in extremely corrosive and odorous gas, or burn the spent caustic in the gaseous phase. The latter is also corrosive and extremely energy-intensive.

When oxidized, spent caustic consists of fully oxidized inorganic salts that are non-reactive and carboxylic organics with a low-molecular weight. As a result, the effluent has no offensive odor and can be safely neutralized. In addition, with the amount of COD dramatically reduced and the organics oxidized to a biodegradable form, the effluent can then be sent to biological treatment for polishing.

Although other oxidation treatment approaches, such as incineration and advanced oxidation, were available, the ZRCC engineering team rejected them because of their high operating and maintenance costs as illustrated below in Figure 1. Incineration requires a lot of costly fuel, while advanced oxidation uses expensive and hazardous liquid oxidizers to oxidize the spent caustic.

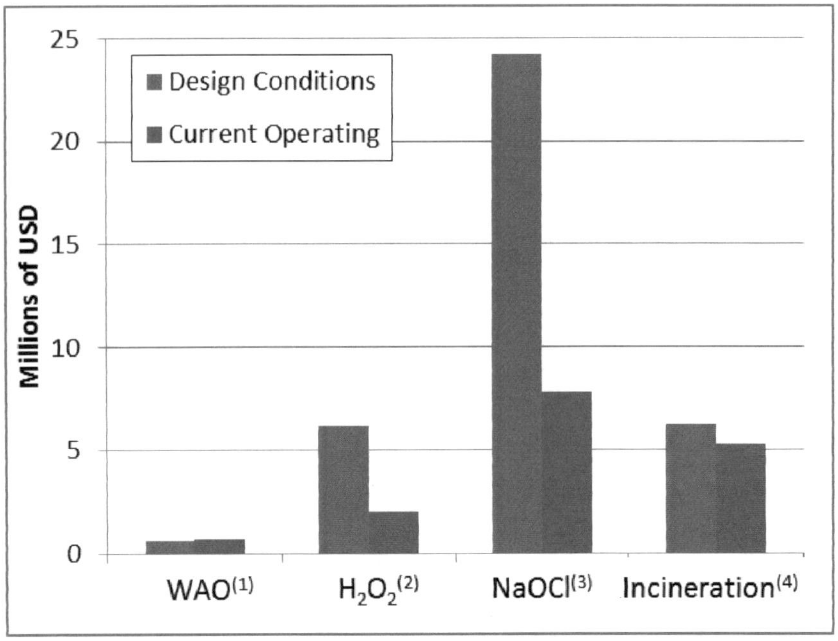

Figure 1. Annual operational expense comparison of WAO technology versus alternative oxidation treatment approaches for ZRCC's ethylene spent caustic at current operating conditions and design conditions. Notes: (1)Includes electricity, steam, boiler feedwater, cooling water, and instrument air; (2)H_2O_2 only; (3 NaOCl only; (4)fuel only.

After evaluating all other options, ZRCC engineers chose to couple a pre-treatment solution with system featuring WAO technology at its core as shown in Figure 2. The former decants the wastewater, then skims corrosive, free-floating red oil from the ethylene spent caustic; the latter then processes the aqueous spent caustic through the WAO system.

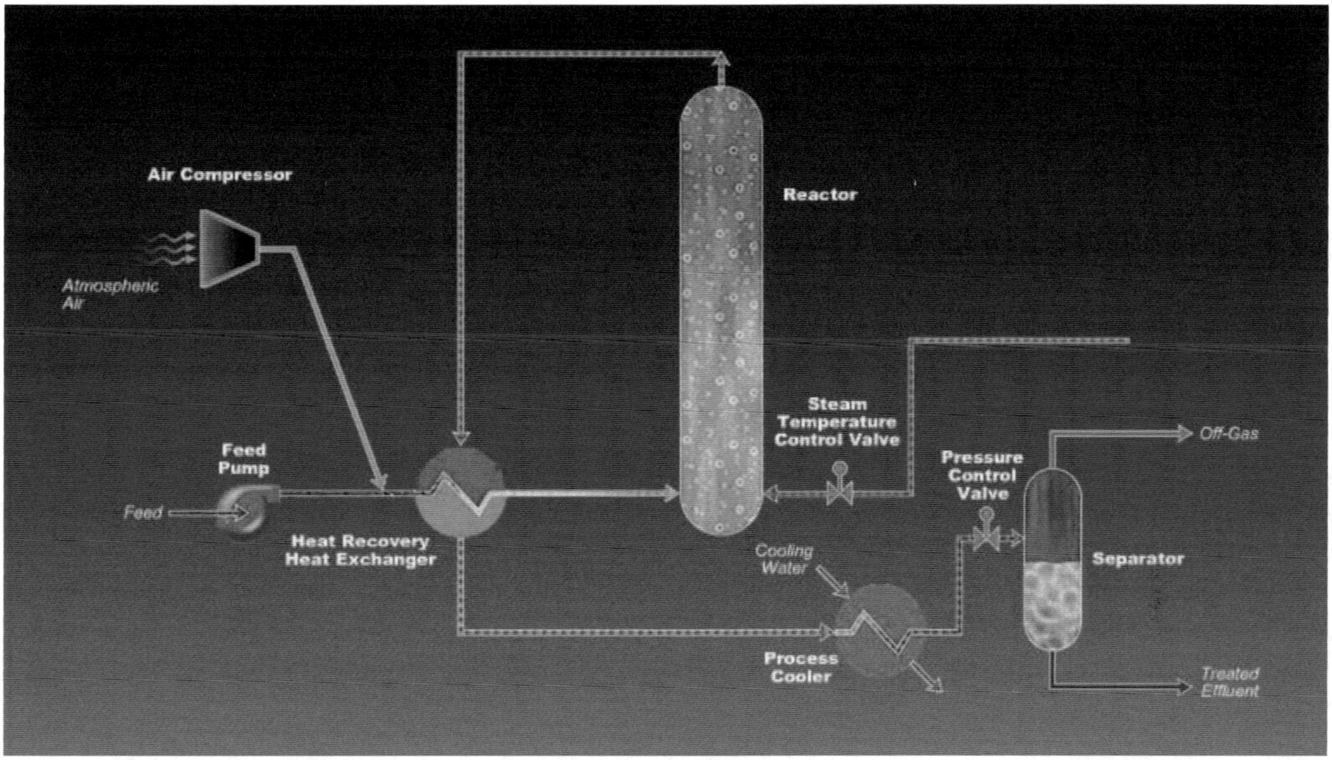

Figure 2. Diagram of the WAO system deployed by ZRCC, designed to process a spent caustic flow of 18.2 m^3/hr.

The ZRCC plant's WAO system is designed to process a spent caustic flow of 18.2 m^3/hr. Inside the system, operating in an aqueous phase at relatively low temperatures, the spent caustic wastewater combines with oxygen from the air and heat, forming hydroxyl radicals that drive the oxidation reactions.

The WAO process converts organic contaminants to biodegradable, short-chain organic acids, such as acetic acid, plus carbon dioxide and water. Inorganic constituents, including sulfides and cyanides, are also oxidized. The process treating ethylene spent caustic involves the following reactions:

- Organics + O_2 → CO_2 + H_2O + RCOOH
- Sulfur Species + O_2 → SO_4^{-2}
- Organic N + O_2 → NH_3 + CO_2 + RCOOH

After the oxidized spent caustic is effectively and safely neutralized, the resulting effluent has a COD load that is a small fraction of what it was before. At this point, it is ready for biological treatment at the ZRCC biological wastewater treatment plant.

The WAO system is designed to reduce odorous and hazardous sodium sulfide from a concentration of approximately 2 percent by weight to less than the analytical detection limit in the oxidized caustic, effectively eliminating odor problems. Total COD in the plant's untreated spent caustic can range as high as 21,450 mg/L.

As shown in Figures 3 and 4, operational data over the course of a month (February 2018) demonstrates that the WAO system is producing a stable effluent with 1,000 to 1,500 mg/L COD, well below the allowable design limit of <5,000 mg/L. This remaining COD in the effluent is in the form of biodegradable organics.

Also, all effluent samples were below the WAO system supplier's guaranteed limits. These limits were selected based upon decades of successful operation of similar plants globally (non-detect which is <1 mg/L of Na_2S as S and <100 mg/L $Na_2S_2O_3$ as S). What's more, all daily effluent phenol readings were under 1 mg/L (detection limit), which is well under the design limit of <20 mg/L.

During the sampling period, the untreated spent caustic contained from 2,300 to 4,900 mg/L sulfides (as S).

The data shows that complete oxidation of sulfides was achieved by the WAO system. At 200°C, the sulfides were destroyed and almost completely oxidized to sulfate. This conversion to sulfate eliminates the noxious odor of the sulfides and the high COD demand associated with sulfides or thiosulfates. As a result, the WAO effluent produces a biologically degradable effluent with no obnoxious odor, which is desirable from a health and safety perspective.

The ZRCC engineering team reports that the WAO system is extremely reliable and has paid for itself many times over, thanks to its lower operating costs compared to other spent caustic oxidation alternatives.

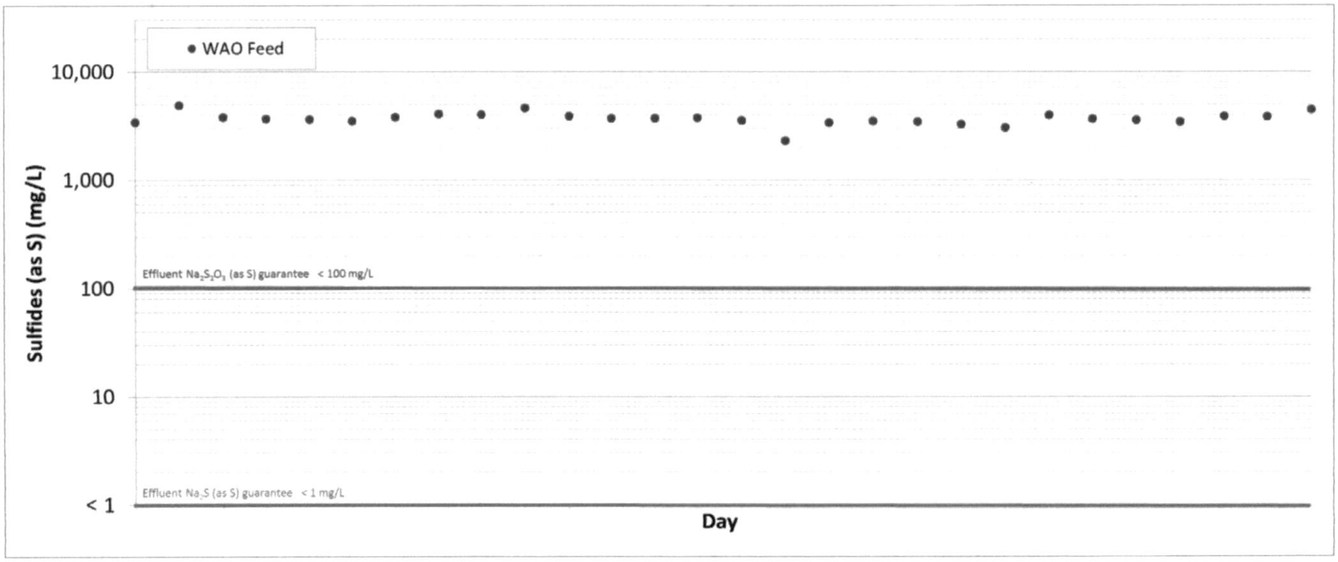

Figure 3: WAO feed and near complete oxidation of sulfides over the course of a month (February 2018).

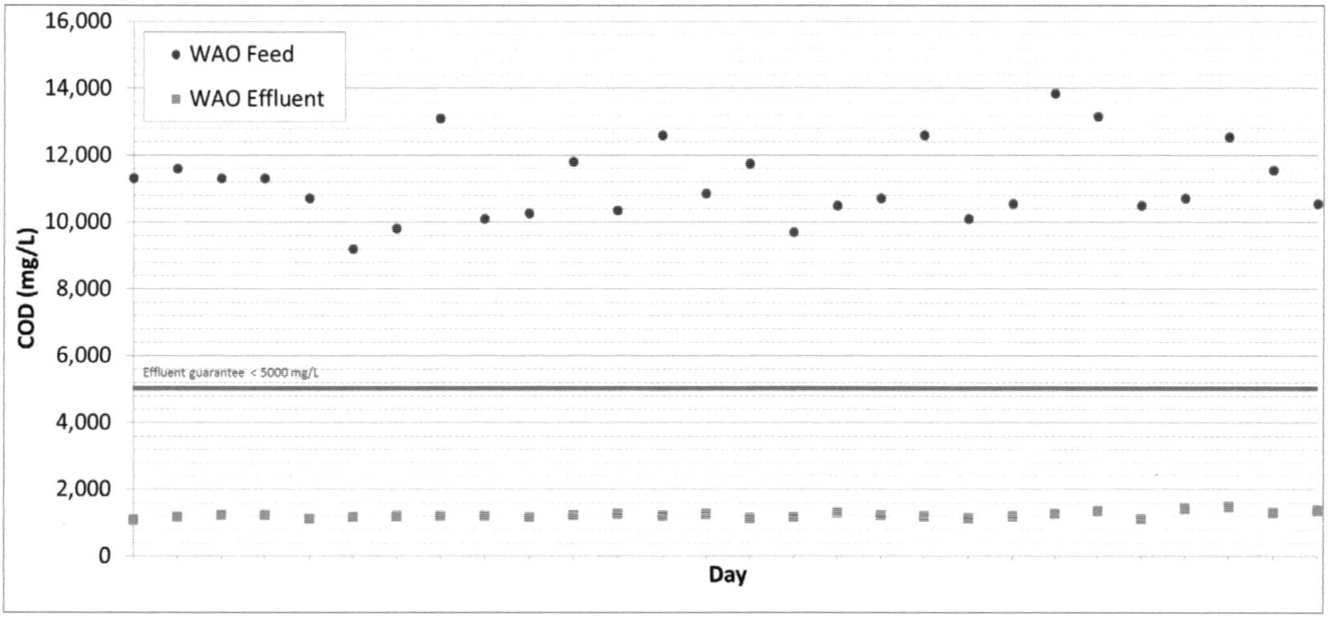

Figure 4: WAO feed and stable effluent COD over one month (February 2018).

PACT© MBR Polishing – A new development

The biggest challenge with the reuse of wastewater in ethylene production is the high concentration of salts in the spent caustic. Once oxidized by WAO, the components in the spent caustic can be treated biologically. However, in conventional activated sludge systems there are limitations on the concentration of salts that can be present in the wastewater while still getting adequate treatment usually thought to be approximately 5%. Through the addition of activated carbon to help reduce toxicity and provide ballasting for biomass solutions with much higher concentrations of salts can be treated. Figure 5 presents data comparing the performance of a conventional active sludge (CAS) bioreactor to one operating as a PACT system at varying salt concentrations in oxidized ethylene spent caustic.

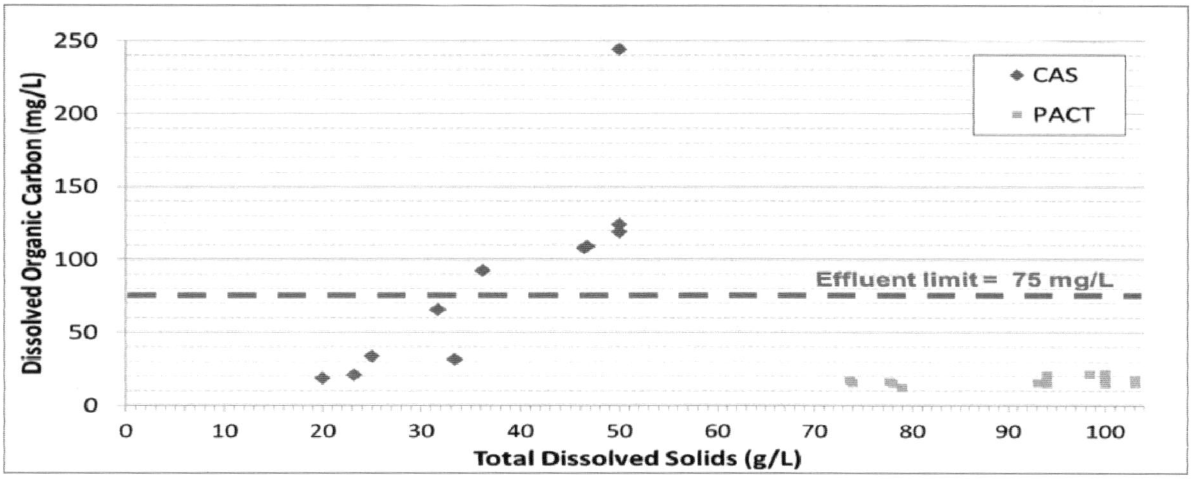

Figure 5: A chart comparing conventional activated sludge (CAS) and PACT for treating oxidized ethylene spent caustic at varying salt concentrations.

By treating the oxidized spent caustic to below discharge limits, no further treatment is required. The added benefit is that the salt no longer needs to be added to the conventional wastewater treatment system and allow for high water recovery.

Conclusion

Under stressful high-sulfide and salt conditions, this total treatment approach is able to deliver effluent quality that alternative wastewater treatment processes typically cannot match, including:

- Guaranteed sulfide and mercaptan levels of less than 1 mg/L
- COD below 150 mg/L
- TOC below 75 mg/L
- TSS below 1 mg/L
- Turbidity below 0.1 NTU.

Equally important, WAO and advanced biological polishing does achieves their wastewater treatment goals at a lower OPEX than alternative ethylene spent caustic treatments and allows for higher water recovery in reuse applications.

References:

1. The Siemens Zimpro® Wet Air Oxidation (WAO) system white paper. 2011. No author byline.

2. "Oil and Gas Wastewater Treatment Simplified — PACT© MBR System" white paper. Published by Siemens. December 2016. No author byline.

3. "Water Recycling Efficiency In Ethylene Spent Caustic Treatment, Part I: Cost," Oil and Gas Online. October 2018. No author byline.

4. "Water Recycling Efficiency In Ethylene Facilities Producing Spent Caustic, Part II: Process," Oil and Gas Online. October 2018. No author byline.

157e

NEW DIRECT FLAME MONITORING TECHNOLOGY TO HELP OPERATORS COMPLY WITH INCREASINGLY STRINGENT FLARING REGULATIONS

Brian Duck, P.E
Global Business Manager
Zeeco, Inc.

Prepared for Presentation at the 2019 Spring National Meeting
New Orleans, LA, March 31 – April 4, 2019

NEW DIRECT FLAME MONITORING TECHNOLOGY TO HELP OPERATORS COMPLY WITH INCREASINGLY STRINGENT FLARING REGULATIONS

Brian Duck, P.E
Global Business Manager
Zeeco, Inc.

Abstract:

Known shortcomings of current flare monitoring methods plus new EPA standards drove the need to develop a new direct flare combustion efficiency (CE) measurement and monitoring method – a technology that directly, autonomously, and continuously monitors flare performance in real time. This patented method, known as VISR, or Video Imaging Spectro-Radiometry, utilizes a multi-spectral infrared (IR) imager to simultaneously measure the relative concentrations of combustion products, carbon dioxide (CO2), and unburned hydrocarbons (HC) at the pixel level. Directly monitoring flare CE eliminates inaccuracies associated with the current practice of monitoring indirect parameters (heating value, velocity, etc.). Because VISR devices can operate autonomously, no aiming or manual data reduction is required. Remote measurement removes the need for contact with corrosive process streams, making VISR devices less costly to maintain and operate over time. This paper will discuss the VISR technology and how it can be used to generate continuous, real-time data on CE and smoke, allowing operators to optimize flare performance in real time.

Introduction

The changing government regulations imposed on refineries, gas processing facilities, petrochemical manufacturers, and OEM vendors demand constant adaptation to meet requirements. Across the globe, facilities are scrutinized for their emissions contributions to our global environment and are soon to be held to even more stringent standards. Since its inception, the United States Environmental Protection Agency (EPA) has been actively involved in various flare system enforcement initiatives to reduce the emissions from flares. Some examples of industrial flares can be seen below in Figures 1 & 2.

Figure 1 – Example of Air Assisted Flare

Figure 2 – Example of Sonic Process Flare

In December 2015, the EPA imposed well-defined flaring regulations regarding emissions in the new Refinery Risk and Technology Review (RTR) Rule. To comply with this updated regulation, flare performance is required to be monitored by strict, defined parameters that currently can only be determined using indirect measurement methods or labor-intensive extractive sampling techniques. Similar regulations are expected to be introduced in 2019 affecting ethylene producers.

Recent testing has shown that the latest EPA methodology for monitoring combustion efficiency (CE) by way of calculating the combustion zone net heating value (CZNHV) varies greatly and invites questions regarding the efficacy of this method. Also, of the few other measurement methods known to measure CE as required to comply with new regulations, none have been credited with being timely, effective in cost or results, or easy to operate and manage. For example, extractive sampling techniques are effective and accurate, but require tremendous effort and labor to produce results. Measuring surrogate parameters (indirect monitoring) requires substantial financial investment in specialized components to ensure compliance (gas chromatographs, BTU analyzers, ultrasonic flowmeters, etc.). Consequently, results do not produce a direct measurement of a flare's performance or any real-time data. A method that will comply with the EPA's newest regulations and optimize flare performance requires a more long-term, simplified, and accurate solution. In this paper, we will discuss how the implementation of new technology utilized in Zeeco's FlareGuardianTM monitoring system can generate real-time continuous monitoring to meet this demand and satisfy compliance with ease.

VISR Technology

Other technology exists to directly monitor flare systems, but most cannot produce results to verify regulation compliance. Those technologies can have tremendous shortcomings as well; for instance, while the use of infrared (IR) flame or pilot monitors can provide imagery capable of determining the presence of a flame, no flare performance or efficiency measurements can be derived with these devices. When "Passive Fourier Transform IR" methods (PFTIR) are employed, they can provide data to comply with regulations but are limited to a single line of sight approximate to wind direction for a flaring event and must be continuously manned and adjusted to monitor CE. With the use of patented Video Imaging Spectro-Radiometry (VISR) technology, achieving real-time measurement of flare efficiency and hydrocarbon (HC) /combustion product emissions is simplified. VISR technology provides a direct flare CE measurement while simultaneously minimizing the chance of calculation errors common to indirect measurement technologies. VISR's multi-spectral imaging system utilizes a specialized micro-lens array and bandpass filtering systems to generate CE and performance information in real time. This capability utilized in conjunction with complex flame behavior algorithms, permits VISR devices to convert the filtered signals identifying chemical species and combustion products into data images to deliver detailed, accurate measurements of a flare's performance. Figures 3a-3d below illustrate the capabilities of VISR to produce filter-specific imaging to achieve this function.

Figure 3a – Image of Smokeless Flaring

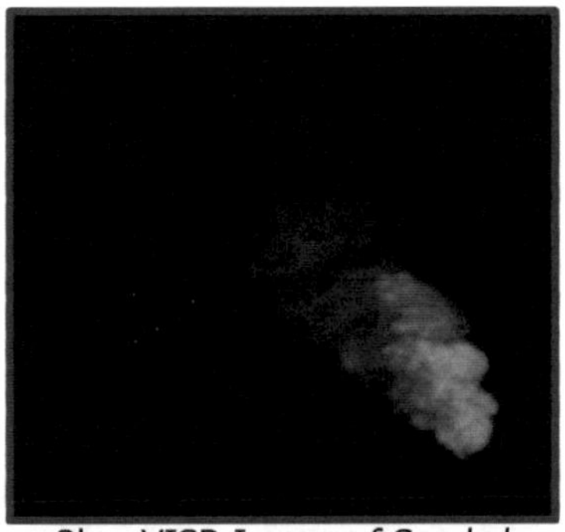

Figure 3b – VISR Image of Smokeless Flare

Figure 3c – Image of Smoking Flare System

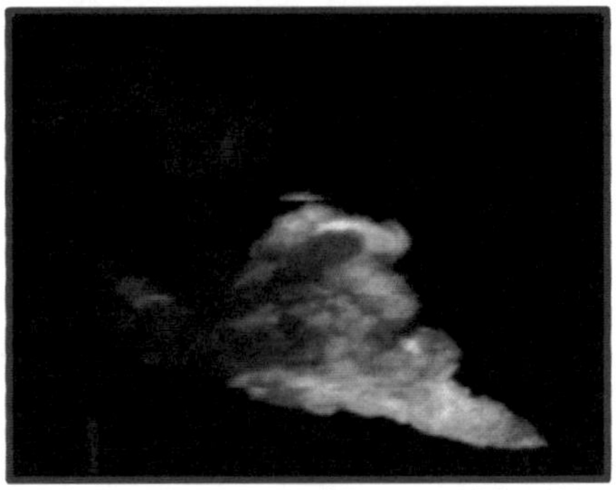

Figure 3d – VISR Reported Image of Smoking Flare

VISR's capabilities to recognize chemical species in distinct wavelength spectrums (CO_2, CO, HC, etc.) at the pixel level produces comparative measurements within the flame envelope. Over a measurement frequency of 30 Hertz (Hz), VISR reports in one (1) second averaged intervals of flare performance and other parameters (smoke production, flame surface area, etc.). Environmental effects such as ultraviolet (UV) and infrared (IR) interference from sunlight in the field of view have essentially no effect on the measurement capabilities of a VISR system. In addition, rain and fog do not prevent this technology from obtaining accurate measurements, ensuring consistent data and performance of the device over various conditions.

FlareGuardian™

To take full advantage of the advancements made by VISR technology, Zeeco has produced a VISR-based product called FlareGuardian to meet the demand for an efficient, accurate, and compliant direct flare monitoring system. An illustration of the testing setup of a Zeeco FlareGuardian unit can be seen in Figure 4.

Figure 4 – Zeeco FlareGuardian Monitoring System in Use During Validation Testing

Table 1 below shows the high-level requirements of the RTR rule in relation to industrial flaring. Because the legislation does not clearly define the methodologies to verify compliance, CE measurement via direct monitoring devices such as FlareGuardian will eliminate many of the hardships faced by companies otherwise forced to employ and substantiate measurement via "alternative options."

Table 1		
EPA Rule 40 CFR, Part 63	**Compliance Requirements**	**Covered by FlareGuardian?**
§ 63.670 (b)	Presence of pilot flame	✔
§ 63.670 (c)	No visible emissions	✔
§ 63.670 (d)	The three requirements are designed to ensure sufficient CE through surrogate parameters	✔
§ 63.670 (e)		
§ 63.670 (f)		

As illustrated, FlareGuardian offers the capability to meet all requirements and ensure compliance while avoiding costly measurement techniques and reducing the time required to adjust in the field – consequently improving flare efficiency. Table 2 explains the full benefits of employing Zeeco's FlareGuardian flare monitoring system:

Table 2
FlareGuardian Design Features and Benefits
-Provides real-time combustion efficiency, smoke index, flame stability, flame footprint, heat release, and pilot status for a complete picture of flare performance
-Autonomous data collection (PLC or DCS) for optimized flare performance
-Simplifies monitoring, reporting, and compliance activities
-Patented technology
-Remote mounted, non-contact monitoring
-More accurate results versus indirect monitoring
-Eliminates the need for monitoring surrogate parameters
-Smoke index (SI) assist in achieving incipient smoke conditions day and night
-Short measurement cycle enables quick response and minimizes cost for supplemental fuel, steam, or air
-Industrial interface allows for closed loop flare operation and control based on direct combustion efficiency and smoke index values
-Easy installation and maintenance, uninterrupted production processes, and no calibrations

Direct measurement for flame CE avoids exposing equipment to corrosive process streams and prevents the use of expensive, unreliable solutions for data collection. An autonomous solution allows for independent monitoring and adjustment to flaring conditions, minimizing the use of operator intervention. VISR devices such as FlareGuardian hold the potential to directly, autonomously, and continuously monitor flare performance in real time. Thus, this data is readily available to review and execute critical adjustment decisions (i.e. adjustments in supplemental fuel, steam, or air) for flaring scenarios to maintain compliance and meet federal and local regulation criteria. Once online, FlareGuardian begins to report directly to the user system performance parameters in a dashboard structured report for ease of interpretation. Filtered imagery of the flame, real-time calculation, and report of CE, indications of smoking, and other parameters can be viewed without an operator needing to attend the flare itself.

In the next section, we will review and discuss the validation of the FlareGuardian device against current approved EPA methods of flaring efficacy, thereby demonstrating that the use of VISR technology can generate unparalleled data quality and further optimize performance of flaring systems in real time.

Testing and Validation

Extensive testing and validation has occurred at Zeeco's testing facility to determine the feasibility of the FlareGuardian system. In November 2014, full scale experimentation was conducted near Tulsa, Oklahoma at Zeeco, Inc.'s testing facility. An example of the testing setup can be seen in Figure 5.

Figure 5 – Example of Testing Setup for FlareGuardian Validation

A total of 39 test points was observed measuring CE for two separate methods (extractive sampling against FlareGuardian VISR monitoring). Validation was performed on three separate Zeeco, Inc. production flare tips as follows. This selection was made to show the diversity of measurement capabilities using FlareGuardian:

- Zeeco Model QFS, 16", steam assisted
- Zeeco Model AFDS, 10", air assisted
- Zeeco Model MPGF, multi-point sonic flare, pressure assisted

For this experiment, flares were evaluated by FlareGuardian at a distance of 300 feet from the base of each flare stack. DRE and CE were measured to validate the results from the FlareGuardian monitor against that derived by extractive sampling. Using the methods consistent with the standard EPA methods for stack testing, a sampling hood was suspended over the flare using a crane setup as depicted in Figure 5. Gas samples were transported back to a stable trailer monitoring facility on site and continuously analyzed for presence of combustion products (carbon dioxide and carbon monoxide), oxygen content, and unburned hydrocarbons.

Results

The results of the testing performed by Zeeco, Inc. are shown graphically in Figure 6. The average CE difference between the two methods for all reported tests was 0.50%. Measurement of CE between extractive sampling and FlareGuardian were highly consistent, varying only in regions where poor combustion efficiency was achieved. The full data set with all measured parameters can be found in Appendix 1.

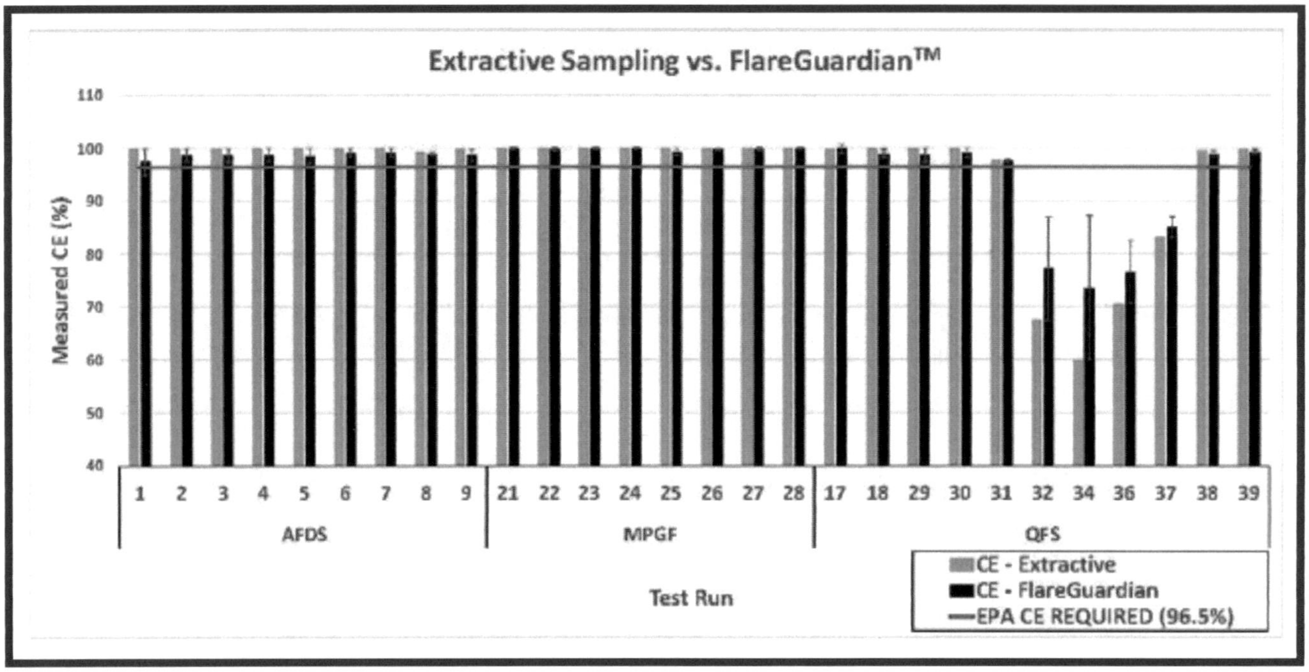

Figure 6 – Flare CE Validation of FlareGuardian vs. Extractive Sampling Method

Based on the individual percent difference against the extractive method reference, the data suggests that FlareGuardian holds a more conservative calculation basis for calculating CE. As federal regulations for compliant CE exist at 96.5%, even in accounting for a conservative calculation, the difference between extractive sampling and VISR are minimal. Based on the data, it can be inferred that excellent repeatability and accuracy exists between FlareGuardian and extractive sampling, thus promoting the use of VISR in flaring applications moving forward.

Discussion

Current measurement methods possess certain pitfalls, such as high capital cost, extensive servicing of all components, constant re-calibration of devices, exposure to corrosive process conditions, and higher labor investment to ensure compliance with mandated regulations. Also, since the EPA RTR rule and expected rules affecting ethylene producers are accompanied by time-dependent monitoring constraints, a problem is created when striving to comply while using methods that do not provide real-time results. Surrogate parameter measurement is dependent on various process components to report data in a multitude of timeframes, which can create prohibitive delays in obtaining results for relative flaring performance. During this composite time delay, flaring process conditions could change dramatically and possibly negate any compliance corrective action, a problem which is not experienced by the real-time data reporting provided with VISR devices. FlareGuardian simplifies monitoring and reduces the involvement of plant operators to ensure compliance and flare efficacy. With the compliance deadline approaching soon, ethylene plants will be required by law to meet these evaluation parameters regardless of the technology employed. More extensive coverage of operations will be in the line of sight for future regulation, such as chemical plants, midstream operations, and further upstream exploration and production efforts. As these regulations progress, so should the technology behind remaining compliant, such as developments like the VISR-based FlareGuardian system.

Conclusions

Available data supports that the use of VISR technology has been validated and will soon emerge not only in the industry but also with regulatory agencies as the best available technology for flare monitoring. VISR-based products such as FlareGuardian will enable operators to reduce flaring emissions, improve flare performance, and troubleshoot flaring systems, giving them a means to control CE and optimize performance in real time.

Appendix

No.	(A)	(B)	(C)	(D)	(E)	(F)	CE – EX	CE – FS	CE % Diff	SI	(G)
1	AFDS	Propane (100%)	7994	33.29	-	259	99.94	97.40	-2.54	2.85	21.13
2	AFDS	Propane (100%)	7994	33.29	-	259	99.99	98.80	-1.19	2.46	19.45
3	AFDS	Propane (100%)	7994	33.29	-	259	99.98	98.70	-1.28	4.58	19.37
4	AFDS	Propane (100%)	6670	39.89	-	221	99.99	98.80	-1.19	2.87	17.63
5	AFDS	Propane (100%)	6670	39.89	-	221	99.97	98.60	-1.37	2.70	18.84
6	AFDS	Propane (100%)	5278	50.42	-	178	99.97	99.20	-0.77	2.66	19.83

7	AFDS	Propane (100%)	5278	50.42	-	178	99.95	99.20	-0.75	2.50	20.03
8	AFDS	Propane (100%)	3063	86.87	-	107	99.33	99.00	-0.33	0.72	20.53
9	AFDS	Propane (100%)	3063	86.87	-	107	99.77	98.70	-1.07	1.44	18.94
17	QFS	Propylene (100%)	4910	-	0.48	1031	99.86	99.90	-0.86	3.99	19.93
18	QFS	Propylene (100%)	4910	-	0.48	1031	99.90	99.00	-0.80	2.24	19.98
21	MPGF	Propane (100%)	5079	-	-	-	100.00	99.90	-0.10	0.24	18.77
22	MPGF	Propane (100%)	5079	-	-	-	100.00	99.70	-0.30	0.27	18.07
23	MPGF	Propylene (100%)	4952	-	-	-	100.00	99.90	-0.10	1.41	17.92
24	MPGF	Propylene (100%)	4952	-	-	-	100.00	99.90	-0.10	1.36	17.38
25	MPGF	Propane/N_2 (50/50)	2448	-	-	-	99.97	99.30	-0.67	0.23	19.48
26	MPGF	Propane/N_2 (50/50)	2448	-	-	-	99.99	99.80	-0.19	0.35	18.19
27	MPGF	Natural Gas (100%)	3300	-	-	-	100.00	99.80	-0.20	0.26	17.03
28	MPGF	Natural Gas (100%)	3300	-	-	-	100.00	99.90	-0.10	0.32	15.76
29	QFS	Propane (100%)	4640	-	0.52	1035	99.99	98.70	-1.29	0.56	19.91
30	QFS	Propane (100%)	4640	-	0.52	1035	99.97	99.10	-0.87	0.70	17.60

31	QFS	Propane (100%)	1879	-	1.25	571	97.75	97.50	-0.25	0.46	19.90
32	QFS	Propane (100%)	1879	-	1.25	571	67.48	77.20	9.72	0.83	20.24
34	QFS	Propane (100%)	1537	-	1.53	489	59.99	73.60	13.61	0.17	19.94
36	QFS	Propane (100%)	1537	-	1.53	489	70.57	76.60	6.03	0.15	18.75
37	QFS	Propane (100%)	1537	-	1.53	489	83.15	85.10	1.95	0.21	18.38
38	QFS	Propane (100%)	3328	-	0.71	850	99.67	99.01	-0.57	0.40	17.38
39	QFS	Propane (100%)	3328	-	0.71	850	99.82	99.40	-0.42	0.46	18.86

Average CE difference between the two methods – all 28 tests: 0.50%
Number of tests with oxygen <19.5% (indication for good extraction): 18
Average CE difference between the two methods – 18 tests with oxygen <19.5%: -0.10%
Where,

(A) = Flare Tip Type (AIR, MPGF, or STEAM)
(B) = Fuel Type Used and Composition
(C) = Fuel Flow Rate, In LB/HR
(D) = Stoichiometric Air Percentage (SA%)
(E) = Steam-to-HC Ratio (lb_{steam}/lb_{HC})
(F) = Combustion Zone Net Heating Value (CZNHV, BTU/FT^3)
(G) = Average O_2 in extracted sample (%)

AIChE Paper Number: 158a

A 15 YEAR FIELD PERFORMANCE REVIEW OF NEW COIL TECHNOLOGIES FOR STEAM CRACKING FURNACES

Dr. Dietlinde Jakobi
Schmidt + Clemens GmbH + Co. KG
51789 LINDLAR/GERMANY

Steffen Heyland
Schmidt + Clemens GmbH + Co. KG
51789 LINDLAR/GERMANY

Dr. Jörg Weigandt
Schmidt + Clemens GmbH + Co. KG
51789 LINDLAR/GERMANY

Prepared for Presentation at the 2019 Spring National Meeting
New Orleans, LA, March 31 – April 4, 2019

A 15 YEAR FIELD PERFORMANCE REVIEW OF NEW COIL TECHNOLOGIES FOR STEAM CRACKING FURNACES

Dr. Dietlinde Jakobi, Steffen Heyland, Dr. Jörg Weigandt
Schmidt + Clemens GmbH + Co. KG, 51789 LINDLAR/GERMANY

Abstract

Radiant coils in steam-cracking furnaces operate under severe and alternating reducing and oxidizing conditions at temperatures even beyond 1100 °C. Therefore, cast alloys installed in the radiant section need to provide high creep strength, a very good corrosion resistance and an excellent resistance against catalytic coking. The coil components (centrifugally cast tubes, cast fittings) must be able to form a dense and continuous oxide layer on the inner surface, free of any catalytic sites active for formation of coke filaments.

Over several decades, iron and nickel based chromia-forming alloys were installed in steam cracking furnaces. 15 years back also alumina-forming alloys were introduced to the petrochemical industry. New insights on the interactions between protective oxide scales formed on such chromia- and alumina-forming alloys and the highly reactive cracking atmospheres allow further understanding of the degradation mechanism as well as of the coking behavior. Sulfur present in the feedstock as natural constituent or added deliberately is also strongly influencing the coking rate and material deterioration. As a conclusion, chromia forming high temperature alloys have a limited resistance against corrosion and also against coking in steam cracking conditions; the limitation being set by thermodynamics. Due to a far better thermodynamic stability, alumina scales like formed on the alloy Centralloy® HT E are favorable as protection in such severe conditions.

Some years back S+C also developed the "SCope" internally profiled tube design and validated its benefits in plant applications. The SCope profile leads to a very uniform, circumferential flow pattern. This improves the heat transfer, balances the heat distribution in the gas phase as well as in the tube shell and leads to reduced tube metal temperatures. These properties contribute to reduce the coking rate. This profile can be individually designed to meet the needs of customers, e.g. regarding increased conversion or ethylene yield, maximum energy efficiency of the furnace, longer maintenance intervals or minimum downtime of steam crackers and additionally an increased coil life.

In order to maximize the operational benefit of the plant, the innovative SCope profile can be combined with the proven Centralloy® HT E tube material.

Performance studies and results from metallurgical investigations carried out on the alumina forming alloy Centralloy® HT E exposed for several years to service conditions in the steam cracking furnace will be presented as well as results from field applications with the SCope profiled tube design.

Introduction

Centrifugally cast tubes and statically cast components installed in the radiant section of steam cracking furnaces are operated at very severe conditions. Alternating atmospheres affect the inner tube surface: Oxidizing atmospheres during furnace start-up or de-coking in steam or in steam-air mixtures, respectively; reducing atmospheres during the normal cracking operation. Tube metal temperatures are rising from 750°C (1382°F) at start of run up to well above 1100°C (2012°F) at the end of each run. Therefore, coil materials need an excellent resistance to corrosion attack (oxidation/carburization/nitridation), good mechanical properties, and should form dense and catalytically inert oxide scale (in order to reduce the rate of coke formation).

Depending on coil design, feedstock and operating conditions, the main damage mechanisms for such components are a combination of carburization and creep load [1]. Carbon can penetrate from the inner surface into the alloy; nitrogen can diffuse from the outer surface into the tube material. Both lead to the formation of brittle precipitates within the alloys matrix (large carbides close to the inner surface and large carbonitrides near the outer tube surface). Increasing degree of carburization and/or nitridation as well as formation of harmful phases during ageing enhances the risk of materials embrittlement with increasing process time. Exposure to operational transients like emergency shutdowns, such aged and embrittled alloys can easily lead to failure by brittle crack formation.

Over several decades, iron and nickel based chromia-forming alloys were the material of choice in steam cracking furnaces. Particularly these alloys are susceptible to the above mentioned failure mechanisms [1]. Therefore, the research work was focused on the development of alumina forming alloys. Alumina scales are thermodynamically much more stable under these conditions; thus, 15 years back such alumina forming alloys were introduced to the petrochemical industry.

A dense and continuous oxide scale, which is stable up to very high metal temperatures, is able to protect coils against carburization and nitridation; this reduces ageing and prolongs tube life. Additionally, properly grown alumina scales do not contain catalytically active sites for coke formation (e.g. Fe- and Ni-particles). These important features reduce the amount of catalytic coking significantly, particularly in case of significant amounts of sulfur in the feed [2, 3, 4].

Such a stable and continuous alumina scale is formed on the inner surface of alloy Centralloy® HT E, which is installed in the meanwhile in the radiant section of more than 330 steam cracking furnaces. Apart from an increased tube/coil lifetime, the alloy is positively affecting the run-length (due to the low coking rates), the fuel consumption, the conversion, etc.

Some years back S+C developed also the internally profiled tube design "SCope" and validated its benefits in plant applications to agree with predictions from fluid dynamic simulations. The SCope profile leads to a very uniform, circumferential flow pattern which improves the heat transfer, balances the heat distribution in the gas phase as well as in the tube shell and leads to reduced tube metal temperatures. As a benefit, all effects contribute to a reduction of coking rate.

Refined Insights on the Performance of Alumina Forming Alloys

Feedback from the market on benefits of the alumina forming alloy Centralloy® HT E has been already reported earlier with respect to improved production capability in terms of run length or throughput of plants [5]. This paper also showed a gain in extra coil life, but in the meanwhile more refined insights are available, especially from metallurgical investigations on aged samples.

In general it can be observed that furnaces being partially upgraded by the alumina forming alloy in the outlet part reflected a prolonged tube life: The material upgrade leaded to an increase in coil life up to 50%. The tubes failed later due to the consumed life of the colder inlet tubes equipped with common chromia forming material. The upgrade of the entire furnace metallurgy to Centralloy® HT E resulted in a life prolongation of up to 100%. However, these individual improvements are depending on the process conditions as well as on the feedstock composition.

Influence of materials choice on Limitation of Coil Lifetime

In different furnace designs (e.g. 2-Pass & 5-Pass Coils) the inlet tubes were equipped in the standard chromium oxide forming alloy 25Cr/35Ni Micro and the outlet tubes (consisting either of 25Cr/35Ni Micro or 35Cr/45Ni Micro) were upgraded to the alumina forming material (Centralloy® HT E). Metallurgical investigations carried out after four years and five years in service, respectively, reflected that the inlet tubes in 25Cr/35Ni Micro failed due to heavy carburization (up to 2 wt.-% carbon), along with strong internal oxidation and finally creep damage – even though these tubes were exposed to much lower metal temperatures. The inner surface of the strongly carburized tubes reflected pits due to the corrosion attack (see Figure 1a).

The hotter outlet tubes in alloy Centralloy® HT E did not show any carburization or internal oxidation; the protective alumina scale on the inner surface was still present and intact after several years in operation and the alloy was still fit for service (see Figure 1b). Figure 2a reflects the typical appearance of the inner surface of such carburized inlet tubes in common chromia forming alloys while Figure 2b shows the cross-section of the none carburized Centralloy® HT E exposed in the same steam cracking coil at higher temperatures.

a) Inlet Tube: 25Cr/35Ni Micro alloy (Centralloy® G 4852 Micro)

b) Outlet Tube: Centralloy® HT E

Figure 1: Appearance of Inner Surface After Five Years in Service (2-Pass Coil)

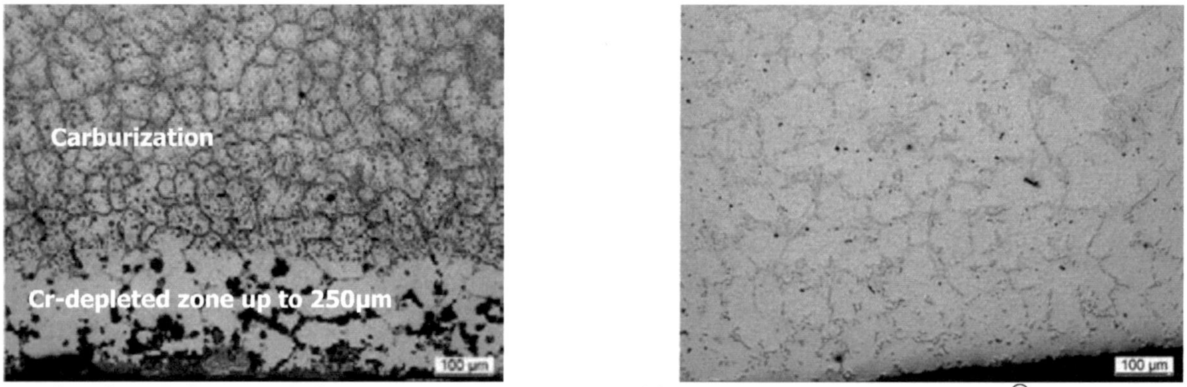

a) Inlet Tube: 25Cr/35Ni Micro

b) Outlet Tube: Centralloy® HT E without carburization

Figure 2: Cross-Section of Inner Tube Surface – Typical Appearance; Samples Removed After Four Years in Service (5-Pass Coil)

Traditionally, the inspection of tubes during shut-downs is focusing on the outlet tube sections because of their much higher metal temperatures. These tubes normally reflect the first damages like bulging, bending and glazed surfaces (indicating the life consumption due to strong carburization and nitridation). After installation of the alumina forming alloy Centralloy® HT E in the outlet part of the furnace, only, the inspection should also focus on the inlet tubes in lower grade alloys in order to avoid unexpected failures.

Impact of Nitridation on Mechanical Properties

Another important aspect of tube ageing is nitrogen pick-up in the outer subsurface zone from the furnace atmosphere (nitridation). This can be pronounced especially at higher temperatures.

The diffusion of nitrogen from the fire box atmosphere into the tube material depends on tube-metal temperature, on gas composition and – most important – on the intactness of a protecting oxide scale on the outer tube surface. The condition of a protective oxide depends on the high-temperature environment. Even evaporation of chromium oxide can occur at high tube temperatures and atmospheres containing water vapor [6]. It is well known that the nitrogen pickup in the standard chromia forming alloy 35Cr/45Ni Micro will lead to the transformation of the chromium carbides to large and brittle chromium carbonitrides (see Figure 3). Therefore, aged common chromium forming alloys – apart from carburization at the inner tube surface – can also suffer from embrittlement of the outer subsurface area. Additionally, niobium carbides in the alloy 35Cr/45Ni Micro can transform to the G-phase ($Ni_{16}Nb_6 Si_7$) by pickup of nickel and silicon from the alloys matrix [7, 8]. G-phase can also become brittle at lower temperatures.

G-phase formation was not observed in the alloy Centralloy® HT E. Generally, due to the ability to form a protective alumina scale, nitrogen diffusion from the flue gas into the alumina forming alloy Centralloy® HT E is less likely than in common chromia forming alloys. Nevertheless, very high tube-metal temperature can cause nitridation; it results in the formation of small aluminum nitrides in the outer subsurface zone (see Figure 4). AlN is not brittle, and no further brittle phases (like G-phase) will form. Additionally, neither chromium carbides nor niobium carbides will be transformed during exposure to service conditions and thus, the alloys mechanical properties remain unchanged.

It can be therefore concluded that the formation of small aluminum nitrides will not negatively impact the ductility of the aged material, even if under certain circumstances (temperature/flue gas composition) nitrogen will diffuse into the material Centralloy® HT E. This was confirmed by welding trials carried out on ex-service material (exposed up to eight years in service). Due to the maintained alloys properties, no special treatment needed to be applied before welding of the aged material.

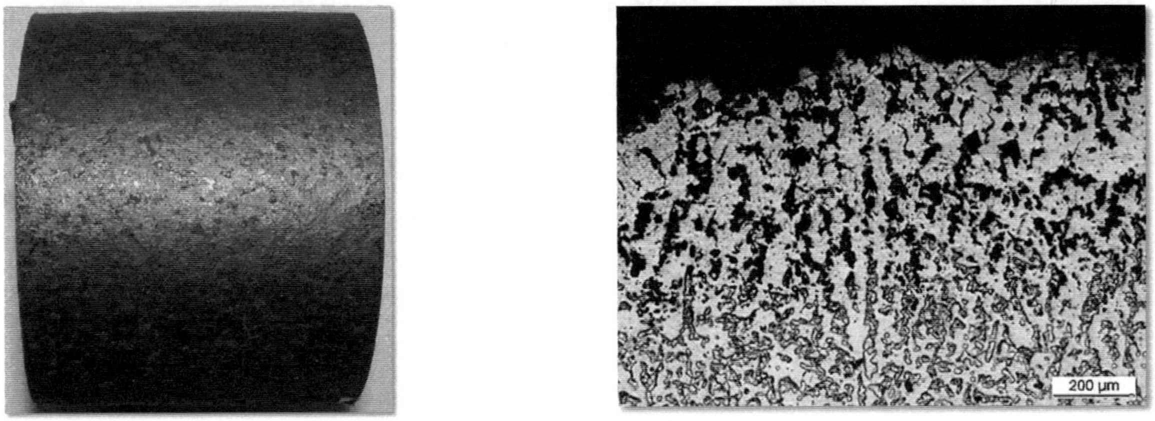

a) Outer surface: Glazed surface due to severe nitridation

b): Outer subsurface area/cross section: chromium depletion & coarse, brittle chromium carbonitrides

Figure 3: Outer Surface of Material 35Cr/45Ni Micro (Centralloy® ET 45 Micro) After Several Years in Service – Cross-Section

a) Outer surface: Typical appearance of aged, nitrided tube

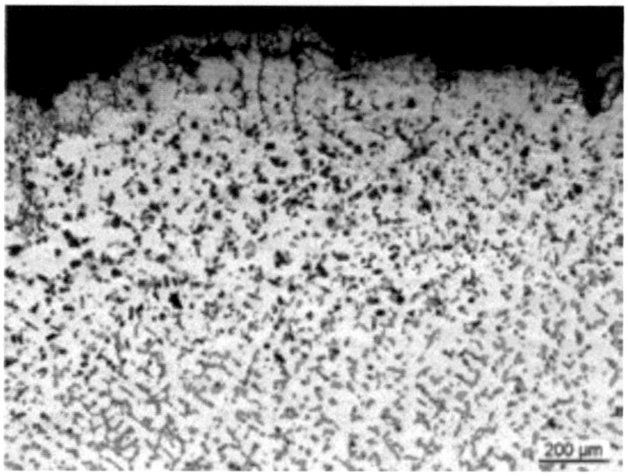

b) Outer subsurface area/cross-section: Nitridation at the outer tube surface

Figure 4: Outer Surface of Material Centralloy® HT E After Several Years in Service

Scheduled Furnace Coil Replacement

Very often a scheduled coil replacement takes place – without any prior tube inspection or metallurgical examination. The time in service might vary between four years and eight years, depending on operating conditions, furnace design and feedstock composition.

After the scheduled replacement of coils in alumina forming technology Centralloy® HT E plant operators and maintenance quite often decide to carry out metallurgical evaluations regarding lifetime consumption and material embrittlement to draw conclusions about weldability of the aged alloy.

In one of such cases the alumina forming alloy was removed from service after four years as this corresponds to the normal schedule for coil replacement. Several samples were taken for metallurgical investigations from this furnace and no corrosion attack (like carburization or strong nitridation) was detected. The very stable, protective oxide scale initially formed on the inner tube surface was still present and it successfully prevented carburization.

In order to determine the remaining coil lifetime, corresponding samples were creep-rupture tested. The material reflected that after four years exposure to service conditions in the steam cracking furnace, just 5% (average) of the minimum creep life was consumed.

Many operators who removed the coils during the scheduled shut-down without prior destructive examination decided to carry out investigations on the weldability of the aged material. This has given maintenance valuable information about the handling of the alumina forming alloy after several years in service. When, like in the case of alloy Centralloy® HT E, neither carburization (inner surface) nor nitridation (outer surface) are influencing the materials properties negatively, the weldability of the aged alloy is comparable to the one of the new material.

Condition of Alumina Scales after Exposure to Service Conditions

Due to the high number of furnace installations with the alumina forming alloy and the huge amount of information obtained from samples removed from different coils/furnace designs it was decided to systematically determine the condition of the oxide scale present on the inner tube surface, especially the oxide layer thickness, and relate it to the operating conditions.

After removal of such aged tube sections from the coils, the inner tube surface normally reflects the original machining marks (corresponding to the appearance of the "as cast" tube material) as the surface is still covered by a thin but very dense and continuous aluminum oxide scale.

The thickness of alumina scales on the inner surface of different ex-service tubes can directly be related to the applied process temperature. Although theoretically alumina scales can grow slowly with time, even under steam cracking conditions (parabolic growth), process temperature has a much stronger influence on the thickness of ex-service tube material than service time (see Figure 5): As more severe the process conditions, as thicker the oxide scale. This explains the tendency in Figure 5 that samples after longer service time had smaller scale thickness. Nevertheless, for the alumina forming alloy Centralloy® HT E the scale formed in the most severe conditions (metal temperatures above 1100 °C) resulted in a thickness of only 4 µm (initially formed oxide scale after start-up of a new coil: 0.15 µm). After comparable service times chromia-forming alloys would have formed non-protective reaction layers with thicknesses of several 10 µm [3, 4].

The common chromia forming alloys like 35Cr/45Ni Micro reflect a completely different behavior. Due to their limited thermodynamic stability the main oxides initially formed during furnace start-up in steam (chromium-manganese-spinel, chromium oxide) tend to transform into less protective carbides during the normal (reducing) stream cracking conditions at high metal temperatures. These carbides once formed can easily grow under cracking conditions leading to a non-reversible damage of the oxide scale. Even inclusions of nickel, iron and other traces become incorporated into this so called "reaction zone". This reaction zone is not protective anymore with regard to carburization and coke formation. With increasing process time the reaction zone deteriorates successively by further chromium oxidation during decoking, and by continued carbide formation during further cracking. The subsurface zone becomes more and more depleted in Cr, Mn and Si, and quite early iron- and nickel-rich oxides form during decoking. These oxides are unstable in the cracking phase and become reduced to catalytically active sites for coke formation.

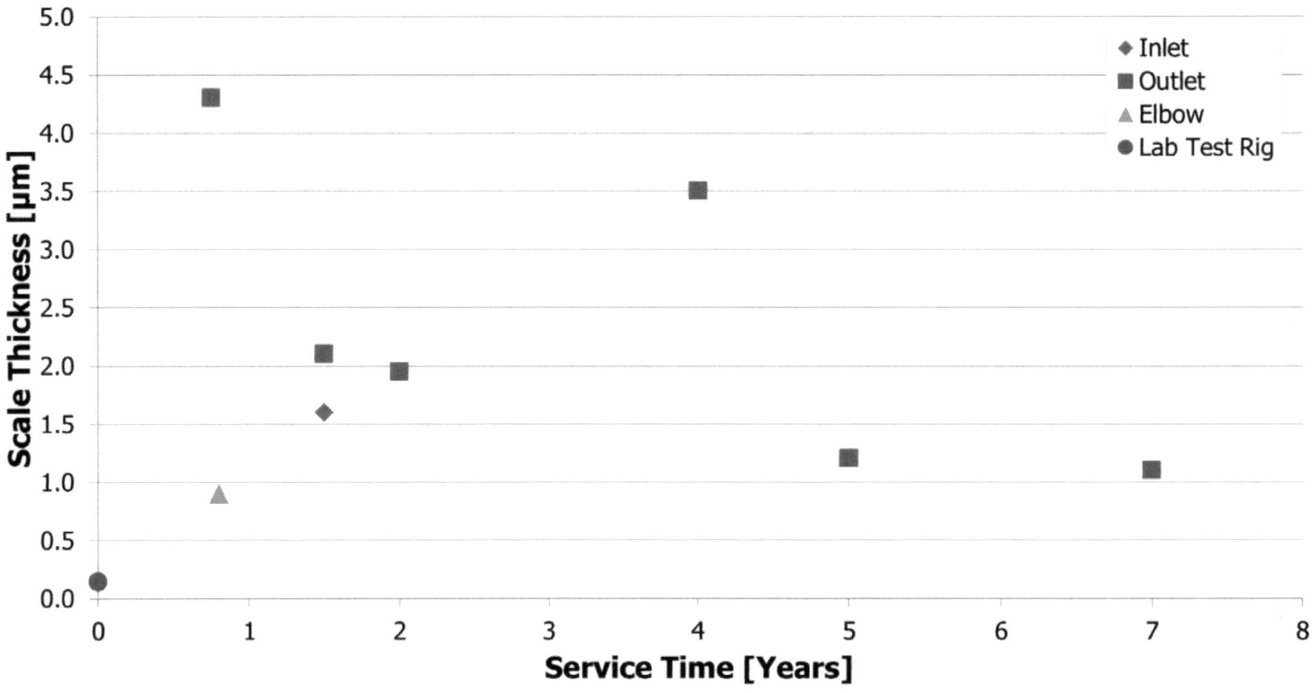

Figure 5: Alumina Thickness of Ex-Service Material

The composition of oxide scales and especially the condition of the oxide scales after ageing is determining the corrosive attack and therefore the tube/coil lifetime, but at the same time it is strongly influencing the coking rates. Particularly when the feedstock contains sulfur additives, the coking rates of chromia forming alloys can increase significantly with the degree of ageing, but sulfur can additionally induce corrosive deterioration of the protective oxide scales. Both effects are strongly depending on the quality and stability of the protecting oxide scale on the inner tube surface. The alumina-forming alloy Centralloy® HT E maintains a dense and uniform α-alumina scale, which withstands sulfur corrosion, as well as phase transformation. These effects have been described in recent publications [3, 4].

S+C SCope Tubes – Dedicated to the Special Needs of the Steam Cracking Process

S+C SCope is a profiling shape modification of the inner tube surface that addresses typical shortcomings of conventional bare tubes. It is aimed to optimize the operation performance of steam-cracker furnaces.

In a steam cracker firebox, furnace tubes are mainly heated by radiation coming from two opposing directions. This leads to an inhomogeneous temperature distribution in the tube shell as well as in the gas phase inside the tube (see Figure 6), i.e. the so called sunny and shady sides of tubes. Furthermore, heat transfer from the burners into the cracking gas is both limited by the maximum allowable tube skin temperature as well as by the heat transfer from the inner tube side into the gas phase. The resulting effects are triggering various issues that negatively influence the steam cracker performance: Sunny sides are causing a local overheating of the tubes, therefore the maximum design temperature is reached faster and consequently the length of the cracking cycle is reduced. Furthermore the local overheating results in temperature gradients from the sunny to shady side. These can induce mechanical stresses across the tube shell, which reduce the tube lifetime. The constrained heat transfer through the tube into the gas phase connected with a rather high gas velocity produces a radial gradient in gas temperature in bare tubes with a minimum along the tube axis (Figure 6a). The consequence is an overheated gas phase diffusion layer at the inner tube surface and a relatively cold temperature in the core flow of the gas phase. Hence, the temperature driven gas phase reactions are resulting in overcracking in this boundary layer. This increases the coking rate on the inner tube wall, whereas chemical activity is low in the cold core flow. As a result, both the furnace cracking cycle length and the product yield of the cracking gas are reduced.

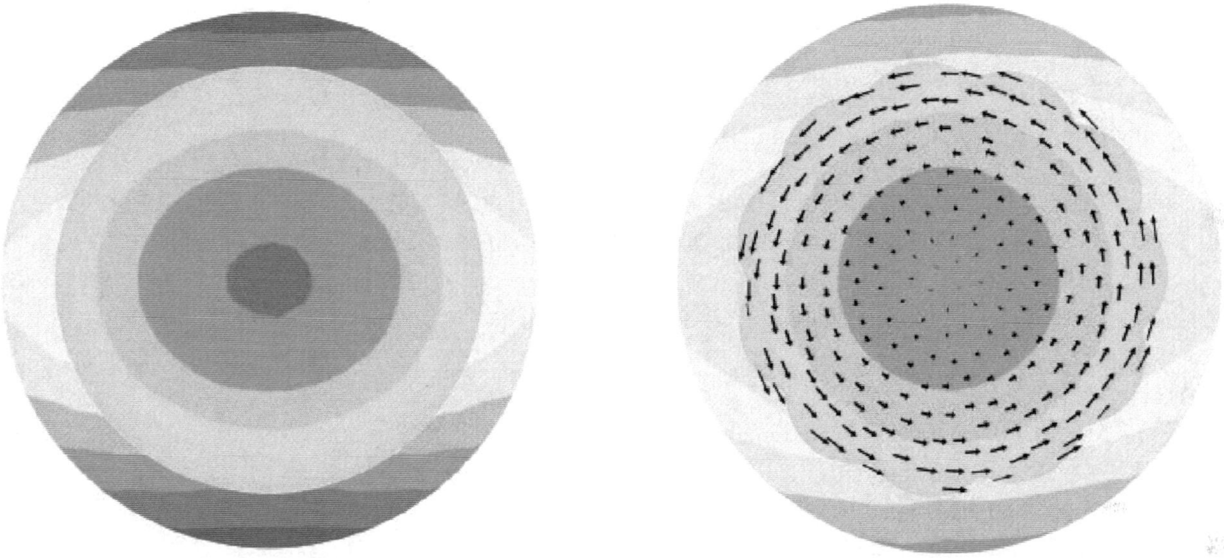

a) bare tube b) SCope

Figure 6: Temperature Distribution and Gas Rotation after 2 m

Taking these physical factors into account, S+C designed the SCope profiled tubes to modify the gas-flow characteristics in a beneficial way to overcome the inhomogeneous gas-temperature distribution. The SCope profile induces a smooth gas phase rotation that increases the heat transfer at the inner tube side, reduces the maximum skin temperature of the tube and balances the temperature gradients in the tube shell as well as in the gas phase (see Figure 7). The SCope profile has got fins with specially designed shape, height and inclination to the tube axis.

The careful adjustment of these profile parameters enables a quick buildup of a fully developed gas phase rotation and prevents elevated coking rates in the valleys of the profile, so that the process impact of the profile is maintained even at progressed run length [7]. Since an increasing pressure drop along the coil is a considerable constraint of process performance, special attention was paid to providing all mentioned beneficial properties at the lowest possible increase in pressure drop to minimize the negative influences.

S+C developed a sophisticated manufacturing method to ensure that surface quality of the profile is equal to the conventional pull-bored tube surface of a bare tube. Therefore and especially with the aluminum oxide former Centralloy® HT E complete protection by the oxide layer provides unrestricted anti-carburization and anti-coking properties.

Figure 7: Gas Rotation

The positive impact of SCope on the steam cracking process has already been proven in operation. The information and data obtained from field installations reflect similar results like the fluid dynamic simulations: lower tube metal temperature levels at start of run open up the production window and additionally a lower slope of temperature increase over time stretches the run length further.

The effect of reducing the over- and under-cracking in the gas phase was also investigated in field applications based on an elaborate and thorough on-site product yield analysis. It reflected an enhancement of olefin yield by more than 1% when operating at naphtha feed. In this regard, SCope can be used either for increasing the run length or the throughput or the cracking severity and even a reduction in fuel consumption can be accomplished.

Special development work has also been performed to keep maintenance procedures, especially related to the field replacement or repair works, unchanged: After applying a special beveling machine, welding of SCope on site can be done with exactly the same welding procedure specification like for bare tubes.

S+C Research & Development Services has access to various tools like CFD, chemical kinetics simulation and statistical operation data analysis to carry out evaluations and advise how to apply S+C tube technologies to target the most promising technical and economical coil configuration for every furnace design on the basis of each customers' specific operation conditions.

Conclusions

Over several decades, iron and nickel based chromia-forming alloys were the material of choice in steam cracking furnaces. Due to the very high stability of alumina scales, the alumina forming alloys were introduced 15 years back to the petrochemical industry.

Such a stable and continuous alumina scale is formed on the inner surface of alloy Centralloy® HT E, which is installed in the radiant section of more than 330 steam cracking furnaces. Apart from an increased tube/coil lifetime, this alloy is e.g. positively affecting the run-length (low coking rates), the conversion rates, and it can decrease the fuel consumption and lower the furnace emissions.

Some years back S+C developed also the internally profiled tube design "SCope" and validated its benefits in plant applications to agree with predictions from fluid dynamic simulations. The SCope profile leads to a very uniform, circumferential flow pattern which improves the heat transfer, balances the heat distribution in the gas phase as well as in the tube shell and leads to reduced tube metal temperatures. As a benefit, all effects contribute to a reduction of coking rate.

In order to maximize the operational benefit of the plant, the innovative SCope profile can be combined with the proven Centralloy® HT E tube material.

References

1. D. Jakobi, R. Gommans (2003), "Typical failures in pyrolysis coils for ethylene cracking", *Materials and Corrosion 54*, pp. 881-887
2. D. Jakobi, C. van de Moesdijk, P. Karduck, A. von Richthofen (2009), "Tailor - Made Materials for High Temperature Applications: New Strategies for Radiant Coil Material Development", *NACE Corrosion 2009*, Paper No. 09155
3. D. Jakobi, P. Karduck, A. von Richthofen (2013), "The High-Temperature Corrosion Resistance of Spun-Cast Materials for Steam-Cracker Furnaces – A Comparative Study of Alumina- and Chromia-Forming Alloys", *NACE Corrosion 2013*, Paper No. 2287
4. D. Jakobi, P. Karduck (2018), "Behavior of High-Temperature Tube Materials in Sulfur-Containing Steam-Cracking Conditions", *NACE Corrosion 2018*, Paper No. 11191
5. D. Jakobi, J. Weigandt (2014), "Benefits from Educated Selection of Advances Materials for Radiant Coils – a 10 Year Field Performance Review of Alumina Forming Alloys" *AIChE Spring Meeting 2014*, Paper No. 152f
6. S.R.J. Saunders, M. Monteiro, F. Rizzo (2008), "The oxidation behaviour of metals and alloys at high temperatures in atmospheres containing water vapour: A review", *Progress in Materials Science 53*, pp. 775-837
7. L.H. de Almeida et al. (2003), "Microstructural characterization of modified 25-Cr-35Ni centrifugally cast steel furnace tubes" *Materials Characterization 5546*, pp. 1-11
8. A. Reihani, R. Derakhshandeh Haghighi (2015), "Failure analysis and weld ability improvement of 35%Cr-45%Ni heat resistant alloy" *Engineering Failure Analysis 52*, pp. 97-108
9. C.M. Schietekat et al. (2014), "Computational Fluid Dynamics-Based Design of Finned Steam Cracking Reactors", *AIChE Journal 60 (2)*, pp. 794-808

550608 The Impact of High Emissivity Coating Systems on Process Tube Coking and Heat Transfer

The Impact on Process Tube Coking and Heat Transfer by High Emissivity Coating Systems

We examined the effects high emissivity coating systems (HECS) have when applied as coating systems to Process Tubes for ethylene steam cracking furnaces. Utilizing tests including carburization, thermal conductivity, and total hemispherical emissivity, HECS were analyzed as a potential method for reducing coking and increasing the heat flux across process tubes. Application of HECS on the external surface of the radiant process tubes could improve the energy consumption[1] in the radiant fire box. Incomplete use of all radiant heat in the fire box along with the reoccurring decoking process represent high costs to the industry every year. One of the most important ways to reduce the energy input in steam cracking furnaces per ton ethylene produced is to increase the absorption of radiant heat by the process tube and to uniformly distribute that heat across the tube with no "hot spots" thus reducing coke formation inside the process tubes[2]. Steam cracking is one of the most energy-consuming processes in the petrochemical industry and globally uses approximately 8% of the sector's total primary energy[3]. Improving the energy efficiency and reduced coking of process tubes has an immediate payback that cannot be overlooked in typical ethane or naphtha-based olefin plants[4].

A key component to the steam cracking process' efficiency and throughput is uniform and efficient heat transfer onto and through the process tubes. One technology that has been shown to enhance heat transfer in high temperature process furnaces is HECS[5].

The HECS performance on process tubes must be viewed as making each tube a new composite process tube structure with improved thermal performance. The thermal performance enhancements can be defined by improvements in 1. Uniform heating of the tube surface thus reducing hot spots that may attribute to increased coking, 2. Increased heat absorption by the process tube with more heat going to product conversion. Through all this the metallurgy of the process tube is not impacted. A new composite process tube concept is created.

To prove or disprove the above HECS concepts studies and research was undertaken on Duraloy process tube MO-RE 10 with the test procedures detailed below.

Results:

1. Carburization Testing:

The impact on carburization on uncoated and HECS coated process tubes was assessed utilizing the NACE International TM0498 test protocol. Control process tubes consisted of thermally aged uncoated tubes being compared with HECS tubes for the rate and degree on carburization. Results of these tests indicate and impact on carburization can be affected by HECS on process tube coking performance. Data will be presented on the HECS effects. Field studies will be underway shortly to verify the laboratory results.

2. Thermal Conductivity Testing:

Thermal conductivity is the ability of a material to conduct heat usually measured in watts per meter-kelvin (W/(m K). ASTM E1530 was chosen as the test procedure to compare control process tubes with HECS process tubes. Thermal conductivity and heat flux are directly related. Thermal conductivity is the ability of a material to conduct heat, heat flux is defined as "a flow of energy per unit of area per unit of time," or a rating of quantity of heat that is conducted across the process tube thickness expressed as (W m−2). How do HECS impact the thermal conductivity must be answered because improved reaction rates in process tubes has been reported from field data where HECS have been utilized on process tubes. So, as thermal conductivity increases, heat flux, or the amount of heat conducted through the tube also increases. Data to identify whether thin filmed HECS lead to increased heat flux will be presented. Laboratory testing will be verified with field heat flux and reaction rate data and compared.

3. Total Hemispherical Emissivity Testing:

Total Hemispherical Emissivity (THE) testing via ASTM C 835-06 has been accomplished. THE testing results indicate that an increase of 0.4 - 0.6 in THE can be seen when HECS's are applied to process tubes. THE increases on the surface of the process tube increases the amount of energy absorbed and distributed across the tube uniformly. Data will be presented on these results and the impact of HECS increase in process tube THE thus leading to positive impacts on the reaction rates and reduced coking.

Conclusion

Can existing process tube performance be improved by making them into composite tubes with HECS on the exterior surface? This study will show what the improvements can be for existing process tube reaction rates and reduced coking. Have we now entered the era of composite process tubes with HECS enhancement?

References

1. Georgios D. Stefanidis, Kevin M. Van Geem, Geraldine J. Heynderickx and Guy B. Marin (2008), "Evaluation of high-emissivity coatings in steam cracking furnaces using a nongrey gas radiation model," Chemical Engineering Journal, pp. 411-421.
2. Andres E. Munoz G., Kevin M. Van Geem, Marie-Francoise Reyniers and Guy B. Marin (2014), "Coking Resistance of Specialized Coil Materials during Steam Cracking of SulfurFree Naphtha," Industrial & Engineering Chemistry Research, pp. 13644-13655.
3. Tao Ren, Martin K. Patel and Kornelius Blok (2008), "Steam cracking and methane to olefins: Energy use, CO2 emissions and production costs," Energy, May, pp. 817-833.
4. Tao Ren, Martin K. Patel and Kornelius Blok (2006), "Olefins from conventional and heavy feedstocks: Energy use in steam cracking and alternative processes," Energy, June, pp. 425-451.
5. Kongkaew, Nithitorn & Juyploy, Pachara & Uengbowontrakoon, Poramate & Sri-U-Thai, Pankam & Srimitrungroj, Theerasak. (2016). CFD simulations of high-emissivity coatings on reactor tubes and furnace walls in steam cracking furnace.

AIChE Paper Number: 158c

SABIC EXPERIENCE FROM ANTI-COKING COILS IN STEAM CRACKERS AND METHODOLOGY OF BENEFITS QUANTIFICATION

Abdullah Alqahtani
Senior Manager
SABIC

Abdulrahman Sarang
Research Engineer
SABIC

Thomas Dijkmans
Senior Scientist
SABIC

Prepared for Presentation at the 2019 Spring National Meeting
New Orleans, LA, March 31 - April 4, 2019

SABIC EXPERIENCE FROM ANTI-COKING COILS IN STEAM CRACKERS AND METHODOLOGY OF BENEFITS QUANTIFICATION

Abdullah Alqahtani
Senior Manager
SABIC
Abdulrahman Sarang
Research Engineer
SABIC
Thomas Dijkmans
Senior Scientist
SABIC

ABSTRACT

Although ethylene plants are well established, there are plethora of challenges that ethylene producers are facing. Furnace run length and availability are on top of this list. As mitigation and in line with its 2025 strategy, SABIC has been deploying Anti-Coking Technologies (ACT) in the recent years in its ethylene plants to maximize production and improve energy intensity. Anti-coking improvements cover developments of chemical additives, flow-modifiers, furnace tube coatings and enhanced alloys. All of these technologies work in different mechanisms to reduce coke formation, which will benefit the plant by extending furnace run length, improving furnace availability, reducing decoking cycles and hence increase profitability. Among these, anti-coking coils (enhanced aluminum based alloys) work in reducing catalytic coke formation and the carburization effects by forming a protective layer, which shields the catalytic sites.

This paper illustrates SABIC experience in evaluating and implementing different technologies, and the methodology of benefits quantification from ACT to justify the investment cost in its furnaces. SABIC has over 100 furnaces in Saudi Arabia where more than 30 furnaces have upgraded coils to different enhanced alloys. Moreover, various anti-coking technologies implemented in SABIC crackers from several vendors have been investigated, analyzed, and compared. Multitudes of benefits have been reported depending on several factors such as alloy selection, furnace limitation, downstream limitation, feedstock type and availability.

INTRODUCTION

Steam cracking is one of the most important petrochemical processes for the production of basic chemicals. At the heart of a steam-cracking plant lies the steam-cracking furnace, which converts the feedstocks into valuable chemicals such as ethylene and propylene. In this furnace, the feedstock is cracked in radiant tubes at high temperatures (~800 – 900°C) to produce the major products. During this process, coke also deposits on the wall of the furnace. The coke layer inhibits heat transfer from the tube to the process gas, which raises tube wall temperature and reduces the cross-sectional flow area of the tube, thus increasing pressure drop which in turn decreases cracking yields.[1] Coke formation remains one of the main challenges in the petrochemical industry and an important area for current developments.[2-6]

Coke formation during thermal cracking is a complex phenomenon. It is typically considered that cokes can be formed by three different mechanisms. The first type of coke formation is the catalytic coke formation where the properties of the tubes and the tube skin material play an important role.[7] The second type of coke formation is formed after the surface has been covered by the first type, known as radical or pyrolytic coke formation. In this phase, poly-aromatic compounds at the gas/coke interface are not yet fully hydrogenated. By abstraction of a hydrogen, radical sites can form on this interface, which can absorb other molecules in the gas phase and thus build up the coke layer.[8] Due to the nature of the coke formation, the first type of coke formation will give you a typical filament like structure as depicted in Figure 1A. The second type of coke formation will give more of a cauliflower like structure like depicted in Figure 1B. The third type of coke formation is typically called condensation cokes. It occurs when heavy poly-aromatic compounds condense on a cool surface (e.g. TLE tubes) where these droplets can further dehydrogenate to form condensation cokes[8] which typically looks like as depicted in Figure 1C.

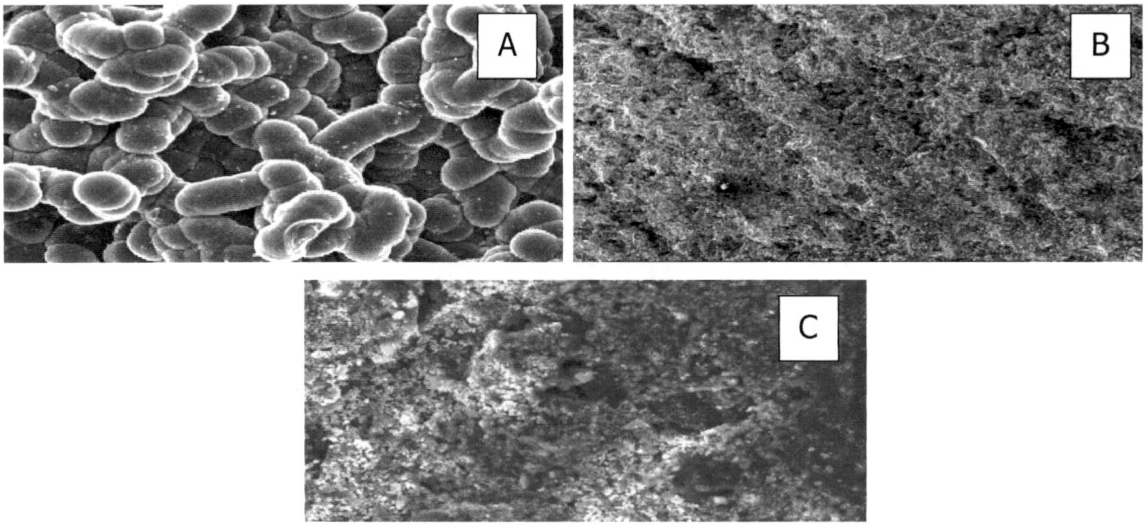

Figure 1: Structure of the different types of coke formation

Depending on the position in the furnace and the feedstock type of the steam-cracking furnace, either of the mechanisms can be the dominant mechanism. The catalytic coke formation is typically dominating at the beginning of the runs due to the high availability of active sites[8, 9], whereas the radical coking will typically start thereafter.[10] The condensation coke formation is typically important in the transfer line exchangers of furnaces cracking heavy feedstocks.[11]

Several types of anti-coking technologies (ACT) exist, which can typically be divided into three groups:[4]

- Surface technologies (e.g. alloys and coatings)
- Three dimensional reactor technologies (e.g. tube finning)
- Feed additives

Depending on the technology, different challenges related to run length are being targeted. For example, the surface technologies and feed additives typically aim at reducing coke formation directly and thus limiting the tube metal temperature (TMT) rise (or pressure drop rise) in the furnaces. The three-dimensional reactors aim at increasing heat transfer and decreasing the laminar boundary layer inside the tube. This reduces the internal and external wall temperature, which both reduces coke formation as well as give more room for the TMT to rise before the temperature limit is hit.

VALIDATION OF ANTI-COKING TECHNOLOGIES

While there are many proven anti-coking technologies available on the market, it remains challenging to pick out the best (economic) technology. In the past, SABIC has developed its own methodology to validate if existing ACT technologies work and to quantify the effectiveness of the technologies in our furnaces. While this work has been described elsewhere[12], a short summary of this work is given in this paper.

At the heart of the methodology lies an in-house developed tool called gCRACKER. gCRACKER is a model to simulate steam-cracking furnaces. A schematic overview of the tool is given in Figure 2. While the mechanistic details of gCRACKER are beyond the scope of this paper, it does contain a detailed coking mechanism required for the validation of existing ACT technologies. Unlike previous works[13-15], SABIC's coking mechanism assumes two mechanisms taking place in parallel, pyrolytic and catalytic coking. More details about the coking mechanism can be found in the referred paper.[12]

Utilizing gCRACKER as a tool, the anti-coking qualification approach consist of four steps.[12] In the first step, a base line performance of a reference furnace (where the ACT technology has already been applied) is obtained by tuning the parameters in the coking model to actual plant data before the ACT technology was installed. In the second step, the base line is repeated for the same furnace, using plant data after the ACT technology was applied resulting in two sets of tuned coking parameters. In the third step, the tuned coking models are used to predict the performance of the technology in the target furnace where the technology will be applied. This is done by simulating the run length using the obtained coking parameters. If required, the model predictions can be compared with existing plant data of the furnace (only before case). In the fourth step, the results of the simulations can be used to quantify the economic benefits and risks when the ACT technology will be deployed. This step will be discussed in more details later in this paper.

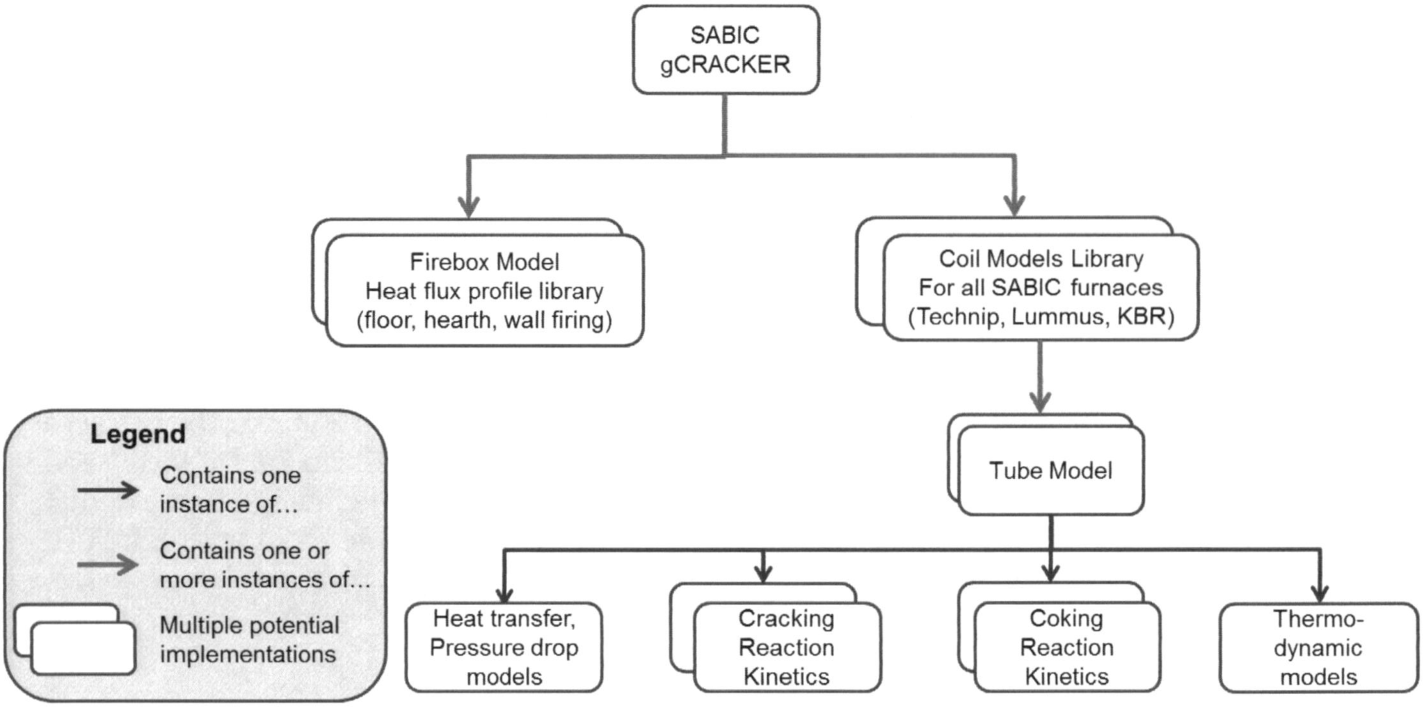

Figure 2: Overview of SABIC gCRACKER Furnace Simulation Tool

SABICS EXPERIENCE WITH EXISTING ANTI-COKING TECHNOLOGIES

In the past few years, several plants within SABIC have already implemented different ACT technologies to help and improve the performance of the plant. The furnaces in these plants can also act as reference furnaces to apply the abovementioned quantification method at other sites within SABIC. Figure 3 shows an overview of three different alloys that have been tested within SABIC at different affiliates using both ethane and propane as a feedstock. In general, all applied technologies have shown an increase in run length at these sites. However, often there are mixed experiences on the actual run length increase (e.g. Alloy 1 on ethane). Often this can be traced back to the operation of the furnaces, as maximizing the benefits of the ACT technology requires careful planning and monitoring of the state of the furnaces by means of accurate tube metal temperature (TMT) measurements and/or pressure drop measurements or venturi ratios. Furthermore, operational issues can not only decrease the lifetime of the coil but also have a negative impact on the performance of the coil. This holds true for both the conventional technologies and the ACT technologies. Thus, good operation practices are still required when ACT technologies are implemented.

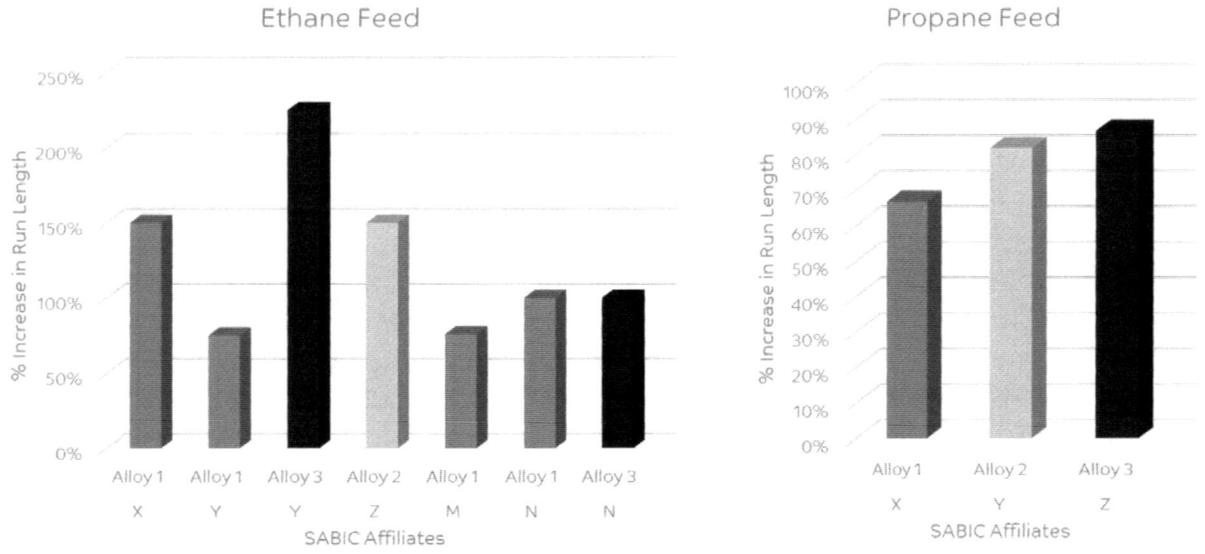

Figure 3: Overview of SABIC experience with different alloys for different feedstocks at different affiliates

BUSINESS CASE DEVELOPMENT

While many of the anti-coking technologies available on the market can be considered proven technologies, it is often difficult to quantify the actual benefits of such technologies. You will often come across several claims that sound very promising such as:

- Increases run length by a factor x
- Increases ethylene selectivity by y
- Increases furnace capacity by z
- Increases coil lifetime by m
- Reduces energy consumption by w%

While these claims are not necessarily false and can be validated with the methodology described before, it is often challenging to meet these expectations in existing steam cracking furnaces of plant, which typically have their own inherent limitations. For example, while increasing the capacity of the furnace (without increasing the run length) sounds promising, if the feed pump of the plant cannot supply the additional feedstock this cannot be considered as a direct benefit. SABIC is operating fourteen steam-cracking plants around the world each with its own inherent limitations. Based on internal experience with existing technologies, a decision tree was developed to help each of SABICs affiliates to decide on how they can use their anti-coking technology to its fullest potential. The decision tree is shown in Figure 4. The figure shows that one of the most important criteria for realizing the maximum benefit of anti-coking technologies is having a furnace bottleneck. If the furnaces are a bottleneck of the plant, an increase in availability of the furnaces can be used to increase the average capacity of the furnace. This capacity can either be utilized by processing more feedstock (if available and can be supplied by the feed pumps) or to lower the average severity of the furnaces (thus increasing the recycle flows). The latter is typically only relevant for ethane, propane and butane furnaces as decreasing the severity of a liquid furnace typically has a large impact on the overall economics of the plant due to a shift in the propylene/ethylene ratio and only a small impact on the recycles being sent back to the furnaces. If the furnaces are not limiting the capacity of the plant, benefits can still be gained from anti-coking technologies though they are significantly smaller. Typically, these benefits are in the form of energy savings. As depicted in Figure 4, energy savings can either be gained by keeping the original run length (before ACT) or by increasing the run length. In the former case, energy will be saved due to less coke formation on the tubes and thus less firing is required on average during a run. In the latter case, the average firing remains more or less the same over a run but less decokes are required which also saves fuel. The last potential benefit, that some of the affiliates in SABIC utilize but very difficult to quantify in terms of economics, is increasing the time available for maintenance. Longer run lengths of the ACT technology allows them to spend additional time on furnace maintenance and thus reducing the unplanned downtime of the furnaces (again increasing availability of the furnace) and thus influencing the economics of the plant. This also reduces the chance that multiple furnaces are out of operation at the same time and that the overall capacity of the plant needs to be reduced at those times. While the latter is an actual economic benefit, it remains challenging to quantify the actual dollars of unplanned but prevented shutdowns.

While most of the major benefits are depicted in Figure 4, other benefits are possible for ACT technology. One such benefit is an increase in the coil lifetime, which decreases the number of times a coil needs to be replaced, which could increase the NPV and the IRR of the coil investment if the CAPEX of the ACT coils is not significantly higher than the conventional coils. Based on the plant limitations and the chosen scenario a business case can be developed for different ACT technologies to determine which technology best fits the plant. It should be noted that when several furnaces have ACT technology implemented, the actual benefits of a new implementation could shift to a different scenario and care should be taken to extend a business case to all furnaces. This is also reflected in the case study that will be discussed later in this paper.

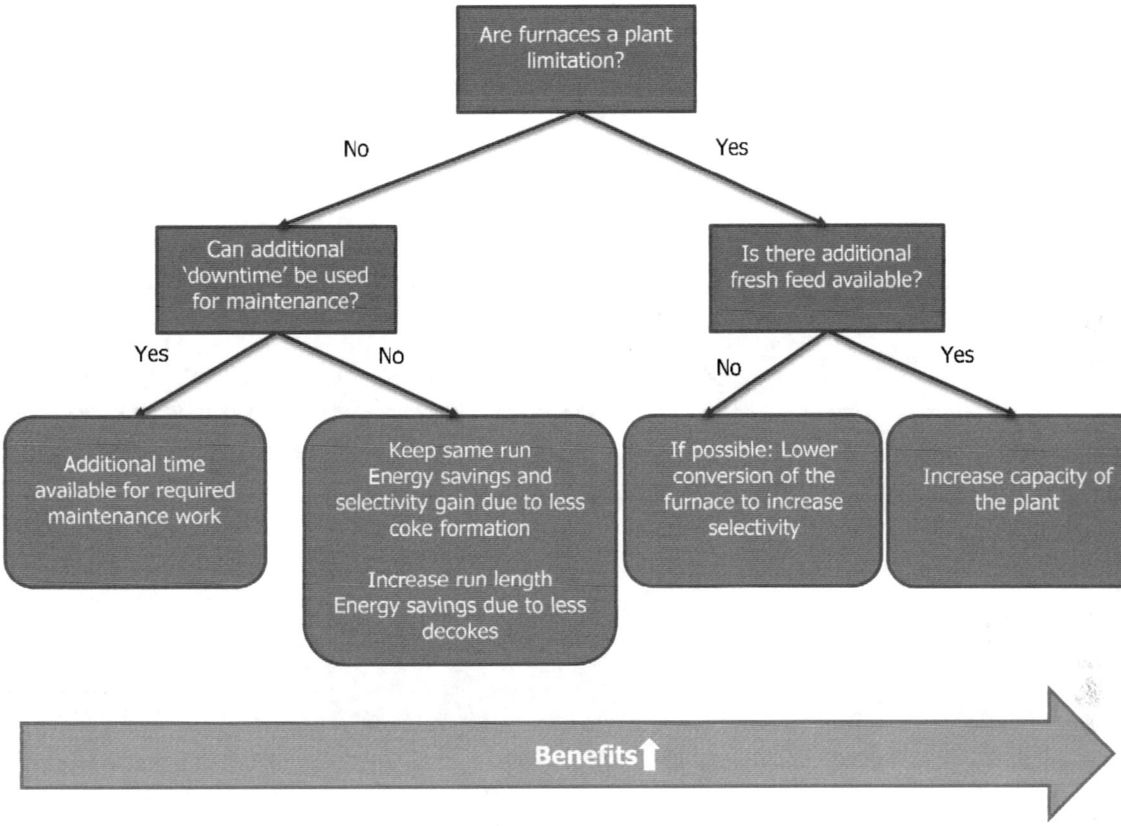

Figure 4: Overview of potential benefits that can be gained from applying anti-coking technologies taking into account the actual plant constraints.

CASE STUDY: ETHANE PLANT

As an example for the methodology, a case study of an existing ethane plant within SABIC is given in this paper. The ethane plant takes in fresh ethane, which is converted in their eight steam cracking furnaces into the valuable products. The unconverted ethane is recycled, sent back to the furnaces, and thus cracked to extinction. The previously described method[12] was used to validate the benefits of different, commercially available ACT technologies and to quantify the potential run length increase for the different technologies. An overview of the results of this study are given in Table 1. The CAPEX of the different technologies that were considered are also given in Table 1.

Table 1: Overview of typical properties for the different technologies that were tested

Technology (-)	Run length increase (%)	Lifetime increase (%)	CAPEX increase (%)
Conventional	+0%	+0%	+0%
Alloy 1	+150%	+25%	+150%
Alloy 2	+100%	+0%	-10%

To analyze the potential benefits of implementing ACT technologies several items were investigated. Firstly, the benefits (compared to the conventional technology) of the three potential scenarios for a single furnace depicted in Figure 4 were quantified regardless of any limitations in the plant. An overview of these results are given in Table 2.

	Units	Capacity	Severity	Energy
		***Alloy* 1**		
Δ C_2H_4 production	X1000 MT / year	6.89	0.325	0.115
ΔEnergy	GJ / year	21,423	21,423	62,866
Yearly benefits	MM$ / year	5.17	0.26	0.16
Payback period	Years	0.15	2.97	4.87
Benefits over coil life	MM$	25.15	0.52	0.021
		Alloy 2		
Δ C_2H_4 production	X1000 MT / year	4.88	0.048	-0.081
ΔEnergy	GJ / year	0	0	32,021
Yearly benefits	MM$ / year	3.73	0.035	-0.019
Payback period	Years	0*	0*	N/A
Benefits over coil life	MM$	14.72	0.2	-0.014

Table 2: Overview of the benefits of both ACT technologies compared to the conventional technology (*Alloy was cheaper than conventional coil)

While all three cases show benefits, there is a huge difference between the increase in plant capacity case as compared to the severity cases and the energy cases, which was already indicated earlier. To quantify the actual benefits it was further investigated if the furnaces are the actual bottleneck of the plant. Based on discussions with the plant engineers as well as analyzing the existing plant data, it was found that the furnaces are running at maximum capacity for most of the year and that typically none of the constraints in the separation section are hit at these times. The next question depicted in Figure 4 is whether additional fresh feed can be supplied to the plant. In this case, this specific plant has an assigned maximum feedstock quota and the question thus becomes if they are consuming their allocated quota. Figure 5 shows the average feedstock consumption for a period of one and half year before any of the ACT technologies were implemented. While it shows that most of the times the feedstock consumption meets the quota and sometimes even (marginally) exceeds the quota there are also times that the quota is not met. The lower consumption of the feedstock typically corresponded to periods were two or more furnaces were out of operation at the same time (either due to decoke and/or maintenance of furnaces). On average during this one and half year, only 98% of the feedstock quota was utilized. Leaving around 2% of the quota not being utilized by the plant. Implementing ACT technologies would allow the plant to loosen up on the planning of decoking. This would decrease the periods of time when more than one furnace is out of operation and thus utilize the remaining 2% of the plant consumption.

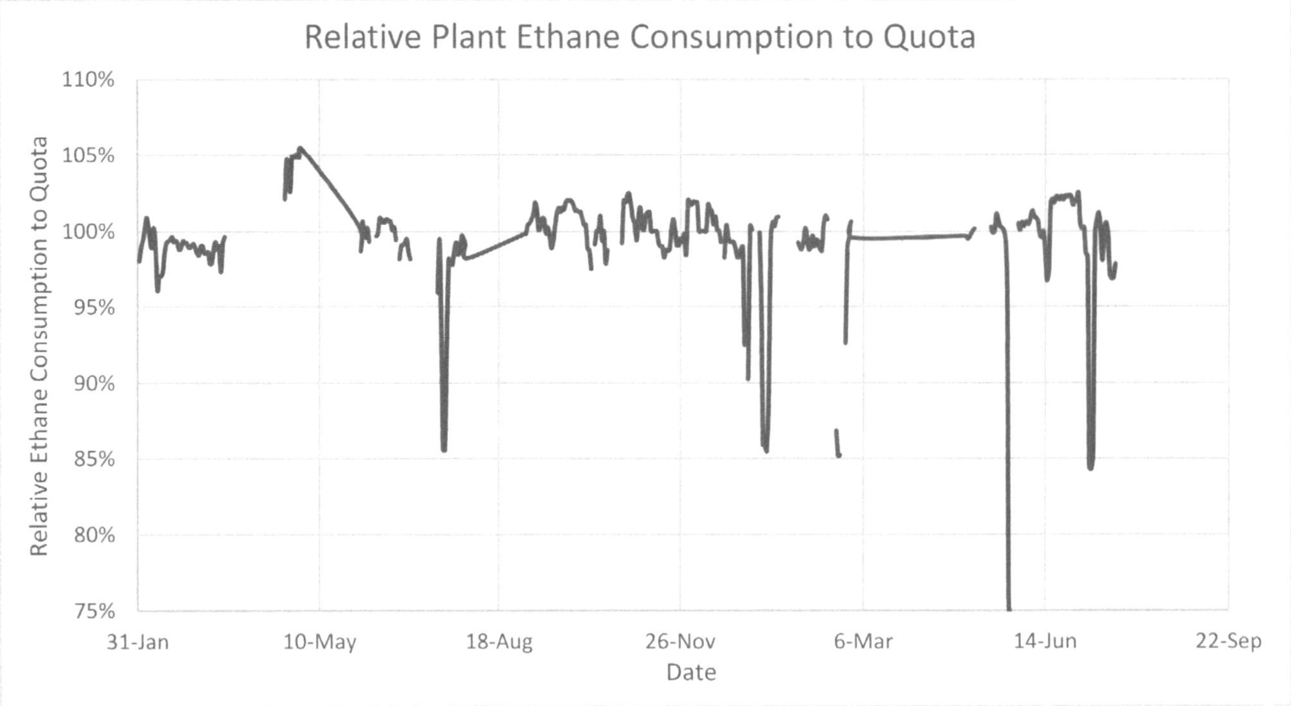

Figure 5: Overview of the feedstock consumption of the ethane plant compared to their feedstock quota

A full implementation of ACT on all furnaces is not required to capture these benefits. To estimate the amount of furnaces required to utilize the remaining 2%, a simplified calculation method has been applied. The number of cycles before and after implementation on a yearly basis is calculated using Equation 1.

$$N_{cycle} = \frac{365 - M}{R + D} \tag{1}$$

With R the run length of the furnace (before or after implementation), D the required decoke time in days and M the maintenance time per year (averaged) in days.

Once the number of cycles is calculated, the availability of a furnace (Av) can be calculated using Equation 2.

$$Av = \frac{R.N_{cycle}}{365} \tag{2}$$

The additional feed in a furnace with ACT can process compared to a furnace without ACT is then calculated using Equation 3.

$$\Delta Feed = Feed_B.\left(\frac{Av_A}{Av_B} - 1\right) \quad (3)$$

With the $Feed_B$ the yearly processed feed on the furnace before ACT implementation and the index A and B indicating the availability after and before implementation of ACT technology respectively. Using the outcome of Equation 3, the number of required furnaces can be calculated using Equation 4.

$$N_{fur} = \frac{Available\ feed}{\Delta Feed} \quad (4)$$

The latter number needs to be rounded to the upper nearest integer. Using the calculations from Equation 1 to 4, it was found that for this specific plant only three to four of the available 8 furnaces would be required to reach maximum capacity of the plant and to alleviate the furnace bottleneck. As such, the business case for the remaining four to five furnaces would need to be adjusted to reflect the change of this constraint. For this specific plant, the additional furnaces would have ACT implemented to alleviate the planning and to allow for more maintenance time for the furnaces as well as energy savings due to less decoke.

CONCLUSIONS

While steam cracking to produce ethylene is a mature technology, producers often face issues with furnace run length and the availability of furnaces. Existing anti-coking technologies (ACT) can help solve these issues. Technology vendors often have promising claims about their technology related to capacity and/or selectivity increase. While these claims are not necessarily false, care has to be taken when calculating the benefits of the technologies of existing plants. These plants typically come with their own inherent limitations that can typically limit the actual benefits obtained from these technologies. In this paper, it was shown that one of the major consideration to take into account, if you would like to achieve the maximum benefits from the anti-coking technologies, is whether your plant has a furnace constraint. If this is not the case, the benefits of the technology will be greatly reduced. With a case study of an ethane cracker, it was furthermore shown that full implementation of ACT technology on all furnaces is not always required and that benefits of reaching maximum plant capacity using the technology can sometimes be achieved by only implementing the technology in a limited number of furnaces.

ACKNOWLEDGEMENTS

The authors are grateful for the support obtained from Process and Energy management group within SABIC as well as the Materials Engineering group. The authors are also grateful to the focal points at SABIC Olefin plants for their support in providing plant data, technical exchanges and encouragement. Special thanks especially to Darren Le Geyt and Dr. Ali Al-Shawaf for their support and discussions, which made this work possible.

REFERENCES

1. Ethylene. In *Ullmann's Encyclopedia of Industrial Chemistry*.
2. General Electric, C. Method and article for cracking hydrocarbon, and method for protecting article against coking during hydrocarbon cracking. 2019.
3. Bao, B.; Wang, Z.; Xu, H.; Liu, J., Anti-coking Effect of MnCr2O4 Spinel Coating during Light Naphtha Thermal Cracking. *Energy Procedia* **2017,** *105*, 4808-4813.
4. Symoens, S. H.; Olahova, N.; Muñoz Gandarillas, A. E.; Karimi, H.; Djokic, M. R.; Reyniers, M.-F.; Marin, G. B.; Van Geem, K. M., State-of-the-art of Coke Formation during Steam Cracking: Anti-Coking Surface Technologies. *Industrial & Engineering Chemistry Research* **2018,** *57* (48), 16117-16136.
5. Anti-Coking catalyst coatings with alumina barrier layer. 2015.
6. Sarris, S. A.; Patil, M.; Verbeken, K.; Reyniers, M. F.; Van Geem, K. M., Effect of long-term high temperature oxidation on the coking behavior of Ni-Cr superalloys. *Materials* **2018,** *11* (10).
7. Wauters, S.; Marin, G. B., Kinetic Modeling of Coke Formation during Steam Cracking. *Industrial & Engineering Chemistry Research* **2002,** *41* (10), 2379-2391.
8. Towfighi, J.; Sadrameli, M.; Niaei, A., Coke Formation Mechanisms and Coke Inhibiting Methods in Pyrolysis Furnaces. *JOURNAL OF CHEMICAL ENGINEERING OF JAPAN* **2002,** *35* (10), 923-937.
9. Hay, G.; Rasouli, G.; Carbognani-Arambarri, L.; Suzuki, R.; Urata, K.; Inoue, M., Reduce coke formation and save operating costs with optimization of DMDS into ethane cracking furnaces. *Hydrocarbon Processing* **2017,** *96* (2), 49-52.
10. Salari, D.; Niaei, A.; Reza Shoja, M.; Nabavi, S., *Coke Formation Reduction in the Steam Cracking of Naphtha on Industrial Alloy Steels Using Sulfur-Based Inhibitors*. 2010; Vol. 8.
11. Geem, K. M. V.; Dhuyvetter, I.; Prokopiev, S.; Reyniers, M.-F.; Viennet, D.; Marin, G. B., Coke Formation in the Transfer Line Exchanger during Steam Cracking of Hydrocarbons. *Industrial & Engineering Chemistry Research* **2009,** *48* (23), 10343-10358.
12. Iliyas, A.; Saudagar, M.; Al-Zenaidi, A.; Al-Dahlous, W.; Al-Zahrani, A.; Karime, M.; Belfadhel, H., New Approach for Anticoking Technologies Qualification for Ethylene Furnaces. In *EEPC*, Czech Republic 2015.
13. Sundaram, K. M.; Froment, G. F., Modeling of thermal cracking kinetics—I: Thermal cracking of ethane, propane and their mixtures. *Chemical Engineering Science* **1977,** *32* (6), 601-608.
14. Sundaram, K. M.; Froment, G. F., Modeling of thermal cracking kinetics. 3. Radical mechanisms for the pyrolysis of simple paraffins, olefins, and their mixtures. *Industrial and Engineering Chemistry Fundamentals* **1978,** *17* (3), 174-182.
15. Sundaram, K. M.; Froment, G. F., Kinetics of coke deposition in the thermal cracking of propane. *Chemical Engineering Science* **1979,** *34* (5), 635-644.

Ethylene Producers Committee

AIChE Paper Number: 158d

Latest possible performance enhancers in radiant coils

Hugues Chasselin

Tech Support & Services Senior Vice-president

Manoir Industries

Heinz Zimmermann

Managing Director

HZ Business Consulting

Manuel Roussel

R&D – Metallurgy and Surface Science

Manoir Industries

Prepared for Presentation at the 2019 Spring National Meeting

New Orleans, LA, April, 03,2019

The radiant coils are one of the most important equipment parts in an ethylene plant in which the cracking reactions take place, transforming the raw materials into the valuable products under precisely controlled conditions of temperature, pressure and residence time. Recent improvements in coil materials and internal geometries offer significant improvements in operation and economics of the cracking furnaces and the whole ethylene plant [1]. Improvements are achieved in furnace run length, throughput and severity by proper utilization of the advanced technologies for radiant coils.

In parallel to the hardware improvements of radiant coils, softwares simulating the cracking reactions have been improved. They are now able to cope with new materials and coil internal geometries and calculate product yields and coil run lengths for different feedstocks.

Correlations of coking rates of new materials in commercial operation and lab tests allow to simulate the coking in the simulations even for new materials with a high accuracy. However, the maximum benefit of the new developments can be achieved by combination of new materials and new internal geometries with a "state of the art" kinetic simulation, being able to handle the hardware improvements. This combination of hardware and software developments is the sound basis for maximizing the benefits of improved technologies.

New materials

For decades cracking coils have been based on high Cr/ Ni alloys which have been cast or wrought into radiant coils. Additions of elements to these alloys have been used to improve the mechanical properties, especially the thermal stability and the mechanical strength [2-4]. Since a couple of years, new alloys containing Aluminum have been developed and used in commercial operation [5-6]. Typical compositions of commercially available materials hereafter referred as 25/35, 35/45 and 4%Al are given in Table 1.

Alloy	C	Al	Cr	Ni	Fe
25/35	0.45	0	25	35	Bal.
35/45	0.45	0	35	45	Bal.
4%Al	0.45	4	25	45	Bal.

Table 1: typical composition of alloys used in radiant coils

Aluminum content in aluminum containing alloys has been optimized over time based on commercial results and currently 4 % Al seems to be the optimum value allowing TMTs up to 1130 °C, offering high corrosion resistance and showing an excellent creep resistance. Creep performances can be seen in figure 1, in which the 100 000 h minimum creep rupture data of standard materials 25/35, 35/45 and aluminum containing material 4%Al is shown.

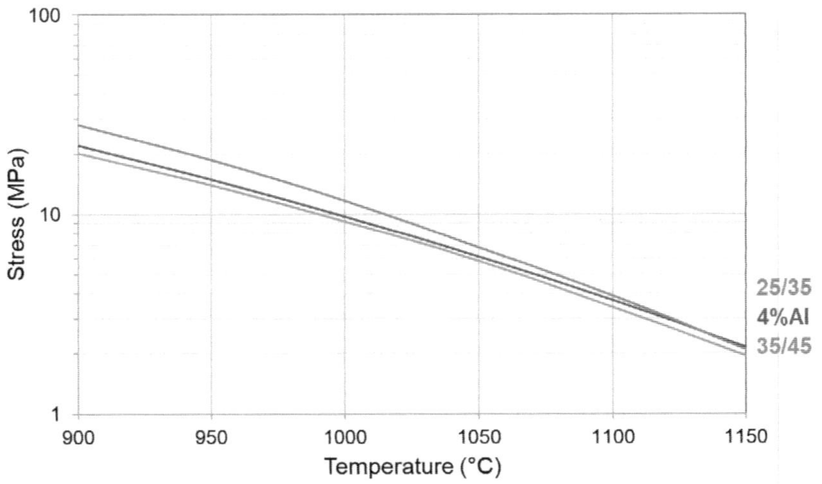

Fig 1 : 100 000h minimum creep rupture data of three commercially available alloys 25/35, 35/45 and 4%Al

One of the most important properties of Al alloys in radiant coils is their surface chemistry. Standard Cr/Ni alloys show a certain catalytic activity due to the Ni content in the oxides they form at their surface during operation whereas Al-containing alloys form a more protective alumina layer on their surface. Ni has a high catalytic activity under the cracking conditions and promotes the "catalytic coke" formation in a cracking coil. Comparison between coke formed on a 4%Al alloy and a 35/45 alloy is shown in figure 2. Coke filaments typically produced by catalytic coking are shown on the surface of the 35/45 alloy in fig. 1.b.

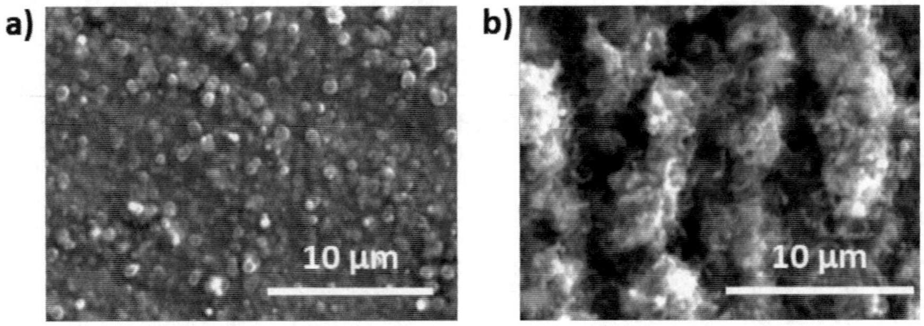

Fig 2: SEM observation of coke deposited on an a) 4%Al sample and b) 35/45 sample after a 1h cracking cycle at 900°C.

Especially in light feedstock cracking, the deposited catalytic coke layers on the internal surface of the coil, together with the pyrolytic coke deposits increase the required Tube metal Temperatures (TMTs) for a constant Coil outlet temperature (COT) and ultimately, when the maximum TMT of the coil material is reached, terminates a coil run and requires a decoke operation. Consequently, Sulphur, most in form of DMDS is used to transfer the catalytically active Ni into Nickel sulfides, in order to decrease the catalytic coke formation.

Carburization is another consequence of the coke deposits and an ageing mechanism for radiant coils. As carbon penetration induces microstructural changes in the tube which ultimately leads to the loss of its creep resistance. Figure 3 shows cross sections of two samples artificially carburized using a pack cementation test at 1200°C for 100h. In stainless steels, the formation of a chromium oxide scale is known to mitigate carburization [8]. Nonetheless, a chromium depleted area due to Cr consumption at the alloy surface during the experiment can be seen on the cross section of the 35/45 alloy. Beneath this Cr-depleted zone, transformed and coarsened chromium carbides evidence carburization of the alloy. The 4%Al alloy shows no Cr-depleted area nor transformed carbides.

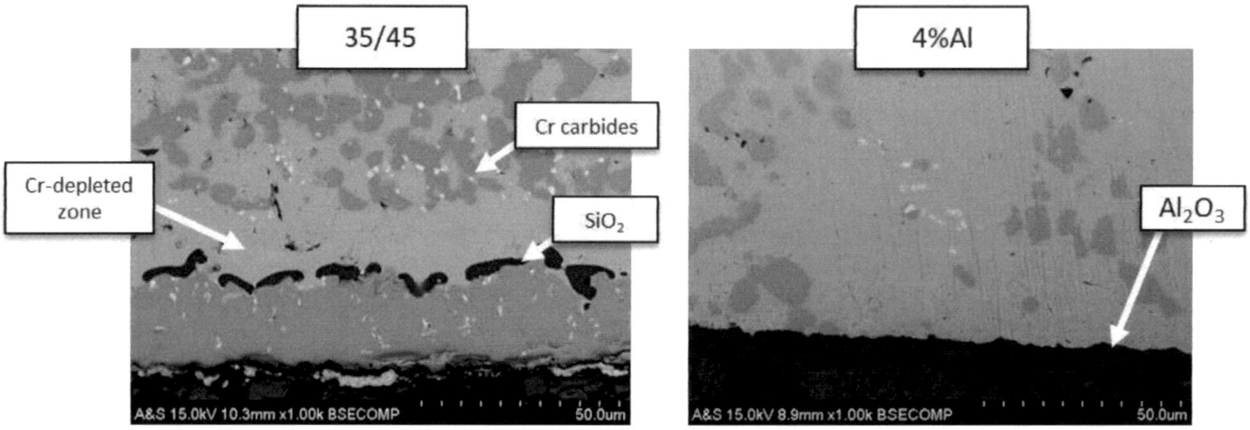

Fig 3 : SEM cross section observation of carburized a) 4%Al sample and b) 35/45 sample after pack cementation for 100h at 1200°C .

Furthermore, Cr depletion from the surface of the alloy causes a significant change of material properties, which leads to the end of a coil life and also increase the coking rates. Cr losses on the outside of the radiant coils have a significant effect in poisoning the SCR catalyst used in cracking furnaces, which requires replacement of the SCR catalysts from time to time. In addition, as Ni and Cr particle emission are drawing more and more attention from the environmental agencies such as the EPA, new rules and regulations could soon lead to new alloy standards and might speed up the change to alumina-forming alloys.

To sum up, new Al containing radiant coils develop an alumina layer on the coil surface, which offers several advantages:

- The catalytic activity of the coil surface is significantly reduced
- There is no Sulphur addition (DMDS) required to protect the radiant coils (probably some of the hottest convection coils need DMDS in smaller quantities)
- No Cr depletion from outer surface to poison SCR catalyst
- Carburization of coils is minimized to extend coil life and improves mechanical properties

It would not be surprising to observe this type of alloy to become a standard over the next coming years for light feedstocks.

New geometries

For quite some time internal fins and corbels are used to improve the heat transfer of the radiant coils in order to achieve sufficient run length or to increase throughput or severity. The mean to obtain them was tested also as the cast materials have very limited elongation properties and cannot be extruded. However, due to increase of pressure drop of those internal structures, yields are reduced, if the cross section of the coil is not increased. With a proper design of the internals, the heat-flux on the inside of the coil can be improved and at constant conditions fuel gas savings of 1 to 2 % can be achieved.

However, a careful design of the internal coil shape is required and needs to be optimized by application of a kinetic simulator. Other parameters, like severity or throughput, require a check of the separation train as well.

As mentioned above, the new developments offer significant performance enhancement, which will be demonstrated by 3 case studies. These case studies are based on data for coking activity of materials in a lab, simulations with a sophisticated Kinetic model and commercial results from industrial furnaces as it is shown in Fig. 4.

The idea in the use of the model is to be able to adjust the set-in coking rates (catalytic & pyrolytic) equations to make them match what lab and full-scale test showed.

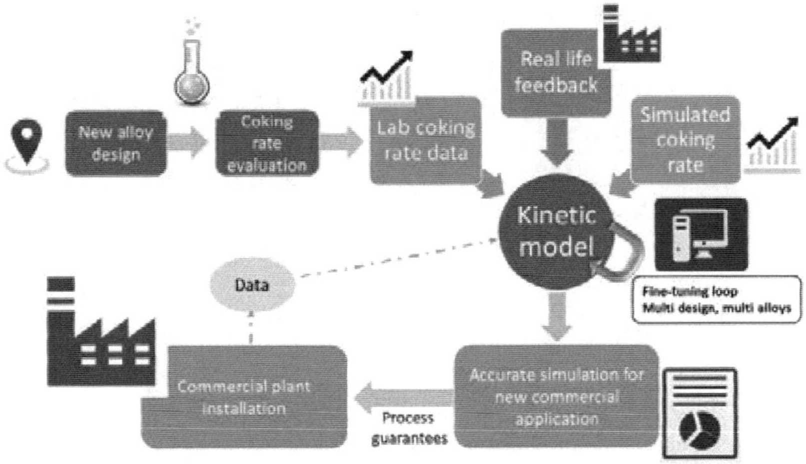

Fig. 4 Scheme for the correlation of LAB data, kinetic simulation and commercial furnace data.

1. Case study 1: Improvements by Al alloy in ethane service

Short residence cracking coils in ethane service suffer in many cases from short run length and a high number of decokes per year, which can lead to a furnace limitation of the plant. In times of high oil prices and therefore also high ethylene prices, this is economically one of the worst scenarios.
Commercial tests have shown that by change of coil material to Al alloy, the run length could be significantly improved as shown in tab.2

The extended run length reduces the number of decokes significantly and can avoid a furnace limitation of the plant, when capacity of the furnace section is reduced since more than one furnace needs to be de-coked at the same time.

#	Case	Number of Furnaces	Runlength / days	No of decokes/ y	Duration of decoke Fee out / feed in (hrs)	Max. No. of Decoke days per year with 1 furnace in decoke	Required decoke days	Furnace limitation
1	base	8+1	14	209	36	243	312 (-69)	yes
2	Mini RL not to be furnace constrained	8+1	20	146	36	243	219 (+24)	No
3	Al bare tubes	8+1	28	104	36	243	156 (87)	No
4	Al tubes finned	8+1	34	86	36	243	128 (115)	No

Tab.2 Effect of run length on plant capacity

As the plant, in case 1 (base), requires 2 furnaces in decoke on 69 days, the plant capacity is reduced to 75% of its design furnace capacity.

Max capacity 243 days *8 furnaces= 1 944 furnace days

Reduced capacity   69 days* 1 furnace = 69 furnace days = 3.5 %

Case 2 shows the scenario where the plant has only 24 days for furnace maintenance, but theoretically is not furnace limited anymore. For this, a technology extending RL from 1' to 20 days is necessary.

Case 3 shows a scenario, using Al material bare tubes with sufficient spare capacity of 87 days for maintenance in the furnace area.

Case 4 shows a scenario with even higher maintenance reserve of 115 days, achieved with straight finned tube and Al alloy

2. Case study 2: Capacity increase by new AL coil material

In a Middle East cracker, Ethane and Propane are cracked in an 8+1 furnaces unit. However, the run lengths are quite different. For ethane cracking, the average run length is 45 days, as for Propane cracking, 85 days are achieved.

With this distribution, 68 decoke days per year are required.

By application of Al alloy coils the run length of the ethane furnaces could be increased to 100 days and for the propane furnaces also to 100 days.

With 100 days run length for both types, the number could be drastically reduced from 68 to only 44 decoke days per year.

This lower number allows to use the spare furnace also as a production furnace during the time, when no decoke operation is required. Theoretically, the plant can use on 365- 44 days = 321 days the spare furnace for production. Allowing 100 maintenance days for the plant, and the need for the spare furnace, the plant could still operate over 200 days with the spare furnace in production.

This gives an advantage in furnace capacity of 6.8 % increased furnace capacity, as the design furnace capacity is 8*365 operation days = 2920 days compared to 8*365+200 days = 3120 operation days.

Capacity increase is 100/ 2920 *200 = 6.85 %. This increase is economically of highest interest should the plant have some spare capacity in the separation train. In this case the payout of the modified material will be much less than1 year. Moreover, the coil life is anticipated be significantly extended by at least 2 years thanks to higher carburization resistance properties, assuming end of run TMT is unchanged.

3. Case study 3: Pressure drop limitation resolved

In an ethane cracker in Middle East, the run length of the furnaces is only 45 days. The furnaces are equipped with a swaged coil with smaller diameters at the 3 legs of the inlet section and larger diameters in the 3 outlet legs.

A simulation of the coil was made, indicating that the limitation was caused by coke deposit in the inlet coils with smaller diameter. By utilization of a combination of Al containing coil material, an increase of the diameter of the inlet tubes and application of straight finned tubes, a significant increase in run length could be achieved. The finned tubes in the inlet section increased the heat flux of the coils in order to achieve a rapid temperature increase of the process gas, which is typically by use of small diameter tubes in the inlet section. With the straight finned tubes, the same effect could be achieved with a much larger cross-sectional area, but at a much lower sensitivity to pressure drop increase. As the Al containing coil has a much lower catalytic coking activity, the coke deposit in the critical inlet section could be reduced. Together with the much higher maximum TMT of 1130 °C for the coil material, a run length increase from 45 to over 100 days could be achieved and the pressure drop limitation was resolved.

4. Next steps

As can be seen from the case studies, the combination of new available technologies and materials, together with a proper simulation, can be extremely useful to resolve operational bottlenecks and to significantly improve the performance of a cracker, by small modifications in the furnaces section. Those modifications do not require any further alteration of structure, or other key equipment around the furnace area.

The next stage of development is to apply active coatings to radiant coils in order to gasify coke deposits. This is also a technology that has been known for quite some time, but the coatings offered so far had the disadvantages to be very expensive, due the method of coating by physical vapor deposition. In addition, the coatings did not last very long. After a few cycles the coating became inactive. Manoir, in exclusive partnership with Suez have developed an affordable coating method for active surfaces, which have demonstrated in lab tests a very good long-term stability.

Based on these lab results and very promising independent tests at the pilot plant at the university of Ghent/ Belgium, a commercial test of this exciting technology has been delivered in Europe and commercial scale results are expected shortly.

The test consists in a full furnace cracking naphta, with outlet tubes coated. This furnace has its identical neighbor uncoated for comparison of data. Those data include TMT measurement, pressure drop increase, CO/CO^2 generation, DMDS injection control, and more.

This new technology can be a very powerful tool in Naphtha cracking, to achieve similar run lengths improvements as with AL alloys in light feedstock cracking.

The use of the model will also provide process benefits generated by this technology.

5. Summary

Summarizing, taking into example some concrete case studies, it can be stated that the combination of kinetic simulation and new coil materials and shapes is a very powerful tool to achieve maximum profitability in commercial furnace operation. The combination of material enhanced properties correlated with lab and industrial data can, in certain cases, alleviate plant constraints, remove furnace limitations, and increase plant output by several percent. Some side effects, all positive, linked with maintenance costs or DMDS consumptions, or more remotely, SCR catalyst poisoning, can also be identified. In these instances, a coil manufacturer offers the benefits of having access to a worldwide range/list of commercial applications and data collections, in each furnace design.

This can bring alternatives and major improvements in radiant coil configurations and retrofit options.

References

[1] H. Zimmermann, R. Walz, 2000, Ethylene in Ullmann's Encyclopedia of industrial Chemistry, Wiley-VCH Verlag GmbH & Co. KGaA

[2] F. Abe, T.-U. Kern, R. Viswanathan, Creep-Resistant Steels - 1st Edition, Woodhead Publishing, 2008.

[3] T. Sourmail, Mater. Sci. Technol, 2001, vol. 17, pp. 1–14

[4] M. Roussel , X. Sauvage , M. Perez , D. Magné , A. Hauet , A. Steckmeyer , M. Vermont , T. Chaise , M. Couvrat, Materialia, 2018, vol.4, pp. 331-339

[5] Y. Yamamoto, M. P. Brady, Z. P. Lu, P. J. Maziasz, C. T. Liu, B. A. Pint,

K. L. More, H. M. Meyer, E. A. Payzant, 2007, vol. 316, pp. 433-436

[6] D. Jakobi, P Karduck, A. Freiherr, 2013, Nace international corrosion 2013 conference & expo, pp. 2287

[7] https://lib.ugent.be/fulltxt/RUG01/001/887/214/RUG01-001887214_2012_0001_AC.pdf

[8] H. J. Grabke, L. Wolf, 1987, Materials Science and engineering, vol.87, pp. 23-33

[9] Jidong Wang †, Marie-Françoise Reyniers*, and Guy B. Marin, Ind. Eng. Chem. Res., 2007, 46 (12), pp 4134-4148

552017 Advancements in Catalytic Coatings for Coke Gasification: Year 11 - Update of Three Full Lifecycle Field Trials of CAMOL Products in Medium Severity Cracking

Hydrocarbon steam pyrolysis is used to produce over 180 and 120 million tonnes annually of ethylene and propylene, respectively. As economics have continued to favour higher severities in steam cracking, the materials supply industry has striven to provide tube alloys / coatings / surfaces / internal profiles to contain and manage the process' severity demands. In the last 20 years alone, medium severity ethane cracking (*primarily defined as ~64-67% ethane conversion, with 0.25-0.35 Steam-to-Hydrocarbon ratio, maximum End-of-Run (EOR) Tube-metal-Temperature (TMT) <1090 °C, and well controlled decoking*) has advanced from an average of 20-50 day runlengths to 60-100 days now being common. Some operators routinely achieve 500+ day runlengths with proper furnace coil alloy and surface technologies, along with good furnace and feedstock controls. Extreme bookends of ethane medium severity runlengths range from 1 day and up to 1,000 days primarily due to the control of the topmost few nanometres of the internal tube surface.

The development of **catalytic (gasifying) furnace coatings** has been advancing since ~2000, driven by the preference and need to mitigate both filamentous and amorphous coke-make. This dual objective cannot be achieved with coil materials and technologies that are engineered solely to provide an internal tube surface that is inert to filamentous coke-make, for example, alumina-formers, silica-formers and spinel-formers. Quantiam's CAMOL^{TM} (*"Catalytically-assisted Manufacture of Olefins"*) catalytic coatings commenced field demonstration trials in 2006, was commercialized in 2012, and now exceeds 10 furnace installations operating at various cracking severities. The trials have mapped and demonstrated the benefits of CAMOL^{TM} across full coil life cycles when remaining reasonably within the **operating envelopes of the base steel and the coatings**, as well as mapping regimes outside of the operating envelopes of the materials systems.

Furnace operation of CAMOL^{TM} products within the operating envelopes has demonstrated runlength improvements of >10X in ethane cracking and >2X in naphtha cracking for furnace-specific coil life cycles ranging from 4.5 to 7.5 years, depending on site-specific operating conditions. These runlength improvements have also been accompanied by energy reductions of up to 6% per unit of production. The first two installations in ethane cracking furnaces (2006-2011 and 2008-2016), and a third installation in naphtha cracking (2010-2016), have now completed one full life cycle. This paper follows up on 1-year and 4-year updates previously presented at the AIChE Ethylene Producers' Conference in 2008 and 2010, and provides a Year-11 update on the three full life cycles completed and now fully autopsied. The discussion will focus on autopsy results from start-of-life to end-of-life, specifically on the steel metallurgy, coating, and surface properties responsible for delivering anti-coking benefits and their correlation with furnace operating conditions and benefits realized or realizable. Finally, a summary of Quantiam's new line of coated products for higher severity cracking that are being advanced based on all learnings to date from the field will be provided with field trials targeting to commence in 2020.

AICHE PAPER NUMBER 158f

Kubota Cracking Tube Technologies

Michael Györffy, Matthew Webb, Eric Hendriks, and Harvey Fraser
Kubota Materials Canada Corp.
Takeo Sato, and Motoyuki Matsubara
Kubota Corp.

Prepared for Presentation at the 2019 Spring National Meeting
New Orleans, Louisiana, March 31st – April 4th, 2019

Kubota Cracking Tube Technologies

Michael Györffy, Matthew Webb, Eric Hendriks, and Harvey Fraser
Kubota Materials Canada Corp.
Takeo Sato, and Motoyuki Matsubara
Kubota Corp.

Abstract

Within the petrochemical market, there is increased demand of more capacity and reliability of steam cracking furnaces. Kubota, a long time supplier of high-quality cracking tubes, has been continuously developing alloys and technologies to improve radiant coil performance. This is realised by different factors of process and tube interactions, such as higher operating temperature, lower coke formation, improved heat transfer, or longer coil life. Technologies have been developed which combine different perspectives of metallurgical design, fluid mechanics and the chemical influence between the process and reactor to meet the industry demands. This approach provides the end user flexibility to address risk and specific technical bottlenecks of the coil's performance. This paper presents a summary of Kubota cracking tube technologies and alloys.

Kubota Materials

Kubota Corporation began as a centrifugal foundry and has evolved to be a leader in the supply of cast heat resistant tubes for cracking coils[1]. Material research and technology development has been a critical component to service the steam cracking tube market. This began with the modification of alloys with a focus on improving creep properties and carburization resistance to validate the supply of standardized replacement materials with alloys of better or equivalent performance. However, a change in the research focus occurred with the understanding that there are interactions between the reactor tubes and the hydrocarbon cracking process. Kubota cracking tube technologies have developed to explore beneficial interactions and to support improved operations and maintenance needs.

The generation of coke in the cracking furnace tube results in higher energy consumption due to reduced heat transfer in the radiant section of the furnace. Coke generation is the cause of carburization the primary tube failure mode. Coke deposits constrict the flow of the product stream throughout the coil, increasing the pressure drop across the coil and insulating coke decreases the heat transfer into the process. Maintenance is then required to assess the level of carburization, coil performance, and planning of a coil turn-around to maintain plant capacity.

Coke is a complex deposit and the result of many mechanisms. The four primary processes are considered to be catalytic coking, pyrolytic coking, condensation of high boiling point components and mechanical deposition from upstream sources[2]. The first is due to the presence of nickel and iron that catalytically increase the kinetics of coking[3]. The coke is characterized by the formation of carbon whiskers with a nickel bearing catalyst tip[4]. Pyrolytic coking is caused by over-cracking of hydrocarbon at high temperature. All forms of coke deposit carbon on the inside surface of the tube, and contribute to the migration of carbon into the tube across that surface. Minimization of coke formation is therefore beneficial to improving tube life.

The first distinctive alloy developed was called UCX, and this material was heavily alloyed with chromium and silicon to provide high levels of carburization resistance. It was noted that the oxide formed had unique properties compared to industry standard alloys. It could resist carburization but also reduce the generation of coke on a lab scale, as shown in figure 1. Understanding that coke was a major contributor to tube failure, future research explored Mixing Element Radiant Tube (MERT) technology[5-9] to lower coking kinetics by lowering tube metal temperatures. Also, ANK 400[10-11] was licenced and developed from NOVA Chemicals Corporation as a low coking surface in a gas feed. These technologies are described and work differently where MERT introduced a static mixing element to increase turbulence in the process fluid and improve radiant heat transfer, while ANK 400 was an inert, stable surface that mitigated catalytic coke formation by eliminating Ni or Fe catalyst particles. The latest technology, called AFTAlloy, is an alumina forming material which provides high temperature internal oxide surface stability, carburization resistance by the formation of an oxide barrier, and in addition to a clean anti-coking surface.

UCX

The alloy UCX was developed in 1990 as a response to the need for carburization resistance of the bulk alloy. It is a 41Cr-48Ni alloy and high in Si; which provides good creep strength with excellent carburization resistance of the bulk alloy. Table 1 shows the chemical composition of this alloy. It was designed for hot applications to localize low carburization kinetics within the firebox. Additional research found that the material was better at resisting the formation of coke. This was due to the formation of a thick oxide scale, and a large reserve of free chromium in the bulk alloy. As a result, it could continuously form and reform an oxide. This is different from conventional materials that experience chromium depletion on the inner surfaces.

Table 1: UCX Composition Range.

Element	Range [wt. %]
C	0.2 ~ 0.5
Si	2.5 max.
Mn	1.5 max.
Cr	40~ 45
Ni	45~ 50
Others	Addition Fe Bal.

From a process perspective the alloy showed 'like new' performance well into the end of life compared to equivalent 35Cr-45Ni alloy performance. Start-of-run pressure drop remained low for the duration of the run length and the coil life could be extended due to improved carburization resistance. The coil applications were implemented where alloy development was only ever originally considered a solution to high local tube metal temperatures within localized portions of the cracking coil.

Figure 1 shows an indexed result of a coking propensity test which compares the resulting mass of coke formed on UCX to conventional 25Cr-35Ni and 35Cr-45Ni alloys. This test is performed at 865°C (1590°F) in a 3:1 ethane to argon feed. The test was performed over three cycles at 1 hour per cycle. It shows that as chromium and nickel content of the alloys increased, the relative amount coke formed was reduced. While this was a lab scale observation, it did highlight the interaction between the process hydrocarbon and the tube metallurgy.

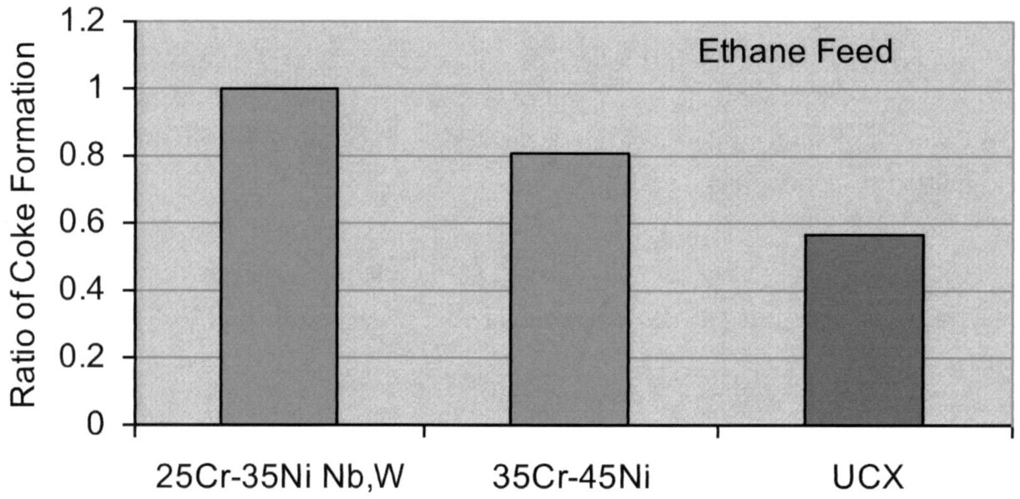

Figure 1: Comparison of lab scale coking formation indexed to a HP alloy. UCX was found to generate less coke than conventional materials.

These benefits were balanced against a few material challenges: the creep strength was acceptable but not improved over conventional alloys and the aged ductility of the material was low. Both of these properties stemmed from the high chromium content. Over 90 orders have been supplied using UCX mostly into Millisecond and Split-U type furnace designs. The high alloy material had immediate results in high tube metal temperatures (TMT) were limiting run length and coil life or there was a very large temperature difference between passes. Therefore, future work focussed on improving these properties but highlighted the material interaction between the coil and the formation of coke.

ANK 400

ANK 400 is an alloy with a stable inert oxide on the inside surface of the tube. Having a surface free of nickel or iron catalytic particles, and a stable and robust oxide is beneficial to reducing coke formation[12]. Elimination of catalytic coking with the use of coatings has been shown to reduce the coking rate by up to 90%[13].The benefit from less catalytic coke formation is that the overall TMT is lower, and therefore kinetically formed coking mechanisms are also less active. The combination of high temperature operation and severe cyclic oxidizing and reducing environments limits the available materials suitable for this process. The ANK 400 material is one such suitably designed alloy and is novel in that it has an engineered oxide surface upon installation.

NOVA Chemicals Corporation and Kubota Corporation have developed a method to reduce both catalytic and pyrolytic coke with ANK 400. This specialized metallurgy separates the heat resistant high nickel tube metal from the hydrocarbon product stream with the inert spinel that is on the tube inner diameter (ID). This spinel layer is formed of manganese chromium oxide and is shown in the scanning electron micrograph in Figure 2.

Figure 2: Scanning electron micrograph of the ANK 400 spinel surface.

Conventional alloys form a Cr_2O_3 oxide which is well documented at 1080°C (1976°F) to decompose into volatile chromium carbides in carbonaceous conditions[7]. This results in a mechanism of chromium loss in the alloy, as well as a loss of oxide protection to carburization. The ANK surface is a stable oxide up to 1100°C (2012°F). Reduced coking rates results in better heat transfer and lower tube metal temperatures. The ANK 400 metallurgy utilizes an industry standard heat resistant 45%Ni-35%Cr alloy base which has known mechanical strength and weldability. The stability of the inner surface provides a reduced rate of surface degradation. The inert spinel has stability at higher operating temperatures than conventional material and acts as a barrier to carbon migration into the substrate metal. It is recognized that oxides are a barrier to carburization[8]. Carbon cannot migrate across an oxide boundary at an appreciable rate.

The performance of three different ANK 400 furnaces is shown in Figure 3. This figure shows the run times using coils with the spinel oxide on the inner surface. The typical furnace run-time prior to the use of this technology was approximately 33 days with conventional materials. Currently, ANK 400 is achieving repeatable run times of over 400 days. Many of these runs were terminated due to plant limitations or reasons not related to coke formation in the coil

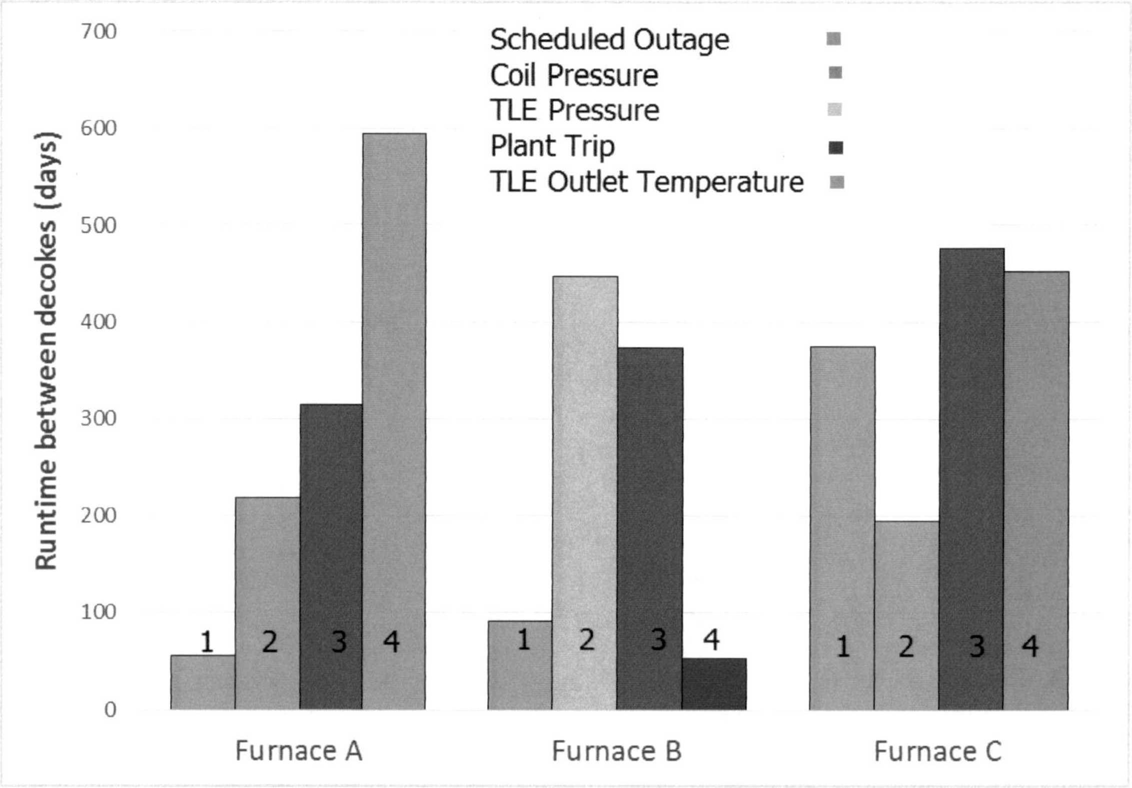

Figure 3: Extended run lengths due to the use of ANK 400. The columns indicate successive runs, and the end of run criteria is indicated by the colour of the column.

ANK 400 is a cracking tube with a continuous oxide on its inner surface that has many beneficial effects to improve the steam cracking process and to reduce maintenance work loads. End-of-life investigations show that intact oxide is an effective barrier to carburization[11]. The effect is not directly causal, in that lower tube metal temperatures, fewer decokes and less coke all have a contributing effect to the end-of-life condition of the tube. This phenomenon has resulted in an achieved 20% improvement in coil life for the spinel oxide users. Less carburization in the coil may result in a possible extension of 50% in coil life based on continuing analysis.

Oxide technologies are being introduced to the cracking tube marketplace that will be able to improve reliability, availability and overall equipment life. From a maintenance perspective, if carburization can be mitigated, it means coil life will be determined by coil creep rates, and the aged ductility of the material. Additionally both of these mechanisms will benefit from a reduction in carburization rates, as the coils will creep less if not subjected to the volumetric expansion of carbide formation. Also, mechanical properties will be improved since there are fewer continuous brittle phases (carbides) in the aged tubes. This allows scheduled planning and coil performance to determine coil turn-arounds.

ANK 400 effectively and repeatably reduces coke deposits in coils during ethane steam cracking. The spinel oxide of ANK 400 is stable up to 1100°C/2012°F. The reduction of TMTs, the improved pressure drop across the coil, the reduced number of coil decokes and the extended run lengths all indicate a significant reduction in coking with to ANK 400 tubes. The technology has been applied to over 223 orders supplied and targets medium to long resistance time gas crackers. It is shown that a robust oxide surface can act on a commercial level to effectively reduce carburization rates, improve coil life and reduce maintenance workloads.

MERT

MERT stands for Mixing Element Radiant Tube. It is a designable three-dimensional tube technology with known heat transfer and frictional resistance parameters. MERT is a centrifugal cast radiant tube with a helical mixing element on the tube inside diameter for use in cracking furnaces for ethylene production. The helical element induces turbulent swirl flow, thereby reducing the stagnant boundary layer of fluid. This increases the inside heat transfer to the fluid cross-section and achieves a more uniform temperature profile. This effect is shown in Figure 6. MERT is distinct from traditional internally finned tubes, in that it does not rely on increased surface area of the helical element in order to achieve improved heat transfer. [5-6] Furthermore, the mixing element does not reduce the effective inside diameter of the tube nor increase its weight significantly. These characteristics, in combination with the performance benefits, have resulted in the great popularity of MERT. Since 1996, MERT technology has been installed in over 630 commercial cracking furnaces, providing the benefits of improved heat transfer, reduced coking, reduced tube metal temperature, increased production rates and increased run times.[8-9]

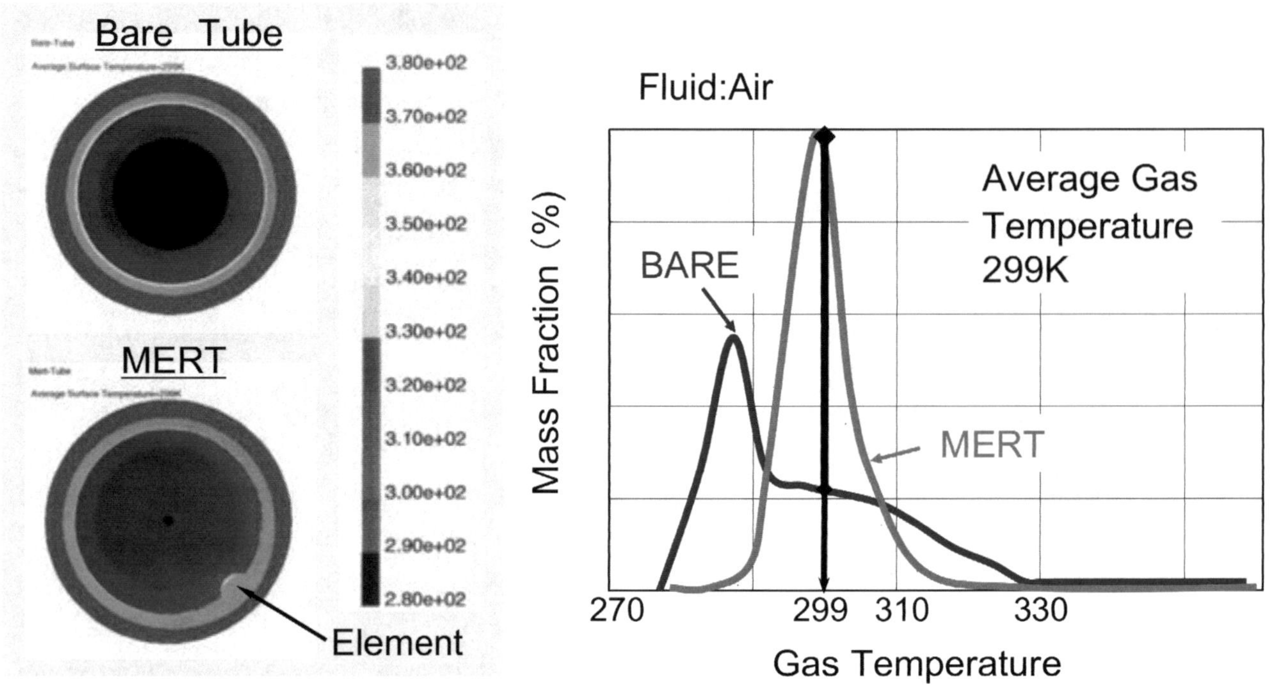

Figure 4: CFD analysis of boundary layer mixing (left) and benefits to make uniform fluid temperature (right).

However, in some cases a continuous weld bead of MERT could not be applied to a particular furnace design due to pressure drop limitations at the plant. To maximize the MERT effect on a furnace, the friction factor needed to be minimized relative to a constant high heat transfer coefficient. [9] In order to address these limitations, Kubota has developed the interrupted helical bead patterns of MERT technology, which provides an improved ratio of heat transfer to friction factor. This provides more opportunities to apply MERT technology in furnaces where there are pressure drop limitations or to provide increase process benefits. The new development has been validated in the laboratory and within multiple commercial furnaces.

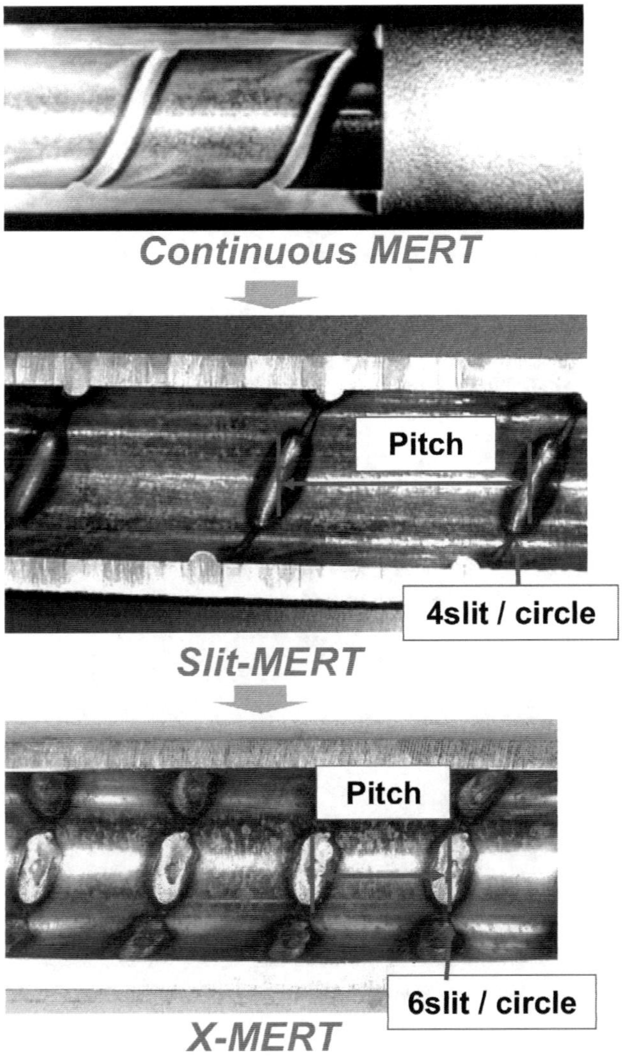

Figure 5: Photographs showing the difference between MERT, Slit-MERT and X-MERT geometries.

Improvements in the element design led to Slit-MERT which has an interrupted helical element. This was a design with 4 breaks in a 360° rotation. Then X-MERT was developed. It is similar to the helical interrupted Slit MERT, except that the element is applied intermittently and at a smaller pitch. The geometry differences between MERT, Slit-MERT and X-MERT are shown in Figure 5. As compared to the standard MERT tube and Slit MERT, the essential geometry is very similar, except for the increased number of small intermittent elements. The heat transfer and friction characteristics of X-MERT tubes have been compared to that of standard bare tubes. This comparison has been performed using CFD (Computational Fluid Dynamics), experimental laboratory experiments, computer simulations and full-scale commercial installations.

Over 630 commercial installations have confirmed the beneficial effects, including reduced tube metal temperature of between 15 to 50°C at start of run (SOR) and reduced coking rates, leading to significantly increased run times. It has been further demonstrated that feed rate and/or severity increases can be consistently achieved without reducing the original run time, due to the high heat transfer capability of MERT technology.

Table 2: Relation between geometry and the pressure drop at constant heat transfer.

	Element Geometry	Heat Transfer relative to Bare Tube	Pressure Drop relative to Bare Tube
X-MERT		1.4	2.1
Slit MERT	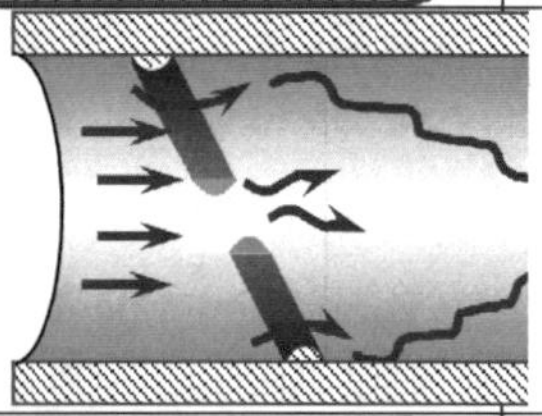	1.4	2.4
MERT	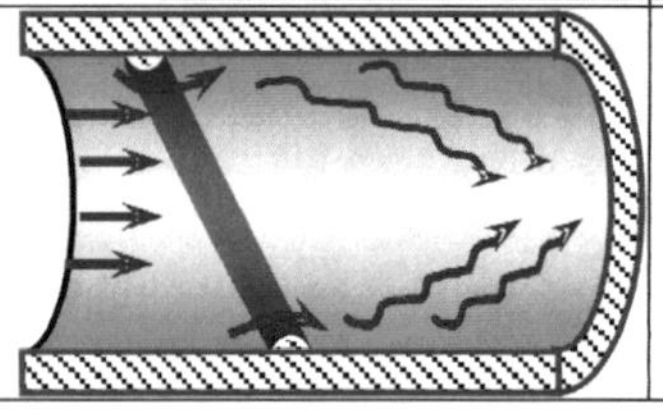	1.4	3.0

The fundamental properties of MERT technology were determined using the Kubota laboratory apparatus.[6] Lab scale data allowed for Kubota to generate a data base relating tube geometry and fluid flow through the tube to the increased heat transfer and friction factor of the X-MERT tube. A summary is shown in Table 2 and this data is then used in a cracking furnace simulation program, to evaluate MERT to specific coil designs, pressure limitations and operating objectives.

AFTAlloy

Alumina forming technology alloys (AFTAlloy) are the next generation of cracking tubes. These materials contain aluminum and generate an alumina surface instead of a chromium oxide surface. The fundamental benefit is a more stable oxide. Chromium oxides are stable in the cracking furnace up to a temperature of 1080°C, above which the oxide breaks down into volatile chromium carbides. This mechanism is harmful in two ways: over the short term the oxide which protects against carburization and coke formation is lost. This causes a change in the tube metal by an increase in carbon content. Then over the long-term the surface is depleted of free chromium and it cannot reform a continuous oxide surface. This loss is due to the previously mentioned volatile loss from the thermodynamic decomposition of the chromium oxide. Additionally, the reduction of free chromium in the austenitic matrix as the volume fraction of carbides in a surface carburized tube increases.

Table 3: Comparison on AFTAlloy composition (wt.%) to conventional chromium oxide forming grades.

	C	Si	Mn	Cr	Ni	Nb	Ti	Al	other
AFTAlloy	0.45	<1.0	<1.0	24	33	<1.0	<0.5	2.0 - 4.0	Add. Fe Bal.
KHR45A 35-45 Grade	0.50	1.8	1.0	31	43	0.5 - 1.0	<0.5	<0.5	— Fe Bal.
KHR35CT HiSi HP Microalloy	0.50	1.8	1.0	25	35	<1.0	<0.5	—	— Fe Bal.

This alloy is developed specifically to form a continuous and robust mono-crystalline layer of α-Al_2O_3. The base material composition is selected to have a reservoir of aluminum dissolved in the austenite for the ability to regenerate oxide if needed, however the initial formation of the oxide is designed to be maintained for the coil operation. For this reason Kubota delivers coils in a pre-oxidized condition for the end-user to ensure a fully formed surface over the entire coil inner surface. The base alloy is a modification of the 25Cr-35Ni micro alloy because this combines high creep strength with the oxidation benefits of the aluminum addition. Since chromium is traditionally added to provide elements for continuous re-oxidation and carburization resistance, an overall lower Cr to Ni ratio can be applied because this function is supplemented by the aluminum additions. A general comparison of bulk alloy chemistry is shown in Table 3.

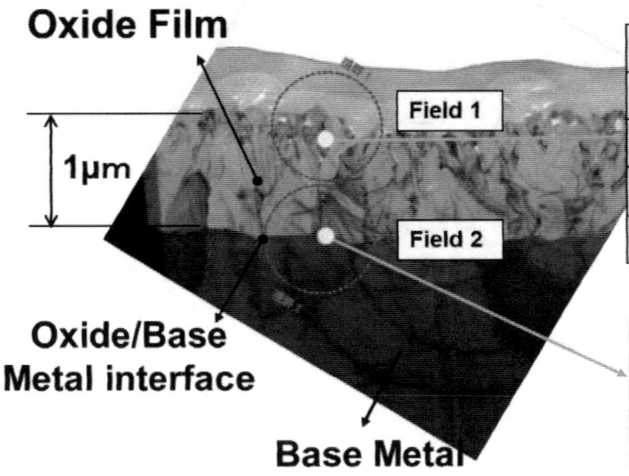

Element	mass%	At,%
O	39.3	54.4
Al	50.4	41.3
Cr	8.0	3.4
Fe	2.3	0.9

Near the base metal

element	mass%	At,%
O	50.0	62.8
Al	50.0	37.2

Figure 6: TEM micrograph analysis of alumina oxide formed on inside tube surface.

The alloy forms a continuous dense crystalline layer of alumina. SEM, TEM and XRD studies confirm the formation of a dense monolayer oxide that is continuously connected to the base alloy interface. This is shown in Figure 6 and demonstrates a mechanism of alumina growth from the base alloy that is different than conventional alloys. This provides several unique properties including exceptional oxidation resistance and carburization resistance as shown in Figure 7 and Figure 8, respectively.

The cyclic oxidation test compares conventional base alloys in a cyclic test to grow and shock the oxide on the material. Since conventional oxides are a multilayered combination of spinel, chromium oxides and silica, the thermal expansion coefficient mismatch can result in spallation and oxide loss with thermal cycling. Alumina formed in AFTAlloy is stable under these conditions due to its chemical composition and microstructure.

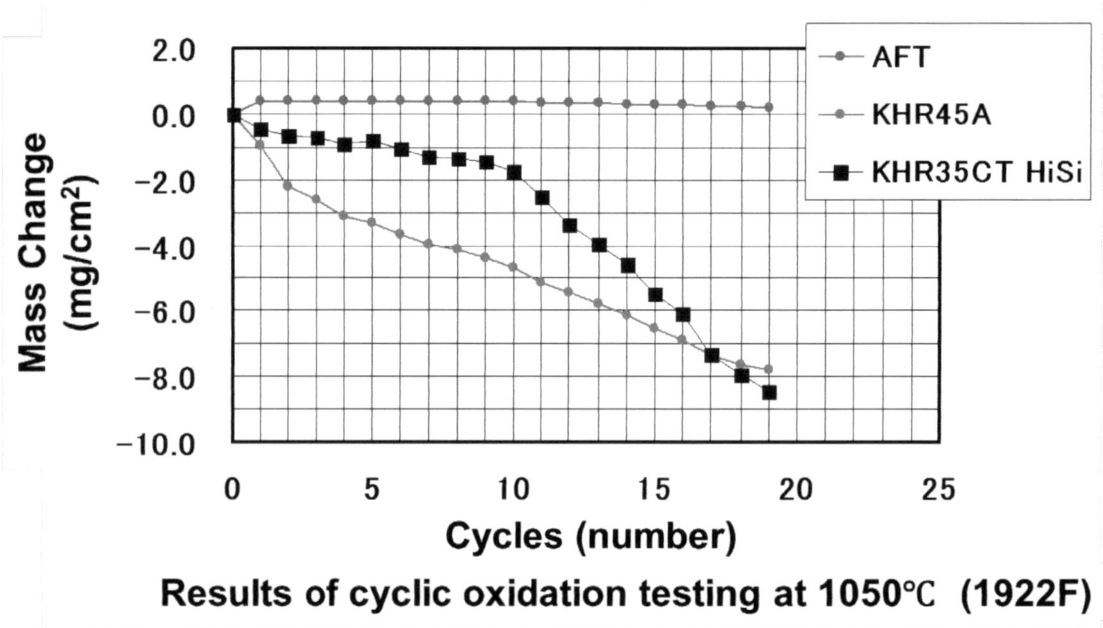

Figure 7: Cyclic Oxidation test of AFTAlloy and conventional chromium oxide forming alloys. Note stable oxide formation and retention.

The stable oxide contributes to exceptional carburization resistance. The alumina oxide is robust and thermodynamically stable in oxidizing and reducing conditions. Therefore it assists in mitigating the migration of carbon from the process gas into the tube metal. Overall less coke is formed due to decreases in catalytic coke formation, and then a barrier oxide promotes longer service life of the coil material. Figure 8 shows an example of this property in a gas carburization test in a reducing atmosphere.

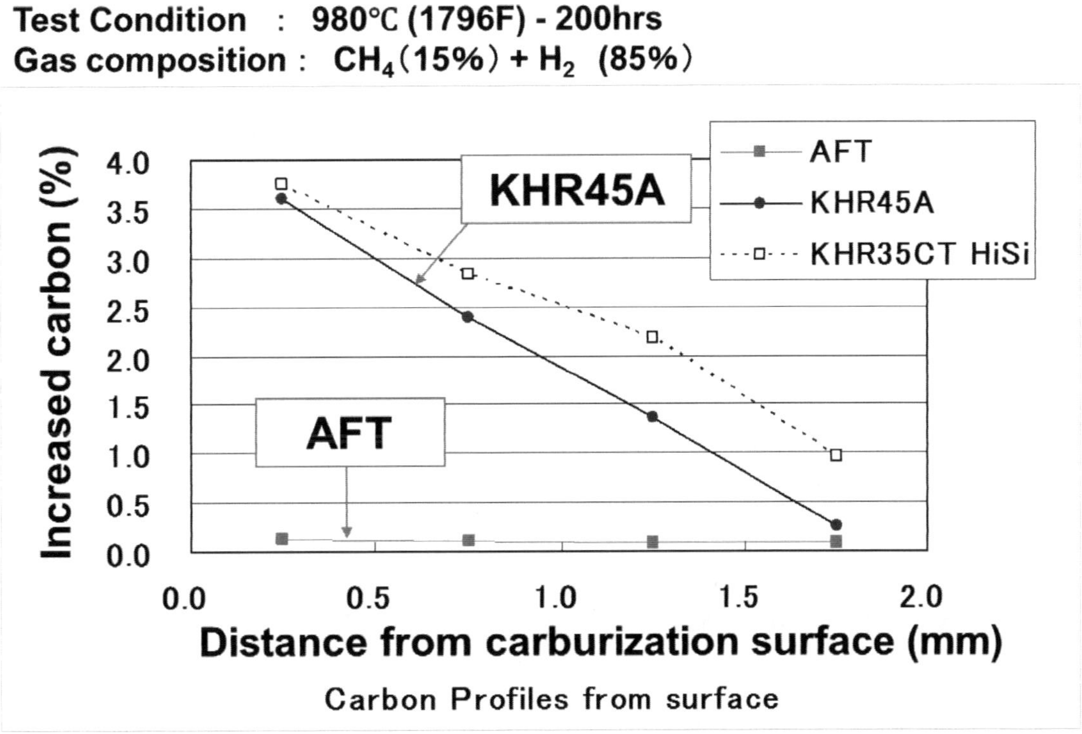

Figure 8: Gas carburization test of AFTAlloy and conventional chromium oxide forming alloys. Barrier oxide and aluminum addition demonstrate a lab scale, step change improvement in carburization resistance.

The AFTAlloy has been applied to over 22 orders since its introduction in 2011. The expectation is that alumina forming technology will provide a reduction in coke generation, and excellent run to run performance due to the stability of the coil's surface. This will increase coil availability due to the need for fewer decokes, increase coil life while allowing an increase in the maximum operating tube metal temperature to above 1150°C (2100°F).

Summary

These technologies employ various mechanisms to improve coil performance. Growing from an understanding of bulk metallurgical anticoking, both the engineered surface chemistry and the process flow dynamics have been developed to optimise a lower coil operating temperature, lower coking rates, improved reliability and furnace availability. Fluid mixing and barrier oxides generated from the tube metal, are a novel and reliable way to improve petrochemical plant operations within the steam cracking furnace and helps to debottleneck issues related to high tube metal temperature limitations or rapid high pressure drop increase within the coil. Additionally, they can be applied to improve coil capacity or cracking severity due to their improved robustness.

References

1. GM Webb, M Györffy, RADIANT CRACKING TUBES–QUALITY FROM THE FOUNDRY, The 2009 Spring National Meeting

2. H. Cai, A. Krywicki, M.C. Oballa, "Coke Formation in Steam Crackers for Ethylene Production", *Chemical Engineering and Processing*, 41, 2002, 199-214.

3. G. Bach, G. Zimmermann, F. Kopinke, S. Barendregt, P. van den Oosterkamp and H. Woerde, "Transfer-Line Heat Exchanger Fouling during Pyrolysis of Hydrocarbons. 1. Deposits from Dry Cracked Gases", *Ind. Eng. Chem. Res.*, 34, 1995, 1132-1139.

4. C. Yin, and W. Tsai, "Nano-sized Carbon Filament Formation during Metal Dusting of Stainless Steel", *Materials Chemistry and Physics*, 82, 2003, 929-936.

5. T. Torigoe, K Hamada, M. Furuta, M. Sakashita, K. Otsubo, M. Tomita, AICHE 1999 Spring National Meeting - Ethylene Furnace Technology and Operations at the 1999 EPC, "Mixing Element Radiant Tube (MERT) Improves Cracking Furnace Performance", March 16, 1999.

6. J. Magnan, K. Hamada, K. Otsubo, K. Mae, E. Sohn, ERTC Petrochemical Conference, "Improved Furnace Efficiency Using High Heat Transfer MERT tubes", Amsterdam, Netherlands, February 20-22, 2002.

7. J. Magnan, K. Hamada, K. Otsubo, E. Sohn, Petroleum Technology Quarterly, "Improved Steam Cracker furnace Efficiency", Summer 2002.

8. J. Magnan, K. Otsubo, S. Matsueda, M. Fujiwara, ERTC Petrochemical Conference, "Next Generation MERT Achieves Further Advances in Cracking Tube Performance." Vienna, 2004.

9. M Gyorffy, M Hineno, K Hashimoto, SH Park, MS You, MERT performance and technology update, AIChE Spring Meeting: Ethylene Producers Conference, Tampa Bay, USA 20

10. M.Gyorffy, L. Benum, N. Sakamoto, "Increased Run Length and Furnace Performance with Kubota and Nova Chemicals' ANK 400 Anticoking Technology; Data from Current Installations as Well as Technology Improvements for Higher Thermal Stability and Decoking Robustness", AiCHe 2006 Spring Meeting.

11. M Gyorffy, M Webb, E Hendriks, H Fraser, CASE STUDY EVALUATION OF END OF LIFE CONDITION OF CRACKING TUBES WITH ANTICOKING BARRIER TECHNOLOGY, AICHE Spring Meeting 2017

AIChE Paper Number: 188A

QUENCH OIL ISOLATION TO PREVENT INCIDENTS

Norman Wilhelm
Process Engineer
LyondellBasell Industries

Prepared for Presentation at the 2019 Spring National Meeting
New Orleans, Louisiana, March 31st to April 4th 2019

QUENCH OIL ISOLATION TO PREVENT INCIDENTS

Norman K. Wilhelm
Process Engineer
LyondellBasell Industries

Abstract:

Quench oil can present many challenges when it comes to isolating and clearing equipment for maintenance. Ethylene plants designed to crack liquids contain equipment in quench oil service that requires frequent maintenance such as feed preheaters, dilution steam generators, process water preheat, quench oil pumps and filters, and quench fittings. Quench oil presents challenges to operators during LOTO and clearing due to its physical properties; it can contain solids, is viscous, and has a high pour point. Failure to isolate and clear this equipment correctly has led to significant process safety incidents and injuries. This paper will review quench oil physical properties, highlight incidents related to quench oil loss of containment, discuss criteria for selecting isolation valves, and make recommendations for design best practices.

Overview

The quench oil loop is a notoriously difficult system for performing maintenance. The physical properties of quench oil can make it problematic when attempting to isolate and clear equipment. Quench oil has a relatively high pour point and contains coke particles from the cracking furnaces. This in turn can create challenges for isolation valves and bleed valves. Loss of primary containment (LOPC) has occurred during line break activities even when Lock Out-Tag Out procedures were correctly followed.

This paper will focus on common pitfalls when isolating and clearing quench oil systems and discuss improvements that can be made to improve operators' success. The causal factors of two incidents will be reviewed as well as a recent installation that incorporates improved design criteria. Selection guidelines for isolation valves will also be discussed.

Quench Oil Incidents

The first incident to be reviewed occurred during maintenance activities on a heat exchanger in the quench oil loop. This exchanger uses hot quench oil on the tube side to pre-heat process water for dilution steam generation. This exchanger is part of a bank of exchangers that are cleaned periodically to maintain heat transfer performance.

To prepare the exchanger for maintenance, operators isolated the quench oil and water lines to the exchanger and placed locks on the isolation valves per the lock out/tag out procedure (LOTO). The quench oil side was cleared by pushing the quench oil back to the primary fractionator with nitrogen, and then by performing a flush on the quench oil side using a light fuel oil. Once the LOTO was completed, operators checked the low point bleed to ensure the system was pressure free and then proceeded to install blinds on the quench oil inlet and outlet lines. It should be noted that the weather conditions were unseasonably cold at this time and ambient temperatures ranged from 27 – 47°F.

The exchanger was out of service for approximately four weeks. During that time, the ambient temperature changed from below average to above average, with a day time high near 80°F. While removing a blind flange on the quench oil inlet line to the exchanger, a LOPC event occurred. The investigation team found that the inlet valve was not completely closed. It was concluded that solids in the quench oil line had prevented the isolation valve from completely closing and that cold ambient temperatures had solidified the quench oil, presenting a false sense that the isolation valve was closed. Figure 1 shows the configuration of the isolation valve – note that the presence of a chain operator on the valve obscures the valve stem from the operators' view.

Figure 1: View of Quench Oil Isolation Valve from Operator's Perspective

The second incident to be reviewed is almost identical to the first; however, it occurred two years prior and at a different site. In this case, a spectacle or figure-8 blind was to be turned from the closed position to the open position after maintenance on a dilution steam generator exchanger. The blind was close-coupled to the outlet isolation valve on the quench oil side (see Figure 2). When the system was isolated nine months earlier, the heat tracing had not been in service. However, at the time of blind removal the heat tracing had been turned back on. Because the blind was close-coupled to the isolation valve, there is no bleed valve, consequently there was no method to make sure the line was free of pressure prior to the line break.

During the removal of the blind, a LOPC event occurred. It was concluded that the gate valve had not been fully seated during the initial isolation and that cold quench oil, due to lack of heat tracing, had provided a temporary seal that allowed the blind to be initially installed.

Clearly, there are some common factors between the two incidents. In both cases, the isolation valves had been reportedly closed and allowed for initial blinding of the system. As cold quench oil or solids settle in the valve seat, the increased resistance can be misinterpreted as the valve being in the closed position. Additionally, cold quench oil provided a temporary "seal" that allowed for initial blinding of the system. It wasn't until ambient conditions warmed or heat tracing was turned on that the temporary "seal" was removed.

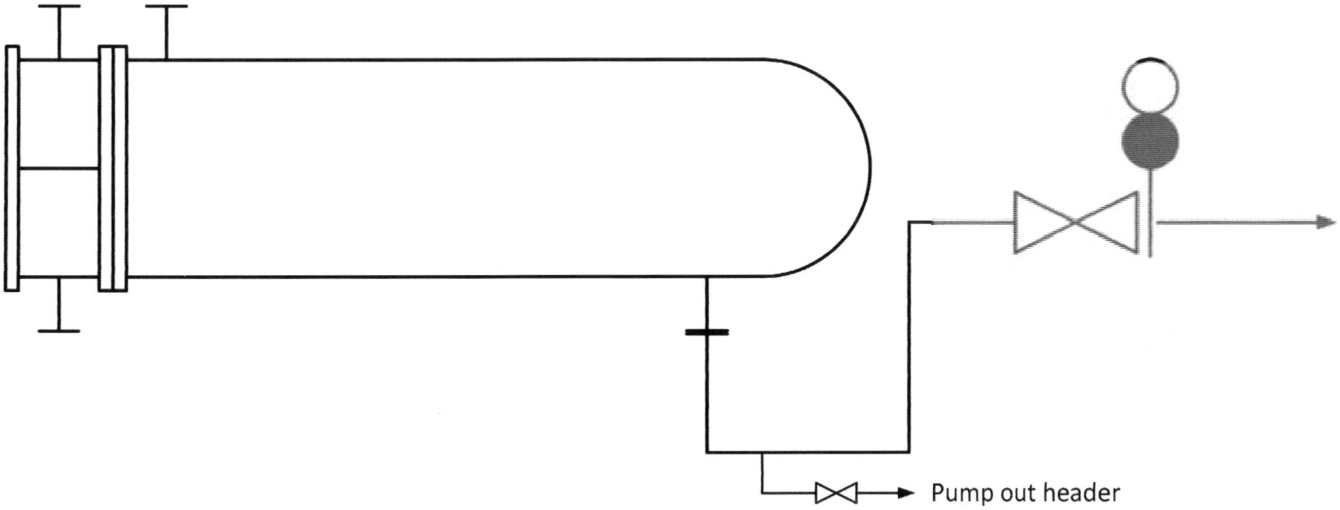

Figure 2: Configuration of Isolation Valve and Figure-8 blind for Incident #2

Quench Oil Properties

The properties of quench oil make it relatively more difficult to clear during equipment maintenance compared to other process streams in an olefins unit. Quench oil is the combination of the heavy oil fraction generated from the cracking furnaces and the accumulation of polynuclear aromatic compounds (PNAs), asphaltenes, and tars[1] due to the long residence time of the quench oil loop. It typically has a boiling range of 500-800°F[2]. The pour point of quench oil was measured to be 45°F at one olefins site operating with a mixture of liquid and gas feeds. Coke from the cracking furnaces can also be present in the quench oil loop and can contribute to plugging in isolation valves and bleed valves.

An uncontrolled release of quench oil has potential to cause severe injury due to high temperature and could result in a fire. The operating conditions of the quench oil loop can vary between plants, but in general, the temperature is between 350°F and 420°F and the pressure is between 100 to 150 psig at the circulation pump discharge. The flash point of quench oil is approximately 200°F and is therefore classified as a combustible liquid per NFPA 30; however, because the operating temperature is above the flash point, it should be assumed that a release of quench oil has the potential to be flammable.

Isolation Frequency of Quench Oil Equipment

The properties of quench oil also lead to frequent maintenance of exchangers, pumps, and filters in the quench oil loop. The risk of loss of containment during isolation of a single piece of equipment is relatively low; however, when considering the frequency at which equipment must come out of service and the quantity of equipment in quench oil service, the opportunity for incidents multiplies. Table 1 shows the typical isolation and clearing frequency of equipment at a liquid cracker. At one LyondellBasell site, the number of planned line-breaks in the quench oil loop was estimated to be at least 80 in one year.

Table 1: Typical Maintenance Frequency of Quench Oil Equipment

Service (# of equipment)	Frequency of Isolation	Size of Isolation Valve
Dilution Steam Generators (12)	Yearly PM	12"
Process Water Preheaters (4)	Yearly PM	14"
Feed Preheaters (4)	Yearly PM	10" - 18"
Furnace Quench Fittings	Bi-annually	10"
QO Circulation Pump	Annually	30"/18"
Coke Removal Equipment	12 weeks	12"

Design of Quench Oil Isolation Systems

A typical isolation scheme for a quench oil user will have the following elements: single inlet and outlet isolation valves, a low point connection to an oil collection header, a connection to a flush oil header, and high/low point bleeds. Figure 3 shows the layout of this system. Some configurations may also include heat tracing on the isolation valves or figure-8 blinds, however that is not universal. Unfortunately, there are some common pitfalls that can make quench oil isolation problematic with this arrangement.

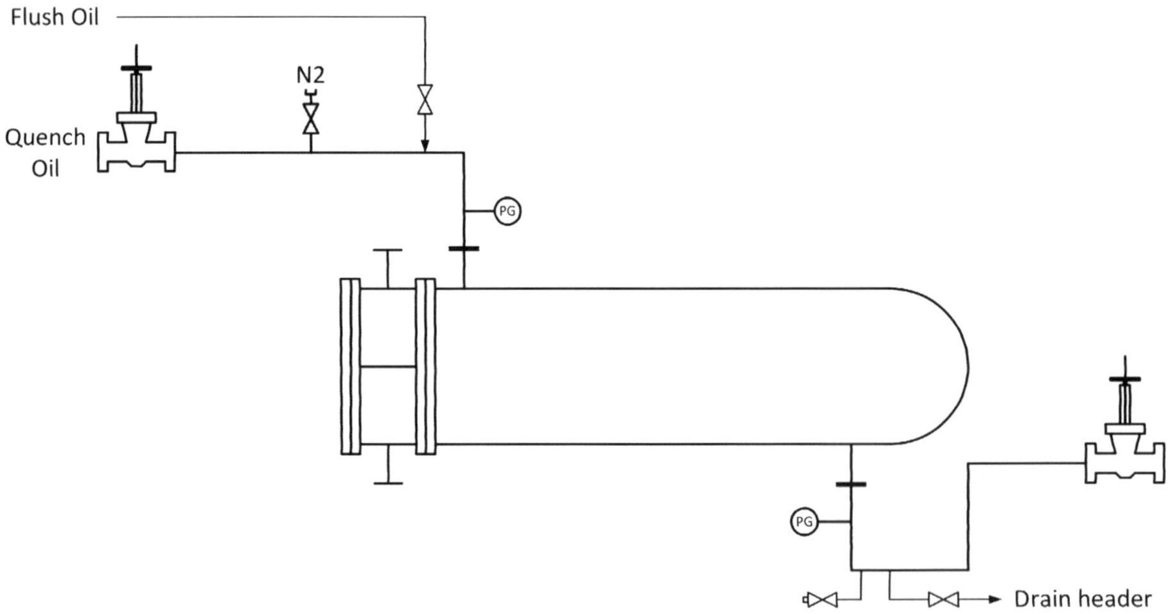

Figure 3: Typical Quench Oil Isolation Scheme

The placement and configuration of low point bleeds is a frequent shortcoming observed in these systems. Bleeds in quench oil service can be plugged with coke particles or solidified quench oil. Bleeds are often placed too close to the ground, which inhibits the use of specialty rod-out tools designed to cut through solids. Bleeds are also commonly configured with a 90-degree elbow to help with ground clearance; however, this design is almost impossible to clear once it becomes plugged and the common "flexible" rod-out tools will likely be ineffective.

Figure 4: Low Point Bleed Configurations

The picture in Figure 4 shows two low point drains on a quench oil exchanger, one of which was original, and one added more recently. The original low point bleed was designed to be low to the ground and included a 90-degree bend. This bleed was frequently plugged and was not adequate to prove the line was pressure-free. A second drain valve was added recently to improve the LOTO and line break practices. The new drain is a larger size (1.5") and does not include any bends.

The selection and configuration of isolation valves is another area that can be problematic. Often, the piping configuration in a quench oil circuit leads to the use of chain operated valves, which have some drawbacks. When used on a typical gate valve, a chain operator will obscure part of the valve stem, removing a visual cue for the operator that the valve is in a fully closed position. Additionally, the use of chain operators does not allow for the operator to have a "feel" for whether the valve is fully seated and can hamper the operator's ability to differentiate between a closed valve and a valve that is pushing against solidified quench oil.

Design Improvements

The design of quench oil isolation systems can be retrofitted to improve isolation and clearing. The layout in Figure 5 is an example of a scheme that was selected for a recent project at one LYB site. The site installed a second gate valve on the inlet and outlet quench oil lines. A piping spool piece was included between the existing and new gate valve, and a bleed was included to ensure no leak-by occurred between the isolation valves. In this case, both the isolation valves and the spool piece are heat traced. Installation occurred during a recent turnaround; however, it should be noted that this type of retrofit is possible outside of a turnaround provided the additional valve and spool piece can fit in the existing piping configuration.

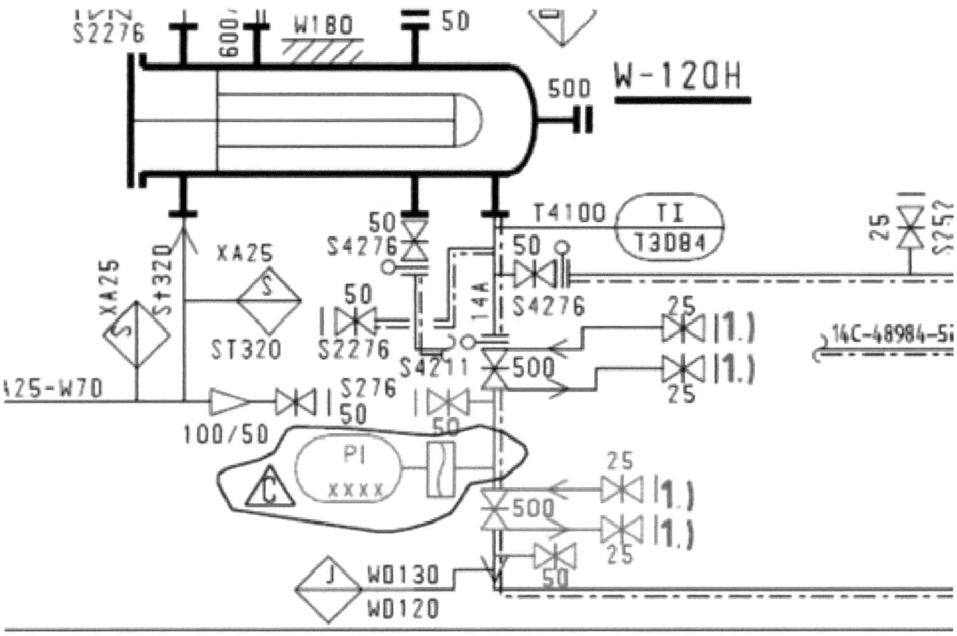

Figure 5: Quench Oil Isolation Retrofit Example

Another option to consider is replacing an existing gate valve with a different style valve with improved isolation performance, such as an articulating plug valve or ball valve (Figure 6). This style of valve can include a flush oil connection which will prevent accumulation of quench oil in the valve cavities and reduce wear of the valve sealing surfaces. This arrangement could be preferred if the piping configuration does not allow for the installation of a double block and bleed.

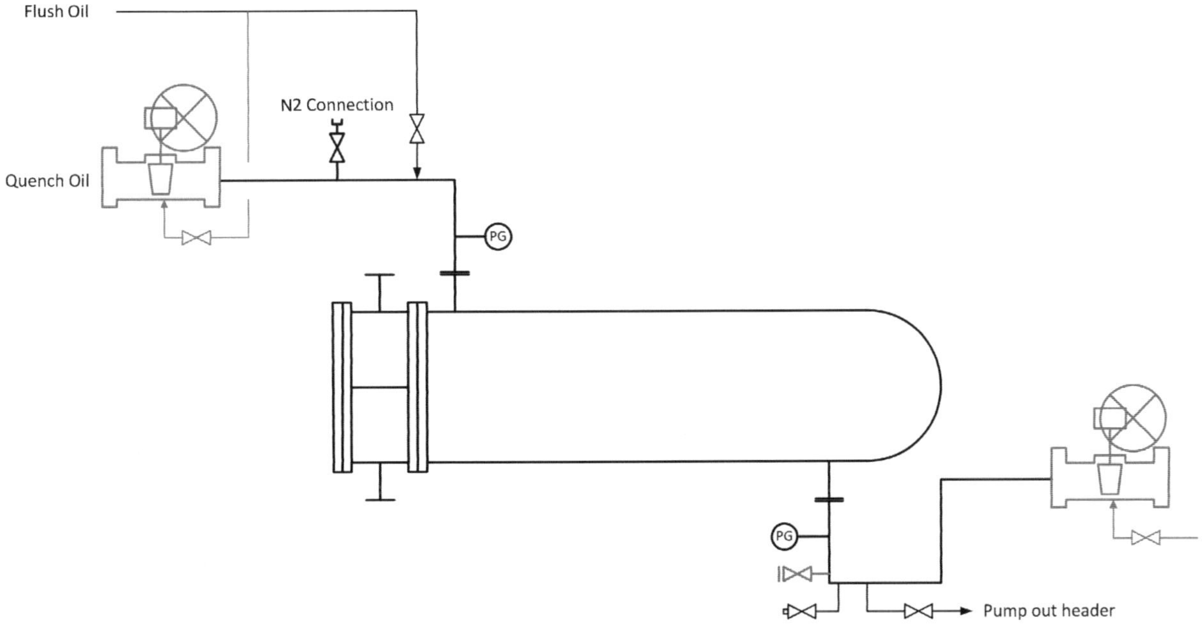

Figure 6: Improved Performance Isolation Valve Example

Another improvement that can be added to an existing system is the inclusion of positive isolation indication. This is especially helpful where the design of valve or use of chain operator obscures the valve stem. At one LYB location, site personnel have fitted an extension to the stem on chain operator valves that allows operators to know if the valve is fully closed. Figure 7 shows an example this indication.

Figure 7: Example of Isolation Indication

Isolation Valve Selection

The choice of valve is another important part of a quench oil isolation scheme. Considerations should be made for how frequently the valve will be used, what orientation the valve will have, the accessibility for operations, and piping support limitations. The following list explores types of valves that have been used in quench oil service. Potential drawback and preferred practices for each type of valve are included. In all cases, it is suggested to include heat tracing on the valve body and to use flush oil into valve cavities to prevent accumulation of solids.

Table 2: Valve Selection Criteria

	Selection Criteria	Potential Problems	Improvements
Wedge Gate Valve	• Inexpensive • Ease of use	• Trapped solids • Chain operators obscure valve stem • Valve Orientation	• Rotate valve – shaft should not be vertical • Add external position indication
Plug or Ball Valve	• QO Solids less likely to inhibit valve movement • Clear position indication	• Relatively Heavy • Frequent use can damage sealing surface	• Flush oil purge will improve seal surface life • Consider use of "lifting" type plug valve to prevent sealing surface damage
Butterfly Valve	• Cost effective for retrofit • Clear position indication • No cavity to trap solids	• Erosion of stem • Stuck in closed position if QO allowed to cool	• Heat trace very important
Sliding gate valve / slab valve	• Less debris in valve body • Effective sealing design	• Relatively Expensive • Difficult for retrofit due to size and weight	• Use in high cycle service • Effective where QO solids are problematic such as inlet to filters

Conclusions

The physical properties of quench oil can make isolation and clearing of equipment problematic and have caused serious LOPC incidents. The typical design scheme for quench oil isolation can be prone to failure due to solidified quench oil. Improvements can be realized by focusing on positive isolation indication and bleed best practices. Heat tracing on isolation valves should be maintained and use of flush oil purge into valve cavities should be considered. A retrofit of the system can frequently be performed online or with minimal scope during a turnaround.

References

1. Gondolfe, Joseph M. and Casey L. Mueller, "The Definitive Solution: Gasoline Fractionator Retrofit Techniques", AIChE Ethylene Producers' Conference 2002.

2. Bernard, Andre', Thomas M. Pickett, M. Beata Manek, Daniel K. Frye, "Flux Oil Stream Import to Quench System Risk and Impacts", AIChE Ethylene Producers' Conference 2011.

AIChE PAPER NUMBER: 188b

REDUCED TOUGHNESS FITTINGS AND POTENTIAL EFFECT ON LOW-TEMPERATURE FITNESS FOR SERVICE

Daniel Ayewah, PE
Senior Associate
Stress Engineering Services

Kannan Subramanian, PhD, PE
Senior Associate
Stress Engineering Services

Ralph King, Retired PE
Subject Matter Expert
Stress Engineering Services

Prepared for Presentation at the 2019 Spring National Meeting
New Orleans, LA, March, 31, 2019

REDUCED TOUGHNESS FITTINGS AND POTENTIAL EFFECT ON LOW-TEMPERATURE FITNESS FOR SERVICE

Daniel Ayewah
Senior Associate
Stress Engineering Services

Kannan Subramanian
Senior Associate
Stress Engineering Services

Ralph King
Subject Matter Expert
Stress Engineering Services

Abstract:

It has been reported in industry that recently manufactured piping and pressure vessel components may have reduced toughness where the component fails the Code -20°F impact exemption. This paper will review the industry reported findings and outline a potential analysis strategy to allow screening out potentially at-risk components that are currently in service. The goal of the proposed analysis approach is to reduce the magnitude of scope of systems that may require advanced analysis or testing. Representative analysis results for various configurations will be presented along with highlighting the need for further industry work to develop a methodology to identify the systems with the highest risk for brittle fracture, if reduced toughness components are within the system.

Introduction

It has been reported by multiple sources that recently manufactured piping, fittings, and flanges have reduced toughness and do NOT meet the American Society of Mechanical Engineers' Boiler and Pressure Vessel Code (ASME BPVC) Section VIII Division 1 -20°F impact test exemption requirements for listed components. All ASME BPVC Section VIII Division 1, Figure UCS-66 and ASME B31.3 Figure 323.2.2A Curve B materials made to the current and previous editions of these codes are considered to be potentially at risk, although the issue has not been found in piping components manufactured from plate materials. Recently, several owner/operators have undertaken toughness audits of new project components to quantify this issue. The results confirm a widespread and endemic problem with a substantial population of supplied components not meeting the toughness requirements for exemption as it is stipulated in the current and previous editions of the design/constructions codes.

Industry response to date has been focused on collecting additional evidence to quantify the scale of the issue and to propose various changes to industry standards and specifications, to mitigate the potential for further issues with new production components. From the design/construction code standpoint, a shift to a warmer temperature of 0°F from -20°F for all ferritic flanges is expected in the upcoming edition of ASME BPVC Section VIII, Divisions 1. In addition, a conservative shift from Curve B to Curve A is expected specifically for standard A105 fittings and a shift from Curve C to Curve B is expected for normalized A105 fittings. Similar changes or warnings are expected to be seen in the upcoming editions of Fitness-for-Service (FFS) standards such as API 579 – 1/ASME FFS – 1. While these changes address the issue for new construction, there currently is no focused industry effort in addressing what is suspected to be a substantial population of currently in-service components having low-toughness.

Despite the low toughness of some recently manufactured fittings and flanges and the understanding that many are currently operating in systems that may present an increased probability of fracture, there have not been a significant number of reported failures. There appears to be some remaining margin between design and operating conditions, resulting in a reduction in the probability of brittle fracture during the normal range of operating conditions. However, the size of this margin is not well understood, making it difficult to identify which systems are fit for continued service and safe to operate without further analysis or testing. There is a need for an FFS screening and analysis approach that can be utilized by industry to identify the most at-risk systems/components and focus the scope of mitigations necessary to avert potential brittle fracture failures.

Understanding Brittle Fracture

Brittle fracture is the sudden and rapid fracture of a component under stress (residual or applied), where the material exhibits little or no ductility or plastic deformation prior to failure but rather is dominated by cleavage fracture. Brittle fracture typically occurs when all of the following criteria are met, as presented in the brittle fracture triangle shown in **Error! Reference source not found.**:

1. Low material fracture toughness (i.e., low resistance to crack initiation and propagation). This may be influenced by the exposure of susceptible steels to low temperatures.

2. The presence of a flaw or crack-like defect. Crack initiation and propagation is influenced by the size and shape of the flaw and any geometric stress concentration effects near the flaw.

3. A sufficient level of stress (typically ~7 ksi for ferritic steels). This may be in the form of residual or applied stresses on the flaw.

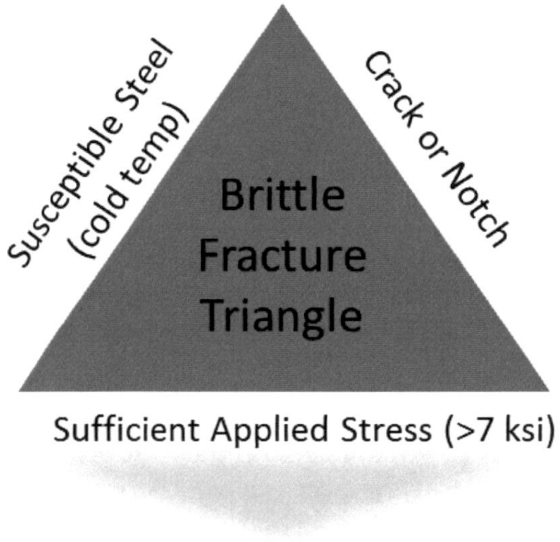

Figure 1 Brittle Fracture Triangle

Fracture Toughness and Temperature

A major influencing factor on the toughness of carbon steel materials is their operating temperature. Carbon steels exhibit a ductile to brittle transition in their toughness with decreasing temperature. The schematic in Figure 2 illustrates a hyperbolic tangent representation of this transition. The parameters of the equation shown can be obtained from a curve fit to actual Charpy test data. Operations involving high temperature exposures (ambient or process temperatures) benefit from upper shelf toughness properties while those at low temperature, as often experienced with light hydrocarbons in ethylene production, move more towards the transition region or lower shelf toughness properties. In addition, it has been observed that different heats of carbon steel exhibit the same transition behavior and differ only by the value of their transition temperature (specified as the 15ft-lb reference temperature for carbon steels). Higher transition temperatures are characteristic of lower toughness materials because the exposure temperature moves closer to the lower shelf. Conversely, lower transition temperatures are characteristic of higher toughness materials.

Increasing Transition Temperatures in Carbon Steels

Pressure vessel and piping materials are grouped into Charpy impact test exemption curves A, B, C and D with increasing toughness in ASME BPVC Section VIII Division 1 Figure UCS-66 and in ASME B31.3 Figure 323.2.2A. Figure 3 shows transition curves representative of each Charpy exemption curve based on calculations in API 579 – 1/ASME FFS – 1 [1]. The steel grades for flanges, fitting and other components of concern for low toughness have historically been categorized as curve B materials. The primary concern with the stated problem of low toughness fittings is the concern that more recently manufactured components are shifting to higher transition temperatures due to changes in what are thought to be a series of interactive and contributing causes in current manufacturing practices (chemistry, thermo-mechanical history, and forming methods). In addition to this effect there may be other mechanisms contributing to the observed low toughness behavior, which must be studied and understood to fully address the problem.

Brittle Fracture Assessment

Common design and FFS methodologies for addressing the issue of brittle fracture aim at establishing a minimum allowable temperature (MAT) at which vessels or piping systems may safely operate. These methods assume a certain level of toughness inherent in the materials from which the components are made, which have historically been based on the Charpy impact test exemption curves to which the material is assigned. Recognizing the recent concern of material non-conformance to their assigned exemption curves, the present problem of low toughness fittings will require a revised FFS approach based on fundamental fracture mechanics methodologies.

While Charpy energies represent a useful quality control indication of toughness in carbon steels, they are not a direct measure of fracture toughness. They are convenient due to their low cost and wide spread availability in pressure vessel and piping systems. Correlations have been developed between Charpy energy and fracture toughness K_{IC}. The most effective in the transition region relates a Charpy reference temperature T_{ref} with a toughness master curve index temperature T_0 [1] to provide an effective estimate of fracture toughness (K_{IC}). The toughness master curve relates fracture toughness to temperature and can be used to establish MATs based on required fracture toughness limits for an assumed reference flaw size.

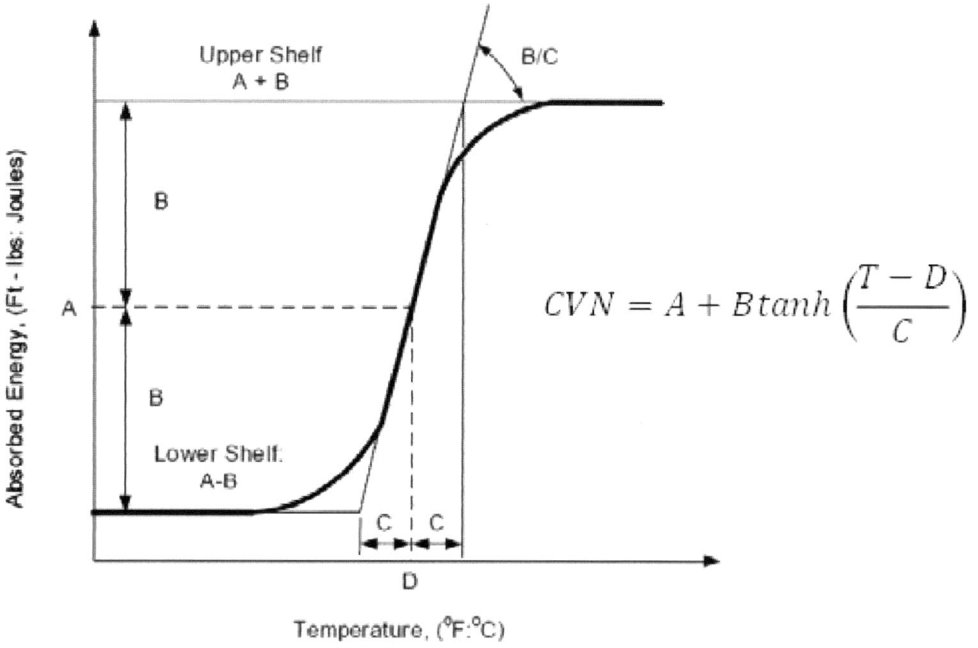

Figure 2 Hyperbolic Tangent Representation of the Ductile to Brittle Transition in Carbon Steels

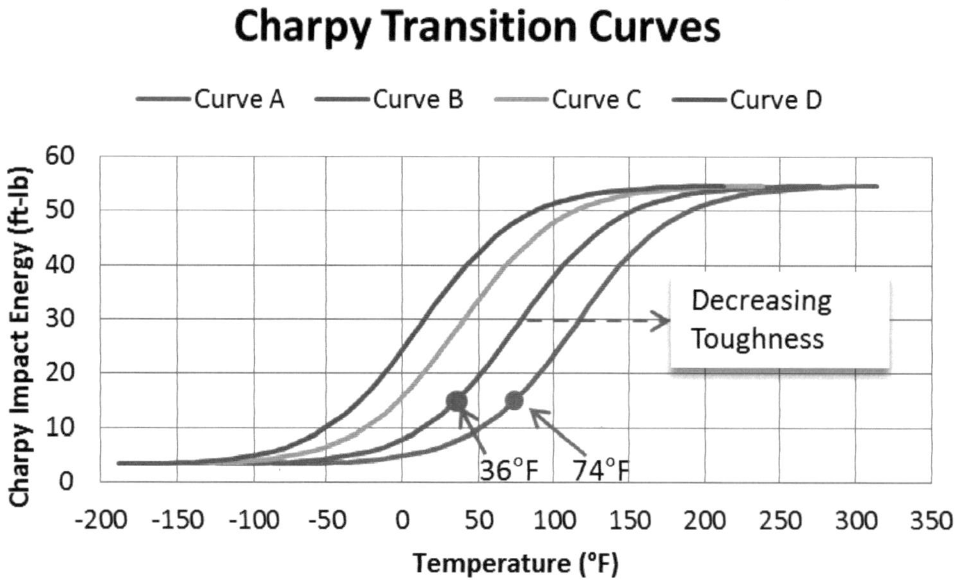

Figure 3 Ductile to Brittle Transition Curves Consistent with the Charpy Exemption Curves Showing the 15ft-lb Reference Temperatures for Curve A and B

Scope of the Problem of Low Toughness Fittings

Over the past 10 to 15 years, the process industry has been plagued by the discovery of substandard materials that exhibit low fracture toughness, even though they meet the requirements (specified composition and reported heat treatment) of current codes and specifications for construction. For example, a non-normalized/as-forged ASME SA-105 fitting (commonly used in the process industry), if designed based on either ASME BPVC Section VIII Division 1 or ASME B31.3, can be rated to a minimum temperature of -20°F without impact testing. The underlying assumption is that this material, if impact tested at -20°F, would exhibit impact energy of at least 15 ft-lbf. However, several users and engineering firms have found that the impact energy of this material can be as low as 3 ft-lbf at temperatures normally expected for processes utilizing this material.

In a recent toughness audit performed on completed piping spool components prior to installation, as many as 30% of the components failed to meet the target Charpy energy requirements. The earliest reported failure involving a substandard component is the failure of a 24-inch ASME SA-105N (normalized) flange in an HDPE (high density polyethylene) plant in Belgium in 1999 [2] that failed during operation. A significant number of other failures occurred during hydro-test. Many of the reported failures within the United States in the last few years can partially be attributed to substandard materials, specifically involving low fracture toughness. It should be noted that of all of the reported failures, the exposure temperatures were well below the calculated MAT, even if the toughness satisfied the Code requirements. Low toughness alone could not be attributed as THE single root cause for the failures.

Materials Affected

The materials affected by the reduced toughness issues are some of the most widely used low carbon and low alloy steel materials. The following are some specific material specifications that have been reported to have reduced-toughness issues:

1. A/SA 105 and A/SA 105N [3]
2. A/SA 234 Gr. WPB [3]
3. A/SA 350 Gr. LF2 [4]
4. A 694 Gr. F60 & F70 [5]

Most of the above are used to make the fittings and flanges used in, for example, the olefins industry, where they experience frequent low-temperature exposure in comparison with the other industries. While many of the owner/operators that experience the reduced toughness issues have issues with almost all of the above materials, issues with commonly used pipe materials such as A106, A53 Gr.B, and API 5L Gr. B have also been reported [3].

Contributing Factors

A major contributing factor to the problem of low toughness fittings is suspected to be a change in steel making practices in recent years. The historical practice of smelting iron ore in a blast furnace is gradually giving way to modern, more energy efficient techniques using recycled steels in an electric arc furnace with some mills running almost entirely on recycle steel with no raw iron ore being utilized. In a few cases, proper quality control procedures were not in place. The use of recycled steel results in increased levels of certain trace elements being detected in the produced carbon steels and is believed to contribute to the observed reduction in toughness properties. Additional factors identified include:

- Changes in hot forming practices
- Inadequate heat treatment
- Thermo-mechanical history during component fabrication

Industry Response

There are multiple industry working groups and Code committee responses underway at this time with proposed modifications of Codes, Standards, and Specifications to address each contributor to the problem and to propose changes to resolve this issue for new construction. Most notable is the decision by ASME to downgrade A/SA-105 from Charpy exemption Curve B to Curve A; and its normalized variant A/SA 105N from curve C to curve B for future versions of UCS-66 [6]. Additionally, it is currently proposed to raise the exemption temperature of forged components from -20°F to 0°F. The changes potentially disqualify some in-service flanges and components that were originally deemed fit-for-service based on ASME BPVC and/or API 579 – 1/ASME FFS – 1, Part 3 procedures [1]. While these changes provide an increased level of protection to the owner/operator for new construction, without the need for additional testing, the resulting disqualification of in-service flanges will need to be addressed from a FFS approach. This paper presents a potential approach to this dilemma.

Alberta Boilers Safety Association (ABSA) issued a warning notice in 2016 and another in 2018 to caution users/designers of the prevailing reduced toughness issue [7]. The warning effectively puts operators on notice as being responsible, but does not provide a recommended approach for addressing the issue. Operators are currently looking for a potential solution, and an FFS approach may be a viable approach for the industry.

Other initiatives to address the reduced toughness issue include the development of advanced NDE methods to determine the grain size, and methods to determine toughness without removing a component from service. Nevertheless, none of those initiatives provide a robust FFS solution for screening large component populations that may be affected by the underlying issue.

An FFS Approach for In-Service Components

In order to ensure safe and reliable operation, the design codes are built with a substantial margin to account for engineering and material property "uncertainties." Clearly, with the reduced toughness values currently being reported, and the reality that very few in-service failures have been reported, it is apparent that in-service components are operating within that margin albeit at a reduced safety margin (i.e., increased risk) to fracture than the original design basis. The ability to quantify that margin across their operating systems would be very valuable to operators with the need to understand and manage their levels of FFS risk.

Detailed stress and fracture analyses can demonstrate under what conditions safe operations may be possible even if a fitting or flange has low impact energy [8]. However, plant operators cannot easily determine the level of safety margin for all of the systems and components currently installed and operating in their facilities. Having a good understanding of the typical variability in operating margins to the design, and how these margins affect minimum allowable temperature predictions, for various piping and vessels, makes it possible to develop an FFS screening approach that considers the reduced-toughness of suspect components. It is expected that such an FFS approach would only eliminate the systems/components least likely to fail due to brittle fracture. However, considering the scope of the problem, such a screening should be of significant value to owner/operators, allowing them to focus mitigation efforts and more detailed assessments on the high-risk equipment.

API 579 – 1/ASME FFS – 1, Part 3 Brittle Fracture Assessment (BFA) includes 3 assessment levels. An increase in level indicates increasing complexity of the method employed. One could apply the various levels to these at-risk components on an individual basis [9]. However, when required to be applied for the numerous piping systems and pressure vessels that may be at risk, such an approach is simply not economical or viable given the scope of the issue. Currently, there is no robust approach to address what appears to be a large number of low toughness fittings and flanges operating at a reduced margin to fracture. An effective screening approach is needed to provide insight into the level of the expected margin available for a large population of specimens. Such an approach should incorporate sound assessment techniques that capture the fundamental fracture mechanics theory behind the issue. The authors have begun investigational development of such an approach and applied it to a number of systems to validate the approach. The two case studies outlined below provide insight to the methodology as applied to a process piping system. The first case study involves an elbow and the second study involves a flange.

Case Study 1

Elbows are often of concern for being some of the more highly stressed components in piping systems. If an elbow is made of substandard material exhibiting low toughness values, the system is at elevated risk of brittle fracture failure. In this case study, such an elbow with a low toughness value is considered. The toughness value assumed was established based on a lower bound Charpy value (~6ft-lb at 20°F), which is representative of data published for systems identified as having low toughness properties. The corresponding Charpy transition curve [1] has a reference temperature of 60°F representing a shift to the right of exemption Curve B (see Figure 3), indicating that the material has a lower toughness value than expected, though better than the Curve A reference temperature.

FFS Level 2 procedures involving piping stress analysis alone cannot be used to assess this elbow since the material does not meet the toughness criteria of its assigned exemption curve to begin with. Therefore, an FFS Level 3 assessment is required. The Level 3 brittle fracture assessment involves the use of a detailed finite element (FE) model (Figure 4a) to determine stresses in the elbow (Figure 4b), based on internal pressure and supplemental loading from the surrounding piping system obtained from a pipe stress analysis model (Figure 4c). This is followed by a fracture mechanics assessment which assumes a reference flaw size and establishes toughness limits using a failure assessment diagram (FAD).

Comprehensive FFS methods for brittle fracture assessments involve the development of a minimum allowable temperature (MAT) curve for varying pressures. An MAT curve, based on the limiting stresses determined for the limiting location, on the sides of the elbow, was developed using a combination of the failure assessment diagram (FAD) approach, by assuming a reference flaw size, and the fracture toughness master curve at 5% (Lower) 50% (median) and 95% (upper) probability bounds. Shown in Figure 5, this type of curve provides an envelope of acceptable operating conditions that fully accounts for the reduced toughness properties in the low toughness fittings. Also shown in Figure 5 are critical exposure temperatures (CET) for two commonly known olefins, based on their flash curves. The fact that the MATs falls below the CET curves indicates that the component may operate safely at temperature at or above the MAT, despite its low toughness.

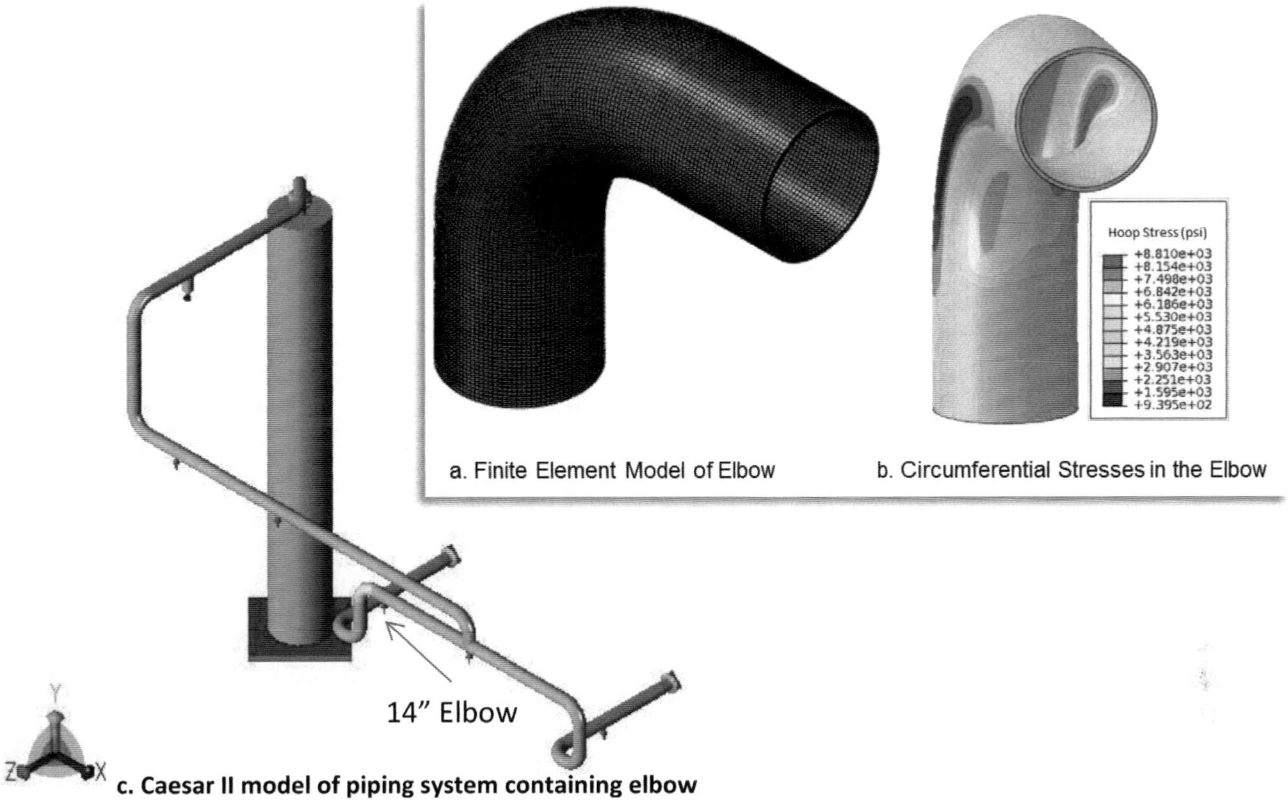

Figure 4 Stresses in an Elbow in an Olefin Piping System.

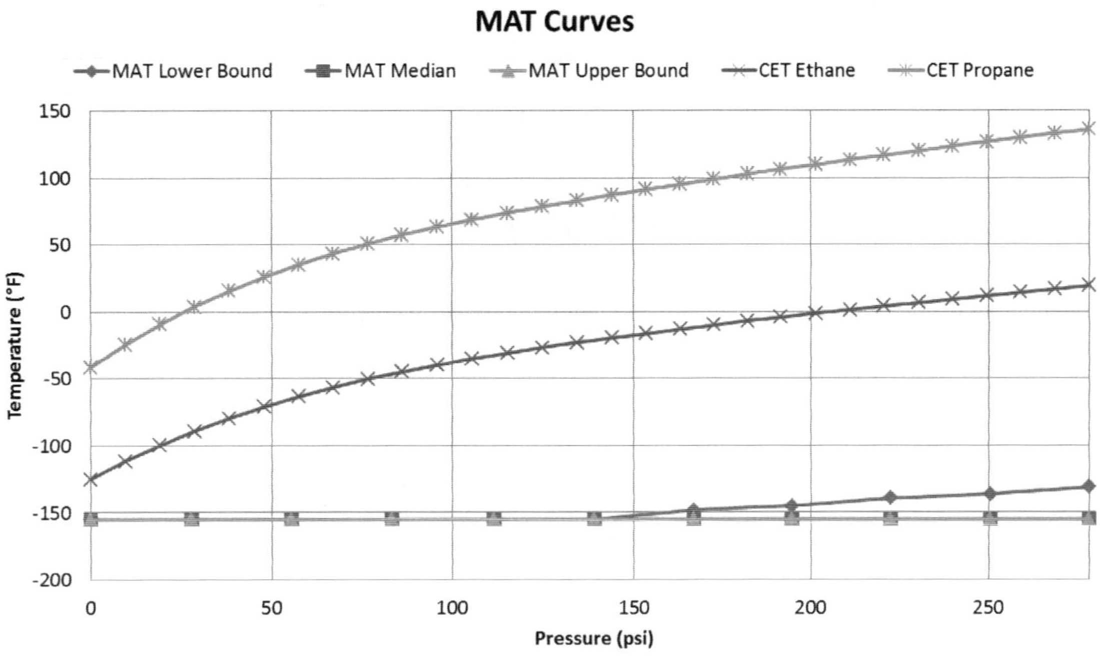

Figure 5 Case Study 1: MAT Curves based on Stresses from the Maximum Stress Location.

Case Study 2

A similar technique was applied to an 18" 300# flange made of A/SA-105 material. As in the case of the elbow considered in the previous case study, a Level 3 assessment requires extensive and detailed modeling of both the bolted flange joint assembly and the piping system. A piping stress analysis was performed to determine piping loads and a finite element model was built, accounting for internal pressure, piping loads and the bolt preloading of the flange assembly. Results (shown in Figure 6) establish stresses at various locations on the flange. Values obtained for axial stresses at the flange neck transition to the pipe, and for radial stresses near the crotch region of the flange were used to establish MAT curves.

MAT curves based on the stresses determined for the limiting locations were developed using the failure assessment diagram (FAD) approach in combination with the toughness master curve at different probability bounds. The resulting MATs are shown in Figure 7. The results provide an envelope of acceptable operating conditions for the flange that is based on fundamental fracture mechanics. It should be noted that the MAT is dominated by the flange bolt preload and is fairly pressure independent as a result, hence the flat MAT curves. There is the potential for further, more complex analyses to fully investigate the interaction of gasket, bolt and flange stress/strain to fully account for stress reductions that occur due to relaxation of the gasket and bolts. Current code rules establishing a -20°F MAT for flanges do not specifically account for the stresses due to bolt preload but they can be considered with this approach. Once again, this analysis exercise is a time and detail intensive process.

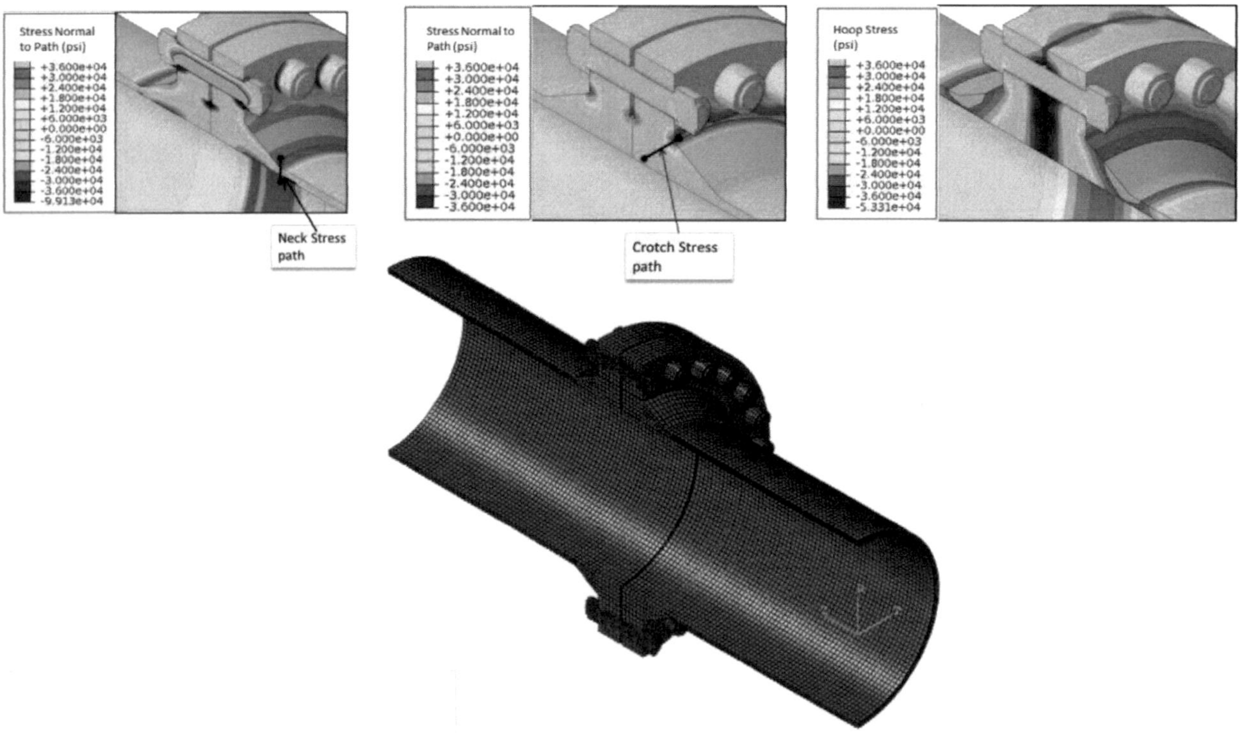

Figure 6 Stresses in a Bolted Flange Joint a Piping System.

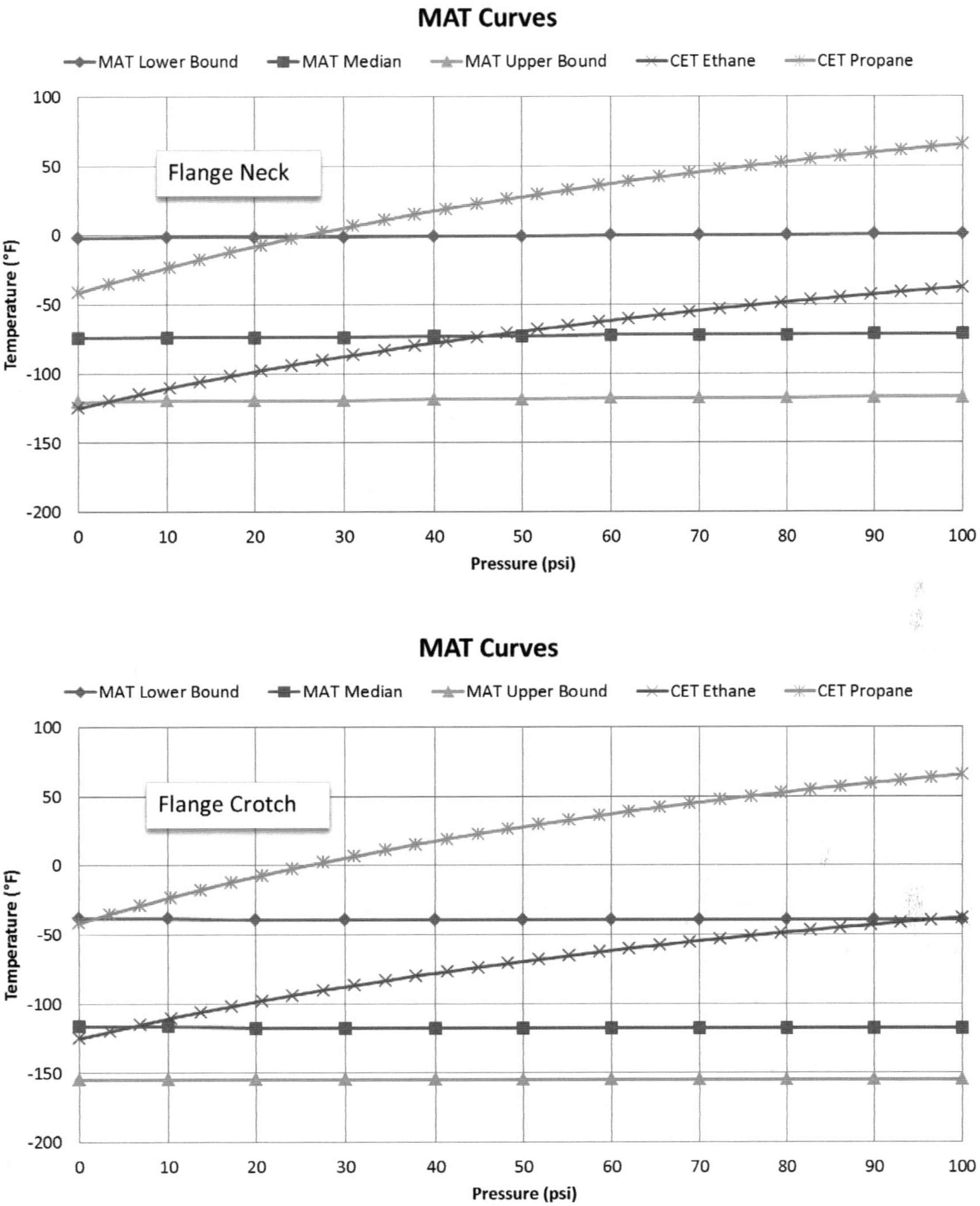

Figure 7 Sample MAT Curve based on Stresses from the Flange Neck Location.

The techniques applied in these case studies fully accounts for the reduced toughness properties in the low toughness fittings by leveraging fundamental fracture mechanics and advanced analysis. They are capable of addressing FFS concerns in components that may previously have been determined to be unacceptable, however these are time and effort intensive techniques that may be impractical for addressing large numbers of components across multiple facilities. There is a need for a more generic approach to address the volume of in-service components and these techniques provide insight into how such a generic approach can be developed.

Need for Further Study

Based on the recent findings, it is clear that there is a substantial population of reduced-toughness fittings that are in-service; however, very few in-service and only a few hydro-test failures and initial inventory/startup brittle fracture failures have occurred. Despite the reduced-toughness, there currently seems to be some remaining margin between design and operating conditions resulting in a low occurrence of these failures. The level of risk and the operating margin to brittle fracture is not currently known. Depending on the potential temperature exposure (operating and auto-refrigeration) and the consequence of failure, sufficient margin may not be available for the reduced toughness components for some systems.

As shown in the case studies above, the operating conditions of loads, temperature, and material toughness in some cases may be sufficient to result in substantially lower MATs than would typically be allowed for the documented design condition. This translates into very few in-service failures. A detailed study to investigate the reduced toughness effects and develop an FFS approach for screening large populations of components is needed. This effort could help identify the available brittle fracture margins between design and current operation and help owner/operators leverage them in a FFS (and documented operating basis reduction) analyses. The objective of such a study should be to:

A. Better understand the typical operating margins to design and brittle fracture predictions for various piping and vessel designs, and

B. Utilize this understanding to develop an effective FFS screening approach.

It is expected that the proposed FFS approach would only eliminate the systems/components least likely to fail due to brittle fracture potentially excluding those systems from a further detailed assessment. This will allow owner/operators to focus more detailed assessments and/or testing on a smaller population of high-risk equipment. Given the extent of the problem, any reduction in the actionable mitigation scope for the risk of brittle fracture failures is thought to be of substantial benefit.

Conclusion

The process and petrochemical industry has substantial concern over the recent findings of reduced toughness fittings and flanges that are at risk of brittle fracture. This is especially a concern for ethylene producing facilities where exposure temperatures are lower than average. There are numerous initiatives that address the metallurgical and manufacturing issues associated with the fittings, and efforts to incorporate changes in the Code to address the issue. However, there is no clear FFS initiative being undertaken to provide a resolution for components that are currently in service.

The authors have studied a significant number of piping systems and applied the advanced analysis approaches presented in this paper, in an effort to investigate an FFS approach that may be applied to large numbers of systems with reduced toughness components. A substantial amount of analysis is required considering "as-designed" versus "as-operated" conditions to fully quantify the statistical distribution of parameters so as to better define the operating margin to the brittle fracture limit for these reduced toughness fittings. Such an approach may be used to categorize the systems based on their risk of brittle fracture and prioritize systems at high risk, for immediate mitigation action. The approach shows promising results as a risk-based screening solution for large numbers of systems. Further development is needed to provide a more definitive FFS solution to the problem.

References

1. API 579 – 1/ASME FFS – 1, Fitness-For-Service, 2016, American Petroleum Institute, New York, NY.
2. Susceptibility to brittle fracture of flanges in steel ASTM A105, CRC/WPS/11/00, Chemical Risks Directorate, Belgium.
3. Susceptibility of Carbon Steel Pipe, Fittings and Flanges to Brittle Fracture, Proceedings of the ASME 2014 Pressure Vessels and Piping Conference (ASME-PVP2017), July 16-20, 2017, Waikaloa, Hawaii.
4. Impact Toughness Deficiencies in ASME SA-350 LF2 Flanges, Proceedings of the ASME 2014 Pressure Vessels and Piping Conference (ASME-PVP2017), July 16-20, 2017, Waikaloa, Hawaii.
5. Mechanical Properties of Vanadium Microalloyed High-Strength ASTM A694 Forgings, Proceedings of the ASME 2014 Pressure Vessels and Piping Conference (ASME-PVP2017), July 16-20, 2017, Waikaloa, Hawaii.
6. ASME: Section VIII Division 1: Rules for construction of Pressure Vessel. American Society of Mechanical Engineering. 2017 Edition.
7. ABSA Information Bulletin # IB16-018, November 25, 2016.
8. Impact Toughness and Brittle Fracture of Carbon Steels, Proceedings of the ASME 2014 Pressure Vessels and Piping Conference (ASME-PVP2017), July 16-20, 2017, Waikaloa, Hawaii.
9. Assessment Methods to Prevent Brittle Fracture failures in Pressure Vessels & Piping Components, AIChE Paper. No.: 178d, Proceedings of EPC Conference, 2015 Spring National Meeting, April 29, 2015, Austin, TX.

549199 Reduced Toughness Fittings and Potential Effect on Low-Temperature Fitness for Service - Part 2

It has been reported in industry that recently manufactured piping and pressure vessel components may have reduced toughness where the component fails the Code -20F impact exemption. This presentation will review the industry reported findings and outline a potential analysis path forward to address fitness for service (FFS) concerns for potentially at-risk components that are currently in service. Representative analysis results for various configurations will be presented along with highlighting the need for further industry work to develop a methodology to identify the systems with the highest risk for brittle fracture if these reduced toughness components are within the system.

AIChE Paper Number: 182d

CURRENT DEFICIENCIES AND PROPOSED MODIFICATIONS TO API 579 PART 3 BRITTLE FRACTURE SCREEENING PROCEDURES

Brian R. Macejko

Consulting Engineer II

The Equity Engineering Group, Inc.

Seetha Ramudu Kummari, Ph.D., P.E.

Senior Engineer II

The Equity Engineering Group, Inc.

Phillip E. Prueter, P.E.

Principal Engineer II

The Equity Engineering Group, Inc.

Prepared for Presentation at the 2019 Spring National Meeting

New Orleans, LA, March 31 – April 4, 2019

CURRENT DEFICIENCIES AND PROPOSED MODIFICATIONS TO API 579 PART 3 BRITTLE FRACTURE SCREEENING PROCEDURES

Brian R. Macejko

Consulting Engineer II

The Equity Engineering Group, Inc.

Seetha Ramudu Kummari, Ph.D., P.E.

Senior Engineer II

The Equity Engineering Group, Inc.

Phillip E. Prueter, P.E.

Principal Engineer II

The Equity Engineering Group, Inc.

Abstract:

Significant inconsistencies and inadequacies exist in the current brittle fracture screening procedures in the various pressure equipment design Codes and Standards as well as the 2016 edition of API 579-1/ASME FFS-1, Fitness-For-Service (API 579) [1]. Results from an evaluation completed using the current published guidance could prove unsafe in some instances. The effects of auto-refrigeration from depressurization of liquefied compressed gases (such as ethylene, ethane, propylene, etc.) can be particularly concerning due to the extremely low equipment metal temperatures (-55°F and below). The ASME/API Joint Committee on Fitness-For-Service (FFS) recently initiated a project to rewrite API 579 Part 3, Assessment of Existing Equipment for Brittle Fracture. The intent of the project is to establish new Level 1 and Level 2 evaluation procedures and acceptance criteria developed using state-of-the-art fracture mechanics consistent with the Part 9 Failure Assessment Diagram (FAD) approach, currently employed in Level 3 FFS evaluations. This paper will provide a comprehensive summary of deficiencies and corresponding concerns with current published methods for evaluating susceptibility of pressure equipment to potential brittle fracture failures. The paper will include a case study demonstrating the benefits of using detailed fracture mechanics and, lastly, the paper will provide a brief introduction to the methodology employed in the recently proposed procedure modifications for Part 3 of API 579.

Nomenclature:

K_{Id}	Dynamic fracture toughness (ksi√in).
σ_{ys}	Material yield strength at temperature (ksi).
T	Assessment temperature (°F).
T_{ref}	Reference temperature used to index material test exemption curves (°F).
A	Fitting parameter for hyperbolic tangent function for material toughness, 1.7.
B	Fitting parameter for hyperbolic tangent function for material toughness, 1.37.
C_R	Fitting parameter for hyperbolic tangent function for material toughness, 66°F.
β_{Id}	Fracture toughness criteria used in development of current published impact test exemption curves.
t	Thickness (inches).
K_I	Stress intensity factor (ksi√in).
K_{IC}	ASME Section XI material toughness (ksi√in).
K_I^P	Stress intensity factor for primary loads (ksi√in).
K_I^{SR}	Stress intensity factor for secondary and residual loads (ksi√in).
Φ	Plasticity interaction factor.
L_r^P	Load ratio for primary stress.
K_{mat}	Material toughness (ksi√in).
P_f	Probability of failure.
L	Crack front length used in Master Curve toughness model (inches).
$T_{20\,ft-lbs}$	20 ft-lbs. reference temperature (°F).
C_V	CVN impact energy (ft-lbs.).
C_{V-LS}	Lower shelf CVN energy (ft-lbs.).
C_{V-US}	Upper shelf CVN energy (ft-lbs.).

R	Stress ratio.
P_a	Applied pressure for the condition under evaluation (psi).
P_{rating}	Pressure rating for the component or assembly under evaluation (psi).
MAT	Minimum allowable temperature (°F).
NDT	Nil ductility temperature (°F).
T_{CVN}	CVN test temperature (°F).
T_0	Master Curve toughness index temperature (°F).

Introduction

API 579-1/ASME FFS-1 (API 579) is the industry accepted standard used to make run, repair, replace decisions for in-service equipment. Part 3 of API 579 is dedicated to the evaluation of existing equipment for brittle fracture. A brittle fracture evaluation may become necessary for any of the following scenarios:

- Brittle fracture was not formally addressed in the original equipment design
- Process hazards review identifies scenarios where process temperatures can fall below original design considerations (e.g. auto-refrigeration, upsets, pressure relief, etc.)
- Change in process operating conditions (re-purposed equipment, etc.)
- Re-rating or FFS for other damage
- Start-up/shutdown optimization
- Hydrostatic pressure test

Significant inconsistencies and inadequacies exist in the current Level 1 and Level 2 brittle fracture screening procedures in API 579 as well as the various pressure equipment design Codes. Results from an evaluation completed using the current published guidance could prove unsafe in some instances. This paper provides a comprehensive summary of deficiencies and corresponding concerns with current published methods and the methodology employed in the recently proposed procedure modifications for Part 3 of API 579 is introduced.

Brittle Fracture

Brittle fracture is the sudden rapid propagation of a crack-like flaw under stress (residual or applied) where the material exhibits little or no evidence of ductility or plastic deformation [2]. Brittle fracture is event dependent, not time dependent. Although rare, the consequences of a brittle fracture are typically catastrophic. The risk for such a failure to occur in an ethylene plant can be significantly higher due to the potential for auto-refrigeration from depressurization of liquefied compressed gases (such as ethylene, ethane, propylene, etc.). Brittle fracture failures experienced within industry have resulted in costly unplanned repairs, extensive production downtime, and most unfortunately loss of life. There can be a large variation in the way owner-users assess their risk for brittle fracture. Owner-users that have previously experienced such a failure may have detailed and comprehensive programs in place to evaluate susceptibility to future issues. Alternatively, owner-users fortunate enough to have avoided such a failure tend to be less cognizant of the potential risk.

There are three (3) key components that drive susceptibility to brittle fracture failure in carbon and low alloy steels:

- Stress (residual and/or applied)
- Material fracture toughness

- Crack-like defect (size and location)

A combination of metallurgical microstructure, weld residual stress, and inherent defects that act as stress intensifiers (e.g. lack of fusion, lack of penetration, porosity, slag inclusions, etc.) make welds and weld heat affected zones critical locations for potential brittle fracture.

Stress

Stress provides the energy necessary to drive a defect to fracture. Typical sources for stress include pressure, weight, and thermal loads, in addition to residual stress from the welding processes. A post weld heat treatment (PWHT) operation will significantly reduce weld residual stress.

Material Fracture Toughness

Fracture toughness is the ability of a material to absorb energy and plastically deform without fracturing. It is a function of both the material strength and ductility. The majority of pressure equipment utilized in the oil & gas and petrochemical industries is constructed of carbon or low alloy steel. For these materials, material toughness is a function of metal temperature. At low temperatures, the material tends to behave in a brittle manner and has a high susceptibility to fracture. At high temperatures, the material tends to behave in a ductile fashion.

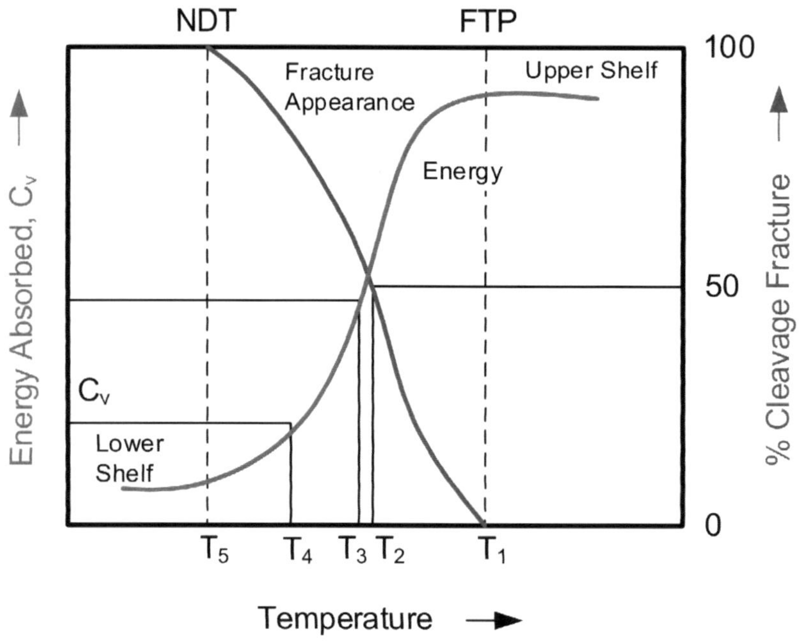

FIGURE 1: Plot Demonstrating CVN as a Function of Temperature

In Figure 1, absorbed energy is plotted against test temperature to demonstrate the transition from brittle to ductile behavior. Absorbed energy is typically measured using a Charpy V-notch impact test (CVN). In the plot, the following transition temperatures are defined:

- T1: Fracture Transition Plastic (FTP); Above T1 fracture is 100% fibrous.
- T2: Fracture Appearance Transition Temperature (FATT) corresponding to 50% cleavage and 50% ductile.
- T3: Average of upper and lower shelf values (often ≈ T2).
- T4: Arbitrary value of CVN (e.g. 15 ft-lb or 20 ft-lb).
- T5: Nil Ductility Temperature (NDT) corresponding to 100% cleavage fracture.

Material chemistry, grain size, and heat treatment also affect toughness. The brittle-to-ductile transition temperature will decrease (i.e., improve) with:

- Decrease in material carbon content
- Increase in manganese-to-carbon ratio
- Decrease in sulfur content
- Decrease in material grain size
- Normalization heat treatment of material through rapid cooling (generally results in fine grain microstructure)

Older steels tend to have a moderate brittle-to-ductile transition slope relative to clean steels. Most of the common steels used to construct pressure equipment in the 1970's, 1960's, and earlier (such as A70, SA-212, SA-201, SA-283, SA-285, SA-515-70) have exhibited relatively high FATT (i.e., low toughness) compared to steels more frequently used today (e.g. SA-516-70, SA-516-70N).

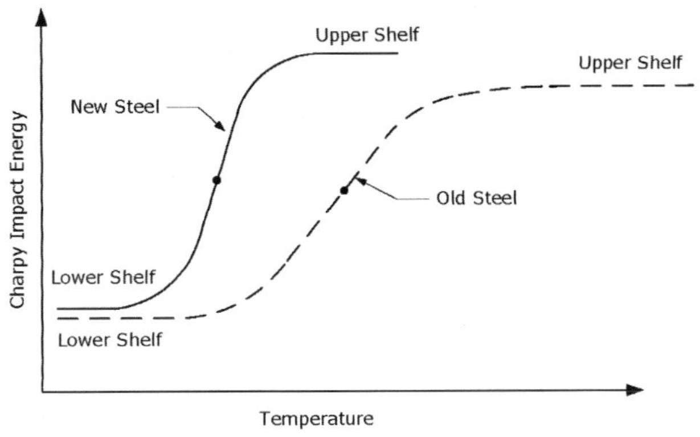

FIGURE 2: Plot Demonstrating CVN Trends for "New" Steels vs. "Old" Steels

A number of environmental issues exist that can degrade material toughness or result in embrittlement (e.g., temper embrittlement in low chrome alloys due to long-term elevated temperature exposure). The adverse effects of any such potential damage mechanisms should be considered in a brittle fracture evaluation.

Crack-Like Defect

Brittle fracture typically initiates at a crack-like defect. Defects can result from environmental damage (such as wet H_2S or caustic exposure), mechanical damage (such as gouges, dents, or fatigue), or from original fabrication (such as laminations, lack of fusion, lack of penetration, slag inclusions, porosity, etc.). Performing a detailed inspection including both surface examination techniques (such as dye penetrant examination or magnetic particle examination) and volumetric examination techniques (such as angled beam ultrasonic examination methods) can be utilized to detect and categorize any crack-like defects present in pressure equipment.

Design Code Procedures for Evaluating Susceptibility to Brittle Fracture

Most industry design codes now have guidance on establishing material toughness requirements or minimum allowable temperature limits. The following is a brief summary of various design Code procedures. The technical basis and background information used to generate these procedures is discussed in a later section of this paper.

ASME Section VIII Division 1 Pressure Vessels

Most pressure vessels operating today were designed to the ASME Boiler and Pressure Vessel Code Section VIII, Division 1 (ASME VIII-1) [3]. The 1987 Addenda to the 1986 Edition of ASME VIII-1 introduced drastic changes to Part UCS-66 due to brittle fracture concerns with carbon and low alloy steels. ASME VIII-1 uses the term Minimum Design Metal Temperature (MDMT) to describe the lowest permissible metal temperature at the vessel Maximum Allowable Working Pressure (MAWP). The MDMT can be established either through destructive testing (i.e. CVN impact testing) or through engineering evaluation.

Prior to 1987, ASME VIII-1 permitted operation of carbon steel equipment down to a temperature of -20°F without any testing or engineering evaluation. The majority of pressure equipment operating today did not receive a formal assessment for susceptibility to brittle fracture at original design.

Impact Testing

ASME VIII-1 permits the use of CVN to establish the MDMT of a component. The CVN impact energy must meet the requirements shown in Figure UG-84.1 of ASME VIII-1 [3] (see Figure 3 below). The required impact energy is a function of the material thickness and minimum specified yield strength for the material.

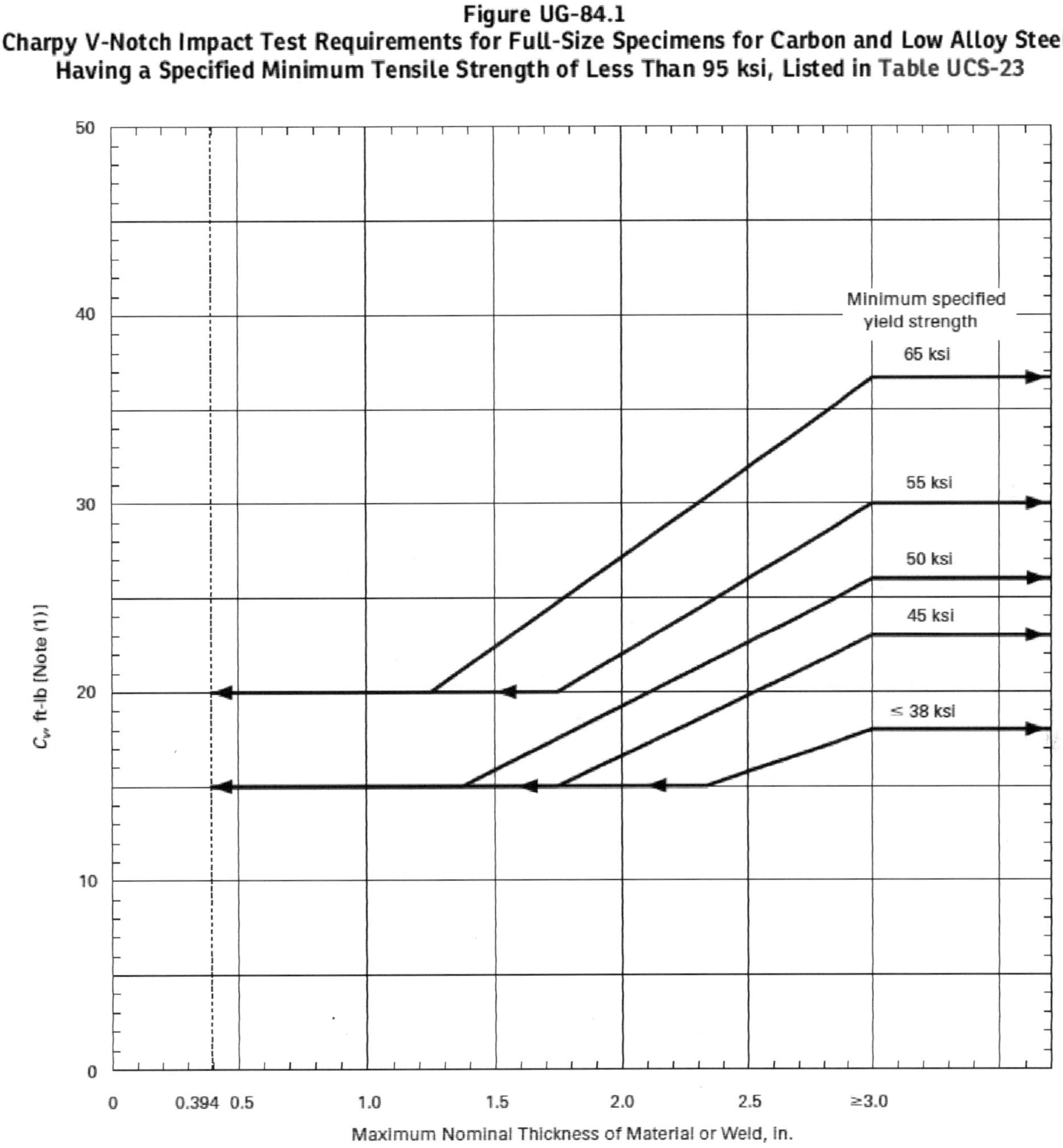

FIGURE 3: Required CVN Impact Energy per Figure UG-84.1 of ASME VIII-1 [3]

For most scenarios, the impact test temperature is assigned as the component MDMT. However, per paragraph UCS-66(g) of ASME VIII-1 [3], materials produced and impact-tested in accordance with the requirements of the materials specifications listed in Figure UG-84.1, general note (c) (e.g. SA-320, SA-333, SA-350, SA-420, etc.) are exempt from impact testing to temperatures 5°F colder than the test temperature required by the specification. Additionally, Table UG-84.4 of ASME VIII-1 (see Figure 4 below) provides additional allowances for assigning an MDMT lower than the actual test temperature.

Table UG-84.4
Impact Test Temperature Differential

Minimum Specified Yield Strength, ksi (MPa)	Temperature Difference, °F (°C) [Note (1)]
≤40 (280)	10 (6)
≤55 (380)	5 (3)
>55 (380)	0 (0)

NOTE:
(1) Impact test temperature may be warmer than the minimum design temperature by the amount shown.

FIGURE 4: CVN Impact Test Temperature Allowance from Table UG-84.4 of ASME VIII-1 [3]

Impact Test Exemption Curves

As an alternative to CVN testing, impact test exemption curves are provided to establish an MDMT based on the material of construction and weld joint governing thickness. An impact test exemption curve (Curve A, Curve B, Curve C, or Curve D) is designated based on the material specification and heat treatment. Figure UCS-66 from ASME VIII-1 [3] is shown below.

Figure UCS-66
Impact Test Exemption Curves

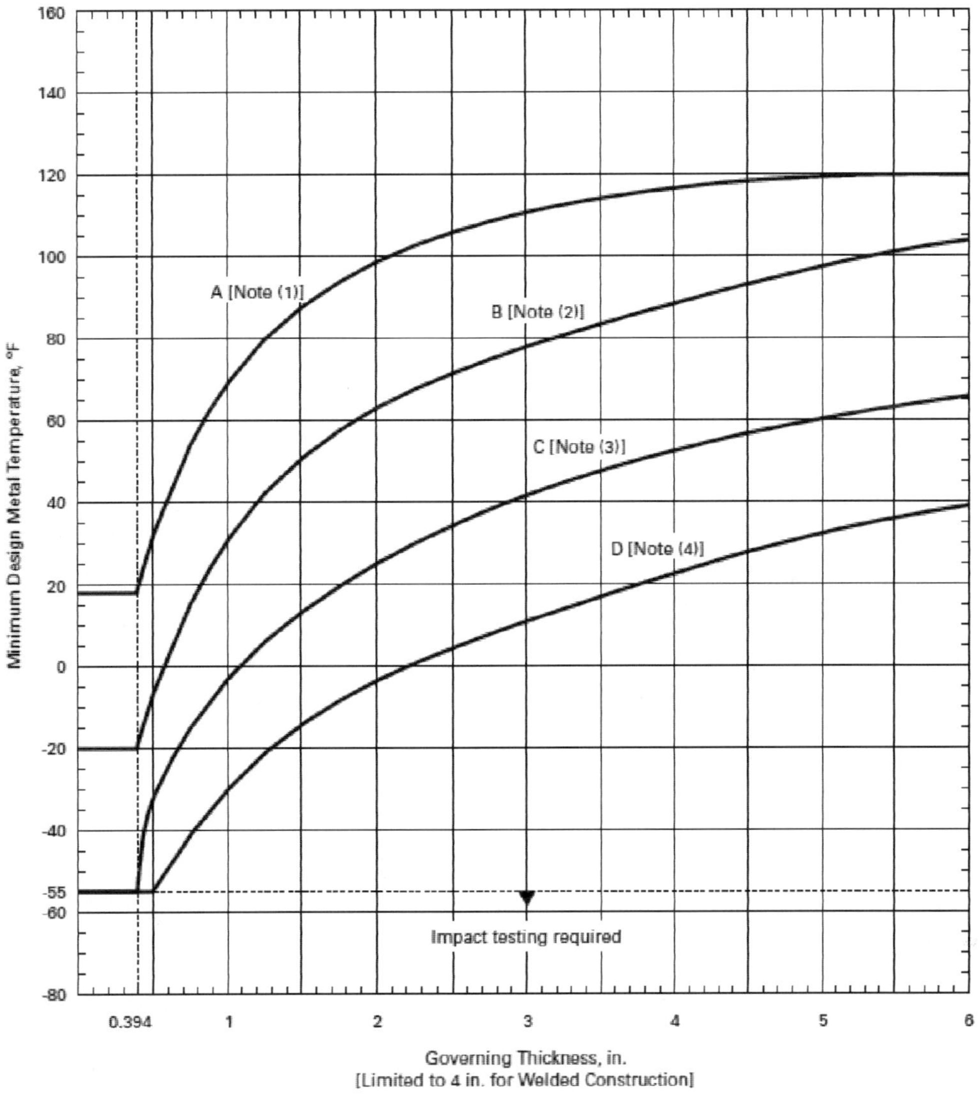

GENERAL NOTES:
(a) Tabular values for this figure are provided in Table UCS-66.
(b) See UCS-66(a).
(c) For bolting and nuts, the following impact test exemption temperatures shall apply:

Bolting

Spec. No.	Grade	Diameter, in. (mm)	Impact Test Exemption Temperature, °F (°C)
SA-193	B5	Up to 4 (100), incl.	–20 (–30)
SA-193	B7	Up to $2\frac{1}{2}$ in. (64), incl.	–55 (–48)
SA-193	...	Over $2\frac{1}{2}$ (64) to 7 (175), incl.	–40 (–40)
SA-193	B7M	Up to $2\frac{1}{2}$ (64), incl.	–55 (–48)

FIGURE 5: Impact Test Exemption Curves in ASME VIII-1 [3]

For ASME P-1 Group 1 and P-1 Group 2 materials with wall thickness less than or equal to 1.5 inches, the MDMT can be reduced by 30°F if a PWHT operation was completed.

ASME VIII-1 Paragraph UG-20(f)

In addition to Figure UCS-66 exemptions, paragraph UG-20(f) in ASME VIII-1 [3] provides an exemption from impact testing for an MDMT of -20°F if all the following requirements are satisfied.

- Material is limited to ASME Code P-1, Group 1 or Group 2, and the uncorroded governing thickness does not exceed the following:
    - 0.5 inches for Curve A material
    - 1.0 inches for Curve B, C, or D material
- Vessel has been hydrostatically pressure tested per ASME VIII-1
- Shock chilling or significant thermal shock loadings are not credible events
- Cyclic loading is not a controlling design requirement

Temperature Reduction Based on Stress Ratio

ASME VIII-1 offers a means to reduce a component MDMT for reduced stress states. This is implemented through the use of a temperature reduction plot as a function of the stress ratio for the component as shown in Figure 6. The stress ratio is the applied tensile membrane stress to the allowable stress for a component.

Figure UCS-66.1
Reduction in Minimum Design Metal Temperature Without Impact Testing

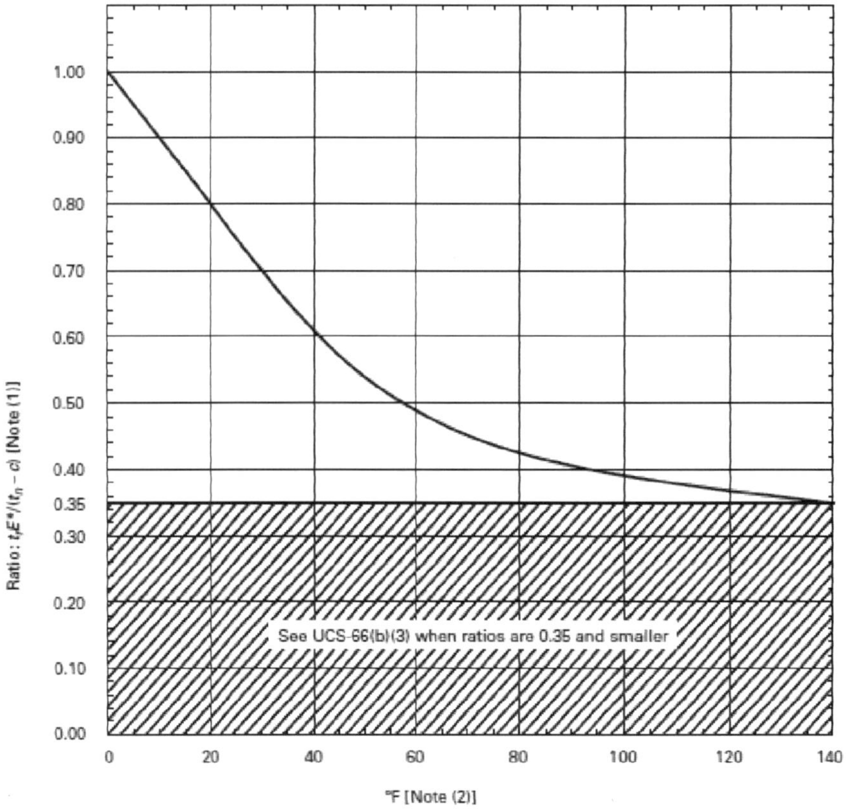

FIGURE 6: Impact Test Exemption Curves in ASME VIII-1 [3]

ASME B31.3 Piping

Most piping operating in process facilities today was designed to the ASME Code for Process Piping, B31.3 (ASME B31.3) [4]. The impact test exemption curves and the temperature reduction curves in ASME B31.3 are almost identical to ASME VIII-1.

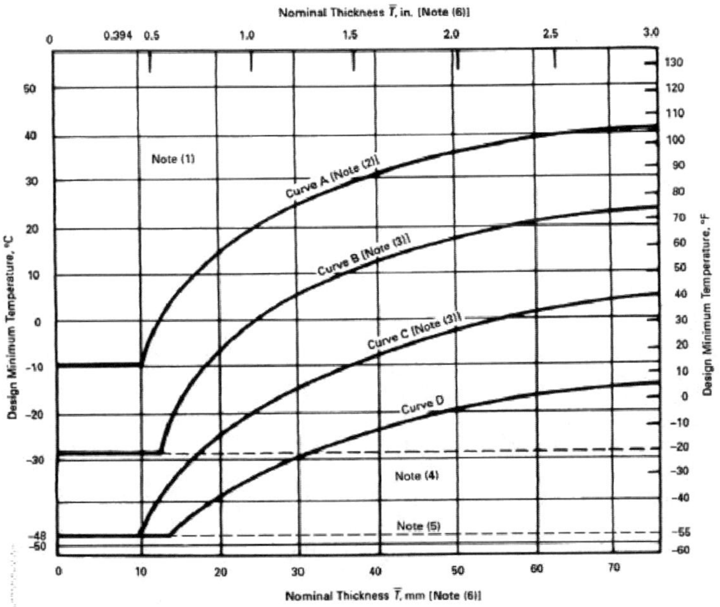

FIGURE 7: Impact Test Exemption Curves in ASME B31.3 [4]

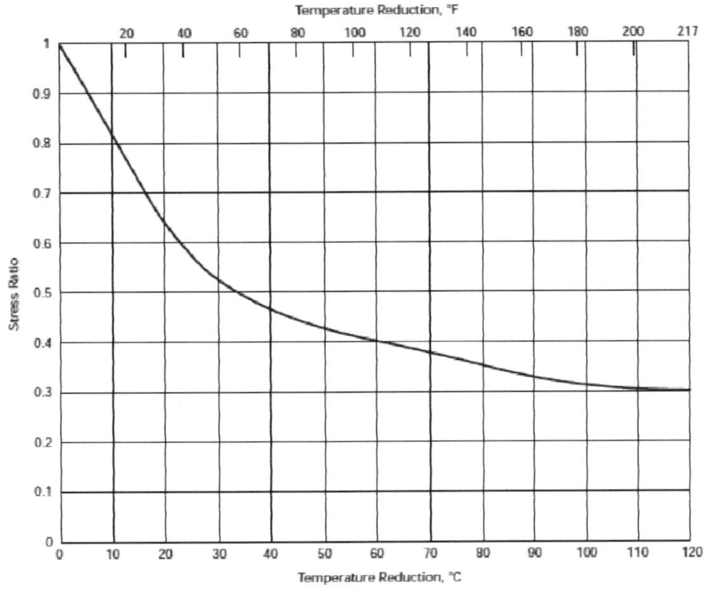

FIGURE 8: Temperature Reduction Curve in ASME B31.3 [4]

A significant difference between ASME B31.3 and ASME VIII-1 procedures for establishing MDMT limits is that ASME B31.3 does not permit the 30°F temperature reduction credit for PWHT (as it is permitted in ASME VIII-1). Additionally, PWHT is no longer a mandatory requirement for any wall thickness in ASME B31.3 provided that multi-pass welding is employed for wall thicknesses greater than 3/16 inch and a minimum preheat of 200°F is implemented for wall thicknesses greater than 1 inch.

ASME B31T Piping

In 2000, the ASME B31 Code for Pressure Piping, Materials Technical Committee (MTC) determined that there was a need to develop a standard set of toughness requirements for piping components that can be adopted by the various piping codes and other Codes and Standards. As a result, ASME B31T Standard Toughness Requirements for Piping (ASME B31T) [5] was published. Currently ASME B31.1 Power Piping [6] references [5] for material toughness requirements, but ASME B31.3 does not. The initial publication of ASME B31T included various differences when compared to ASME B31.3. An example is shown in Figure 9 below.

2010 Edition of ASME B31T:

				Material		Weld		
T-Number Group	Nominal Thickness, in.	Notes	Low-Temp., °F, Service Limit	Minimum Material Permitted Temp., °F	Minimum Temp., °F, Without Impacts	Minimum Weld Permitted Temp., °F	Temp., °F, Without Impacts SR > 0.3	Temp., °F, Without Impacts SR ≤ 0.3
Carbon Steels (Cont'd)								
CS B	≤ 0.394	Max. possible Charpy $t < 2.5$ mm [Note (5)]	−20	−55	−55	. . .	−20	−155
CS B	≤ 0.394	Max. possible Charpy $t \geq 2.5$ mm [Note (6)]	−20	. . .	−20	. . .	−20	−155
CS B	≤ 0.47	. . .	−10	. . .	−10	. . .	−10	−155
CS B	≤ 0.57	. . .	0	. . .	0	. . .	0	−155
CS B	≤ 0.68	. . .	10	. . .	10	. . .	10	−155

<u>*2015 Edition of ASME B31T:*</u>

T-Number Group	Nominal Thickness, in.	Notes	Low-Temp., °F, Service Limit	Material: Minimum Material Permitted Temp., °F	Material: Minimum Temp., °F, Without Impacts	Weld: Minimum Weld Permitted Temp., °F	Weld: Temp., °F, Without Impacts SR > 0.3	Weld: Temp., °F, Without Impacts SR ≤ 0.3
Carbon Steels (Cont'd)								
CS B	≤ 0.394	Max. possible Charpy $t < 2.5$ mm [Note (5)]	−20	−55	−55	. . .	−20	−155
CS B	≤ 0.394	Max. possible Charpy $t \geq 2.5$ mm [Note (6)]	−20	. . .	−20	. . .	−20	−155
CS B	≤ 0.47	. . .	−20	. . .	−20	. . .	−20	−155
CS B	≤ 0.57	. . .	−7	. . .	−7	. . .	−7	−155
CS B	≤ 0.68	. . .	9	. . .	9	. . .	9	−155

FIGURE 9: Example Comparison of ASME B31T 2010 Edition to 2015 Edition [5]

However, it appears based on a cursory review that the current guidance in the 2015 Edition of ASME B31T [5] has been updated to more closely align with current ASME B31.3 [4] requirements. The authors of this paper are unaware of the basis behind the modifications that were made between the various editions of ASME B31T.

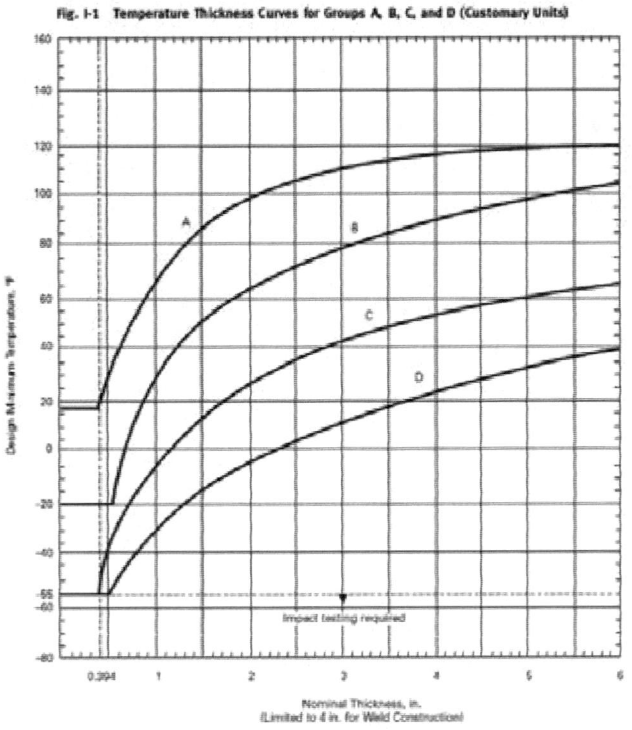

FIGURE 10: Impact Test Exemption Curves in ASME B31T, 2015 EDITION [5]

CURRENT FFS PROCEDURES FOR EVALUATING SUSCEPTIBILITY TO BRITTLE FRACTURE

API 579 [1] provides procedures and criteria to evaluate in-service equipment and justify run, repair, or replace decisions. Part 3 in API 579 is dedicated to the assessment of existing equipment for brittle fracture. The Part 3 Level 1 and Level 2 Method A procedures used to screen for the propensity for brittle fracture are based upon the ASME VIII-1 [3] Part UCS-66 design philosophy. The Level 3 method (which references Part 9 of API 579) includes a detailed fracture mechanics evaluation. For all assessment Levels, a critical exposure temperature (CET) is compared to a minimum allowable temperature (MAT) to determine acceptability. The CET represents the driving force for fracture and consists of the lowest potential metal temperature from all operating, upset, or atmospheric conditions. The CET can consist of a single temperature and coincident pressure or an envelope of temperature and pressure combinations. The MAT represents the material resistance to fracture. The MAT for a pressure vessel or piping system is the highest (i.e. warmest) value of the MAT for any of its components.

Level 1 FFS Evaluation

A Level 1 FFS evaluation is typically completed by an inspector or plant engineer. It is the quickest and easiest assessment and requires the least amount of information. Similar to an ASME VIII-1 MDMT, a Level 1 MAT is established either through engineering evaluation techniques or it assigned based on destructive testing (i.e., CVN). For materials that are not impact tested, an impact test exemption curve (Curve A, Curve B, Curve C, or Curve D) is designated based on the material specification and heat treatment. Using the weld joint governing thickness along with the applicable impact test exemption curve, the Level 1 MAT is obtained from the following figure (Figure 11) which is basically identical to the ASME VIII-1 Figure UCS-66 (Figure 5).

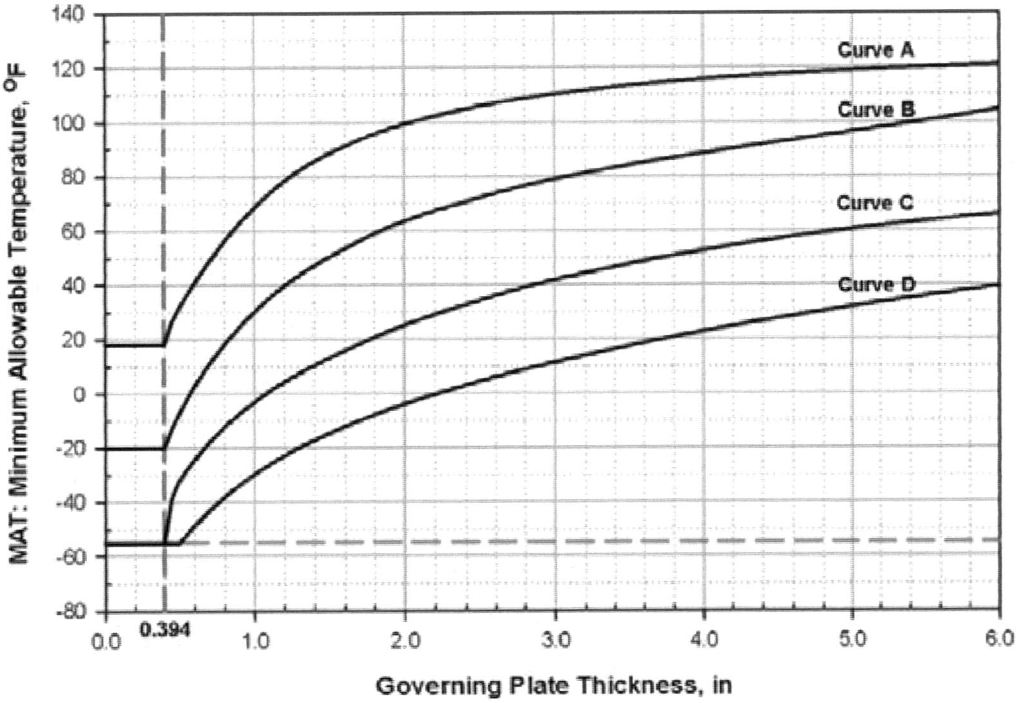

FIGURE 11: API 579 Part 3 Level 1 MAT Curves [1]

For ASME P-1 Group 1 and P-1 Group 2 materials with wall thickness less than or equal to 1.5 inches, the Level 1 MAT can be reduced by 30°F if a PWHT operation was completed and no post-construction repairs or alterations were performed that would diminish the effectiveness of the PWHT.

In a Level 1 assessment, the MAT is a single temperature assigned to all potential operating conditions. There is no flexibility for the analyst to modify the Level 1 MAT based on flaw size, material toughness, or stress.

Level 2 FFS Evaluation

Level 2 FFS evaluations are typically completed by a plant engineer. Three (3) analysis options are provided:

- *Method A:* Calculation of a safe envelope based on an evaluation of actual stress compared to design allowable stress (i.e. trade lower stress levels for lower permissible temperatures). The starting point (i.e. baseline MAT) for a Method A analysis is the MAT from a Level 1 evaluation and therefore it inherently includes assumptions on flaw size, material toughness, and stress. Based on a calculated stress ratio (computed using the defined thickness basis, stress basis, or pressure-temperature rating basis) for the condition under consideration, Figure 12 below is used to permit a temperature reduction from the Level 1 MAT. Similar to a Level 1 evaluation, an additional 30°F reduction is permitted for ASME P-1 Group 1 and P-1 Group 2 materials with wall thickness less than or equal to 1.5 inches so long as a PWHT operation was completed and no post-construction repairs or alterations were performed that would diminish the effectiveness of the PWHT.

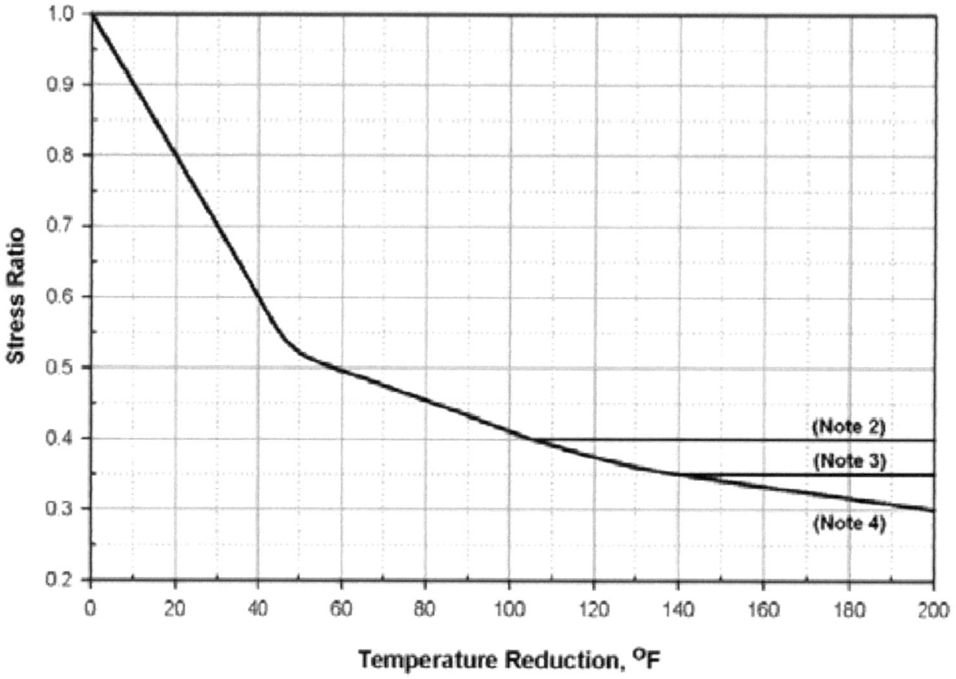

FIGURE 12: API 579 Part 3 Level 2 MAT Temperature Reduction Curve [1]

- *Method B:* Determine the MAT based on the metal temperature at the time of hydrotest. This method is not practical or feasible for many instances, and substantial risk may exist for fracture during the hydrotest.

- *Method C:* Determine the MAT based on a proof test argument (i.e. grandfathering approach). This method assumes that if it can be shown that the worst-case combination of applied stress, temperature, and flaw size has already occurred and did not result in a failure, other less severe conditions will not result in a failure. This method requires detailed historical operating, inspection, and repair records that are not typically available in many instances.

In a Level 2 assessment, the MAT can consist of a single temperature and coincident pressure or an envelope of temperature and pressure combinations.

Level 3 FFS Evaluation

A Level 3 FFS evaluation is the most reliable and accurate method for establishing an MAT because it involves a detailed fracture mechanics evaluation and all critical input parameters (particularly stress, material toughness, and potential crack-like defects) are user-controlled and shall be based upon information and assumptions specific to the equipment or component under evaluation. A Level 3 evaluation should be completed by an engineering specialist with extensive experience in FFS. The API 579 Part 9 FAD approach (as shown in the figure below) is employed in a Level 3 brittle fracture evaluation.

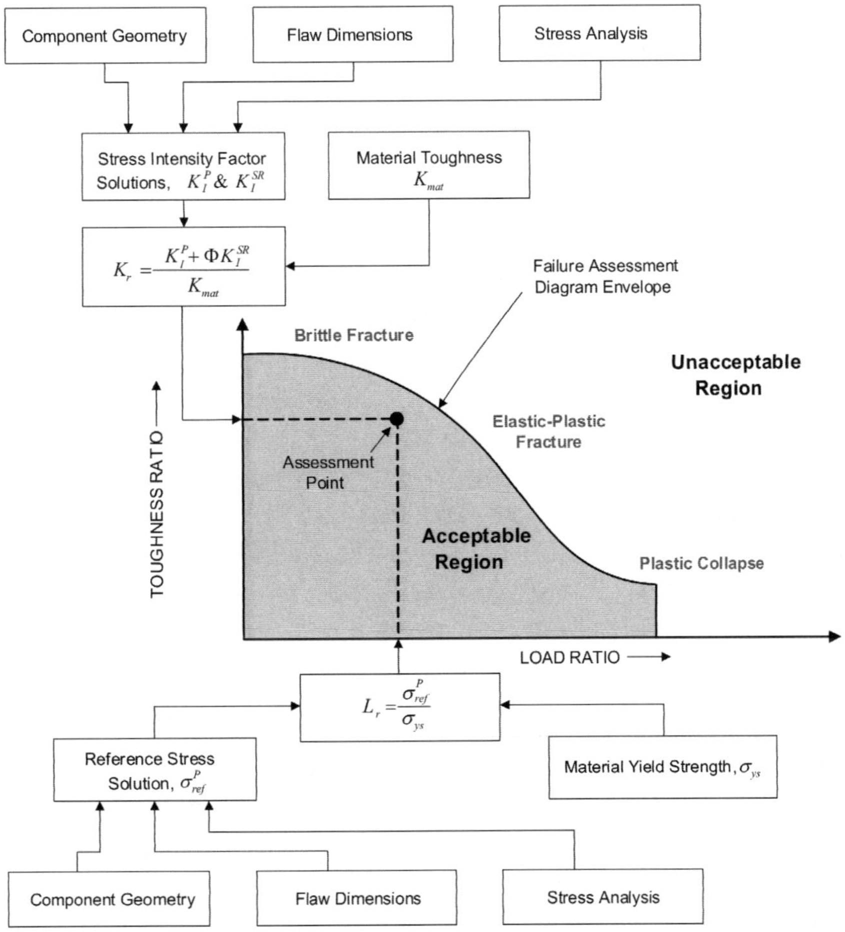

FIGURE 13: API 579 Part 9 Failure Assessment Diagram [1]

Level 3 evaluations are required when a crack-like flaw is known to exist or if the equipment operates in a service where material toughness degradation or an active cracking mechanism exists. A significant advantage of a Level 3 evaluation is that a focused inspection plan can be coupled with the FFS evaluation to optimize inspection costs while minimizing risk.

Similar to a Level 2 assessment, the MAT established through a Level 3 evaluation can consist of a single temperature and coincident pressure or an envelope of temperature and pressure combinations.

DEFICIENICIES AND DISCRPANCIES WITH CURRENT CODE PROCEDURES

A number of deficiencies and inconsistencies exist with the current design and FFS procedures that are used to establish minimum permissible temperature limits [7]. Although an API 579 Level 3 evaluation represents the most reliable and most accurate means for evaluating susceptibility to brittle fracture, results from a Level 3 evaluation can prove more limiting than current design Code requirements and/or Level 1 or Level 2 FFS results. Such an outcome is out of compliance with the intent of Level 1, Level 2, and Level 3 FFS evaluations. When developing FFS procedures, Level 3 technology is typically coupled with conservative assumptions to develop more simplified procedures for implementation as Level 2 analyses. Additional assumptions and simplifications are then applied to establish a conservative set of Level 1 rules. The current ASME VIII-1 design rules and the API 579 Part 3 Level 1 and Level 2 methods are based on obsolete fracture mechanics and "judgement." Each are based on assumptions and technology that are inconsistent with current state-of-the-art (i.e. Level 3), but also inconsistent with each other. In 1998, Welding Research Council (WRC) Bulletin 435 [8] was published to provide insight into potential issues associated with reducing design margins for ASME VIII-1 pressure vessels. One of the final recommendations published at that time was "*Notch toughness requirements should be reviewed and updated.*" Since then, the material toughness requirements in ASME Section VIII Division 2 (ASME VIII-2) [9] have been updated, but ASME VIII-1 has not changed. The following is a summary of many of the deficiencies and discrepancies known to exist with the current procedures.

Impact Test Exemption Curves

The full background, assumptions, and technology used to develop the ASME VIII-1 impact test exemption curves (which also serve as the Level 1 MAT curves for non-impact tested components) are documented in [10] and reproduced in [11]. A summary of some key aspects is provided herein.

The following relationship was used to estimate material dynamic fracture toughness as a function of temperature for carbon and low alloy steels [12]:

$$K_{Id} = \sigma_{ys}\left(A - B\tanh\left[\frac{T - T_{ref}}{C_R}\right]\right) \qquad (1)$$

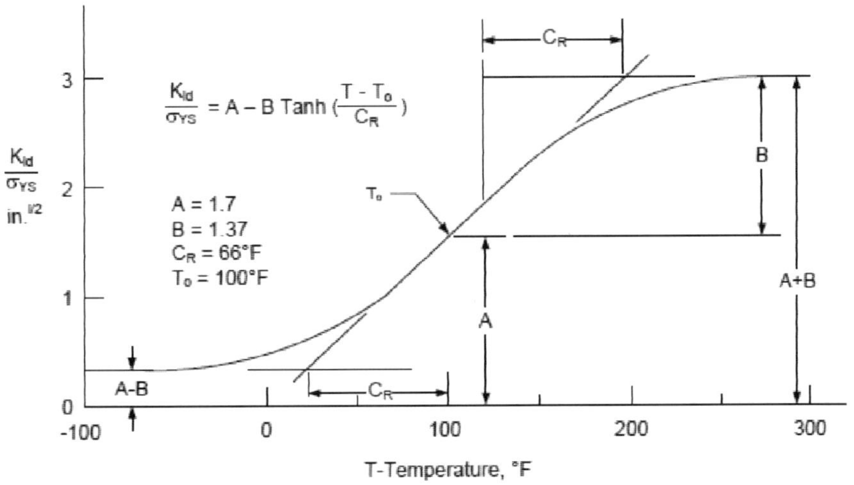

FIGURE 14: Hyperbolic Tangent Function for Material Toughness [11]

The brittle fracture criterion relating dynamic fracture toughness, material yield, and component thickness was established as follows [13][14]:

$$\beta_{Id} = \frac{1}{t}\left(\frac{K_{Id}}{\sigma_{ys}}\right)^2 = 1 \qquad (2)$$

This criterion was recommended "*as a level that insures significant crack tip plastic deformation prior to onset of crack extension*" and because it takes account of the following factors that increase tendency toward non-ductile fracture:

- Triaxial constraint increases at the crack tip with increased thickness.
- The potential for larger flaws increases as the thickness is increased.
- Variability of metallurgical differences increases as the thickness is increased.

Combining equation (1) and equation (2), Corten [10] developed the following thickness-temperature relation.

$$t^{\frac{1}{2}} = A - B\tanh\left(\frac{T - T_{ref}}{C_R}\right) \qquad (3)$$

Based on fits of experimental data, A=1.7, B=1.37, and C_R=66°F. Solving for minimum allowable temperature, the equation is rewritten as follows:

$$T = T_{ref} + 66\left(\frac{a\tanh\left[\sqrt{t} - 1.7\right]}{1.37}\right) \qquad (4)$$

Four reference temperatures were used to index materials of interest with T_{ref} set as:

- 114°F for Curve A materials
- 76°F for Curve B materials
- 38°F for Curve C materials
- 12°F for Curve D materials

The current Level 1 MAT curves for non-impact tested materials were generated by plotting equation (4) for a range of thickness values and then engineering judgement and experience were used to make modifications as shown below [10] [11].

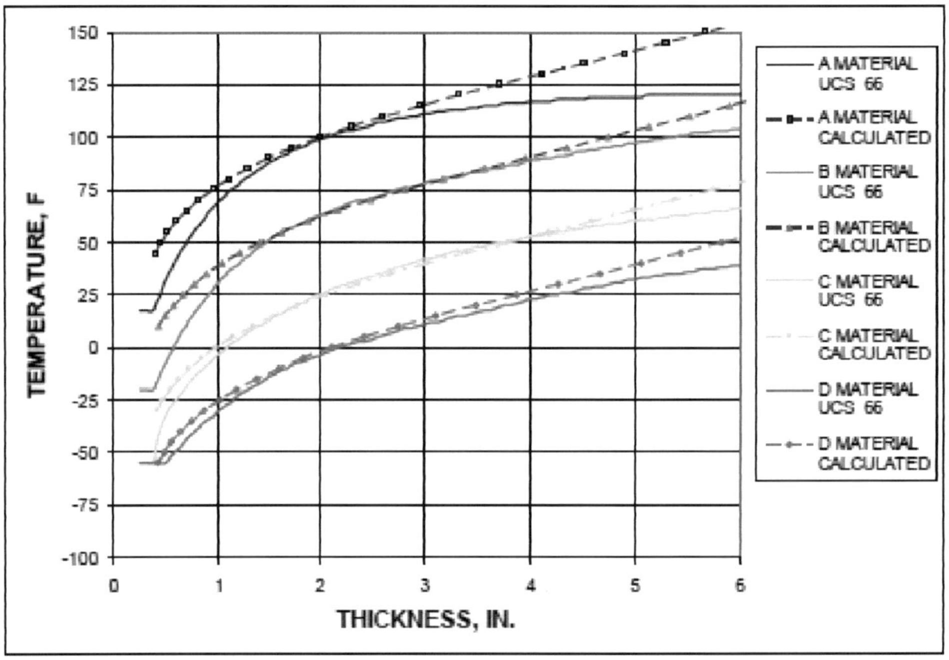

FIGURE 15: Plot of Original Formulation Used to Establish Impact Test Exemption Curves [10]

Temperature Reduction Curves for Reduced Stress

Although the baseline MAT obtained from a Level 1 assessment serves as the starting point for a Level 2 FFS evaluation, the temperature reduction curve that is used to justify a reduced temperature as a function of reduced stress is based on completely different technology and a different set of assumptions.

The temperature reduction curve was established using obsolete fracture mechanics assuming a postulated reference crack-like flaw with a depth equal to 25% of the component thickness (i.e. t/4 flaw) and material toughness as a function of temperature per ASME Section XI [15].

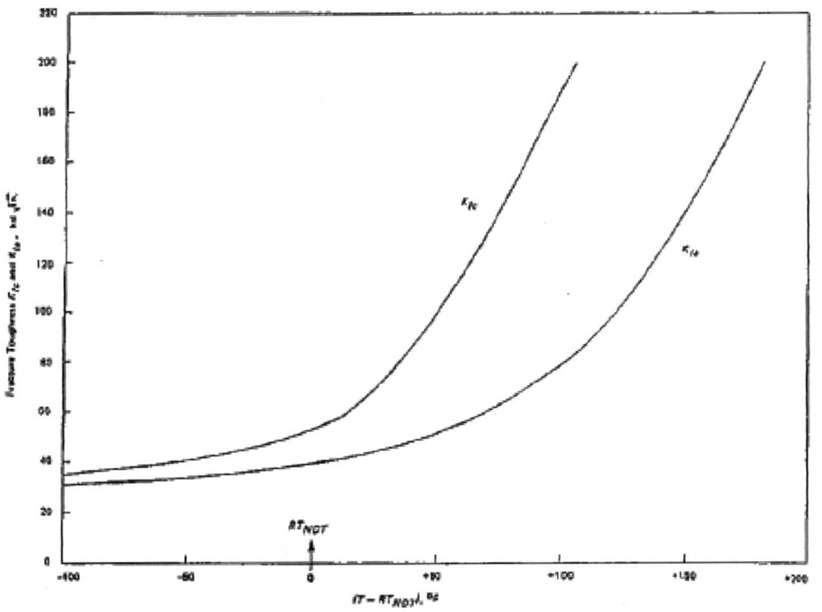

Fig. A-4200-1 Lower bound K_{Ia} and K_{Ic} test data for SA-533 grade B class 1, SA-508 class 2, and SA-508 class 3 steels

FIGURE 16: Original ASME Section XI Material Toughness [15]

The temperature reduction curve was developed assuming the ratio of applied stress intensity to material toughness must be maintained constant as membrane stress is reduced [16].

$$\frac{K_I}{K_{IC}} = \frac{K'_I}{K'_{IC}} \qquad (5)$$

Assuming a starting point that corresponds to the NDT plus 60°F, the relation of temperature reduction as a function of reduced stress intensity (i.e. applied stress) is developed using equation (5). Results for a few reduced stress states are summarized in Table 1 below.

TABLE 1: Example of How Temperature Reduction Curve Values Were Established

K_I/K_{IC}	K_{IC} (ksi√in)	Reference Temperature (°F)	Temperature Delta (°F)
1	115	NDT+60	0
0.75	86.25	NDT+35	25
0.5	57.5	NDT	60

As noted previously, the technology implemented in the Level 2 evaluation method is inconsistent with current state-of-the-art (i.e. Level 3) and inconsistent with that used to develop the Level 1 procedures. Particularly, there does not appear to be any documented technical basis for assuming a t/4 reference flaw at the material NDT plus 60°F would be stable at the Level 1 MAT (which is the starting point for a Level 2 Method A evaluation) for all potential loading scenarios. Another major deficiency with the Level 2 Method A procedure is that the temperature reduction curve does not address the significance that weld residual stress has on the susceptibility to brittle fracture. Weld residual stress does not relax with a reduction in applied stress.

Fracture Arrest and Applied Stress Ratio Thresholds

API 579 [1] currently defines the CET as the lowest (coldest) metal temperature derived from either the operating or atmospheric conditions at the maximum credible coincident combination of pressure and supplemental loads that result in general primary tensile stress (including any stresses due to net section bending) greater than 8 ksi. ASME B31.3 [4] does not require evaluation for susceptibility to brittle fracture for applied stress values less than 6 ksi. Similarly, ASME VIII-1 [3] Part UCS-66(f)(1), ASME B31.3 Figure 323.2.2B, and the API 579 Part 3 Level 2 Method A procedures permit the MAT of a component to drop to -155°F if the applied tensile stress ratio falls below a pre-defined threshold. The applied tensile stress ratio threshold is approximately 10% of the material ultimate tensile strength (UTS). This is typically 5-8 ksi for most materials of interest. However, as noted in [11], 8 ksi limit in the CET definition and the 10% UTS applied stress ratio threshold are not appropriate for pressure vessels and piping systems in the as-welded condition. Failures of non-PWHT weld joints have occurred due to weld residual stress alone. Fracture can occur regardless of the source of stress (e.g. residual stress from the welding process or applied stress from pressure).

The basis for the applied stress threshold is work completed over 50 years ago by T.S. Robertson [17] and later expanded upon by W.S. Pellini of the Naval Research Laboratory [18]

[19] [20] where experimental testing results indicated the lower stress limit for unstable fracture propagation was found to be 5-8 ksi. Pellini's crack arrest temperature (CAT) curve developed from drop-weight testing is shown below.

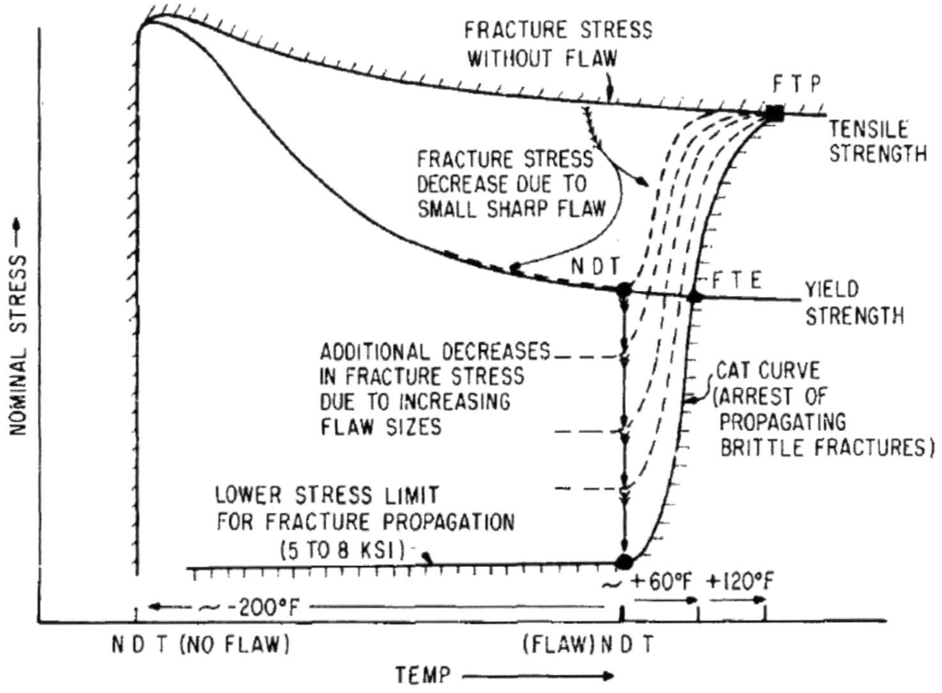

FIGURE 17: Crack Arrest Temperature Diagram [18]

It is critical to note the only welded samples included in the testing were longitudinal weld seams loaded in the longitudinal direction (parallel to the weld seam) with cracks/defects oriented in the circumferential direction (perpendicular to the weld seam) [19]. It is well documented that residual stress attenuation occurs within 1 to 2 times the plate thickness from a weld [1]. As crack propagation occurs for such a scenario, the crack tip extends outside the weld residual stress field such that the driving force for brittle fracture is significantly diminished.

In [19], Pellini concludes that fracture can occur "*at any applied stress level, depending on the level of the residual stress that is present in the weldment.*" In [18], he states that "*if residual stresses are present, it is necessary to add the residual stress component to the load stress component.*" Further, in [20], Pellini specifically highlights concerns for potential brittle fracture in pressure vessels and states, "*fractures may develop for PD/2t stresses of extremely low or zero value (as for the case of spontaneous fracture) because the additional general level of "locked in" stresses may provide for propagation, i.e., may exceed 5 to 8 ksi minimum required by the CAT curve for propagation at temperatures below the NDT.*"

To further illustrate the significance weld residual stress can have on susceptibility to fracture, the plot below shows Level 3 MAT results calculated from fracture mechanics (API 579 Part 9) for SA-515-70 (impact test exemption Curve A designation) plate with a thickness of 1.125 inches considering loading due to weld residual stress alone for non-PWHT joints. No consideration for any applied primary loads (pressure, weight, etc.) was included. Two different postulated reference flaw sizes were analyzed, both with a 6:1 length-to-depth aspect ratio:

- 0.281 inches deep (t/4 flaw) – plotted in red
- 0.141 inches deep (t/8 flaw) – plotted in blue

The MAT ranges are plotted based on the worst-case residual stress distributions for flaws parallel to welds and worst-case residual stress distributions for flaws perpendicular to welds (per Annex 9D of API 579 [1]). For both flaw sizes evaluated, the calculated MAT ranges are substantially higher than the Level 2 MAT of -155°F. For reference purposes, boiling point curves for various hydrocarbon processes are also plotted in the figure to provide perspective on potential metal temperatures that may be achieved during auto refrigeration events. As the plot show, excluding the effects of weld residual stress can prove unsafe, particularly for non-PWHT welds.

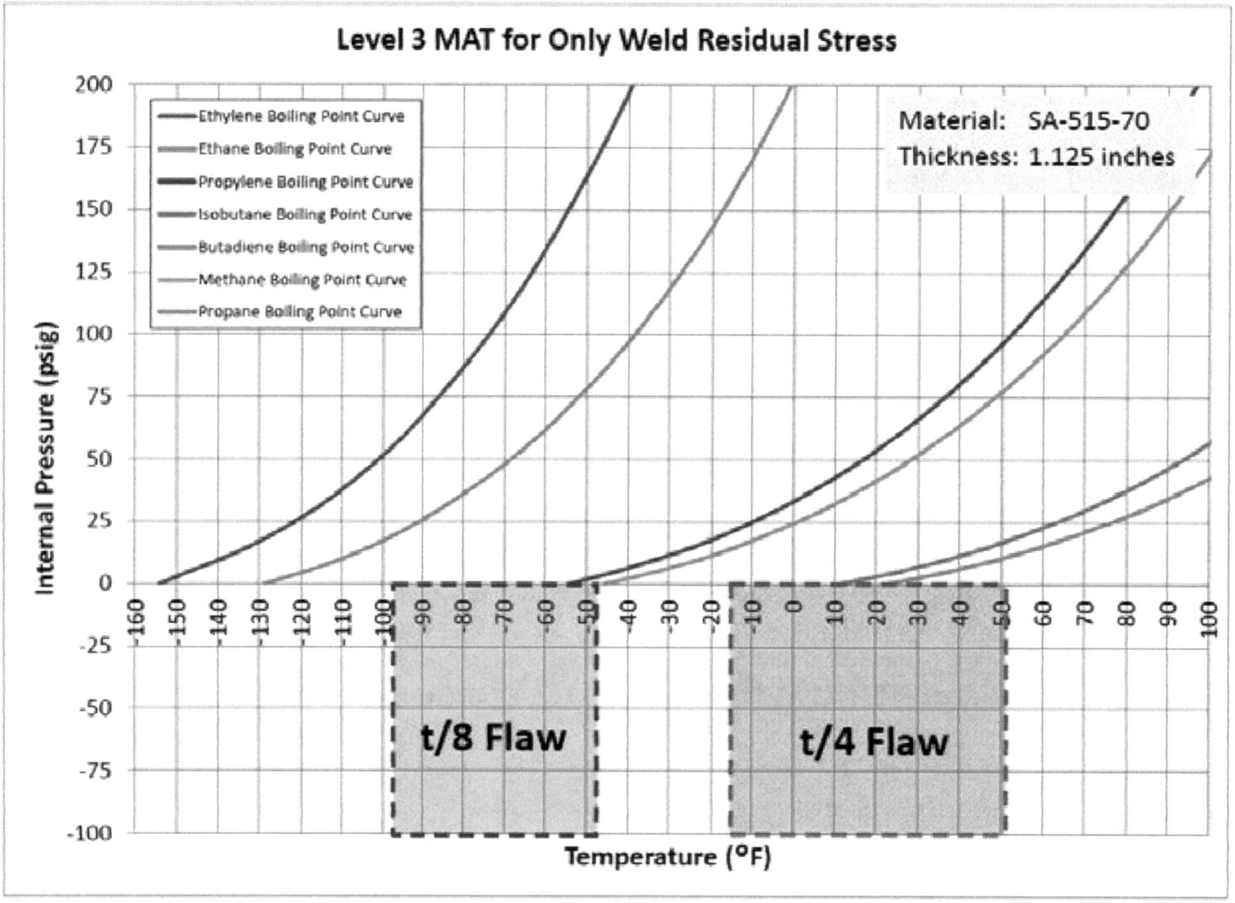

FIGURE 18: Plot Demonstrating Level 3 MAT Calculated Considering Only Weld Residual Stress Loading

PWHT Temperature Credit

Current ASME VIII-1 and API 579 Level 1 and 2 procedures permit a 30°F temperature reduction for PWHT of ASME P-1 Group 1 and P-1 Group 2 materials with wall thickness less than or equal to 1.5 inches. The technical basis for the 30°F is not documented and not consistent with explicit fracture mechanics assessments using weld residual stress profiles as published in Annex 9D of [1]. Results from a Level 3 fracture mechanics evaluation would indicate a much greater benefit can typically be justified (i.e. more than 30°F delta compared to as-welded components).

Impact Tested Materials

As an alternative to the impact test exemption curves, the FFS Level 1 MAT (or design MDMT) for a component can be established through destructive testing. As noted in [21] (and also shown in Figure 3 specifically for ASME VIII-1), current Code required CVN energy can vary anywhere from 10 ft-lbs to 40 ft-lbs depending on material specification, thickness, and the ASME/API standard. As noted in [10], the CVN impact energy requirements were established based on "*employing the accumulated experience among some of the major users of Section VIII, Division 1.*" Subsequently, the MAT assigned using CVN testing can vary significantly from the MAT established using fracture mechanics (i.e. a Level 3 evaluation) for the corresponding material fracture toughness. To address this discrepancy, many international Codes (including PD5500, EN 13445, and EN 13480) provide CVN test temperature vs. target MAT curves or correlations. An example from [22] is shown below.

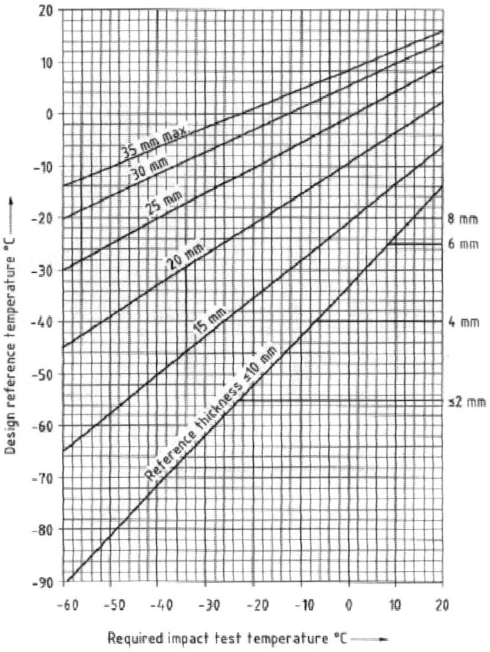

FIGURE 19: Minimum Design Reference Temperature as a Function of Impact Test Temperature per PD5500 [22]

Based on Figure 19, the CVN impact test temperature may need to be higher or lower than the target MAT.

ASME VIII-1 Paragraph UG-20(f)

The impact test exemptions in paragraph UG-20(f) of ASME VIII-1 [3] are inconsistent with results obtained using all other design Code and FFS procedures (including results obtained from detailed fracture mechanics). These exemptions were removed in the 2016 edition of API 579 due to the lack of technical basis.

Minimum Temperature for Hydrotest

Many brittle fracture failures have occurred during hydrotest conditions. Current industry guidance is inconsistent and the technical basis for the temperatures is not well documented. The following is a summary of current guidance from various design and post-construction Codes:

- 2016 Edition ASME VIII-1 [3]: MDMT + 30°F
- 1965 Edition ASME VIII-1: 60°F
- NBIC [23]: 60°F
- API 510 [24]: MAT + 10°F ($t \leq 2''$), MAT + 30°F ($t > 2''$)
- ASME PCC-2 [25]: MDMT/MAT + 30°F
- API 579 [1]: MAT + 30°F

PROPOSED MODIFICATIONS TO FFS PROCEDURES

To mitigate the numerous discrepancies and deficiencies with the current design Codes and the Level 1 and Level 2 FFS procedures in API 579 Part 3, it is proposed that new brittle fracture screening procedures and acceptance criteria are developed that are anchored in state-of-the-art fracture mechanics consistent with Part 9 of API 579 and as outlined in WRC Bulletin 562 [26]. All critical inputs (particularly, stress, toughness, and flaw size) shall be explicitly noted such that an analyst can clearly determine applicability to specific equipment and applications. The following recommendations are provided and examples implementing the procedures based on these assumptions are presented. A comprehensive and detailed explanation of the methodology and assumptions will be published in [27].

Level 1 MAT Curves

Level 1 MAT curves shall be generated by solving the limit state equation below iteratively for a given range of input parameters.

$$\frac{K_I^P + \Phi K_I^{SR}}{\left(1.0-\left(L_r^P\right)^{2.5}\right)^{0.2}} - K_{mat} = 0 \qquad (6)$$

Primary Stress

To maximize the applicability of the Level 1 acceptance criteria across equipment constructed from various design Codes, it is recommended that the applied general primary membrane stress used to establish the Level 1 MAT is set equal to 2/3 of the material minimum specified yield stress.

Secondary Plus Residual Stress

Residual stress from welding must be included in generation of the MAT screening curves. Through-thickness weld residual stress distributions are a function of the weld joint geometry (single V, double V, etc.) and weld joint orientation with respect to flaw orientation (parallel to weld vs. perpendicular to weld). Separate curves shall be provided for the as-welded condition and PWHT condition. The following recommendations are provided:

- To maximize the applicability of the Level 1 procedures (including applicability to repair welds), weld residual stress for the as-welded condition shall be equal to the material minimum specified yield stress + 10 ksi.
- For the PWHT condition, weld residual stress shall be set equal to 30% of the as-welded residual stress distribution.

Material Fracture Toughness

It is recommended that the Wallin Master Curve [28] be utilized with a probability of failure, P_f , of 5% to estimate material lower bound static fracture toughness as a function of temperature as follows:

$$K_{mat} = 18 + \left(9.9 + 70.1\exp\left[0.0106\left(T - T_0\right)\right]\right) \cdot \left(\ln\left[\frac{1}{1-P_f}\right]\left(\frac{1}{L}\right)\right)^{0.25} \qquad (7)$$

The Wallin Master Curve incorporates probability of failure, it is supported by ASTM Standard E1921 [29], and it has been adopted by the technical community as the best approach for predicting toughness, particularly in the ductile-to-brittle transition region. Guidance is provided in [26], [30], and [27] for establishing T_0 for both impact tested materials and non-impact tested materials.

Flaw Size

Acceptance criteria shall be provided for postulated surface-breaking crack-like flaw depths equal to 25% of the component governing thickness (t/4) and 12.5% of the component governing thickness (t/8) with a minimum reference flaw depth equal to 0.0625 inches and a maximum reference flaw depth equal to 1 inch. The postulated reference crack length shall be six times the crack depth (i.e. 6:1 aspect ratio). Historically, a t/4 reference flaw has been used for evaluation of equipment based on WRC Bulletin 175 [31] when no information is available regarding previous inspections for crack-like defects. However, based on the extent of inspection and the type and sensitivity of non-destructive examination (NDE) performed on the component, the analyst shall be offered the flexibility to determine which reference flaw size to use in the screening evaluation.

Example Level 1 MAT Screening Curves

Based on the assumptions and methodologies presented, new Level 1 MAT screening criteria can be established for impact tested components as well as non-impact tested components [27]. Separate screening curves shall be provided for both as-welded and PWHT conditions.

The resulting Level 1 MAT curves for a t/8 reference flaw for non-impact tested 40 ksi yield strength materials in the as-welded and PWHT conditions are shown below.

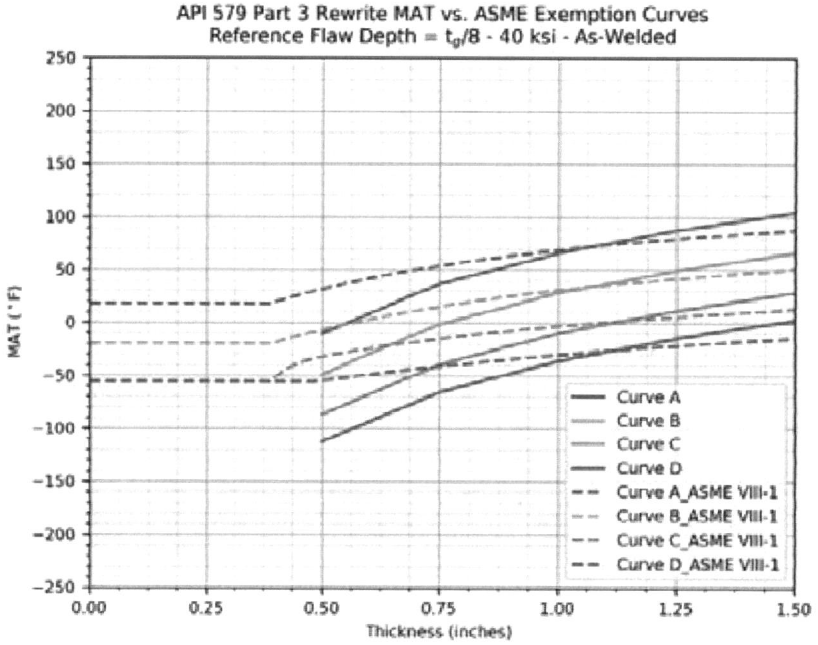

FIGURE 20: Example Comparison of Proposed Level 1 MAT vs. Current Level 1 MAT, t/8 Flaw (As-Welded)

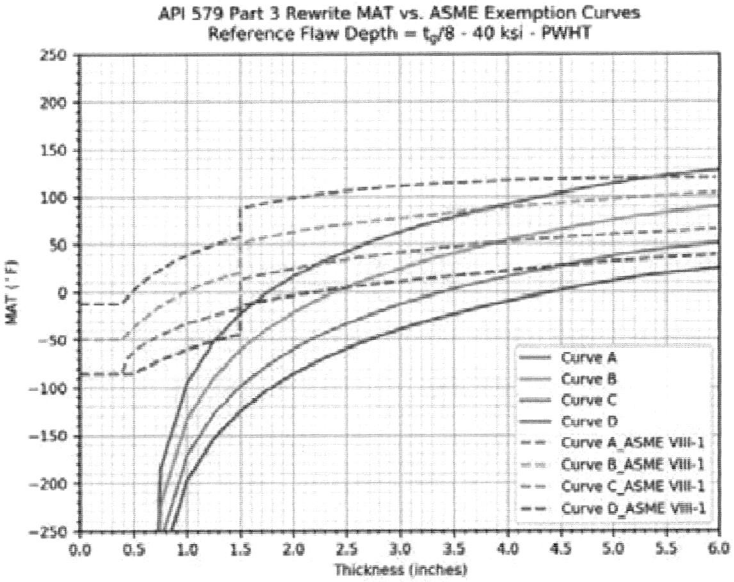

FIGURE 21: Example Comparison of Proposed Level 1 MAT vs. Current Level 1 MAT, t/8 Flaw (PWHT)

In Figure 21 above, the current published API 579 Level 1 MAT curves are shifted downward by 30°F for thickness below 1.5 inches to demonstrate the current procedure allowance for PWHT credit. For comparison, similar plots for a t/4 reference flaw are shown below.

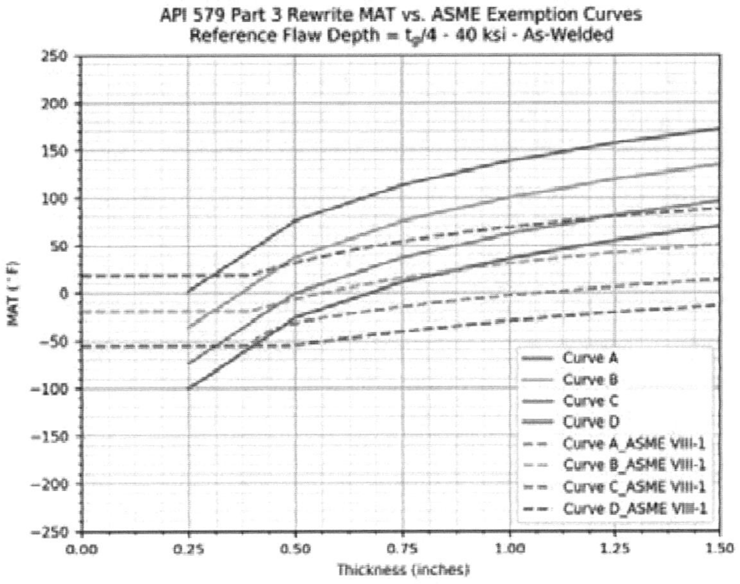

FIGURE 22: Example Comparison of Proposed Level 1 MAT vs. Current Level 1 MAT, t/4 Flaw (As-Welded)

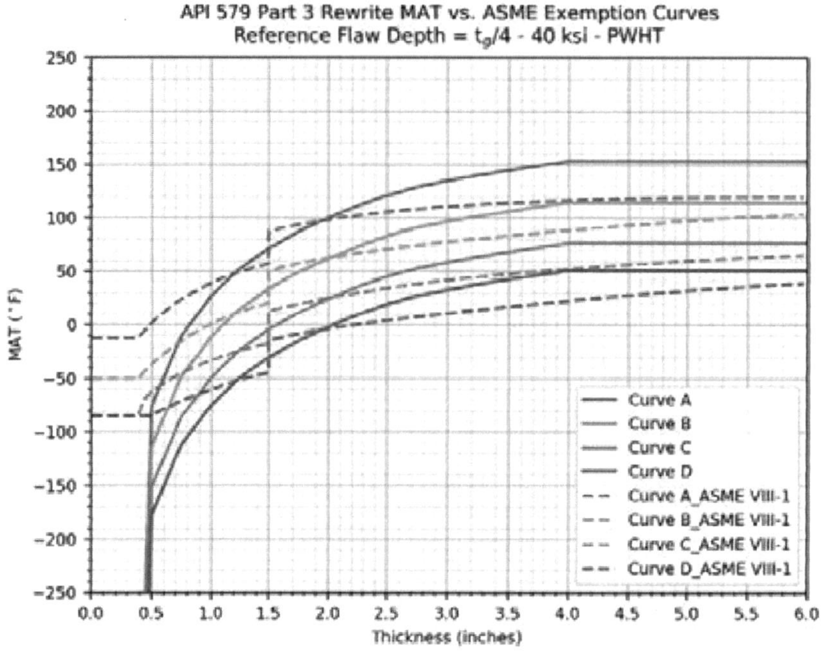

FIGURE 23: Example Comparison of Proposed Level 1 MAT vs. Current Level 1 MAT, t/4 Flaw (PWHT)

Temperature Reduction Curves for Reduced Stress

Using the same technology implemented in the proposed Level 1 screening curves, Level 2 temperature adjustment curves shall be generated to enable an analyst to modify the MAT of a component for applied general primary membrane tensile stress conditions that are lower or higher than the 2/3 of the material minimum specified yield stress assumption that was employed to develop the Level 1 screening curves. Aside from the adjustment on general primary membrane stress, all other assumptions used in the proposed Level 1 screening curves shall remain constant in development of the Level 2 temperature adjustment curves. It is recommended the curves shall be developed to enable an analyst to evaluate stress ratios ranging between 0 to 1.5 (with a stress ratio of 1.5 corresponding to the material specified yield stress). Extending the Level 2 temperature adjustment curves for stress ratios greater than 1, a Level 2 assessment can be utilized to determine the appropriate MAT for hydrotest conditions and/or short-term operating excursions such as those permitted in paragraph 302.2.4 of ASME B31.3 [4] piping systems. Level 2 temperature adjustment curves shall be provided for both as-welded and PWHT conditions.

The Level 2 MAT is calculated as follows:

$$MAT_{Level\ 2} = MAT_{Level\ 1} - T_A \qquad (8)$$

Example Level 2 temperature adjustment curves for 40 ksi yield strength material in the as-welded condition for a t/4 reference flaw are provided below [27]:

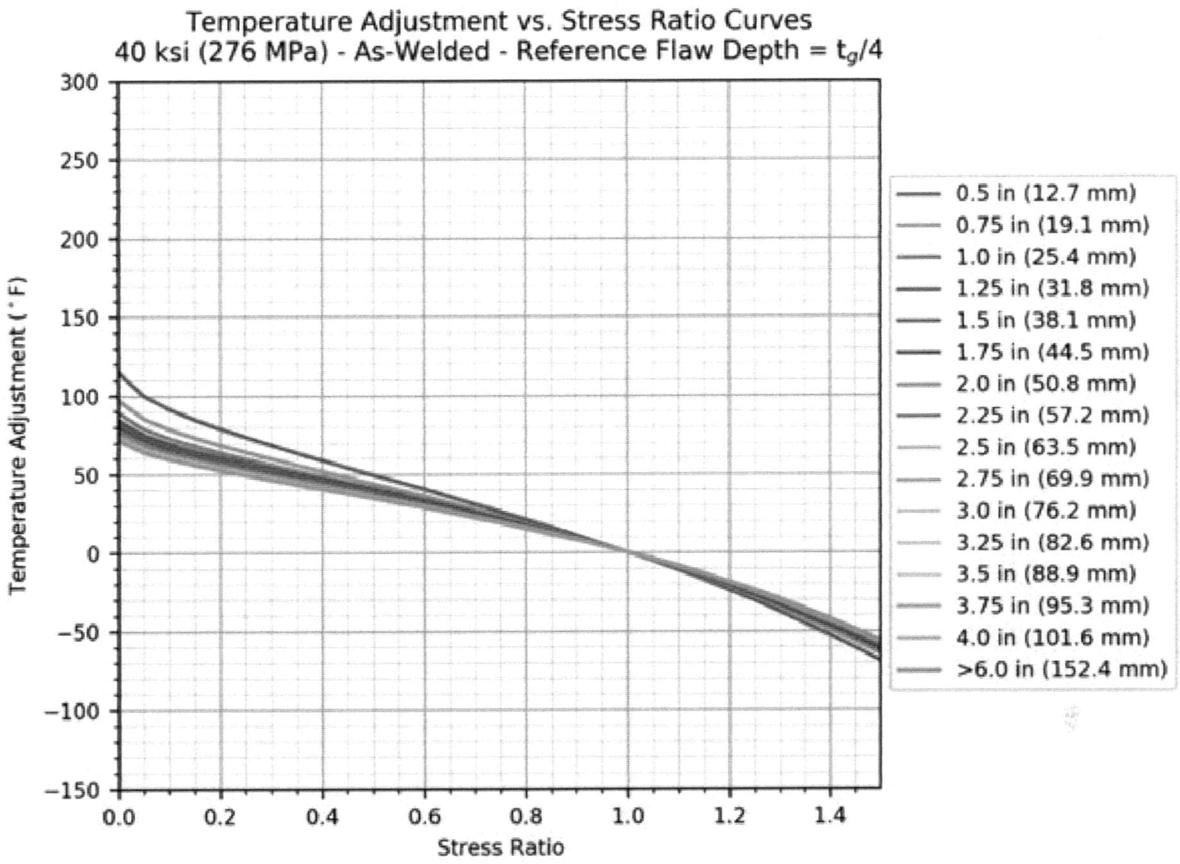

FIGURE 24: Proposed Level 2 Temperature Adjustment as a Function of Stress Ratio

As an alternative to using temperature adjustment curves to refine the Level 1 MAT, mathematical expressions can be used to directly calculate the Level 2 MAT. An example is provided in [27]. Either approach will yield the same Level 2 MAT result.

EXAMPLES

The following examples are provided to illustrate results obtained using the proposed modifications to the Level 1 and Level 2 procedures and acceptance criteria of API 579.

Example 1: Non-Impact Tested Pressure Vessel Shell

Calculate the MAT for a non-impact tested pressure vessel cylindrical shell section.

- Construction Code: ASME VIII-1
- Component: Cylindrical Shell
- Material: SA-516 Grade 70 (MSYS = 38 ksi, Curve B)
- Inside Diameter: 60 inches
- Nominal Thickness: 1.0 inches
- PWHT: No
- Design Pressure: 600 psig
- Impact Tested: No
- Radiography: 100%

Results:

- Level 1 MAT
  - t/4 Reference Flaw 100°F
  - t/8 Reference Flaw 28°F

  Current Part 3 Level 1 MAT 31°F

- Level 2 MAT at 600 psig
  - t/4 Reference Flaw 74°F
  - t/8 Reference Flaw -3°F

  Current Part 3 Level 2 MAT 22°F

Note: The evaluation must be repeated for all components of the pressure vessel and the MAT for the equipment is the maximum (warmest) MAT from all components.

Example 2: Impact Tested Pipe

Calculate the MAT for an impact tested section of pipe.

- Construction Code: ASME B31.3
- Component: Pipe
- Material: SA-333 Grade 6 (MSYS = 35 ksi)
- Inside Diameter: NPS 18
- Nominal Thickness: 1.375 inches
- PWHT: No
- Design Pressure: 1058 psig
- Impact Tested: 13 ft-lbs (avg) at -50°F per Material Specification
- Radiography: 100%

Results:

- Level 1 MAT
  - t/4 Reference Flaw 39°F
  - t/8 Reference Flaw -30°F

Current Part 3 Level 1 MAT -50°F

- Level 2 MAT at 1058 psig
    - t/4 Reference Flaw -16°F
    - t/8 Reference Flaw -99°F

Current Part 3 Level 2 MAT -55°F

Note: The evaluation must be repeated for all components of the piping system and the MAT for the system is the maximum (warmest) MAT from all components.

Example 3: Comprehensive Level 3 FFS Evaluation

A high-pressure, heavy-walled heat exchanger was identified to have the potential for tube side components to reach metal temperatures as low as 17°F at full pressure. Calculate the MAT for the tube side components of the exchanger.

- Construction Code: ASME VIII-1
- Tube Side Design Pressure: 2250 psig
- Tube Side Design Temperature: 34°F-150°F
- Channel Diameter: 27.75 inches
- Channel Thickness: 2.0625 inches
- Channel Material: SA-105-II
- PWHT: Yes
- Minimum Operating Temperature: 17°F

Finite Element Analysis (FEA) was completed to quantify through-wall stress distributions at locations of interest (nozzle junctions, flange junction, welds, etc.).

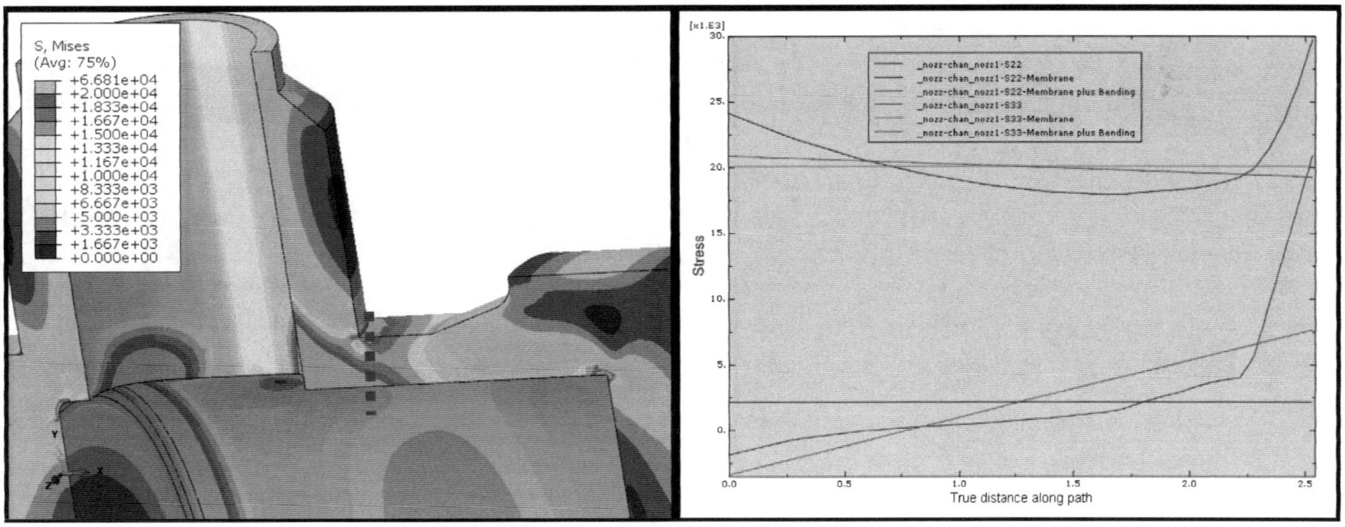

FIGURE 25: Stress Results Obtained Through FEA

As shown in Figure 26, results obtained using current API 579 Part 3 Level 2 procedures as well as results from a Level 3 evaluation using a t/4 postulated reference flaw depth do not meet the API 579 acceptance criteria to qualify the exchanger for the target minimum operating temperature of 17°F for pressures above 1400 psig.

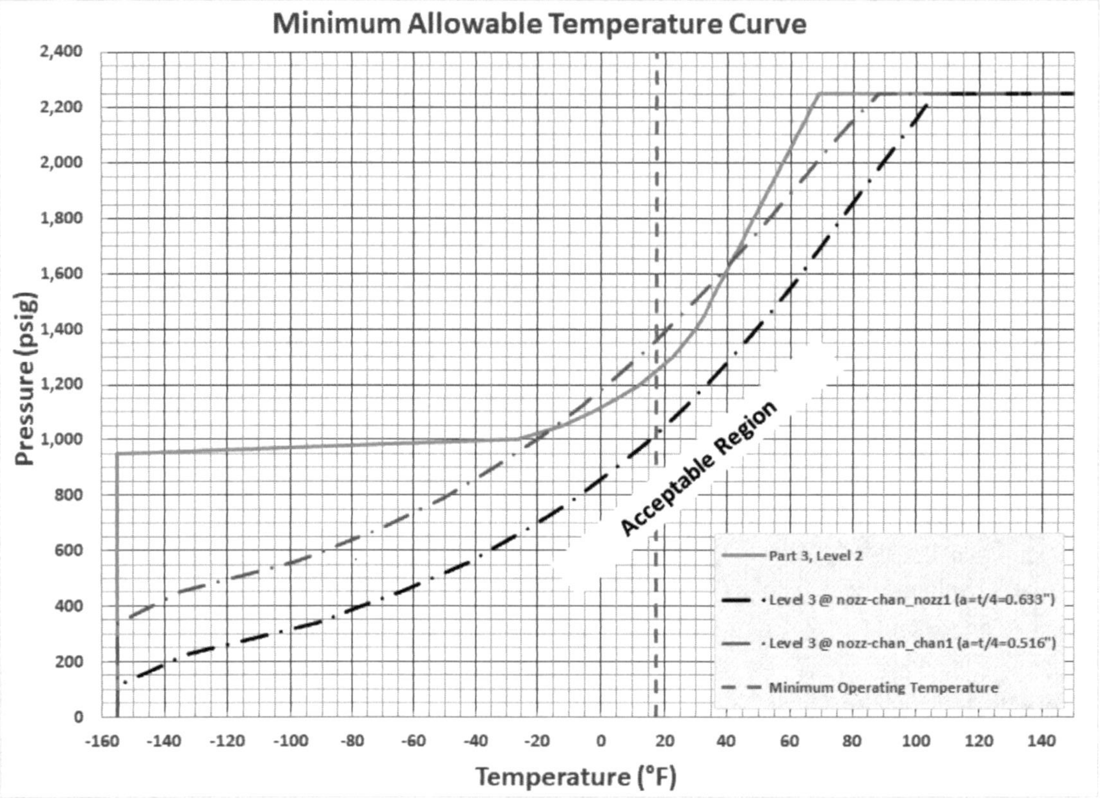

FIGURE 26: Current Part 3 Level 2 FFS Results and Level 3 Results for t/4 Reference

Supplemental Level 3 fracture mechanics calculations were performed iteratively reducing the postulated reference flaw size until the calculated MAT for all locations was within the target design temperature at full pressure. The results from the Level 3 evaluation were used to identify critical locations and optimize the associated inspection methods to ensure adequate sensitivity for detection of potential flaws.

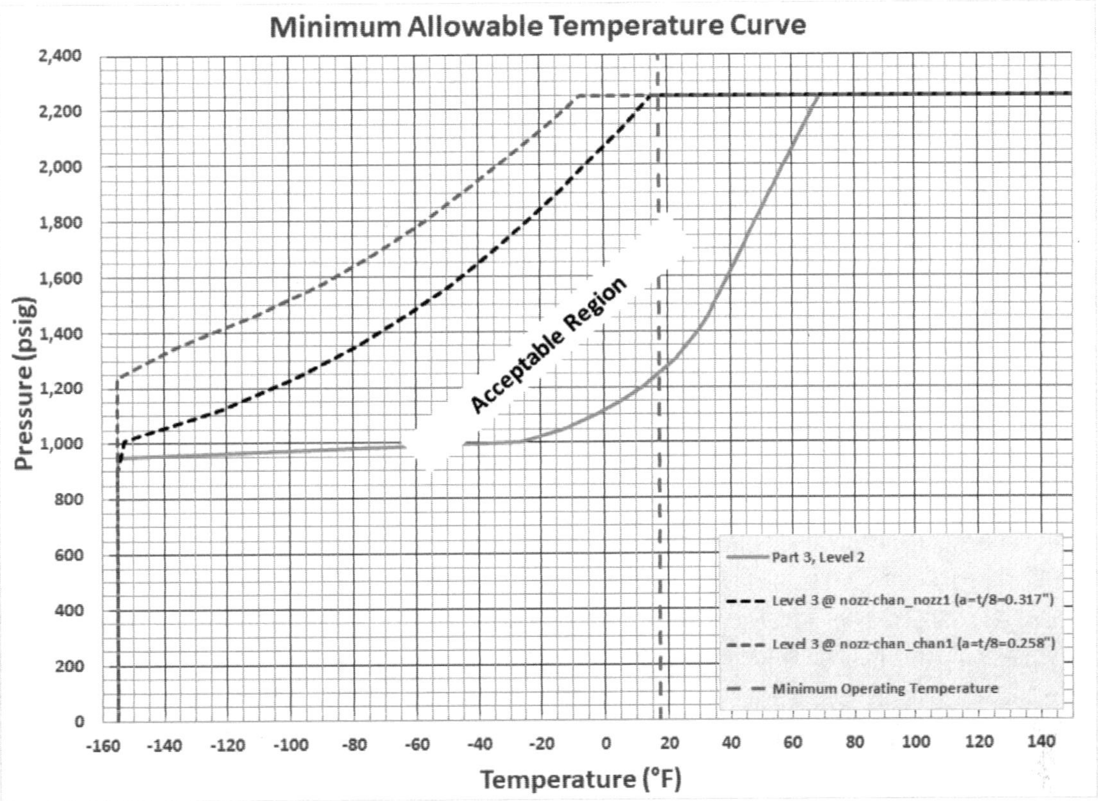

FIGURE 27: Level 3 FFS Evaluation Results of Different Reference Flaw Depths

DISCUSSION

MAT at Nozzle Junctions

In addition to the deficiencies previously noted, current design Code and API 579 Part 3 Level 2 evaluation results can be misleading with regard to locations or components with the highest risk for potential brittle fracture. For example, unless a nozzle material has a lower impact test exemption curve designation, it rarely will limit the MAT for a piece of equipment. However, brittle fracture failures tend to occur at such discontinuities, and results from a Level 3 FFS evaluation frequently identify nozzle junctions as limiting due to the elevated bending stresses local to the discontinuity as shown in the figure below.

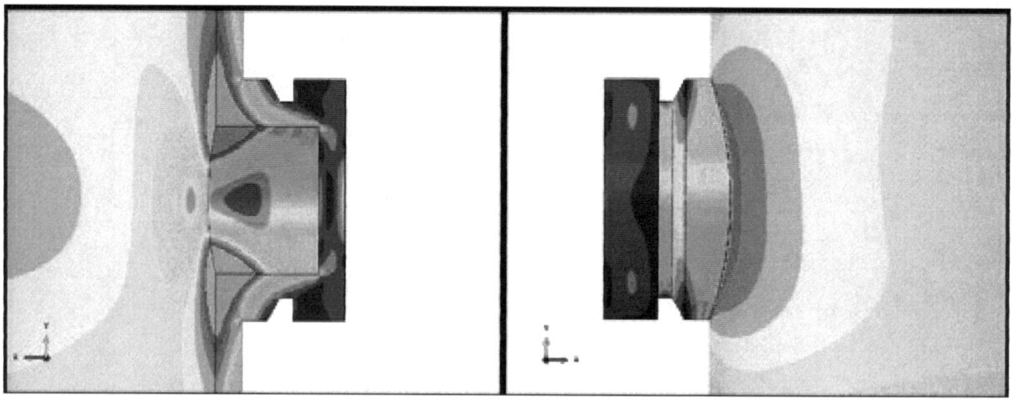

FIGURE 28: Example Stress Provide at Nozzle Junction

To formally qualify flaw tolerance at a nozzle junction, FEA is typically performed to quantify the through-wall stress distribution, and subsequent fracture mechanics is performed using the local stress at the location of interest (such as demonstrated in Example 3). In an effort to address this issue without adding significant complexity to the procedure, it is recommended that the stress ratio used to determine the temperature adjustment at a nozzle junction in a Level 2 evaluation is based on the following:

$$R = \frac{P_a}{P_{rating}} \qquad (13)$$

In formula (13), P_a is the applied pressure for the condition under evaluation and P_{rating} is the nozzle assembly rating calculated using the procedures in 2C.3.10 of API 579 [1]. Although not completely accurate, modifying the temperature adjustment stress ratio based on this procedure may directly improve the ability to identify locations at highest risk for brittle fracture without completing a full Level 3 evaluation. A similar stress ratio modification for flange junctions may also be considered with P_{rating} calculated using the procedures in 2C.3.12 of API 579 [1].

NBIC Alternative Weld Methods

The National Board Inspection Code (NBIC) [23] and the ASME Post Construction Code Repair of Pressure Equipment and Piping, PCC-2 [25] provide alternative welding methods to reduce the potential for cracking without PWHT. These include various combinations of preheat temperatures, weld interpass temperatures, and temper bead procedures. Although these welding methods may help to slightly temper or soften the material, they do not significantly reduce weld residual stress when compared to a proper PWHT stress relief operation as shown in [32]. Therefore, the MAT in a brittle fracture evaluation shall be evaluated in the as-welded condition (i.e. without PWHT credit) if one of these alternative welding methods are utilized.

Toughness Degradation Mechanisms

A number of environmental issues exist that can degrade material toughness or result in embrittlement [2]. The adverse effects of any such potential damage mechanisms should be considered in a brittle fracture evaluation, but currently a Level 3 evaluation is required. Future modifications to the Part 3 Level 1 and Level 2 acceptance criteria may be established to consistently incorporate the effects of such damage mechanisms (such as hydrogen embrittlement, temper embrittlement, etc.) so long as their impact on material fracture toughness can be quantified. Guidance for incorporating such modifications is presented in [26] and an example of application in industry is published in [33].

Constraint Effects

Future modifications to the Part 3 Level 1 and Level 2 procedures and acceptance criteria may be established to incorporate constraint (or "T-stress") effects on susceptibility to brittle fracture [34]. However, to incorporate any such modifications in Part 3, those procedures must initially be included in Part 9 of API 579 in order to ensure consistency with technology leveraged across evaluation Levels.

ASME Section VIII Division 2

As an alternative to the design by rule ASME VIII-1, a pressure vessel may be constructed in accordance with ASME VIII-2 [9]. Prior to 2007, the MDMT rules in ASME VIII-2 were very similar to ASME VIII-1. The CVN impact energy requirements, the impact test exemption curves, and the temperature reduction curves were almost identical to those in post-1987 edition of ASME VIII-1. However, in 2007, ASME VIII-2 was completely rewritten and the design margin on material ultimate tensile strength was reduced from a factor of safety of 3 to a factor of safety of 2.4. To accommodate the reduced design margin, the material toughness requirements were completely updated using explicit fracture mechanics and the Material Property Council (MPC) material fracture toughness model for estimating fracture toughness as a function of temperature for carbon and low alloy steels. The basis for the material fracture toughness rules in the current ASME VIII-2 is thoroughly documented in [11]. Similar to ASME VIII-1, the MDMT for an ASME VIII-2 component can be established either through destructive testing (i.e. CVN impact testing) or through engineering evaluation, and temperature reduction is permitted for reduced stress states. A major enhancement in ASME VIII-2 was the introduction of separate requirements for as-welded materials compared to PWHT or non-welded components. Figure 29 and Figure 30 show the ASME VIII-2 [9] required CVN impact energy for as-welded and PWHT components, respectively. Figure 31 and Figure 32 show the ASME VIII-2 impact test exemption curves for as-welded and PWHT materials, respectively. Figure 33 and Figure 34 show the temperature reduction curves for as-welded and PWHT materials, respectively.

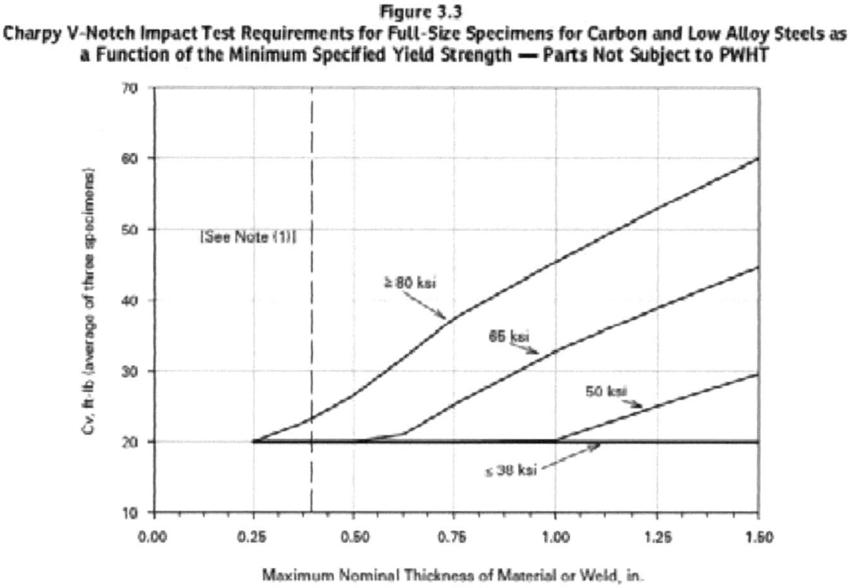

FIGURE 29: Required CVN Impact Energy for As-Welded Components per ASME VIII-2 [9]

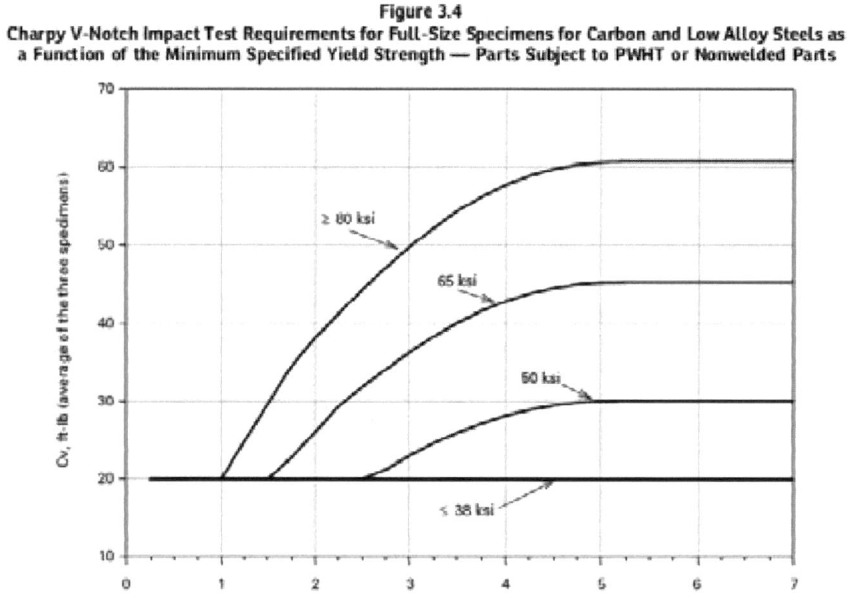

FIGURE 30: Required CVN Impact Energy for PWHT or Non-Welded Components per ASME VIII-2 [9]

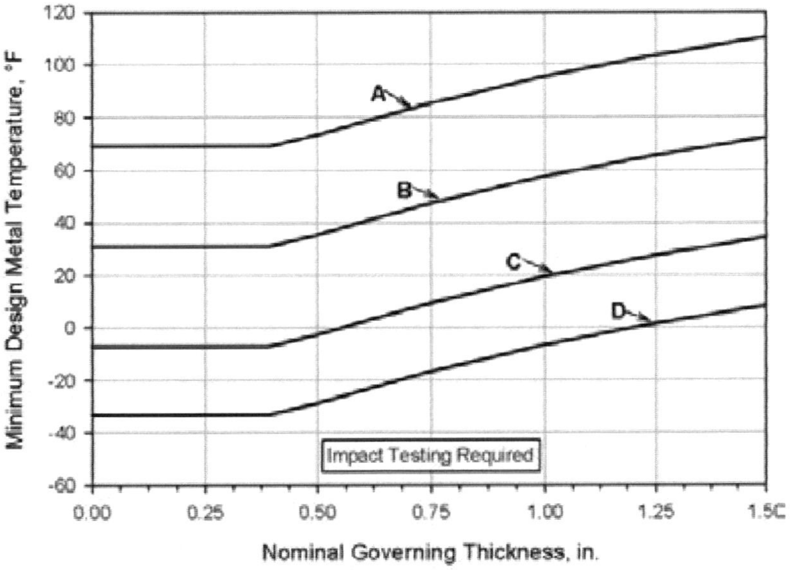

Figure 3.7 – Impact Test Exemption Curves – Parts Not Subject to PWHT

FIGURE 31: Impact Test Exemption Curves for As-Welded Components per ASME VIII-2 [9]

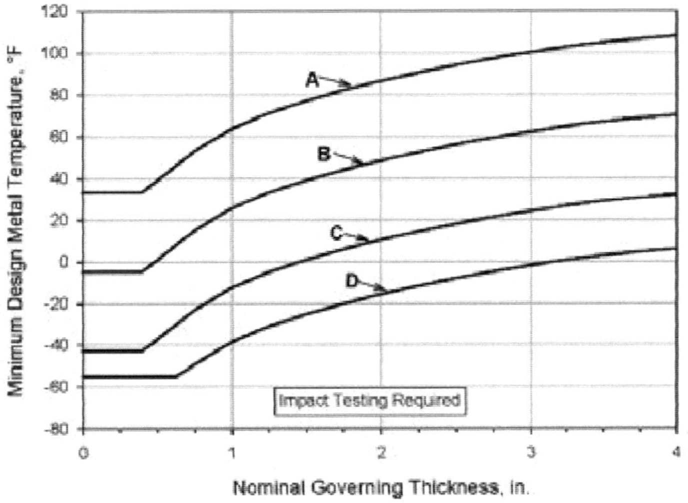

Figure 3.8 – Impact Test Exemption Curves - Parts Subject to PWHT and Non-welded Parts

FIGURE 32: Impact Test Exemption Curves for PWHT or Non-Welded Components per ASME VIII-2 [9]

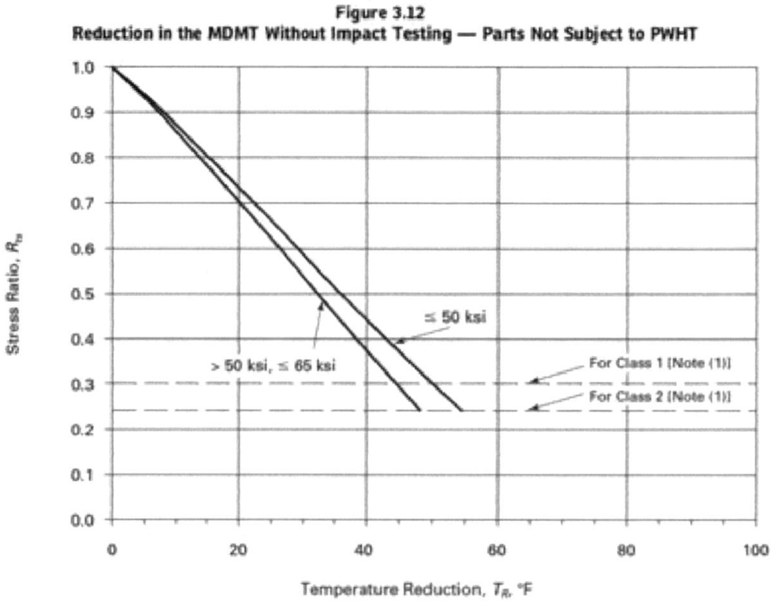

FIGURE 33: Temperature Reduction Curves for As-Welded Components per ASME VIII-2 [9]

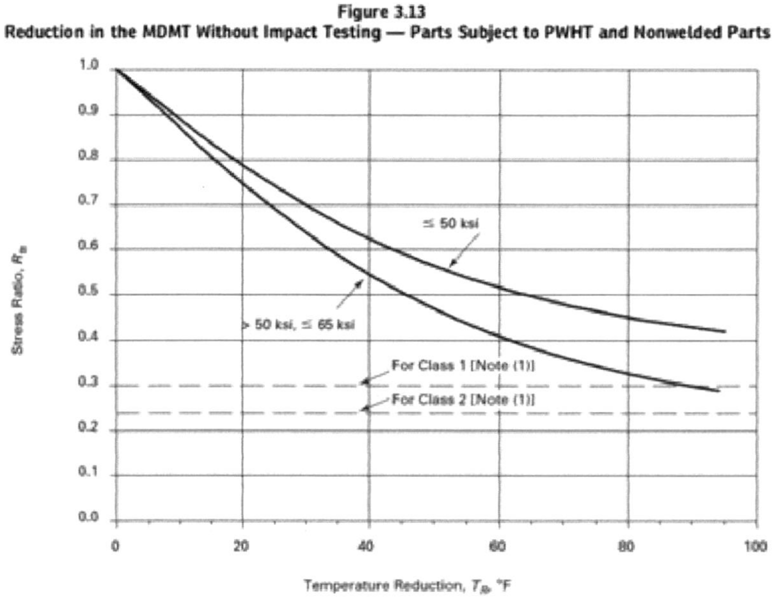

FIGURE 34: Temperature Reduction Curves for PWHT or Non-Welded Components per ASME VIII-2 [9]

A significant difference between the current ASME VIII-2 guidance and the proposed modifications previously discussed is that ASME VIII-2 toughness rules were developed using the MPC material fracture toughness model. The MPC model was developed to obtain a better prediction of lower shelf toughness behavior for carbon and low alloy steels [11]. However, the Wallin Master Curve [28] has been adopted by industry as a more accurate prediction for material toughness in the transition region and it is recognized as state-of-the art. For comparison, a plot of material toughness as a function of temperature for the ASME Section XI [15], the MPC [11], and the Wallin Master Curve [28] toughness models is shown in Figure 35 below.

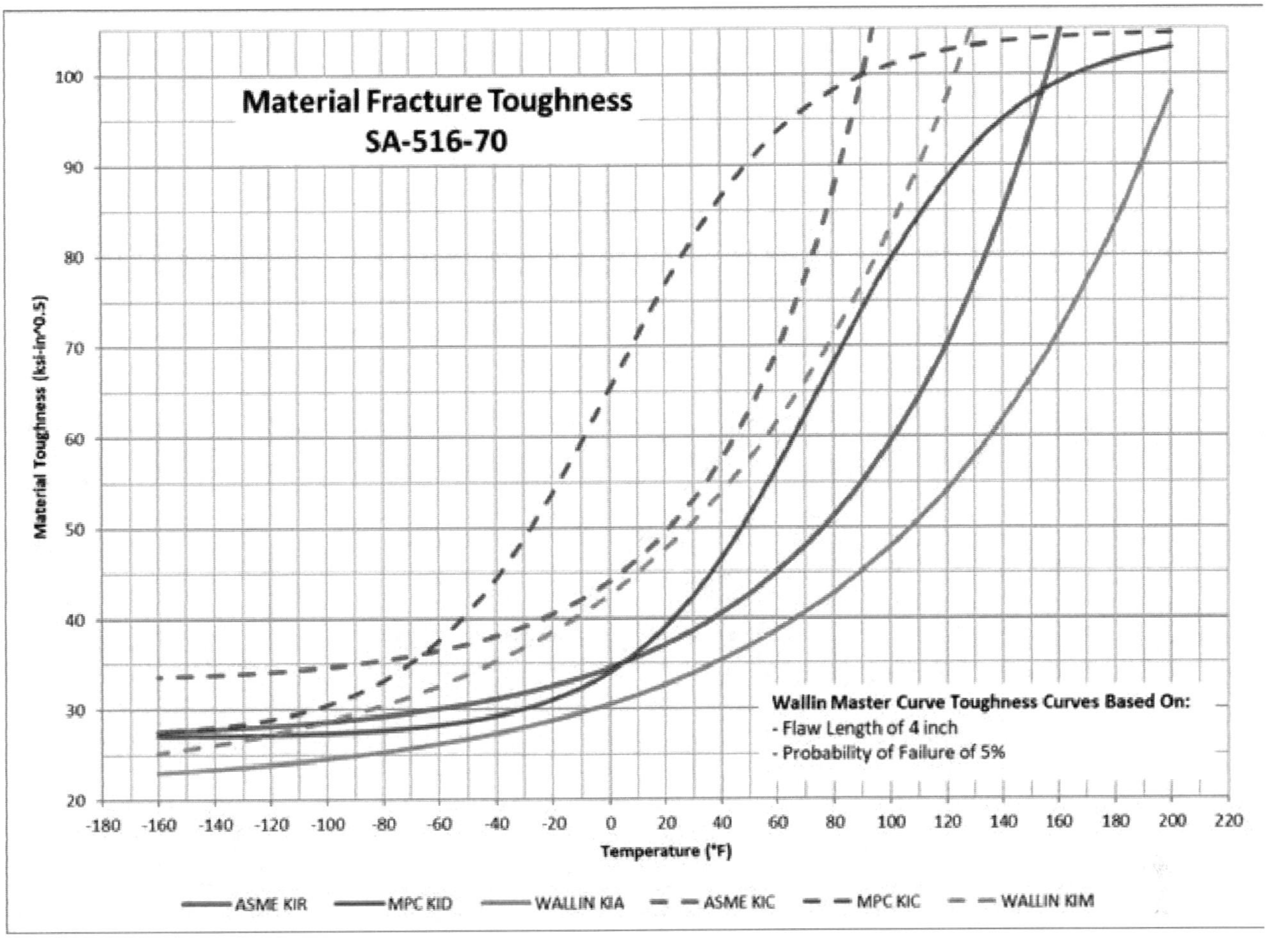

FIGURE 35: Various Material Toughness Model Predictions Showing Toughness as a Function of Temperature

SUMMARY

This paper provides a detailed summary of the numerous inconsistencies and inadequacies that exist in the brittle fracture screening procedures currently published in pressure equipment design Codes as well as the API 579 [1] *Fitness-For-Service* Standard. Results from an evaluation completed using the current published guidance could prove unsafe in some instances. Although relatively rare, the consequences associated with brittle fracture failures are typically catastrophic. A complete rewrite of Part 3 of API 579 is necessary to establish clear and safe direction for analysts. An API 579 Part 3 Level 3 FFS evaluation is the most accurate and most reliable prediction of susceptibility to brittle fracture. Proposed modifications to mitigate the issues with Level 1 and Level 2 screening procedures have been discussed and example results are presented. As demonstrated in this paper, the new procedures and acceptance criteria shall be anchored in state-of-the-art fracture mechanics technology consistently leveraged across Level 1, Level 2, and Level 3 evaluations.

REFERENCES

1. API/ASME, 2016. API 579-1/ASME FFS-1 Fitness-For-Service. The American Petroleum Institute and The American Society of Mechanical Engineers, Washington D.C./New York.
2. API, 2011. API Recommended Practice 571, Damage Mechanisms Affecting Fixed Equipment in the Refining Industry. The American Petroleum Institute, Washington D.C.
3. ASME, 2017. Boiler and Pressure Vessel Code, Section VIII Rules for Construction of Pressure Vessels, Division 1. American Society of Mechanical Engineers, New York.
4. ASME B31.3, 2016. ASME Code for Pressure Piping: Process Piping. The American Society of Mechanical Engineers, New York.
5. ASME B31T, 2015. Standard Toughness Requirements for Piping. The American Society of Mechanical Engineers, New York.
6. ASME B31.1, 2018. ASME Code for Pressure Piping: Process Piping. The American Society of Mechanical Engineers, New York.
7. Macejko, B., Kummari, S.R., Prueter, P., E2G Technical Report 125-2018.
8. Upitis, E., Mokhtarian, K.., 1998. Welding Research Council Bulletin 435: Evaluation of Design Margins for Section VIII, Div. 1 and 2 of the ASME Boiler and Pressure Vessel Code. Welding Research Council, New York.
9. ASME, 2017. Boiler and Pressure Vessel Code, Section VIII Rules for Construction of Pressure Vessels, Division 2 Alternative Rules. The American Society of Mechanical Engineers, New York.
10. Corten, H.T., Fracture Toughness Considerations Underpinning New Toughness Rules In Section VIII, Division 1, of the ASME Code.
11. M. Prager, D.A. Osage, J. Staats, B. Macejko, 2010. WRC Bulletin 528, Development of Material Fracture Toughness Rules for the ASME B&PV Code, Section VIII Division 2. The Welding Research Council, New York.

12. Bamford, W., Oldfield, W., and Marston, T., "An Improved Reference Fracture Toughness Procedure for Pressure Vessel Steels," Proc. Fifth International Conference on Pressure Vessel Technology, ASME, held in San Francisco, CA, September 1984.

13. Irwin, G. R., Kraft, J. M., Paris, P. C., and Wells, A. R., "Basic Aspects of Crack Growth and Fracture," NRL Report 6598, November 21, 1967.

14. Rolfe, S. T. and Barsom, J. M., Fracture and Fatigue Control in Structures Applications of Fracture Mechanics, Prentice-Hall, Inc., NJ, 1977.

15. ASME, 2015. Boiler and Pressure Vessel Code, Section XI Rules for Inservice Inspection of Nuclear Power Plant Components. American Society of Mechanical Engineers, New York.

16. Selz, A., 1988. New Toughness Rules in Section VIII, Division 1 of the ASME Boiler and Pressure Vessel Code, American Society of Mechanical Engineers Pressure Vessels and Piping Division Conference, 88-PVP-8, New York, N.Y.

17. T.S. Robertson, 1955. The Mechanism of Brittle Fracture. Naval Construction Research Establishment. Ship Structure Committee. Serial Number SSC-97.

18. Pellini, W.S., Puzak, P.P., 1963. Fracture Analysis Diagram Procedures for the Fracture-Safe Engineering Design of Steel Structures, U.S. Naval Research Laboratory, March 15, 1963, Washington, D.C.

19. Pellini, W.S., Pruzak, P.P., 1963. Welding Research Council Bulletin 88 Fracture Analysis Diagram Procedures for the Fracture-Safe Engineering Design of Steel Structures." Welding Research Council, New York.

20. Pellini, W.S., Puzak, P.P., 1963. "Practical Considerations in Applying Laboratory Fracture Test Criteria to the Fracture-Safe Design of Pressure Vessels," NRL Report 6030, U.S. Naval Research Laboratory, November 5, 1963, Washington, D.C.

21. Kummari, S.R., Macejko, B.R., Prueter, P.E., "Proposed Methodology Changes to Determine Minimum Design Metal Temperature of ASME/API Impact Tested Materials Based on Fracture Mechanics." Proceedings of the ASME 2018 Pressure Vessels and Piping Conference, July 15-20, 2018, Prague, Czech Republic, PVP2018-84795.

22. BSI, 2016. Specification for Unfired Fusion Welded Pressure Vessels, PD5500. British Standards Institution.

23. NBIC, 2017. National Board Inspection Code (NBIC), Part 3 Repairs and Alterations. The National Board, Ohio.

24. API, 2017. Pressure Vessel Inspection Code: In-Service Inspection, Rating, Repair, and Alteration, API 510. The American Petroleum Institute, Washington D.C.

25. ASME, 2015. Repair of Pressure Equipment and Piping. ASME PCC-2. American Society of Mechanical Engineers, New York.

26. Osage, D., Spring, D., Anderson, T., Kummari, S., Prueter, P., and Wallin, K., 2017. Welding Research Council Bulletin 562: Recommendations for Establishing the Minimum Pressurization Temperature (MPT) for Equipment. Welding Research Council, New York.

27. Macejko, B., Kummari, S., and Prueter, P., "Proposed Modifications to API 579 Part 3 Brittle Fracture Screening Procedures." Proceedings of the ASME 2019 Pressure Vessels & Piping Division Conference PVP2019-93207, July 14-19, 2019, San Antonio, TX, United States.

28. Wallin, K., 2011. Fracture Toughness of Engineering Materials. EMAS Publishing, United Kingdom.

29. ASTM, 2016. ASTM E1921 Standard Test Method for Determination of Reference Temperature, To, for Ferritic Steels in the Transition Range. American Society for Testing and Materials, Conshohocken, Pennsylvania.

30. Kummari, S.R., Macejko, B.R., Shipley, K.S., Prueter, P.E., Evans, G., Jamal, N., "Establishing Fracture Mechanics Based Minimum Allowable Temperatures for Low Temperature Applications of ASME B31.3 Piping." Proceedings of the ASME 2018 Pressure Vessels and Piping Conference, July 15-20, 2018, Prague, Czech Republic, PVP2018-84797.

31. PVRC Ad Hoc Group on Toughness Requirements, 1972. WRC Bulletin 175, PVRC Recommendations on Toughness Requirements for Ferritic Materials. The Welding Research Council, New York.

32. Prueter, P.E., Smith, K.J., Macejko, B.R., Shipley, K.S., "Commentary on Recent Changes in ASME B31.3 Post Weld Heat Treatment Requirements and the Effectiveness of Weld Preheat." Proceedings of the ASME 2017 Pressure Vessels and Piping Conference, July 16-20, 2017, Waikoloa, Hawaii, PVP2017-65859.

33. Prueter, P.E., Kummari, S.R., Osage, D.A., September 2016. A Fracture Mechanics-Based Minimum Pressurization Temperature Assessment on a 2¼Cr-1Mo Hydrotreater Reactor using Finite Element Analysis [Powerpoint Presentation], NACE CTW16 State-of-the-Art on MPT/Brittle Fracture Meeting in Houston, TX.

34. Wang, X., Bell, R., "Elastic T-stress Solutions for Semi-Elliptical Surface Cracks in Finite Thickness Plates Subject to Non-uniform Stress Distributions," Engineering Fracture Mechanics 71, 2004.

AIChE Paper 188f

Ethylene Decomposition:
A Tutorial

Steven T Coleman
Consultant

Prepared for Presentation at the 2019 Spring National Meeting
New Orleans, Louisiana
April 3, 2019

Ethylene Decomposition: A Tutorial

Steven T Coleman

Consultant

Abstract: Ethylene decomposition is a rather subtle hazard that can occur when ethylene is exposed to elevated pressure and temperature. This paper will describe situations where ethylene could decompose, with reference to specific events. Suggested actions to prevent initiation of a decomposition will be presented. A web-based tutorial on ethylene decomposition is being developed and will hopefully be complete and introduced as part of the presentation.

Introduction

The decomposition of ethylene to primarily methane and carbon is a reaction which may be initiated under conditions of elevated pressure and temperature. The reaction is highly exothermic, and if allowed to proceed, will in most instances result in a loss of containment and release of hydrocarbon to the atmosphere.

There have been a number of incidents of ethylene decomposition documented in the open literature, including a review of 4 incidents presented 5 years ago at the Ethylene Producers' Conference (1). The focus of this paper is to examine how incidents such as these may have been prevented, and to facilitate getting this information to those whose actions could be instrumental in preventing the initiation of an ethylene decomposition.

Ethylene Decomposition

Under conditions of high pressure, typically greater than 1000 psig, and elevated temperature, typically over 600 ^{o}F, ethylene can decompose with a significant release of energy in the form of heat. The primary decomposition products are methane and carbon, and the corresponding heat release is 30.4 Kcal / mole of ethylene reacted. The decomposition of ethylene to carbon and hydrogen also takes place, to a lesser extent, and with a lower heat release, 12.5 Kcal / mole of ethylene reacted. For more information on the conditions under which ethylene may decompose, see the paper by Laurence Britton, Alex Taylor and Donald Wobser entitled "Thermal Stability of Ethylene at Elevated Pressures" (2), which summarizes results of a study by Union Carbide to better understand the decomposition. Included in this publication are tools to predict where ethylene can decompose as a function of temperature, pressure and geometry (shape and size), as well as a prediction of ignition delay as a function of wall temperature and pressure. Initiating decomposition mechanisms are described, as are the effects of other gases and contaminants in the ethylene stream on the onset of decomposition.

Ethylene Decomposition: Initiating Events

To initiate an ethylene decomposition, three things are necessary: ethylene, elevated pressure and elevated temperature. This can be visualized as a triangle, similar to the fire triangle, where all three sides must be present for the reaction to take place. There is a fourth element, the time which ethylene can exist at the elevated pressure and temperature before the decomposition begins, and is a function of the temperature and pressure.

To prevent a decomposition from initiating, remove one or more of the three legs of the decomposition triangle. Which one is selected depends on the condition leading to ethylene being present at elevated pressure and temperature. Most reported ethylene decompositions are initiated by one of the following:

- External fire; fire impinging on high pressure ethylene pipeline or vessel
- Sudden compression of gas
- Exothermic chemical reaction (oligomerization, hydrogenation)

Considering the decomposition triangle, under normal circumstances two of the legs of the triangle already exist: ethylene is present and the system is at elevated pressure. These conditions can be found in pipelines transporting ethylene, which can operate at pressures ranging from 1000-2000 psig (70-140 barg) and are present from the ethylene product pump to a high pressure storage facility, such as a salt cavern, or to the downstream process receiving the ethylene. Meter stations and high pressure ethylene heaters, as may be found in polyethylene plants or salt cavern storage facilities with ethylene dryers, should receive particular attention. It is the third leg of the triangle, heat, which when added typically completes the triangle and initiates the decomposition.

In the following sections, measures or actions to be taken to prevent a decomposition from occurring for each of these potential initiators are discussed, with references to incidents where decompositions have occurred.

Initiator: Fire

On July 27, 1996 at the ARCO Chemical Company olefins complex in Channelview, TX (6), a gasket failure in a high pressure ethylene pipeline resulted in a release of ethylene which, approximately 1 hour after detection, ignited. The flame jetting from the leak impinged on the pipeline, initiating an ethylene decomposition. Within 10 minutes of the vapor cloud ignition, the pipeline failed with a "massive fireball accompanying the sudden large gas release" (1).

In this incident, the actions taken prior to the fire included isolation of the leaking section of pipe. As it was deemed unsafe to sent personnel into the meter station where the leak occurred, pipeline personnel were notified to close isolation valves on the two outgoing pipelines. By the time the persons arrived at the isolation valves and closed them, the vapor cloud had already ignited and the decomposition was underway.

How might this decomposition have been prevented?

1. Isolate and depressure the leaking section of pipeline more quickly. *Actions resulting from the incident investigation included a procedural change enabling plant operations personnel to isolate a leaking pipeline in an emergency under direction from the pipeline company personnel. In addition, a stainless steel depressuring header to the plants' two dry flare drums was added.* It should be noted that when ethylene is depressured from 1800 psig (125 barg) and ambient temperature to below 700 psig (48 barg), it becomes a 2-phase mixture. Piping used for depressuring must be designed both for the 2-phase flow as well as the low temperature that ethylene can reach as pressure is further reduced (minus 151°F, or minus 102°C if depressured to atmospheric pressure.)

2. Establish (or continue) flowing ethylene through the pipeline with the leaking gasket. This seems counter-intuitive while attempting to isolate and depressure the leaking ethylene line. Britton, et. al. (2) reports a time-temperature relationship for "predicted time to explosion" for a pure ethylene system:

   $$t = (T_0^3/P)\exp[(19{,}980/T_0) - 36.4]$$

   where

   t = ignition delay, seconds
   To = vessel wall temperature, deg K
   P = system pressure, psia

   With no flow, in a pipeline at 1800 psia, when the wall temperature reaches 500°F, the time for a decomposition to initiate is roughly 2.8 days. At 750°F, the time drops to 3.5 minutes, and at 1000°F, a decomposition would be initiated in 2 seconds. Reducing the pressure also increases ignition delay (inversely proportional) and has the added benefit of reducing the temperature of the ethylene in the pipeline. With water spray on the external pipe wall and flowing ethylene carrying heat away from the internal pipe wall, the time to initiate a decomposition could be significantly extended by reducing the temperature of ethylene in the pipe.

3. Displace the ethylene with another hydrocarbon. Ordinarily there would not be another hydrocarbon stream connected to the ethylene pipeline, and if it were, it would likely be blinded off. And the ethylene pipeline would have to be depressured to a point where another hydrocarbon could be introduced to displace the ethylene. Propylene is one candidate, and while it also could decompose, the conditions required are much more severe than those for ethylene (11). Ethane is another

candidate, less reactive than propylene, but again requires depressurization of the ethylene pipeline and/or ethane at high pressure for it to displace the ethylene.

4. Have a well-defined action plan in the event of a fire engulfing a high pressure ethylene pipeline, and operations personnel knowledgeable of the hazard of high pressure ethylene at elevated temperature to carry out the plan.

It bears repeating that the time from when the fire first impinges on the pipe and the loss of containment is short (less than 10 minutes.) Therefore, when fire is the initiating event, the response must be quick, with predetermined actions that can be rapidly executed.

Initiator: Sudden Compression

On August 20, 1976 at ARCO Chemical Company's Channelview, TX olefins complex, an ethylene decomposition was initiated when a meter run on the outgoing ethylene pipeline was suddenly pressurized from ambient pressure to 1300 psig when the valve isolating the meter run from the pipeline was opened (4,5).

Adiabatic compression of a gas such as described above is accompanied by a rise in temperature according to the equation:

$$T_a / T_b = (P_a / P_b)^{1-1/K}$$

where
T = Temperature, °R
P = Pressure, psia
a and b subscripts indicate after and before compression
$K = C_p / C_v$
= 1.4 for most diatomic gases
= 1.15 – 1.3 for many light hydrocarbons
= 1.24 for ethylene

The meter run, out of service for installation of an orifice box, contained air at atmospheric pressure. Had the meter run been inventoried with pure ethylene at atmospheric pressure, the temperature after compression from the above equation would have been approximately 830 °F. With heat loss to the cold pipeline, this may not have been sufficient to initiate and sustain an ethylene decomposition. However, the pipeline contained air, a diatomic gas, and the temperature rise calculated for the compression of air is roughly 1,500 °F, sufficient to initiate the decomposition.

How might this decomposition have been prevented?

1. Avoid sudden compression. Gradual pressurization allows the heat from compression to be dissipated to the colder pressurizing gas and vessel or pipe walls. One option, which reinforces the awareness of the hazard of sudden compression, is to install a small diameter line for pressurizing high pressure ethylene systems.

2. Air-free the segment of line to be pressurized with nitrogen, slowly raising the pressure in the line as high as practical with nitrogen. Displace the nitrogen with ethylene vapor, then bring the system to operating pressure with ethylene. Note that the heat of compression for ethylene is significantly less than that for the diatomic gases nitrogen and air.

3. Increased awareness of the heat generated and associated temperature rise accompanying sudden compression of gases by ethylene being sufficient to initiate an ethylene decomposition, particularly by those involved in pressurization of high pressure ethylene equipment.

A more recent event involving sudden compression as the initiator of an ethylene decomposition occurred at the DuPont Sabine River olefins unit in Orange, TX (7). The incident occurred on May 15, 2008 as a meter run, inventoried with nitrogen at low pressure, was being placed in service. Ethylene at 1,100 psig was introduced through a large gate valve, and the sudden pressurization initiated a decomposition. In this case, thanks to the timely response by personnel on site, there was not a loss of containment. Specifically, the smoking paint was quickly recognized, and while flow was initially stopped, it was reestablished within minutes to provide cooling. The propagating decomposition appeared to be quenched by high efficiency paper cartridge filters, but a "low order" non-propagating decomposition continued at a hot spot in the header. Reaction appeared to be continuing at the hot spot even after 4 hours and with the header depressured to below 60 psig. The area was considered safe only when the header was completely depressured and cleared of ethylene. The aforementioned recommendations for prevention are applicable for this case as well, and the addition of a small diameter pressurizing line and additional training for operations personnel on the hazards from sudden compression were both implemented.

Initiator: Overheating – Chemical Reaction (Hydrogenation, Oligomerization)

There have been incidents where the apparent conditions were not sufficiently severe as to result in an ethylene decomposition, yet one was observed. Loss of flow to a high pressure ethylene heater could result in an initiation of oligomerization, an exothermic reaction which could then elevate the temperature sufficiently to initiate a decomposition. Such an incident took place at the Lyondell Polymers High Density Polyethylene plant in Victoria, TX on September 27, 1996 (8). In this case, temperature of the reactor feed (hexane, ethylene and hydrogen at 2500 psig) was controlled by bypassing a portion of the feed around the reactor feed heater (550 psig steam, 480°F). At the time of the incident, most of the flow was bypassing the heater.

Without warning, the 6" elbow on the feed line to the reactor feed heat exchanger ruptured. A jet of black "sooty" vapor formed a cloud roughly 100 feet high. The vapor cloud ignited, but without an explosion and accompanying blast effects, and was extinguished within 5 minutes. Subsequent analysis indicated that the failure of the wall was temperature-induced and failed at approximately 1,100 °F. Carbon buildup in the heater and soot accompanying the initial hydrocarbon release support ethylene decomposition as the cause of the failure.

Based on the process conditions available to the operators, an ethylene decomposition should not have occurred. Britton, et. al. (2) proposed the use of an extension of the Frank-Kamenetski model to fit experimental data relating pressure and vessel geometry to the critical wall temperature for initiation of an ethylene decomposition:

$$P = \{T_c^4 C/(Er^2)\exp[(E/RT_c) - 38.52]\}^{0.5}$$

where

P = system pressure, psia
T_c = critical explosion temperature, °K
C = geometry constant (2.0 for an infinite cylinder, 3.32 for a sphere)
R = radius of pipeline, meters
E = activation energy = 40,000 cal/mole
R = gas constant = 1.987 cal/mole °K

Assuming the decomposition was initiated in the exchanger inlet head (characterize as a 24" diameter sphere), the predicted wall temperature to initiate a decomposition is 550 °F (290°C), well above the 480°F (250°C) heater wall temperature.

From information collected during the investigation, it is believed that rust (iron oxide) present on the heater wall catalyzed an exothermic ethylene hydrogenation reaction. The temperature increase from this reaction further reduced the already low flow rate to the heater, allowing the temperature to reach that required to initiate the decomposition.

How might this decomposition have been prevented?

1. Add the hydrogen and ethylene downstream of the feed heater, which would then only be heating the hexane portion of the feed.

2. Use a lower pressure steam as a heating medium when producing products requiring a lower reactor temperature.

3. Add a minimum flow limit on the flow to the reactor feed heater to prevent low flow through the heater.

4. Passivate the process side of the reactor feed heater to deactivate rust, if present, which could act as a catalyst for oligomerization or hydrogenation.

5. Increased training of operations personnel on the hazard of high pressure ethylene at high temperature, with particular emphasis on stagnant or very low flow of high pressure ethylene through a heater.

Note that several of these recommendations were implemented to prevent a recurrence of this incident.

Operating Close to the Edge: High Pressure Ethylene Dryers / Treaters

During the regeneration of high pressure (>1000 psig) ethylene dryers, which may be found at high pressure salt cavern storage facilities, conditions are such that all three legs of the decomposition triangle (ethylene, pressure and heat) are present. Under normal circumstances, the regeneration heater wall temperature and ethylene residence time are not sufficiently high or long enough for a decomposition to initiate. However, events such a stoppage of flow through the heater while in operation, or a control failure resulting in overheating the ethylene, could result in conditions where a decomposition could occur. Rust on the inside of the heater wall could initiate oligomerization, which would in a low or stagnant flow situation provide the additional heat needed to initiate a decomposition. A review of the controls and safeguards associated with high pressure ethylene dryer regenerations, with particular focus on ensuring that conditions severe enough to promote a decomposition cannot be reached, is recommended, with appropriate training to convey awareness of the hazards for those responsible for the dryer regeneration operations.

Conclusions

The initiation of an ethylene decomposition can and in many cases does result in a loss of containment of hydrocarbon with accompanying fire, damage to equipment, and in some cases injuries and loss of life. The decomposition reaction is highly exothermic, resulting in high temperature induced metal failure, and the time from initiation of the decomposition to the loss of containment is brief, typically 5-10 minutes. It is therefore prudent to prevent the decomposition from starting. In some cases, there are design changes which will eliminate or reduce the likelihood of a decomposition. In others, procedural changes can, if followed, prevent a decomposition from occurring. In all cases, training of operations personnel on the conditions which can result in a decomposition and how to respond in the event of an initiating event, with clearly defined actions determined beforehand. Scenario drills, particularly for a fire-initiated event, are one option for determining whether the action plans can be executed in the short time available before initiation of a decomposition.

To aid in the training aspect, the Ethylene Producers' Committee has commissioned development of a web-based tutorial on ethylene decomposition. The tutorial will be available for use by member companies, with the intent of reducing and hopefully eliminating ethylene decomposition incidents through better understanding of the conditions which could lead to initiation of an ethylene decomposition and steps to prevent its occurrence and mitigate its impact.

Additional Reading

For information on more ethylene decomposition incidents, "Loss Case Histories in Pressurized Ethylene Systems" by Laurence Britton (9) and the articles presented in the 1983 Texas Chemical Council on Ethylene Decomposition (10) as well as the Proceedings of the Ethylene Producers' Conferences (1,3) are recommended for your consideration.

Acknowledgements

Thanks to:

The Ethylene Producers Committee, whose Proceedings provided much of the information related in this presentation, as well as additional information on the subject.

Texas Chemical Council, for sponsoring the 1983 Symposium on Ethylene Decomposition (10) which provided information for this presentation and prompted additional research into ethylene decomposition.

The work by Union Carbide (now Dow) to better understand ethylene decomposition, and in particular the published work by Laurence Britton, Alex Taylor and Donald Wobser, which is perhaps the most widely referenced publication on this subject.

The companies who took the time and effort to document and share their ethylene decomposition experiences such that these incidents may be avoided in the future.

References:

1. Coleman, S. T., "Ethylene Decomposition: A Review", Process Safety Issues in Ethylene Plants Item 4, 26th Ethylene Producers' Conference Proceedings, New Orleans, LA; April, 2014.

2. Britton, L. G., Taylor, D. A. and Wobser, D. C., "Thermal Stability of Ethylene at Elevated Pressures", Plant/Operations Progress (now Process Safety Progress), vol. 5, No. 4, October, 1986.

3. Melching, J. S., "Ethylene Decomposition", Process Safety Issues in Ethylene Plants Item 4, 19th Ethylene Producers' Conference Proceedings, Houston, TX, April, 2007.

4. McKay, F. F., Worrell, G. R., Thornton, B. C. and Lewis, H. L., "If an Ethylene Pipe Line Ruptures…", Hydrocarbon Processing, Vol. 56, No. 11, November, 1977.

5. McKay, F. F., "A Case History of an Ethylene Decomposition in a Pipeline", Texas Chemical Council Symposium on Ethylene Decomposition, May 17-18, 1983.

6. Coleman, S. T. and Knickerbocker, R. L., "Ethylene Pipeline Decomposition Incident", Paper 81a, 13th Annual Ethylene Producers' Conference, Houston, TX; April, 2001.

7. White, J. D., "Ethylene Decomposition in Meter Header – Sabine River Olefins Unit, Orange, TX", Paper 105b, 22nd Annual Ethylene Producers' Conference, San Antonio, TX; March, 2010.

8. Black, D. E., "An Ethylene Decomposition Event at Lyondell Polymers' Victoria TX High Density Polyethylene Facility", CCPS 1997 Int'l Conference and Workshop on Risk Analysis in Process Safety, Atlanta, GA, October, 1997.

9. Britton, L. G., "Loss Case Histories in Pressurized Ethylene Systems", Process Safety Progress, Vol. 13, No. 3, July, 1994.

10. Proceedings from the Texas Chemical Council Symposium on Ethylene Decomposition, May 17-18, 1983.

11. Russell, F. R., and Mueller, R. H., "Explosive Decomposition of Propylene", 12 Int. Congress of Pure and Applied Chemistry, New York; Sept, 1951.

550638 Tutorial: The Fundamentals of Dilution Steam Generator / Quench Water System Fouling

Dilution Steam Systems (DSS) separate and recover ethylene quench water from hydrocarbons, recover heat, and generate steam for pyrolysis furnaces. Dilution steam is essential to reduce hydrocarbon partial pressure, promote the formation of ethylene, reduce the formation of undesirable heavier compounds, and reduce coke formation in the furnace tubes. The DSS incorporates a number of individual functions including process water recovery, hydrocarbon stripping, and dilution steam generation. Each function is closely linked to changes in plant operation; i.e., cracking severity, feedstock, and imported or recycle streams into the plant. Typically, a DSS is comprised of the following units:

- Quench Water Tower (QWT)
- Quench Water Drum Separator (QWDS)
- Filtration, Fine Filtration, and Coalescer (DOX unit)
- Process Water Stripper (PWS)
- Dilution Steam Generator (DSG) and/or Feed Saturator

The dilution steam system reliability is extremely important when it comes to the ethylene plant's energy balance.

The problems associated with the operation of the DSS are:

- Emulsion formation
- Fouling
- Corrosion
- Foaming
- Sodium carryover to furnaces

The paper will focus primarily on emulsions, fouling and corrosion and how to find the balance between the three through chemical, operations and mechanical means.

550640 Tutorial: The Fundamentals of Fouling in the Acid Gasremoval System and Fractionation Train

This tutorial presentation reviews the fundamentals of process-side fouling in an ethylene unit's acid gas removal system and fractionation train. The general mechanisms of fouling are first discussed, with a particular focus on differences between liquid- and gas-heavy feed slates. Common techniques for anticipating and monitoring the fouling of equipment are then presented. Finally, an assortment of proven ways to mitigate fouling – equipment design, process treatment, operational adjustments, and maintenance practices – are discussed.

549493 Designing Ethylene Plants for Fouling - a Primer

The designers, owners and operators of ethylene plants must consider the reality that foaming and fouling will occur. Knowledge is required to understand how these problems occur. Experience and expertise is needed to select design features of the olefin plant to deal with the impact of these unwanted actions.

For crackers to operate safely and profitably, the design must incorporate features to minimize or eliminate fouling. These range from process scheme selection and equipment design details to operating and maintenance procedures. This paper presents fundamental principles for the design of these systems such as performance objectives, configuration choices and design procedures. This paper is focused on providing basic knowledge to people relatively new to the ethylene business.

550637 Insoluble Polymer Fouling in CGC Systems

Insoluble polymer control is critical for avoiding unplanned shutdown in compressor systems. This paper describes fundamentals of insoluble polymer formation and field experience with mitigating solutions, including: 1) Mechanism of insoluble polymer formation, 2) Challenges in distinguishing and quantifying insoluble polymer formation and deposition, 3) Difficulties in predicting rate of deposition and impact of deposits on run length targets, and 4) Field experience connecting fundamentals with outcomes for various treatment strategies.

Alternative treatment strategies are evaluated for effectiveness in controlling fouling due to insoluble polymer formation. Advanced strategies that are more proactive and reliable for control of insoluble polymer formation are presented.

A case history on a liquid cracker and a case history on a gas cracker illustrates similarities and differences in fundamentals and treatment strategies for different feedstocks.

550052 Olefins Plant Cracked Gas Compressor Fouling Tutorial

Olefins plant cracked gas compressors frequently foul, resulting in loss of efficiency that increases energy consumption, and may threaten operational reliability, and ultimately cause throughput limitations. Critical factors in controlling compressor fouling are: 1) monitoring performance to identify and value impact of fouling, 2) identify fouling mechanism, 3) apply mitigation appropriate to the mechanism, 4) continue monitoring to validate mitigation effect. There are a lot of options available to the operator for monitoring, servicing and treating the cracked gas compressor. In this session we will discuss the various options to help the unit engineer understand the benefits and limitations of each.

AIChE Paper Number 192e

Advanced Monitoring of Fouling in Gasoline Fractionators

David Hood

Senior Product Applications Engineer

SUEZ – Water Technologies & Solutions

Conrad Teran

Senior Product Applications Engineer

SUEZ – Water Technologies & Solutions

Norman Wilhelm

Olefins Process Engineer

LyondellBasell

Prepared for AIChE 2019 Spring Meeting – 31st Ethylene Producers' Conference

New Orleans, Louisiana, April 4, 2019

Advanced Monitoring of Fouling in Gasoline Fractionators

Authors: **David Hood**, Senior Product Applications Engineer, SUEZ – Water Technologies & Solutions, Tomball, TX

Conrad Teran, Senior Product Applications Engineer, SUEZ – Water Technologies & Solutions, Tomball, TX

Norman Wilhelm, Olefins Process Engineer, Global Engineering & Turnarounds, LyondellBasell, Channelview, TX

Abstract:

The gasoline fractionators in the two sister ethylene units at LyondellBasell's Channelview facility historically ran from turnaround to turnaround without issue. However, after the units' feed slate shifted towards appreciably more gas, fouling in the towers became more problematic.

This paper will briefly review the general mechanisms of fouling in the gasoline fractionators, with a particular focus on the impact of moving from a primarily liquid to a primarily gas feed slate. The process treatment program will then be described, with a primary focus on the monitoring strategy.

The gasoline fractionator is a rather complex system with many correlated process variables. As a result, it is generally difficult to discern an increase in the fouling rate with conventional monitoring techniques until it becomes truly problematic and potentially irreversible. The use of sophisticated multivariable regression modeling and a modified statistical process control strategy allows us to monitor the performance of the tower and clearly distinguish between common cause variation (variation attributable to normal changes in the various operating parameters) and special cause variation (variation due to a statistically significant, unexpected event such as fouling). As a result, we can immediately respond to subtle changes in the fouling rate.

Introduction

LyondellBasell operates two olefins units at their Channelview, Texas manufacturing facility. Designated OP1 and OP2, these units were commissioned in 1976 and 1977, respectively, and were originally designed to process heavy liquid feedstocks. The two units have been through two major debottlenecks in 1989 and 1995 and have included upgrades to allow for NGL processing. Historically, operation of the gasoline fractionators in the OP1 and OP2 units had been relatively problem-free until 2011, when fouling of the OP2 gasoline fractionator was suspected going into a turnaround that year.

Process inspection during the OP2 turnaround revealed a significant change to the amount and location of fouling inside the tower than what had been observed previously. Figures 1 and 2 show the fouling that had collected on the top tray of the tower and adhered to the tray valves underneath the tray. The trays were cleaned during the turnaround and the tower was returned to service. Unfortunately, the tower performance began to deteriorate shortly after the turnaround, and a decision was made to clean the tower one year later during a site-wide outage. The fouling observed in 2012 was nearly identical to what was found during the 2011 inspection. Analysis of the solids from OP2 indicated that the material was primarily polymers of styrene, indene, and cyclopentadiene. This is consistent with the monomers normally found in the gasoline fractionator reflux.

Source: LyondellBasell

Figure 1: *OP2 gasoline fractionator, top tray, 2011 turnaround*

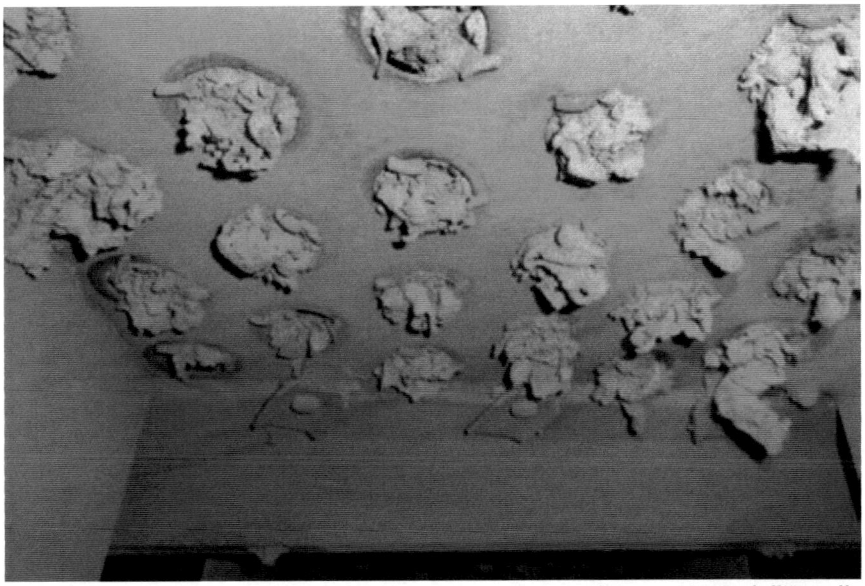

Source: LyondellBasell

Figure 2: *OP2 gasoline fractionator, underside of top tray, 2011 turnaround*

The OP2 tower was returned to service in 2012 after cleaning the trays, and again, the performance of the tower declined in the months following the turnaround. There were several observations that signaled the fouling problem was becoming more severe. The first indication was rising pressure drop, which suggested that part of the tray open area was beginning to plug. Another indication was the decline in fractionation performance, which was observed through high gasoline end-point and high quench oil viscosity. After one year of operation, the tower became difficult to operate. As furnaces were brought on- and offline, large shifts in gasoline inventory between the gasoline fractionator and quench tower were observed, and frequently led to temporary rate reductions to stabilize the system.

LyondellBasell began working with SUEZ in 2014 to develop a chemical treatment program for the OP1 and OP2 gasoline fractionators. The program proposed by SUEZ had three primary objectives: (1) apply chemical treatment to the two towers that allowed for operation between turnarounds without unscheduled outages, (2) provide a monitoring program that indicated changing conditions so that treatment dosage and operating targets could be changed before problems occurred, and (3) provide technical expertise in fouling mechanisms so that underlying causes of fouling were better understood.

Fouling Mechanisms

Multiple fouling mechanisms are potentially of concern in a gasoline fractionator. This is especially true after a unit's feed slate shifts towards appreciably more gas.

Fouling in the quench section of a gasoline fractionator and its ancillary heat exchangers is generally attributable to the accumulation and agglomeration of coke fines and polynuclear aromatics (PNA's)[1]. These heavies tend to become more problematic with a lighter feed slate because their residence time in the system can increase by more than an order of magnitude[2]. There was some evidence of this fouling mechanism in the OP1 and OP2 gasoline fractionators, as evidenced by episodes of elevated quench oil viscosity. However, these episodes were readily addressed by importing an external light cycle oil as needed.

Fouling in the rectification sections of the OP1 and OP2 gasoline fractionators was clearly more problematic. Fouling in this area is largely attributable to two mechanisms: (1) heat-induced and, in some cases, metal-catalyzed *in situ* free-radical polymerization of conjugated dienes such as styrene and indene; and (2) Diels-Alder cycloaddition reaction of conjugated dienes and dienophiles[1]. Once again, these mechanisms become more problematic with a lighter feed slate due to the increased residence time of the reactants[2].

Internal inspections of the OP1 and OP2 gasoline fractionators determined that fouling was consistently most pronounced on the top tray. It got progressively better as one moved down the tower before becoming largely inconsequential in the lower half of the rectification section. This fouling profile suggests a contribution from existent polymer in the reflux, which consists of both internally produced and imported raw pyrolysis gasoline.

Finally, the industry has encountered accelerated fouling in the rectification section due to inadequate wetting of the trays when cracking a lighter feed slate[3]. The higher volumetric flow of process gas is certainly a factor. Additionally, some operators have not adjusted their targeted temperature profile in the column, despite the process gas having a greater propensity of more volatile compounds. This can lead to an excessive reduction in the reflux flow rate. The OP1 and OP2 gasoline fractionators never approached dryness on the trays; however, monitoring did detect upticks in the fouling rate at low reflux-to-feed ratios.

Chemical Treatment

A SUEZ chemical treatment program started up in the OP1 and OP2 gasoline fractionators in late 2014. Given the various fouling mechanisms that are generally in play in the gasoline fractionators, the program provides multiple actives, each with a different functionality. A highly efficacious antioxidant, or chain-stopper, is used to mitigate thermally induced free radical polymerization. A metal deactivator is included to address metal-catalyzed free radical polymerization. Finally, a highly efficacious dispersant ensures that any polymer particulates that are still able to form do not agglomerate and foul the trays and downcomers.

The multiple actives are delivered in two products, which are both injected into the reflux stream. The actives are not volatile at the gasoline fractionators' operating conditions, so they move down the column, providing protection throughout. Ultimately, they exit the system in the pyrolysis fuel oil byproduct stream. Some applications require supplemental dosage in the quench oil circuit, but that is not the case in OP1 and OP2.

Monitoring Strategy

Given that the OP1 and OP2 were experiencing greatly accelerated fouling with a lighter feed slate, with run-lengths measured in months instead of years, it was critically important to develop a monitoring strategy that could accurately and precisely quantify fouling as it happened. This would facilitate the identification of cause-and-effect relationships that would otherwise be masked by operating conditions that are nearly always in flux as the cracking furnaces cycle in and out of service, the feed slates and severities are optimized, the quantity and quality of the imported raw pyrolysis gasoline change, etc. Furthermore, an issue could be identified and addressed in a timely manner, before the fouling became intractable.

Conventional Techniques

Naturally, when one is concerned with fouling in a tower, the differential pressure across the tower is monitored. Generally, the differential pressure is flow-corrected to eliminate noise due to changes in the operating rate. However, even then, this parameter lacks the sensitivity to detect anything more than a long-term fouling trend (e.g., month to month). This is especially true when the taps span a large number of trays.

Some other indicators of fouling are monitored, most notably the temperature profile throughout the tower and the separation efficiencies of key components. Gamma scans of the rectification section can also be incorporated as need be. However, as with the differential pressure, these techniques provide a view of only long-term trends.

A number of fouling-related risk factors are also included as feed-forward components to the monitoring plan: the cracking feed slate, the concentrations of reactive monomers in the system, the concentrations of existent and heat-induced gums in the pyrolysis gasoline, and the reflux-to-feed ratio, just to name a few. Based upon experience across SUEZ's entire global portfolio of gasoline fractionator applications, the risk factors are weighted and then compiled in order to project the appropriate dosage target for each chemical treatment active.

Multiple Regression Analysis

Conventional monitoring techniques certainly provide critical insight into the condition of each gasoline fractionator; however, they fall short of the ultimate goal: accurate and precise quantification of fouling as it happens. Often the next step in monitoring is a simple multivariable regression analysis (MRA). MRA is a compilation of statistical techniques that are used to develop a mathematical model that describes the behavior of a dependent variable (e.g., differential pressure in a gasoline fractionator) as a function of a set of independent variables or predictors (e.g., flow rates, compositions).

Postulating an MRA linear model for a dependent variable, Y, and a series of predictors, X_n, yields the following equation, where a and b are constants to be predicted by the regression algorithm:

$$Y = a + b_1X_1 + b_2X_2 + b_3X_3 + b_4X_4 + ...$$

The correlation coefficient, R^2, is a measurement of the accuracy of the model from zero to one, with one being a perfect fit. The adjusted correlation coefficient is similar, but it takes into consideration the degrees of freedom in the model (there is an increasing penalty as the number of independent variables increases).

When developing an MRA model, there are several practical considerations:

1. Instantaneous values for the various process variables of interest are harvested from the data historian at a prescribed frequency (e.g., every eight hours).
2. Time is included as an independent variable that is reflective of fouling. It is usually expressed as "days of run."
3. Statistical outliers in the process data set must be objectively identified and removed. This is typically done with a statistical technique known as the Box Plot Method.
4. Variables of dramatically different magnitudes should be centered and standardized. Otherwise, regression techniques can lead to a biased model that inappropriately places more significance on the variables with significantly greater magnitudes.
5. The process data should be randomly divided into two groups. The first, typically consisting of 70 to 80% of the data, is used to train the MRA model. The balance is used to validate the MRA model once it is complete.
6. There are many MRA tools in the marketplace, such as the one within Microsoft Excel's Analysis ToolPak add-in.
7. Predictor variables in the MRA model should only be used if they have a statistically significant impact on the dependent variable, as determined via a t-test. A predictor variable may be clearly relevant from a first-principles perspective, but may not be deemed statistically significant by the t-test (this

typically happens when the variable is relatively constant throughout the training dataset).

Multicollinearity and the Need for Advanced Regression Methods

Given the complexity of the gasoline fractionator and its operation, it should come as no surprise that MRA models are not always entirely adequate. Beyond the practical limitations of some regression packages (e.g. the Excel add-in limits the number of predictor variables to twelve), the most significant and insidious issue when using MRA methods on this type of system, is the presence of multicollinearity – the moderate to high correlation between predictor variables in a regression model. The process variables that are ostensibly independent from each other, do in fact affect one another. Furthermore, variables in a continuous process do have varying degrees of inertia, if you will: they are autocorrelated.

The presence of multicollinearity in MRA models, can switch the signs, increase the variance, and increase the sensitivity of the predictor coefficients, making the model highly unstable and unreliable over time. As result, more advanced methods are in many cases required.

Many methods have been developed over the years (Ridge, Lasso, PCS, PLS and others) to deal with multicollinearity. However, there is one such method utilized by SUEZ that has proven to be very reliable and robust across many types of systems, the so-called "Correlated Component Regression" or CCR algorithm, developed by Dr. Jay Magidson several years ago[5].

CCR utilizes correlated components (S_k), which are exact linear combinations or composites of the predictors (X_p) such that $k<p$, to predict the dependent variable (Y). Simultaneous variable reduction is achieved via a step-down algorithm, removing the least important predictor at each step (importance is defined by the absolute value of the standardized coefficient). Cross-validation is used to determine the values of k and p, the two tuning parameters for the model. By tuning k, we prevent overfitting and uncontrollable or extraneous effects (confounding effects) of high predictor correlations. By tuning p, we eliminate irrelevant predictors. Selection of k is driven by practical experience and selection of p, by the automatic step-down process. The tuned model is then converted back to a linear equation identical to that of an MRA model, but much more reliable and robust.

The first three considerations for building a CCR model, are the same as those required for building an MRA model. Beyond that, CCR models are scale invariant, so there is no need to transform or standardize the predictor variables. Whenever possible, datasets are divided into 70% training and 30% validation subsets. SUEZ uses an internally assembled algorithm to execute the CCR methodology. It is important to note that whenever tuned model fails to satisfy statistical thresholds, we take an additional step to capture potential non-linear effects via the use of genetic algorithms and other machine learning tools, but discussion of these methods is beyond the scope of this paper.

The CCR models for the OP1 and OP2 gasoline fractionators are represented in Figure 3 and Figure 4. The dependent variable in each case is the differential pressure across the rectification section. Predictor variables include, but are not limited to: time, total feed, liquid fraction in the cracking feed slate, feed composition, reflux rate, operating temperatures.

Figure 3: *CCR model for OP1 gasoline fractionator*

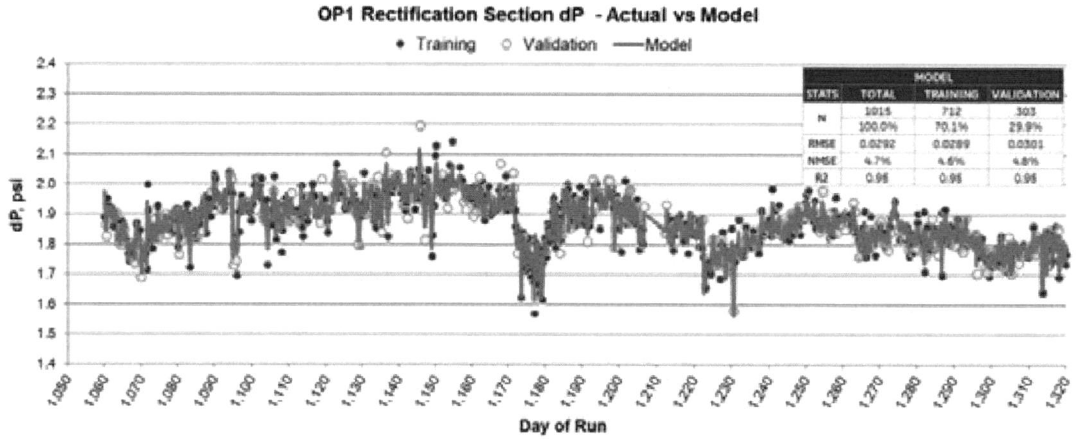

Figure 4: *CCR model for OP2 gasoline fractionator*

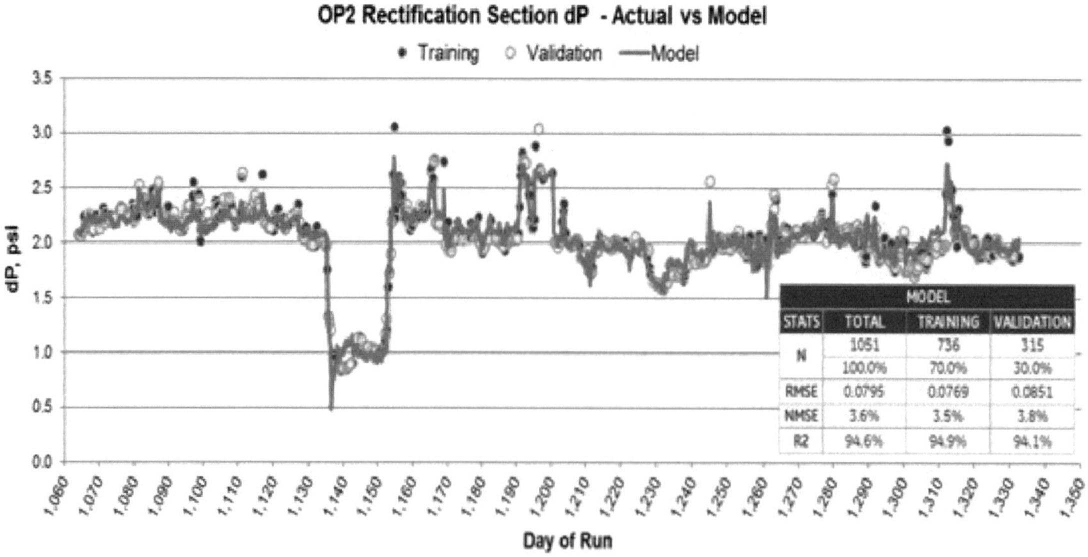

Modified Statistical Process Control Charts

Statistical process control (SPC) strategies are used to migrate a regression model into a chart of the dependent variable, specifically differential pressure, over time, one that enables the user to clearly distinguish between common cause variation (variation attributable to normal changes in the various operating parameters) and special cause variation (variation due to a statistically significant, unexpected event such as changes in fouling rate).

The first step in creating a modified SPC chart is the normalization of the model. All the independent variables other than time (X_1) are normalized to a set of standard conditions. These conditions are usually the values at the beginning of the run, the averages during the monitored period, or any set of values within the range of the training and validated data sets.

- The normalized dependent variable (Y_N) at chosen standard conditions (denoted by the subscript S) is defined as follows:
  $Y_N = a + b_1X_1 + b_{2S}X_{2S} + b_{3S}X_{3S} + b_{4S}X_{4S} + \ldots b_{KS}X_{KS}$

- Since $a + b_{2S}X_{2S} + b_{3S}X_{3S} + b_{4S}X_{4S} + \ldots b_{KS}X_{KS}$ = a constant (C), the equation is reduced to $Y_N = C + b_1X_1$, where X_1 represents the time variable.

The second step is the correction of the model. The corrected dependent variable (Y_C) describes the deviation of an actual measurement of the dependent variable (Y_A) at any point in time from the corresponding estimated value from the model (Y_E) at the standard conditions used for normalization.

- $Y_C = Y_N * [1 + (Y_A - Y_E) / Y_E]$

An SPC chart naturally requires upper and lower control limits. In conventional SPC methodology, these limits are defined as the mean of the dataset for the variable of interest ±3 times its standard deviation. In this modified methodology, the upper and lower control limits (UCL_X and LCL_X) are calculated at each point in time (X) using the model root mean square error (RMSE) as a substitute of the standard deviation as follows:

- $UCL_X = Y_{N,X} + 3 * RMSE_X$
- $LCL_X = Y_{N,X} - 3 * RMSE_X$

Now, one simply plots the normalized values (Y_N), corrected values (Y_C), upper control limits (UCL) and lower control limits (LCL). As the modified SPC chart is populated, the corrected values are automatically evaluated to determine if any are out of statistical control.

Since it is difficult to detect non-obvious out-of-control situations just by looking at an SPC chart, we incorporate an algorithm based on control zones, which automatically detects and tags out-of-control points. This method guarantees the strict implementation of the traditional out-of-control detection rules but is modified to minimize false positives. The implementation is as follows:

- The control chart is divided into control zones, with weights assigned to each zone:

Zone	Range	Weight
4	$> Y_N + 3.0*RMSE$	8
3	$Y_N + 2.0*RMSE$ to $Y_N + 3.0*RMSE$	4
2	$Y_N + 1.5*RMSE$ to $Y_N + 2.0*RMSE$	2
1	Y_N to $Y_N + 1.5*RMSE$	1
0	Y_N (centerline)	0
-1	Y_N to $Y_N - 1.5*RMSE$	-1
-2	$Y_N - 1.5*RMSE$ to $Y_N – 2.0*RMSE$	-2
-3	$Y_N – 2.0*RMSE$ to $Y_N – 3.0*RMSE$	-4
-4	$< Y_N - 3.0*RMSE$	-8

- Values of Y_C that fall within a zone get assigned the corresponding weight. The weight is accumulated and reset to zero every time the centerline (Y_N) is crossed.
- Accumulated values ≥8 or ≤-8 indicate an out of statistical control situation.
- For easy visual detection, out-of-control points outside the natural statistical control limits of +/-3*RMSE are tagged as "Alarms" and are colored in red.
- Out-of-control points within the control limits are colored in orange and, while in statistical terms they should be considered alarms, from a practical standpoint we tag them as "Warnings" (lack of normality in the distribution of the data, instrumentation errors), with the understanding, that accumulated warnings may eventually lead to a violation of the control limits.
- Out-of-control points indicate a greater than a 99.7% likelihood that a special cause is contributing to the variation in the dependent variable. In other words, the dependent variable is being impacted in a way that the model cannot explain given the dataset that was used for training and validation. The impact may be desirable (the differential pressure is better than expected) or not (the differential pressure is worse than expected). Regardless of the effect, the special causes should be identified if possible, with the goal of eliminating it or mitigating it by incorporating a beneficial cause into normal operating practices.

In the case of the OP1 and OP2 gasoline fractionators' differential pressures, some of the special causes that have been encountered are shifts in the balance of the gas feeds (ethane, propane, and butane), changes in the number of furnaces in hot-steam standby, and "shock" dosages of dispersant active.

Modified SPC charts for the OP1 and OP2 gasoline fractionators' differential pressure during specified time periods are depicted in Figures 5 and Figure 6.

Figure 5: *Modified SPC chart for OP1 gasoline fractionator*

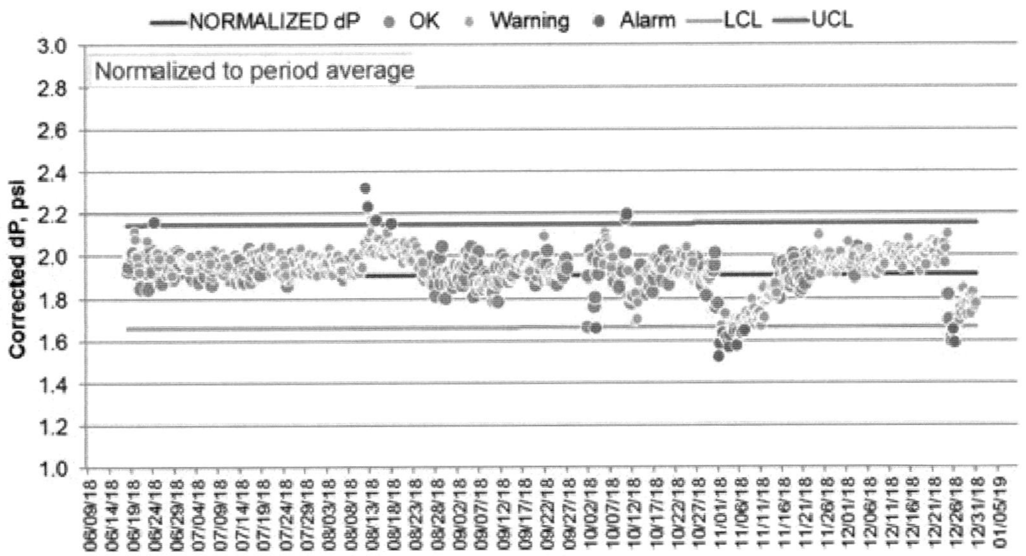

Figure 6: *Modified SPC chart for OP2 gasoline fractionator*

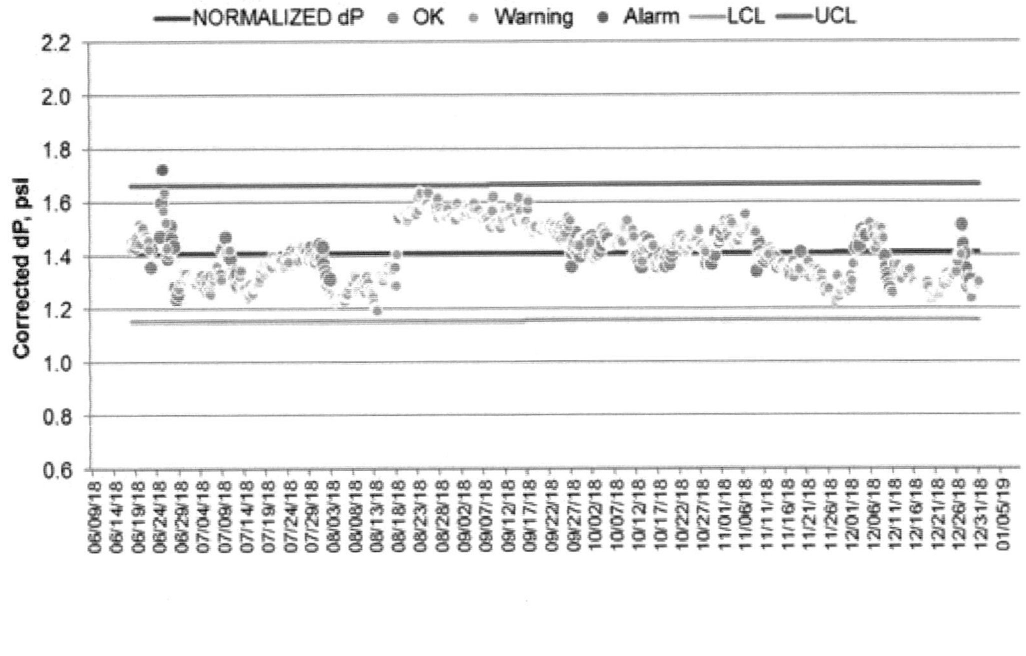

Conclusion

SUEZ's chemical treatment program has successfully controlled fouling in the OP1 and OP2 gasoline fractionators, thereby enabling the units to achieve their targeted run-lengths. The program delivers multiple actives – namely, an antioxidant, metal deactivator, and dispersant – to address the various fouling mechanisms.

Fouling-related risk factors are monitored to provide feed-forward control of product dosages. Additionally, CCR modeling and modified SPC charts are used for feedback. Special causes of variation are identified in a timely fashion – typically, within days. This facilitates the identification of cause-and-effect relationships and ensures that an issue is quickly addressed, before fouling becomes intractable.

One final note: while it has not been addressed in this paper, the predictive power of these models is such that forecasting is also possible.

References

1. Hong, Mike and Patrick Lucas (2011), "Fouling Mitigation Program Recovers Operating Efficiency and Extends Run Length for Ethylene Plant Oil Quench Tower," proceedings from the 23rd Ethylene Producers' Conference.

2. Bernard, Andre, Feridoun Fahiminia and Grace Kim (2013), "Challenges of Primary Fractionators on NGL Feed," proceedings from the 25th Ethylene Producers' Conference.

3. Quadro, Emerentino B., Amanda de Carvalho Costa, Flavio Weissheimer and David Hood (2015), "Minimizing Fouling in Ethylene Primary Fractionators," proceedings from the 27th Ethylene Producers' Conference.

4. Teran, Conrad (1989), "Multiple Regression Analysis: An Efficient Combination for Process Modeling and Optimizing," proceedings from the 1989 NPRA Computer Conference.

5. Magidson, Jay (2012), "Introduction to Correlated Component Regression", proceedings from 7th International Conference on Partial Least Squares and Related Methods.

550639 Fouling and Foaming in a Caustic Tower

The Caustic Tower in an ethylene plant is subject to fouling and foaming. Both can cause significant operational issues for the operator. Fouling is caused primarily by the base catalyzed condensation of acetaldehyde with itself. Foaming can be a result of fouling.

Across the industry, there appears to be an increase in caustic tower operational issues as plants crack lighter and shale ethane is brought into the plant. An increase in differential pressure ("delta P" or DP) across the caustic tower is one of the key performance indicators of fouling and foaming, and it can sometimes be difficult to determine which processing issue is occurring. Generally, a continuous dP increase is attributed to fouling, while a dumping and loading effect to the dP is attributed to foaming. The caveat to this is the stable foam that builds and stays. High salts, including rust, can help stabilize foam as well as high temperatures and throughput.

The paper will focus on fouling and foaming mechanisms in a caustic tower, changes in the industry, how to monitor a caustic tower and troubleshoot problems. As part of the troubleshooting discussion, a recent foaming event at a gulf coast ethylene producer will be discussed, including the steps taken to identify foaming vs fouling and ultimately how the foaming was controlled operationally instead of chemically at this plant.

AUTHOR INDEX

AUTHOR INDEX